Evolutionary Biology – New Perspectives on Its Development

Volume 9

Evolutionary biology has been a remarkably dynamic area since its foundation. Its true complexity, however, has been concealed in the last 50 years under an assumed opposition between the "Extended Evolutionary Synthesis" and an "Alternative to the Evolutionary Synthesis". This multidisciplinary book series aims to move beyond the notion that the development of evolutionary biology is structured around a lasting tension between a Darwinian tradition and a non-Darwinian tradition, once dominated by categories like Darwinian Revolution, Eclipse of Darwinism, Evolutionary Synthesis, and Post-Synthetic Developments.

The monographs and edited volumes of the series propose an alternative to this traditional outlook with the explicit aim of fostering new thinking habits about evolutionary biology, a multifaceted area composed of changing and interacting research entities and explanatory levels. Contributions by biologists and historians/philosophers are welcomed. Topics covered in the series span from (among many other possibilities):

- An Overview of Neutralist Theories in Evolutionary Biology
- Developmental Biology: From Reductionism to Holism and Back
- Selection Theories Beyond Hard and Soft Inheritance
- Divergent, Parallel, and Reticulate Evolution: Competing or Complementary Research Programs?
- The Rise of Molecular Biology: Between Darwinian and Non-Darwinian
- Biologizing Paleontology: A Tradition with Deep Historical Roots
- The Darwinian Revolution and the Eclipse of Darwinism: Blurring the Historiographical Lines
- Darwinism, Lamarckism, Orthogenesis: Can We Really Define Them by Their Hard Explanatory Cores?
- The Evolutionary Synthesis: A Fabricated Concept?
- The Opposition to the Evolutionary Synthesis: Criticizing a Phantom?
- A Reversed Perspective: Approaching Charles Darwin from the Pre-1859 Period
- The Long Development of the Multilevel Paradigm in Evolutionary Biology
- Self-Organization: A Research Tradition from Morphology to Cosmology
- Human Evolution: Sociobiological or Sociocultural?

All chapters are systematically reviewed by the series editor and respective volume editor(s).For monographs, the editor of the book series reach out to two competent reviewers. The editor ensures that reviews are fair and relevant.For edited volumes, the volume’s editor selects two competent reviewers for each chapter, ensuring that reviews are fair and relevant. The series editor oversees the whole evaluation process.

Carlos Ochoa

Hierarchical Evolutionary-Developmental Theory

Towards a New Theory of Evolution

Springer

Carlos Ochoa
Department of Evolutionary Biology
Faculty of Science
Universidad Nacional Autónoma de México
Mexico City, Mexico

ISSN 2524-7751 ISSN 2524-776X (electronic)
Evolutionary Biology – New Perspectives on Its Development
ISBN 978-3-031-88722-2 ISBN 978-3-031-88723-9 (eBook)
https://doi.org/10.1007/978-3-031-88723-9

This Springer imprint is published by the registered company Springer Nature Switzerland AG
The registered company address is: Gewerbestrasse 11, 6330 Cham, Switzerland

For my parents, Silvia and Eligio

Acknowledgments

The idea of developing a new evolutionary theory may seem ambitious. However, all of us devoted to this field agree that the framework in which we were trained is in urgent need of restructuring. This work represents an effort to step away from the orthodoxy that has dominated evolutionary thought for decades. Throughout this journey, I have been fortunate to have the right resources and the support of remarkable individuals who helped lay the groundwork for this book. It has been a truly monumental undertaking, and I can confidently say that this project is one of the most meaningful milestones in my academic career. While I hope you explore much of this work, for now, I sincerely appreciate your interest in it.

I wish to express my deepest gratitude to Richard G. Delisle, not only for his invitation and invaluable role in the development and publication of this book but also for his unwavering personal support since my initial collaboration in his book series. I am also sincerely grateful to the Springer staff, especially Sabine Schwarz, Srinivasan Manavalan, and Rekha Kumar for their dedication and meticulous oversight throughout the various stages of this publication process. My appreciation extends to the Universidad Nacional Autónoma de México (UNAM), which has been my professional home for many years, as well as to its academic community and invaluable database, which provided essential resources for this work. I also deeply appreciate the financial support provided by the Secretaría de Ciencia, Humanidades, Tecnología e Innovación (SECIHTI) through my appointment to the Sistema Nacional de Investigadoras e Investigadores (SNII). I am grateful to the reviewers for their suggestions, which improved this work. I also wish to recognize those who, in various ways, played a role in its development. Luis Ramírez Trejo provided vital guidance in understanding the philosophical aspects of natural kinds and evolutionary individuals during the early stages of this project. Laura Elliott carefully reviewed the English text, offering thoughtful comments and suggestions to enhance the writing. Aldi De Oyarzabal Salcedo created magnificent illustrations that resulted from extensive discussions and the considerable effort invested in their creation. Oscar Morales Arteaga contributed with his cartoon-style illustrations. Julio Ernesto Velasco Olmos improved many of the images featured in the book. Keila Nunes Purificação provided me with the photos of the shiny cowbird. Additionally, I would like to thank José Uriel Gante Escudero, Mariana Martínez Peláez, Rafael Paredes Montesinos, and Laura de Lourdes Cárdenas Flores for their invaluable support and willingness to assist during the search for images of *Lacandonia schismatica*. In this regard, my special thanks go to Esteban Manuel Martínez Salas, who kindly received me and provided exclusive photos of this iconic plant. Finally, I extend my heartfelt thanks to my students and colleagues at the Faculty of Sciences at UNAM, with whom I have shared countless reflections on the themes explored in this book. I am also deeply grateful to my family, especially to Mariana Martínez Peláez, to whom I owe much of this book. Certainly, this book is also dedicated to her and Luna, in appreciation of their unconditional support and companionship throughout this journey.

Competing Interests The author has no competing interests to declare that are relevant to the content of this manuscript.

Contents

Introduction: Hierarchical Evolutionary-Developmental Theory (H-Evo-Devo Theory)

1

Abstract

This book introduces a new evolutionary model called *Hierarchical Evolutionary-Developmental Theory* or *H-Evo-Devo Theory*. This theory proposes that the organism's internal properties reflect entities that are highly integrated in developmental terms that evolve hierarchically. In particular, it emphasizes the importance of "developmental constraints." These constraints operate at different scales, producing distinct states of morphological stability reflected as subspecies, species, higher taxa, and homologues, all representing units of evolution. Microevolution is the subspecies level. Macroevolution is the species level. Megaevolution is the higher taxa level. Modular-evolution is the homologue level. Each of these levels exhibits patterns of origin, stability, and extinction. Thus, the study of evolution aims to explain the mechanisms involved in these patterns, and the H-Evo-Devo Theory tackles these challenges. In short, the H-Evo-Devo Theory establishes a new synthesis between systematics, morphology, and evolution, suggesting a reinterpretation of the taxonomic structure and the evolutionary processes overall. Importantly, this model significantly diverges from the achievements reached by the Modern Synthesis, and it attempts to explain the phenomenon of evolution from a completely different angle.

Keywords

H-Evo-Devo Theory · Developmental constraints · New theory of evolution · Units of evolution · Subspecies · Species · Higher taxa · Homologues · Linnaean hierarchy · Levels of evolution · Modular-evolution · Microevolution · Macroevolution · Megaevolution · Hierarchy theory of evolution · Natural System

C. Ochoa, *Hierarchical Evolutionary-Developmental Theory*, Evolutionary Biology – New Perspectives on Its Development 9,
https://doi.org/10.1007/978-3-031-88723-9_1

Since the publication of Charles Darwin's *On the Origin of Species* (1859), biological evolution has stood out as one of the most debated topics in science. While the fact of evolution has been accepted, the issue revolves around how evolution has occurred. How can the patterns that establish biological disparity and diversity be explained? (Gould 2002). In the mid-twentieth century, the theoretical framework of the Modern Synthesis attempted to address this issue by positing a supposed agreement between Darwin's theory and the knowledge of genetics (Mayr 1980). Despite its status as the prevailing framework in contemporary evolutionary biology, this model focuses on genes as the primary actors in the evolutionary process, relegating the organism's internal properties and its role as an active agent in the environment (Wagner 2015; Walsh 2015). In addition, it alludes to the idea that the entire hierarchical structure of the phylogenetic tree is built from a single level and bottom-up processes, i.e., genetic mechanisms with small-scale effects are sufficient to explain the origin of large taxonomic groups (see Adams 2021).

Evolutionary Developmental Biology (Evo-Devo) is a multidisciplinary field that explores the relationship between embryonic development and evolution (Laubichler 2007; Love 2024). Many of its achievements are related to studying the organism as an active entity constructed by a hierarchical relationship between its parts that responds—either favorably or unfavorably—to different environments (Gawne et al. 2017; Baedke and Fábregas-Tejada 2023). Consequently, throughout its history, it is unsurprising that this discipline was not integrated into the framework of the Modern Synthesis (Hamburger 1980). As a result, many scholars, particularly in the 1980s, revisited the insights provided by embryonic developmental phenomena to comprehend the mechanisms of evolution (Love 2015).

From these discussions, contemporary researchers have attempted to construct novel evolutionary representations that move beyond the framework of the Modern Synthesis. For instance, one group of scientists proposes the so-called Extended Evolutionary Synthesis, a model incorporating specific causal factors from Evo-Devo to enrich our understanding of the evolutionary process (Laland et al. 2015; Lala et al. 2024). However, as I discuss in detail in another publication, this framework requires more rigorous elaboration and clarity regarding its meaning (Ochoa 2024). Alternatively, another group of scientists advocates for a Hierarchy Theory of Evolution (e.g., Tëmkin and Eldredge 2015); nonetheless, many of its foundational principles lack proper integration of Evo-Devo insights.

The question is, given that it is now recognized that Evo-Devo involves developing models that extend beyond a gene-centered perspective (i.e., only genetic mechanisms account for evolutionary change) and the extrapolationist premise (i.e., the sum of microevolutionary processes explains macroevolution), why does the Modern Synthesis framework continue to be upheld? Is a new theory of evolution needed? Although most researchers wager that such an extreme view is unnecessary, this research attempts to demonstrate that it is a worthwhile endeavor. This book aims to establish a new synthesis between systematics, embryology, and evolution, which entails a general reinterpretation of the taxonomic structure and evolutionary theory. In particular, it is written for all those scientists who disagree with maintaining the traditional frameworks and need a new model to interpret biological data with more clarity.

1.1 What Is H-Evo-Devo Theory?

In this book, I present a new evolutionary model that I call *Hierarchical Evolutionary-Developmental Theory* or *H-Evo-Devo Theory*, which proposes that evolution cannot be reduced to a single level but occurs at different levels, each operating under its own causes and effects, resulting in a complete perspective of the phenomenon of biological evolution. This theory focuses on the organism—rather than solely on genes—as the subject of study, i.e., it is based on an organism-centered perspective. It posits that the internal properties of the organism reflect entities that are highly integrated in developmental terms that evolve hierarchically. In particular, it highlights the importance of the concept of "developmental constraints," understood as a mechanism indicating that the interaction of the developmental processes limits morphological variation by maintaining it at a specific position within the morphospace (Alberch 1982). These developmental constraints operate hierarchically, both up and down, producing distinct states of morphological stability, which can be recognized as subspecies, species, higher taxa, and homologues, all representing units of evolution.

It is important to note that the H-Evo-Devo Theory significantly deviates from the principles of the Modern Synthesis and traditional systematics, offering an entirely different perspective to explain evolutionary phenomena. Likewise, although this model involves natural selection operating at multiple levels, it is not a traditional multilevel selection theory. Rather, it is a more complex theory that considers various processes acting hierarchically, emphasizing developmental constraints as the key to identifying units of evolution.

Usually, many traditional biologists define evolution as "inheritable changes" occurring at the population level. While they separate evolution into microevolution and macroevolution, the scale difference is seen as merely a matter of length of time. For example, they define microevolution as processes observable within a human lifetime and macroevolution as the longer-time scale that accounts for the evolutionary innovations resulting from those observable processes (e.g., Gingerich 1987; Bromham 2016). However, under the H-Evo-Devo Theory, evolution involves not only the pattern of "change" but also the patterns of "stability" (spatiotemporally recognizable) and "extinction." Thus, evolution can be defined as the science that studies the processes involved in the patterns of origin, stability, and extinction of subspecies, species, higher taxa, and homologues. In addition, evolution must be understood hierarchically, necessitating separation into different scales according to morphology, i.e., not based on the length of time but on the difference between the scales of complexity of the characters involved.

In accordance with the H-Evo-Devo Theory, homologues are at the highest level, where developmental constraints operate. Thus, homologues refer to quasi-autonomous parts of the organism that are highly integrated in developmental terms and phylogenetically maintained for a long time. This conception also leads to interpreting the Natural System in this light, which is reflected in how developmental constraints act on the character states of homologues, thus shaping the taxonomic structure of the phylogenetic tree. According to the H-Evo-Devo Theory, the Linnaean hierarchy is expressed as a kind of hierarchy of types. The concept of "type" or "body plan" refers to the common pattern of highly conserved characters shared by a group of closely related organisms. Hence, the ranks or categories represent body plans identified by different scales of morphological disparity, each maintaining stability. In short, taxonomic categories represent a series of body plans arranged in a nested compositional hierarchy.

In this context, species (or genera) represent the least inclusive body plan within a hierarchy of body plans. Given that there are taxonomic categories above the species (or genus)—such as family, order, class, and phylum (or division in plants)—I will use the term "higher taxa" to simplify these categories. Therefore, higher taxa are groups representing more inclusive body plans than the species body plan within a hierarchy of body plans. Finally, subspecies are entities that do not represent a body plan; rather, they are variations of the species body plan. That is, subspecies are groups that share the species body plan and are separated from each other by small morphological gaps of simple characters (e.g., color, pattern, size, proportion, etc.) as well as by the fact that they live in different environments.

Meanwhile, in the framework of the H-Evo-Devo Theory, units of evolution play different roles than those assigned by the traditional view. For instance, homologues are no longer merely elements that infer a common ancestry (such as molecules, behavior, or even the absence of a character); instead, they are considered units integrated into all levels that contribute to the evolution of body plans over geological time. Additionally, higher taxa are no longer seen as just historical records of species; they are now interpreted as real, active, and autonomous entities that shape the structure of the phylogenetic tree at higher levels. Under this view, higher taxa can undergo their own evolutionary processes that are not reducible to those at the lower level. Similarly, species are no longer recognized solely by their extrinsic properties (e.g., interbreeding); their role now resembles that of higher taxa, as they are considered body plans. However, they differ from higher taxa in that they represent the least inclusive body plan within the hierarchy. In short, higher taxa and species are considered in studying the role that body plans play in evolution. Finally, subspecies are no longer seen as incipient species or models for the origin of species; they are entities that adapt species to different environments, thereby protecting them from extinction.

Once the units of evolution have been identified, this model reveals that each of these entities resides at a specific level through which their patterns of origin, stability, and extinction can be recognized. For example, microevolution is the level that studies patterns of the origin, stability, and extinction of subspecies. To identify these patterns, the terms *subspeciation*, *substasis*, and *subextinction* are used, respectively. Macroevolution is the level that studies the origin, stability, and extinction of species (or genera); the proposed terms are *speciation*, *specific stasis*, and *specific extinction*. Megaevolution is the level that studies the origin, stability, and

Table 1.1 *According to the H-Evo-Devo Theory, subspecies, species, higher taxa, and homologues are units of evolution*

Units of evolution	Levels of evolution	Origin	Stability	Extinction
Subspecies	Microevolution	Subspeciation	Substasis	Subextinction
Species	Macroevolution	Speciation	Specific stasis	Specific extinction
Higher taxa	Megaevolution	Supraspeciation	Suprastasis	Supraextinction
Homologues	Modular-evolution	Morphological novelty	Morphological stability	

These entities reside at different levels, which can be studied through certain recognizable patterns: origin, stability, and extinction. A definition of evolution must encompass not only the origin of the entities but also their persistence over time and their extinction. Therefore, evolution involves studying the patterns or modes of subspecies, species, higher taxa, and homologues. The challenge lies in investigating the causes involved in these patterns at different levels

extinction of higher taxa, suggesting the terms *supraspeciation*, *suprastasis*, and *supraextinction*. Finally, modular-evolution is the level that studies the origin and stability of homologues, corresponding to the terms *morphological novelty* and *morphological stability* (see Table 1.1). Thus, the true challenge in evolutionary theory is to explain each of these evolutionary patterns or modes through the mechanisms proposed by Evo-Devo.

In this regard, the H-Evo-Devo Theory proposes the following mechanisms concerning the modes of origin, stability, and extinction of the units of evolution: Subspeciation is explained through what will be referred to as *Darwinian mechanisms* (dispersal model). In contrast, speciation and supraspeciation are explained by *supraDarwinian mechanisms* (prospecies model). Regarding modular-evolution, homologues arise from the co-option of elements (e.g., gene networks, cellular tissues, or even submodules) of a preexisting homologue. It is worth noting that these events are indeed so irreducible to the processes observed in simple characters that they must be studied as independent units.

Regarding stability, the modes of substasis, stasis, suprastasis, and morphological stability are all explained through developmental constraints. The difference is that substasis focuses on the stability of simple characters (e.g., color, pattern, and size), whereas specific stasis and suprastasis focus on the stability of complex characters allowing for the recognition of an underlying body plan. Meanwhile, morphological stability is the mode that denotes the most cohesive part of the system as a whole; although it falls within the category of complex characters, it does not represent body plan stability.

As for extinction, since subspecies are adapted to their respective local environments, the primary causes of their decline relate to factors altering their local habitat and their population structure (e.g., environmental stochasticity, habitat fragmentation, competition, population size fluctuations, etc.). In contrast, the extinction of body plans involves a hierarchical view. For example, when the extinction of a body plan depends on the decline of its lower-level units, it is referred to as the *loss of parts*. In contrast, when the extinction of a body plan occurs due to global catastrophes, it is exclusive to its level and associated with the concept of *mass extinction*. In this regard, I will explore the close relationship between the origin and extinction of evolving entities. Understanding them through a hierarchical view helps in visualizing the taxonomic structure of the phylogenetic tree at different points in time.

In terms of scientific scope, the H-Evo-Devo Theory introduces a new way of thinking in evolutionary biology, potentially offering a predictive approach to studying biological phenomena. Moreover, it would encourage new lines of research that have not yet been explored due to limitations imposed by the traditional conceptual framework (e.g., studying higher taxa as real and active entities in the field of evolution). In fact, the H-Evo-Devo Theory impacts evolutionary theory in several ways. Here are some of them:

1. *History of evolutionary thought*. Since much of the historiography of evolutionary thought is biased toward highlighting the achievements of the Modern Synthesis, the H-Evo-Devo Theory offers an alternative narrative that brings to light key insights overlooked in its account of evolution. For example, in the traditional narrative, typology was considered an obstacle to understanding evolutionary thought. However, in the historiography of the H-Evo-Devo Theory, typology was essential for synthesizing taxonomy, morphology, and evolution. This integration facilitated the formation of the realism of the Natural System.
2. *A reinterpretation of the Natural System*. Taxonomy is among the earliest disciplines of biological practice. Its use corresponds to the recognition of the natural order of biological groups that depends on the theoretical framework assigned to it. For example, according to contemporary interpretation, the natural order is built from the bottom up, assuming that microevolutionary processes account for macrotaxonomic units by extrapolation. In other words, population divergence gradually gives rise to higher taxonomic categories over time. Additionally, taxonomic units such as subspecies, species, and higher taxa are framed within a genetic context of genealogical relationships. However, the H-Evo-Devo Theory proposes a new approach to systematics, reinterpreting these Linnaean categories from an organism-centered perspective. Its structure is largely construed through a hierarchy of body plans, suggesting that the phylogenetic tree is constructed in reverse—from the top down rather than the bottom up, i.e., the body plans of the higher levels emerge first, followed sequentially by those of lower levels.
3. *A resolution to the ontological problem regarding units of evolution*. Given the previous point, the H-Evo-Devo Theory sets aside the ambiguity of recognizing the units of evolution, which is typically established from ontological pluralism. Instead, it opts for a monistic representation of the Natural System by employing ontological criteria such as natural kinds and evolutionary individuals to highlight the realism of such units. For example, one of the most significant challenges in biology is the species problem, which involves defining species based on multiple criteria without a clear and consistent definition of

what they are. Rather than proposing a new species concept, the H-Evo-Devo Theory offers a new perspective on evolution, through which the species category acquires a new meaning.

4. *A distinct interpretation of tempos and modes in evolution.* According to the traditional view, gradualism is considered the predominant mode for the origin of taxonomic units (e.g., species). Importantly, this view of gradualism is independent of the speed at which lineages change over geological time, i.e., these rates can occur either rapidly or slowly. In contrast, the H-Evo-Devo Theory posits that the most prevalent mode of change is through saltational evolution, which is measured in generations rather than geological time. Additionally, supported by paleontological and embryological evidence, the H-Evo-Devo Theory suggests that the evolutionary patterns in the units of evolution are discrete, characterized by a punctuated origin, a stable form, and a well-defined end. In short, this view challenges the mode of gradualism as the intermediacy of transitional forms.
5. *A strictly macroevolutionary view.* Even though many scholars have developed supposedly macroevolutionary models (e.g., punctuated equilibrium, species selection, etc.), in reality, they continue to maintain the extrapolationist premise since they assert that there are no distinctive macroevolutionary mechanisms other than those provided by microevolution. That is, although they recognize macroevolutionary patterns, they explain them based on microevolutionary processes. In contrast, the H-Evo-Devo Theory suggests that microevolution and macroevolution are entirely decoupled, considering there are differences in the application of mechanisms involved in the patterns of origin, stability, and extinction, as well as of the type of variation being studied (e.g., simple or complex characters). In fact, the H-Evo-Devo Theory offers conceptual tools for studying the higher units of the hierarchy. For example, it assigns a central role to higher taxa and homologues as primary initiators in the construction of the phylogenetic tree.
6. *A hierarchical perspective on the processes involved in the origin of the units of evolution.* By challenging the classic modes of geographical speciation, which lack an adequate integration of embryonic development phenomena, the H-Evo-Devo Theory introduces distinct models of divergence from a hierarchical view. For example, subspeciation is explained by the *Darwinian mechanisms* or dispersal model. That is, when dispersing into new environments, individuals within a subspecies undergo evolutionary change due to the direct interaction with the new environment. This process promotes divergence at that level. In contrast, the origin of body plans is explained through the *supraDarwinian mechanisms* or prospecies model, popularly known as the hopeful monster model (saltational evolution). That is, a new body plan emerges through a radical discontinuous change. Once it is established morphologically and ecologically, it generates lower-level entities to diversify.
7. *A hierarchical perspective on the processes involved in the stability of the units of evolution.* Although many evolutionary factors have been proposed as potentially contributing to the stasis of evolutionary entities (e.g., stabilizing selec-

tion), the H-Evo-Devo Theory posits that developmental constraints are the sole cause of such stability. Indeed, these constraints act hierarchically, enabling their phylogenetic recognition spatiotemporally. Considering this, while the mode of stasis has been recognized at the species level, subspecific and supraspecific stasis are rarely acknowledged by researchers. Nonetheless, some of these patterns have been studied despite the lack of consensus, e.g., supraspecific stasis has been explored through the concept of "living fossils." Furthermore, although various researchers have attempted to understand the mechanisms underlying the persistence of homologous characters over time, little attention has been paid to the stability of their variations as character states.

8. *A hierarchical perspective on the processes involved in the extinction of units of evolution.* The traditional evolutionary model has devoted little attention to extinction events. However, the H-Evo-Devo Theory not only focuses on these episodes but also emphasizes their hierarchical nature and their interrelation with other evolutionary processes. For example, while extinction at lower levels can influence extinction at higher levels, the latter also follows its own rules. Moreover, although extinctions negatively impact diversity and disparity, they also promote the emergence of new groups at different levels. In fact, this evolutionary dynamic between the processes of origin and extinction across distinct scales has shaped the taxonomic structure of groups over time.
9. *A structure focused on a mechanism distinct from natural selection.* Developmental constraints are a central mechanism in this model since they account for the recognition of units of evolution through time. This does not mean that natural selection is not important; yet, it posits that its role is not particularly "creative" as established in the traditional model.
10. *Reintroduction of the "organism" in evolutionary biology.* By supporting a gene-centered perspective, the Modern Synthesis framework regarded the organism as irrelevant in evolutionary processes. Under this view, the organism is seen as a passive entity whose sole function is to bear genes. By adopting an organism-centered perspective, the H-Evo-Devo Theory emphasizes that the development of the organism is crucial for understanding the origin of variations. Additionally, it supports the idea that the organism is an active entity that responds positively to its environment, chooses where it lives, actively shapes its own evolution through its behavior, and so on.
11. *A structure based on the concept of morphological modules (homologues), which are considered to be units that explain macroevolutionary inheritance and variation.* In the traditional view, inheritance is defined as the transmission of genetic information, while variation refers to the changes in that information. However, the H-Evo-Devo Theory challenges this perspective by emphasizing the concept of "modules" or quasi-autonomous elements that make up the bodies of multicellular organisms. Because these modules are highly integrated systems due to their developmental constraints, they can remain unchanged across generations in both their character identities and their character states. This helps explain the cohesion of the units of evolution. Furthermore, the ori-

gin of these modules, along with changes in their variational modalities, accounts for the origin of variation from an embryological point of view.

Given these considerations, this book will provide a detailed description of this model from different perspectives: historical (how it is historically structured), philosophical (how the reality of units of evolution is established), and biological (what mechanisms are involved in the patterns of origin, stability, and extinction of evolutionary units). To achieve this, it is divided into five parts. The first (Chaps. 2 and 3) addresses the history of the H-Evo-Devo Theory. The second (Chaps. 4 and 5) explores the relationship between evolution and hierarchy. The third (Chaps. 6, 7, and 8) offers a philosophical discussion on the metaphysics of units of evolution. The fourth (Chaps. 9 and 10) challenges some traditional evolutionary assumptions. Finally, the fifth (Chaps. 11 and 12) examines biological mechanisms that operate at different levels of evolution.

Most of the book is organized according to a methodology that contrasts an earlier theory with a new one: first outlining the problems and limitations of the previous theory and then presenting the proposed solutions of the new theory. However, I recommend starting with Chap. 5, followed by Chaps. 11 and 12, for a more direct understanding of this new proposal. Below, I will provide a summary of each chapter.

1.2 Part I: The History of the H-Evo-Devo Theory

Chapter 2 presents an alternative narrative on the history of taxonomy and evolution, challenging the traditional "essentialism story" and emphasizing the significance of the type concept and structural constraints in the Natural System's foundation. This chapter contends that types reflect structural constraints underlying the organism's internal properties. It traces how the use of taxonomic practices to recognize types beginning in the eighteenth century, along with advances in morphology and the theory of evolution in the nineteenth century, all contributed to the development of the Natural System's realism. This view was reflected in the emergence of "morphological systematics" in the twentieth century. Additionally, this chapter questions the idea that the type concept was replaced by that of the common ancestor, demonstrating that typology and evolution are compatible doctrines. This is supported by exploring Darwin and Owen's views, as they used the type concept within an evolutionary framework. Ultimately, this chapter concludes that typology facilitated a synthesis between taxonomy, morphology, and evolution.

Chapter 3 explores the history of the H-Evo-Devo Theory from the late nineteenth century to the mid-twentieth century. Firstly, it emphasizes the significance of anti-Darwinian theories such as neo-Lamarckism, orthogenesis, and saltationism, which focus on an organism-centered perspective. It argues that population geneticists, based on a gene-centered perspective, failed to address the problems of Darwinism, as an organism-centered perspective was needed to understand the phenomena of variation and inheritance. Secondly, it shows that both the Modern

Synthesis architects (Dobzhansky, Mayr, and Simpson) and the Unmodern Synthesis architects (Goldschmidt and Schindewolf) formulated their evolutionary ideas under a hierarchical view with common points such as a hierarchy of types, decoupled evolutionary levels, and a phylogenetic tree constructed from the top down.

1.3 Part II: Hierarchies and Evolution

Chapter 4 challenges traditional hierarchical theories of evolution. It begins by offering an overview of hierarchy theory and traces its early applications in the evolutionary framework from the 1960s to the 1980s. Subsequently, it critiques two main models: the *Hierarchical Theory of Selection*, particularly species selection, and the *Hierarchy Theory of Evolution* (*sensu* Eldredge and Salthe), which relies on an outdated replicator-interactor framework. However, this chapter focuses mainly on challenging the Hierarchy Theory of Evolution. Overall, it argues that these models fail to properly integrate Evo-Devo insights.

Chapter 5 provides a more detailed description of the structure of the H-Evo-Devo Theory. By integrating hierarchies into the Evo-Devo context, this model highlights developmental constraints as the main mechanism. These constraints are established through a procedural hierarchy, reflecting interactions within the developmental process, and a compositional hierarchy, identifying evolving entities in a phylogenetic context. The model proposes that developmental constraints operate hierarchically, producing distinct states of morphological stability that result in evolutionary entities. Homologues represent the highest entities, and their character states within the whole organism generate steady states at different scales, interpreted through the Linnaean hierarchy as subspecies, species, and higher taxa.

1.4 Part III: The Units of Evolution—Philosophical Criteria

Chapter 6 focuses on a philosophical discussion by resorting to a metaphysical argument to identify units of evolution. Firstly, it examines the philosophical argument regarding whether species are better understood as natural kinds or individuals. In this context, it defends the position of natural kinds in their traditional sense, emphasizing that the essentialism story is challenging and that the fundamental attributes of natural kinds can be supported. While treating evolutionary entities as individuals has been considered contrary to the philosophy of natural kinds, this chapter upholds that both categories are metaphysically compatible. For example, it posits that developmental constraints are the causal mechanism that shapes entities as natural kinds and, at the same time, as evolutionary individuals. Secondly, this chapter also suggests some criteria of individuality that help in spatiotemporally recognizing units of evolution based on specific patterns and modes.

Chapter 7 aims to argue in favor of subspecies and species as units of evolution. Initially, it explores the difficulties of recognizing subspecies and species in their traditional sense. For example, subspecies are difficult to delimit, while species have a wide array of concepts, making it impossible to reconcile them with a general notion of "species." However, through the H-Evo-Devo Theory, subspecies and species acquire a new ontological reality. Subspecies are groups separated by small morphological discontinuities in simple characters, whose function is to adapt species to different environments, whereas species are the least inclusive body plan within a hierarchy of body plans.

Chapter 8 argues in favor of higher taxa and homologues as units of evolution. Initially, it explores the difficulties of recognizing higher taxa and homologues in their traditional sense. For example, higher taxa are conceptualized as entities reflecting the historical record of species, lacking evolutionary processes, while homologues are defined as any element that allows the inference of a genealogical nexus. It is important to note that, under the H-Evo-Devo Theory, higher taxa and homologues acquire a new ontological reality. Higher taxa are interpreted as more inclusive body plans than the species body plan within a hierarchy of body plans. Meanwhile, homologues are defined as parts of the organism that are highly integrated in developmental terms and maintained phylogenetically over a long time.

1.5 Part IV: Challenging Evolutionary Statements

Chapter 9 challenges traditional evolutionary models, particularly the model of gradualism, which posits that the accumulation of small variations over geological time explains macroevolution. It questions the terms "cladogenesis" and "anagenesis" and demonstrates that punctuated equilibrium is similar to phyletic gradualism in terms of evolutionary rates. By challenging methods that attempt to quantify evolutionary rates through the premise of gradualism, the chapter argues that discontinuous rates of change better explain the origin of body plans and complex characters. The H-Evo-Devo Theory proposes that units of evolution undergo their processes at discrete rates and in different modes: an instantaneous origin, a period of stasis, and punctuated extinction. These patterns are supported by fossil records and embryology.

Chapter 10 challenges traditional models of speciation and the extrapolationist premise. Firstly, it analyzes and criticizes the theoretical foundation of allopatric speciation. It demonstrates that the original framework by which it was postulated is obsolete from a contemporary point of view. Moreover, it claims that evolutionary divergence is not caused by geographical barriers or the interruption of gene flow. Rather, divergence is explained through two simultaneous processes: variation and adaptation to local environments. Secondly, this chapter examines whether macroevolution is more than the sum of microevolutionary processes. It reviews the history and contemporary meanings of the terms "microevolution," "macroevolution," and "megaevolution." Three approaches to resolve this dilemma are explored, concluding that the irreducible approach is deemed the most plausible. The H-Evo-Devo Theory offers a new perspective on evolutionary divergence at multiple evolutionary levels and posits a qualitative separation between different levels of evolution.

1.6 Part V: The Levels of Evolution—Biological Processes

Since the H-Evo-Devo Theory supports the view that macroevolution (species level) is decoupled from microevolution (subspecies level), Chapter 11 explores the biological processes related to the origin, stability, and extinction of subspecies and species. Subspecies arise through Darwinian mechanisms, while species emerge through supraDarwinian mechanisms. The stability of both levels is due to developmental constraints acting on simple and complex characters, respectively. Subspecies extinction results from factors altering their local habitat and their population structure, whereas species extinction is due to the loss of parts and mass extinction events. This chapter represents the first modern effort to clearly delineate evolution into two levels: microevolution and macroevolution.

Chapter 12 focuses on the biological processes that occur in megaevolution (the level of higher taxa) and modular-evolution (the level of homologues). The origin of higher taxa also occurs through supraDarwinian mechanisms. However, this chapter highlights that the taxonomic structure of the phylogenetic tree is built from the top down, i.e., the body plans of higher levels originate first, followed sequentially by the body plans of lower levels. The sudden emergence of these entities at various taxonomic levels, described as hopeful monsters, is emphasized as well. In addition, higher taxa undergo stasis, which has been indicated in other works through the concept of "living fossils." This phenomenon is explained solely by developmental constraints. Finally, the extinction of higher taxa is caused by the loss of parts and mass extinctions. Regarding homologues, they arise from the phenomenon of co-option, and the cause of their stability is explained by developmental constraints acting at the deepest level of the hierarchy.

1.7 A Final Thought

Keep in mind that this book should be read from an objective perspective, as it might not align with your views on systematics and evolution. Furthermore, if you think radical change is needed in evolutionary theory, this book presents a viable alternative. Perhaps you might notice that many biological phenomena are difficult to understand within the traditional framework. But if this model proves useful to you and works well in your scientific practice, then we are on the right track.

References

Adams MB (2021) Little evolution, BIG evolution: rethinking the history of Darwinism, population genetics, and the "synthesis". In: Delisle RG (ed) Natural selection: revisiting its explanatory role in evolutionary biology. Springer, Cham, pp 195–229

Alberch P (1982) Developmental constraints in evolutionary processes. In: Bonner JT (ed) Evolution and development. Springer, New York, pp 313–332

Baedke J, Fábregas-Tejada A (2023) The organism in evolutionary explanation: from early twentieth century to the extended evolutionary synthesis. In: Dickins TE, Dickins BJ (eds)

Evolutionary biology: contemporary and historical reflections upon core theory. Springer, Cham, pp 121–150
Bromham L (2016) Testing hypotheses in macroevolution. Stud Hist Philos Sci A 55:47–59
Darwin C (1859) On the origin of species by means of natural selection, or preservation of favored races in the struggle for life. John Murray, London
Gawne R, McKenna KZ, Nijhout F (2017) Unmodern synthesis: developmental hierarchies and the origin of phenotypes. BioEssays 40. https://doi.org/10.1002/bies.201600265
Gingerich PD (1987) Evolution and the fossil record: patterns, rates, and processes. Can J Zool 65:1053–1060
Gould SJ (2002) The structure of evolutionary theory. The Belknap Press of Harvard University Press, Cambridge, MA
Hamburger V (1980) Embryology and the modern synthesis in evolutionary theory. In: Mayr E, Provine WB (eds) The evolutionary synthesis: perspectives on the unification of biology. Harvard University Press, Cambridge, pp 97–112
Lala KN, Uller T, Feiner N, Feldman MW, Gilbert SF (2024) Evolution evolving: the developmental origins of adaptation and biodiversity. Princeton University Press, Princeton/Oxford
Laland KN, Uller T, Feldman MW, Sterelny K, Müller GB, Moczek A, Jablonka E, Odling-Smee J (2015) The extended evolutionary synthesis: its structure, assumptions and predictions. Proc Biol Sci 282:20151019. https://doi.org/10.1098/rspb.2015.1019
Laubichler MD (2007) Evolutionary developmental biology. In: Hull DL, Ruse M (eds) The Cambridge companion to the philosophy of biology. Cambridge University Press, New York, pp 342–360
Love AC (2015) Conceptual change and evolutionary developmental biology. In: Love AC (ed) Conceptual change in biology: scientific and philosophical perspectives on evolution and development. Springer, Dordrecht, pp 1–54
Love AC (2024) Evolution and development: conceptual issues. Cambridge University Press, Cambridge
Mayr E (1980) Prologue: some thoughts on the history of the evolutionary synthesis. In: Mayr E, Provine WB (eds) The evolutionary synthesis: perspectives on the unification of biology. Harvard University Press, Cambridge, pp 1–48
Ochoa C (2024) Deconstructing the extended evolutionary synthesis: do we need a new theory of evolution? In: Delisle RG, Esposito M, Ceccarelli D (eds) Unity and disunity in evolutionary biology: deconstructing Darwinism. Springer, Cham, pp 403–439
Tëmkin I, Eldredge N (2015) Networks and hierarchies: approaching complexity in evolutionary theory. In: Serrelli E, Gontier N (eds) Macroevolution: explanation, interpretation and evidence. Springer, Cham, pp 183–226
Wagner GP (2015) Reinventing the organism: evolvability and homology in post-Dahlem evolutionary biology. In: Love AC (ed) Conceptual change in biology: scientific and philosophical perspectives on evolution and development. Springer, Dordrecht, pp 327–342
Walsh DM (2015) Organisms, agency, and evolution. Cambridge University Press, Cambridge

Part I

The History of the H-Evo-Devo Theory

The Role of Typology and Constraints in the Formation of the Natural System

2

Abstract

This chapter presents an alternative narrative for the history of taxonomy and evolution. It challenges the "essentialism story" and stresses the significance of the type concept and structural constraints in the foundation of the Natural System. The projection of the type reflects the structural constraints operating hierarchically within an organism's internal properties. Naturalists developed a more realistic view of the Natural System by interpreting Linnaean ranks as types. The recognition of types began in the eighteenth century within taxonomic practices. Subsequently, during the eighteenth and nineteenth centuries, morphology contributed to achieving greater precision in their recognition. Finally, in the nineteenth century, the theory of evolution bolstered the realism of types by explaining them through genealogical relationships. This collective scientific progress contributed to the Natural System's realism, culminating in the formation of "morphological systematics" in the twentieth century by establishing a hierarchy of types. Moreover, this chapter challenges the assumption that the type concept was replaced by that of the common ancestor, arguing that typology and evolution are not opposing doctrines but compatible ones. To illustrate this, it delves into Charles Darwin's and Richard Owen's inquiries into morphology. Upon scrutiny, it emphasizes that the type concept served at least two distinct purposes in the evolutionary context: First, the Archetype (general type) was a concept used to identify homologous structures; second, the *Bauplan* (specific type) was utilized to indicate the morphological stability of characters at different scales. Thus, this chapter concludes that typology facilitated a synthesis between taxonomy, morphology, and evolution.

C. Ochoa, *Hierarchical Evolutionary-Developmental Theory*, Evolutionary Biology – New Perspectives on Its Development 9,
https://doi.org/10.1007/978-3-031-88723-9_2

Keywords

H-Evo-Devo Theory · Typology · Type · Archetype · *Bauplan* · Linnaean hierarchy · Natural system · Homologues · Richard Owen · Charles Darwin · Taxonomic realism · Constraints · Essentialism story · Typological thinking · Species fixity

Since the eighteenth century, naturalists have aimed to find a classification that reflects order in nature. The pursuit of uncovering the Natural System is one of the most ambitious endeavors in biology. Yet, a critical question arises: Does such a taxonomic structure really reflect the natural order, or does it merely cater to the practical needs of human convenience? This chapter reinterprets the history of taxonomy by challenging the traditional narrative of essentialism. It presents a historical alternative that stresses the significance of types and the structural constraints in the foundation of the Natural System. In so doing, it appreciates how typology has positively influenced the scientific progress of taxonomy and evolution. This line of reasoning contributes to recognizing units of evolution as postulated by the H-Evo-Devo Theory.

Section 2.1 emphasizes that the type concept arises not from a particular philosophy but from a natural phenomenon: It reflects the structural constraints operating hierarchically within an organism's internal properties. By interpreting Linnaean ranks as types, it will become clear that this phenomenon allowed naturalists to develop a more realistic view of the Natural System. Types were initially recognized through taxonomic practice in the eighteenth century. Subsequently, during the eighteenth and nineteenth centuries, morphology played a crucial role in attaining a higher degree of precision when identifying them. Finally, in the nineteenth century, evolutionary thought strengthened the realism of types by explaining them through genealogical relationships. This scientific progress collectively influenced the realism of the Natural System, culminating in the foundation of "morphological systematics" in the twentieth century through the development of a hierarchy of types.

Section 2.2 challenges the prevailing notion that the type concept was replaced by that of the common ancestor. It concludes that typology and evolution are not opposing doctrines but rather compatible ones, and an in-depth exploration of Charles Darwin and Richard Owen's inquiries into morphology illustrates this point. To this end, it becomes evident that the type concept served at least two distinct purposes within the evolutionary framework: first, the Archetype (general type) was employed to identify homologous structures; second, the *Bauplan* (specific type) was used to indicate the morphological stability of characters at different scales. Considering this, it becomes clear that evolutionary theory did not fundamentally alter the morphological program. This leads to the conclusion that typology facilitated a synthesis between taxonomy, morphology, and evolution.

2.1 Typology and the Linnaean Hierarchy

According to the taxonomic ideal, the Natural System reflects a natural arrangement that is inherently hierarchical. Evolutionary biologists agree that this system represents genealogical relationships among organisms. However, the problem lies in interpreting this supposed hierarchy of natural organization. For example, in order to illustrate the taxonomic structure, the Linnaean hierarchy organizes organisms into ranks or categories, e.g., subspecies, species, genus, family, order, and so on. Philosophical debates on the meaning of these categories center specifically on their ontological value, i.e., do subspecies, species, genera, families, etc., exist? Now, most biologists assume the reality of species category, while the other categories are artificial (subspecies, genus, family, etc.). Still, the issue is that species have different definitions, resulting in various interpretations of the hierarchy.[1]

An essential point in this debate revolves around the concept of "type." Also known as *Bauplan* (body plan) or Archetype,[2] the type suggests that closely related groups of organisms are abstractly recognized by a highly integrated common pattern. Recognition of types stems from the observation that the underlying structures of the group in question seem to remain without significant changes, as no transformations beyond certain structural constraints are observed.[3] This interpretation contrasts with the unscientific and anti-evolutionary view of the type often portrayed in traditional historiography. This section demonstrates that types—reflected as categories in the Linnaean system—were initially recognized through taxonomic practice. Naturalists observed structural constraints and, under these criteria, rationally organized groups of organisms. Subsequently, these recognition practices were refined and empirically professionalized through the morphological program. During this process, types started to be distinguished as natural patterns and, therefore, as real entities. Contrary to the traditional viewpoint, the theory of evolution did not rule out the type concept; instead, it bolstered it by highlighting the genealogical connections between groups. Ultimately, I will claim that this scientific progress collectively influenced the realism of the Natural System, which is reflected in the foundation of morphological systematics in the twentieth century through a hierarchy of types.

The idea that the type concept hindered the progress of taxonomy finds its roots in the historiography of the Modern Synthesis. It focuses on "evolutionary systematics," advocating the Biological Species Concept and interpreting higher taxa as the historical record of species lacking evolutionary relevance. Referred to as the

---

[1] See these discussions in detail in Chaps. 7 and 8, Sect. 8.1.

[2] It is important to recognize that while the terms "body plan" and "*Bauplan*" are interchangeable, the concept of the Archetype holds a different meaning. While the body plan, or *Bauplan*, denotes a stable pattern of characters at different scales, the Archetype is used to identify the same structures (homologues) in a higher group of organisms without hierarchical classification. See below: Sect. 2.2. See also: Chap. 8, Sect. 8.2.2.

[3] As detailed in Chap. 5, this phenomenon is explained through developmental constraints in the negative sense.

"essentialism story" by Mary P. Winsor, various historians and philosophers have promoted such a historical narrative since the mid-twentieth century. According to the essentialism story, the notion of type emerged from either Platonic or Aristotelian essentialism, which dominated taxonomic methods in the pre-Darwinian era. Following this line of thought, species were defined by essential, unchanging, and timeless characters, influencing Carl Linnaeus' foundation of the Natural System. Unfortunately, this perspective also endorsed a creationist and fixist stance regarding species. Yet, thanks to Darwin and the advancements of the Modern Synthesis, this "typological thinking" was replaced by "population thinking," aligning more closely with evolutionary principles (see Mayr 1982).

Although the essentialism story has been popularized for a long time—in sources ranging from scientific papers to textbooks—several historians have challenged its historical accuracy (e.g., Winsor 2003, 2006; Amundson 2005; Wilkins 2009a, b, 2013; Richards 2010; Müller-Wille 2011; Witteveen 2020; Wagner 2024). For example, the publications constructing the essentialism story lack rigorous historical analysis (Winsor 2006). None of the pre-Darwinian naturalists can be accurately described as "essentialists," except Louis Agassiz, despite having been a contemporary of Darwin (Amundson 2005). Also, typology and essentialism are not comparable lines of reasoning (Wilkins 2009b). Remarkably, Ernst Mayr played a significant role in promoting the essentialism story, devoting much time to writing the history of taxonomy within the Modern Synthesis framework. His writings underscore the famous dichotomy between typological thinking and population thinking. According to Mayr (1959), typological thinking traces its roots to Plato's *eidos*, while population thinking is grounded in Charles Darwin's ideas.

Nevertheless, some scholars assert that such a dichotomy serves merely as a rhetorical construct devised by Mayr to support his view on species and speciation theories. Furthermore, they demonstrate numerous inconsistencies within this historical dichotomy (e.g., Chung 2003; Müller-Wille 2011; Witteveen 2015, 2016a; Wagner 2024). Unfortunately, this portrayal of history has led to the type being unfairly stigmatized; as a result, many researchers associate it with species fixity as well as other metaphysical notions (e.g., a preexisting idea in God's mind). The resulting negative perception has become so ingrained that discussing typology within the context of evolution is often viewed as taboo. As Winsor (2003, p. 396) states: "Mayr's campaign against 'typological thinking' has been so successful that many people think they know what a terrible thing 'types' were. Yet we have no hope of understanding the history of systematics if we cannot domesticate that word." Thus, rehabilitating the type concept necessitates the development of a historical framework that enables us to recognize its significance in the development of the Natural System.

Like today's researchers, early researchers perceived the type as an abstract projection of shared characters observed within a closely related group, indicating a highly integrated common morphological pattern (see Chap. 8, Fig. 8.3). As will become clear, this perception did not stem from an adherence to any specific philosophical doctrine (whether Platonic, Aristotelian, or otherwise); instead, its distinctiveness is a natural attribute of biological entities that reflects an underlying stability in their internal properties. Hence, while early naturalists recognized types, they were not committed to explaining their causes, even when referring to them in a metaphysical context (Amundson 2005). For instance, Farber (1976, p. 112) points out that Agassiz conceived the type, not because it was rooted in theological arguments about God's design, but because its recognition was based on a solid scientific foundation, i.e., it emerged from meticulous comparative anatomical studies. In this regard, it is noteworthy that the type does not align directly with Plato's ideas and that its formulation was shaped by diverse methods, ranging from the development of taxonomy's nomenclature and classification rules to the empirical research programs in morphology (Farber 1976; van der Hammen 1981; Winsor 2003).

For example, in order to delineate the different research methods applied in identifying the type, Farber (1976) utilized the following categorization:

1. *Classification type concept.* This concept highlights taxonomic practices in which specimens serve as models to establish their placement in the classification system. For example, a species might be used as a type to characterize the genus, a genus as a type for a family, and so forth.
2. *Collection type concept.* This concept pertains to the collection of name-bearing specimens in museums that are used to identify specific taxa.[4]
3. *Morphological type concept.* This concept is characteristic of the morphological program, where naturalists recognized an underlying "Unity of Type."

Considering this, it is worth mentioning that the conceptualization of the type progressed alongside the development of taxonomy methods (classification type and collection type), and its identification was subsequently rectified with the advancements in morphology (morphological type). Amundson (2005), for example, discusses that during the establishment of the Natural System in the eighteenth century, most naturalists were taxonomic nominalists, i.e., they did not believe that categories reflected an objective reality. However, taxonomic realism—i.e., the idea that taxonomic categories are real—gained increasing credibility over time. Morphology played a significant role in this endeavor and influenced the discovery of evolution. Accordingly, he asserts that morphological typology has "been a necessary step toward evolutionary thinking because it contributed to the development of taxonomic realism, that is, to the establishment of the Natural System as a real objective structure in nature" (p. 81).

Regarding the above, my hypothesis posits that the projection of the type, viewed as a natural phenomenon (due to structural constraints), played a pivotal role in fostering realism in the Natural System, primarily through early taxonomic practices (mainly through nomenclature methods). Subsequently, the discipline of morphology—whose research program lies in comparative anatomy and embryology—professionalized the study of types with greater precision. Eventually, types came to be regarded as more real once defined in evolutionary terms,

[4] However, Witteveen (2016b) argues that it is more appropriate to employ the term "nomenclature type-concept."

portraying them as historical patterns shaped by a genealogical nexus. Later, this evolutionary interpretation of the type contributed to establishing a hierarchy of types, which supposedly reflected a real structure of the Natural System. The following discussion delves into this process by examining the realism of categories within the Linnaean system.

2.1.1 Species as Types

Under the essentialism story, types have been characterized as immutable, timeless, and abstract entities. This is precisely how Linnaeus's conception of species was allegedly interpreted as unchanging entities with a metaphysical origin (e.g., created by God). Consequently, Kunz (2012, p. 62) points out, for example: "Species essences cannot be reconciled with the theory of evolution… The idea of the species as a type cannot be reconciled with variation due to evolution." Conversely, Amundson (2005) states that typology was not implicated in the initial attempts to classify organisms in a Natural System. He asserts that timeless types do not imply species fixity unless species themselves were types, but "species are not types." Supporting this assertion, Winsor (2006, p. 159) and Wilkins (2009b, p. 118) contend that types have only been described in cases of higher taxa. According to them, typologists never focused on lower levels, i.e., differences between species or individuals. Considering this, I maintain that, contrary to the consensus, on some occasions naturalists did describe species as types, including Linnaeus. Nonetheless, my conclusion differs from the essentialism story.

As mentioned earlier, rather than an ideology, the type represents an abstraction brought from nature: the projection of the type indicates the presence of underlying structural constraints within the biological entities. During the eighteenth century, in their quest to uncover the Natural System, early naturalists classified groups based on the stability of their phenotypic structures (see Wilkins 2009a). Although these methods were initially rudimentary—i.e., despite inaccuracies in species diagnosis during early taxonomy endeavors—advancements in morphology later allowed for a more precise determination of groups, which, in turn, reinforced the realism of species. Let's take a closer look at this point.

Amundson (2005) maintains that before the foundation of the Natural System, most Western naturalists believed in transformism, albeit not in the modern sense of the term but through myths and legends. However, Linnaeus and his contemporaries challenged this view with their taxonomic notion of species, which implied the persistence of essential characters within groups. Amundson states that this perspective, popularly known as species fixity, "is a fact of nature, not of metaphysics" (p. 40). Indeed, species fixity is based on a solid empirical foundation. For instance, when plants were exchanged between horticultural gardens in Europe, naturalists noted that they exhibited changes in their traits when exposed to varying climates in new regions. Yet upon returning to their native habitats, these plants reverted to their original form. In light of this, it can be deduced that early naturalists' belief in species fixity was due to their recognition of structural constraints at the species level.

Likewise, the systematics of the time was based on naming species by comparing various individuals with type specimens (collection type). While this task might appear straightforward, it was far from simple, as it demanded the development of specialized techniques for accurate identification and recognition. In 1753, Linnaeus wrote: "It is no easy task to establish the essential characters for a SPECIFIC NAME; it requires a thorough knowledge of several species, a meticulous examination of their parts, the careful selection of differences, and, finally, the precise application of technical terms to ensure they are as concise and reliable as possible" (translated from Latin). In this regard, Witteveen (2016b) notes that although Linnaeus never explicitly used the term "type," he implicitly recognized it; in fact, he also employed this notion to establish the boundaries of categories above species level (classification type). In conclusion, species as types began to be perceived from early taxonomic practices.

Nevertheless, the problem of recognizing species as types is due to the ambiguity—both in the past and present—surrounding the distinction between a true species and a subspecies or variety of the species, since the lowest denominations in the Linnaean hierarchy might not represent types but rather variations of the type (see Farber 1976, p. 118). However, during the eighteenth and nineteenth centuries, as morphology progressed, many naturalists implicitly identified species through the morphological type, particularly when they emphasized the stability of certain characters, and such constancy was evidence of their reality. For example, in 1756, Louis-Jean-Marie Daubenton asserted that each species possessed a set of invariable characters. In order to specify their species recognition, he disregarded variable traits such as color. In 1782, he wrote: "Specific characteristics, that is those which distinguish species, are essential to the individuals, they do not depend on the wish of the naturalist. They are not faulty, and in this regard only, the systems are infallible because they represent nature" (Daubenton cited in Farber 1975, p. 71). Similarly, considering the advancements in the morphology of his time, in 1832, Charles Lyell affirmed that species could adapt to any environment, yet their characters remained constant:

> The entire variation from the original type, which any given kind of change can produce, may usually be effected in a brief period of time, after which no farther deviation can be obtained by continuing to alter the circumstances, though ever so gradually,—indefinite divergence, either in the way of improvement or deterioration, being prevented, and the least possible excess beyond the defined limits being fatal to the existence of the individual.[5] (Lyell 1832, p. 65)

Given these considerations, "it appears that species have a real existence in nature, and that each was endowed, at the time of its creation, with the attributes and organization by which it is now distinguished" (Lyell 1832, p. 65). It is evident that, by examining Lyell's quote through the lens of the essentialism story, species fixity and creationism could be perceived as interconnected concepts. However, the phenomenon of structural constraints is a natural reality that prompts the recognition of

[5] This assertion was likely influenced by Georges Cuvier's principle of the correlation of parts.

types, suggesting their genuine existence. Yet, although Lyell recognized types, he was not committed to explaining the causes of their origin. As Amundson points out regarding early naturalists:

> They were committed to the reality of a kind of thing (a natural group, sometimes a taxon, sometimes a type) even though they did not pretend to understand its deeper nature. In fact, this cautious realism is prevalent in the history of science. Galileo was realistic about planetary orbits even though he did not know what caused or maintained them. Mendeleev was realistic about the periodicity of the elements even though he did not know what caused that periodicity. (Amundson 2005, p. 33)

In a nutshell, if types are manifestations resulting from structural constraints (invariable characters), if they are a "fact of nature," and if species are types, then species are indeed real entities in nature. Put differently, in line with the viewpoints of certain naturalists from the eighteenth and nineteenth centuries, species may have been occasionally perceived as types, suggesting that they are entities that exist in nature.

2.1.2 Higher Taxa as Types

Interestingly, although naturalists such as Linnaeus believed that species (or even genera) were an objective reality of nature while establishing the Natural System, they considered the other taxonomic categories in the Linnaean hierarchy—e.g., family, order, and class—as merely conventional arrangements.[6] In fact, as previously mentioned, Amundson (2005) states that taxonomic nominalism (i.e., the belief that the groups forming species do not exist) was prevalent among pre-Darwinian naturalists and that the idea of the Natural System truly reflecting the natural order only became possible with the advancement of the burgeoning science of morphology: "The development of the morphological type, along with its hierarchical structure of subtypes and its associations with embryological development and the fossil record, was a major contribution to this increased realism about the Natural System, and so ultimately to an evolutionary view of life" (pp. 81–82). According to Amundson, the morphological type influenced taxonomic realism, thereby fostering the acceptance of evolution. However, I propose that taxonomic realism gained greater acceptance once the morphological type was interpreted within the framework of evolutionary theory. This consolidation would not occur until the early twentieth century, culminating with the advent of morphological systematics, as explored below.

Taxonomy According to Farber (1976), during the eighteenth century, the notion of type was implicitly used in taxonomic practices (e.g., classification type). Nonetheless, in the first half of the nineteenth century, with the proliferation of specimens, the recognition of types helped naturalists objectively organize the

[6] Nevertheless, according to Pavlinov (2023, p. 135), by the end of his life, Linnaeus might have considered the higher categories as real as species.

material. For example, in 1847, the renowned British philosopher William Whewell described these classification practices through the "Method of Type." He mentioned that naturalists, engaged in the progress of classification, organized groups naturally rather than artificially. In order to facilitate this process, they "have a Type for our direction." Whewell describes this procedure below:

> A Type is an example of any class, for instance, a species of a genus, which is considered as eminently possessing the characters of the class. All the species which have a greater affinity with this Type-species than with any others, form the genus and are ranged about it, deviating from it in various directions and different degrees. Thus a genus may consist of several species which approach very near the type, and of which the claim to a place with it is obvious; while there may be other species which straggle further from this central knot, and which yet are clearly more connected with it than with any other. (Whewell 1847, pp. 494–495)

However, although taxonomic practices facilitated the recognition of types and their arrangement implied a degree of naturalness, many naturalists were not compelled to make an ontological commitment to types. For example, Hugh Strickland—widely distinguished for establishing standardized rules in zoological nomenclature—expressed the following in 1845:

> We may obtain a great amount of fixity, in the position at least, if not in the extent of our groups, by invariably selecting a type, to be permanently referred to as a standard of comparison. Every family, for instance, should have its *type-subfamily*, every subfamily its *type-genus*, and every genus its *type-species*. But it must not be supposed, with some theorists, that these types really exist as such in nature; they are merely examples or illustrations selected for convenience to serve as permanent fixed points in our groups, whatever be the extent which we may give to their boundaries. (Strickland 1845, p. 219; his emphasis)

Morphology Establishing a hierarchy of types as a true reflection of the Natural System did not progress linearly. Johann Wolfgang Goethe introduced the term "morphology" to designate the science devoted to studying the theory of organismal form. Despite its roots in an ancient tradition, this discipline primarily focuses on analyzing the similarities and differences among organisms, with particular attention to the organization and function of their constituent organs. Building on this foundation, in the late eighteenth- and early nineteenth-century Germany, figures from the *Naturphilosophie* school—such as Goethe, Lorenz Oken, and Carl Gustav Carus—developed a type concept unrelated to the study of classification (morphological type). For them, the notion of type was not employed to "naturally" cluster organisms. Instead, they resorted to this concept—called "Unity of Type"—to identify the same structures (homologues) within a higher group of organisms (e.g., animals). This research methodology was independently pursued in France, mainly by Étienne Geoffroy Saint-Hilaire (1818), and much later in England by Richard Owen through the notion of the Archetype (1848). Conversely, Georges Cuvier espoused a different perspective on morphology, which gained prominence in

France in the early nineteenth century, positing that organs were shaped by their functions. Within this framework, Cuvier proposed four body plans for animals, asserting that they lacked any comparative relationship.[7] Nevertheless, although this version of the type stands apart from the taxonomic hierarchical arrangement, it nonetheless exerted a nuanced and intricate influence on its structuring.

In 1828, embryologist Karl Ernst von Baer used the type concept to challenge the prevailing theory of recapitulation. According to his observations, for example, all vertebrates in their early developmental stages exhibited common characteristics. Yet there was no evidence of any progressive sequence of characteristics from supposedly inferior adult invertebrates during the early developmental stages of vertebrates. For von Baer—as for Cuvier—animals displayed four body plans, each apparently without any connection of similarity. Building on these observations, von Baer formulated the idea that type characters appear first, followed by class characters, then order characters, and so on. In conclusion, embryology stands out as a process within the type, occurring from the general to the specific.

Under this framework, the implication of taxonomic realism began to take shape. For example, thirty years later, approaching it from an embryological standpoint,[8] Louis Agassiz doubted whether the Natural System merely served as a tool to facilitate the study of Natural History:

> Is not this in itself evidence enough that *genera, families, orders, classes, and types* have the same foundation in nature as *species*, and that the individuals living at the time have alone a material existence, they being the bearers, not only of all the different categories of structure upon which the natural system of animals is founded, but also of all the relations which animals sustain to the surrounding world—thus showing that species do not exist in nature in a different way from the higher groups, as is so generally believed? (Agassiz 1859, p. 8; emphasis added)

Note that for both von Baer and Agassiz, the term "type" solely refers to the designation for the higher group of organisms—like Cuvier's classification of animals into four body plans—and the taxonomic categories below the type are seen as varying degrees of differentiation of the type. Here, it is interesting to highlight that, as far as I know, Agassiz was among the first naturalists to admit the reality of the higher taxa. Nevertheless, the ultimate cause of their reality stems from an argument rooted in natural theology. For example, he views the type and the taxonomic categories as "translations into human language of the thoughts of the Creator" (p. 9).

Evolution Although the term "evolution" was initially associated with the idea of the embryological alteration of types, it gradually came to be linked to the transfor-

---

[7] These different schools in morphology mark the well-known debate between formalism and functionalism (see Russell 1916; Asma 1996; Gould 2002; Amundson 2005; Ochoa and Barahona 2014).

[8] In contrast to von Baer, Agassiz embraced the theory of recapitulation. Despite rejecting von Baer's law, Agassiz believed that embryo offspring could only recapitulate the adult stages of lower animals within the confines of the type (see Gould 1977).

mation of types throughout their phylogenetic history (Richards 1992). As the following section will delve into detail, with the advent of the modern theory of evolution by natural selection proposed in 1858 by Charles Darwin, the type concept gained an evolutionary framework, i.e., types were no longer perceived as an abstract metaphysical construct, but rather as evidence of genealogical connections. Thereby, types obtained a sort of material reality. I suspect this shift likely indirectly impacted the realism of the Linnaean hierarchy, particularly regarding the construction of a hierarchy of inclusive types.

For example, in early twentieth-century Germany, Goethe's morphology (along with his concept of the Unity of Type) and its evolutionary interpretation (by figures such as Ernst Haeckel and Carl Gegenbaur, who adopted Darwin's interpretation of the type during the late nineteenth century)[9] paved way for the emergence of morphological systematics or "idealistic morphology," epitomized primarily by the Swiss zoologist Adolf Naef. It is worth noting that this version of systematics differs in many aspects from the systematics later developed by Willi Hennig,[10] as it was more realistic regarding the structure of the Natural System. In broad terms, morphological systematists, such as Naef, claimed that the phylogenetic tree was structured around a series of types organized hierarchically from greater to lesser inclusiveness, i.e., it was represented by a hierarchy of types (see Fig. 2.1.). This arrangement was not arbitrary but based on a comprehensive research program encompassing morphology, embryology, and paleontology, all aimed at elucidating the Natural System (Rieppel et al. 2013). As Sokal (1962, p. 240) holds: "Taxa are real in the system of the typologist, at least as real as the types."

As is evident, the typology within the Natural System, manifested as a hierarchy of types, neither stemmed from ancient ideas (rooted in Plato's concepts or Linnaeus's creationism) nor represented a metaphysical approach to taxonomy. Rather, it emerged relatively recently in the twentieth century under evolutionary considerations, arising from a rigorous collective research effort spanning approximately 184 years (see Fig. 2.2). Thus, this view not only contributed to synthesizing taxonomy, morphology, and evolution but also upheld a realistic interpretation of the Natural System. At its peak, one of the leading figures in modern typology, paleontologist Otto Schindewolf, wrote in 1950: "[W]e regard types as not just abstract patterns of form or merely idealized archetypes of taxonomic categories but, rather, as the *products of concrete, evolutionary processes*, as *particular evolutionary entities*" (Schindewolf 1993, p. 205; his emphasis).[11]

---

[9] See Sect. 2.2.2

[10] Particularly because Hennig formulated his method as a proposal far removed from the consideration of types. This resulted in a distorted view of the hierarchical nature of the Linnaean system. See Chap. 5, Sect. 5.1.3 and Chap. 8, Sect 8.1.1.

[11] Regarding Schindewolf's theory, see Chap. 3, Sect. 3.2.1.

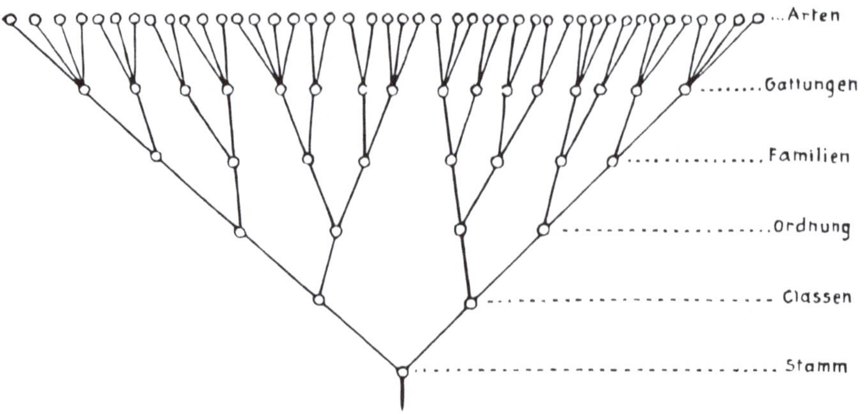

Fig. 2.1 *An ideal phylogenetic tree representing a hierarchy of types according to Adolf Naef.* Under this view, each node represents a type (category), and the branching lines represent the connection between inclusive types. Note that Naef arranges the categories of the same level on a horizontal line from the bottom up, indicating differences in scale and implying that evolution occurs at the boundaries of the general type (Stamm). Notice the similarity of this tree with Hennig's phylogenetic tree, which is shown in Fig. 8.2 in Chap. 8. However, as other chapters will explore, Hennig's tree is a false hierarchy of inclusion. In Hennig's scheme, the nodes are not aligned horizontally to represent categories; instead, they are positioned at varying heights, indicating points of divergence at the species level. (Modified from Naef 1919)

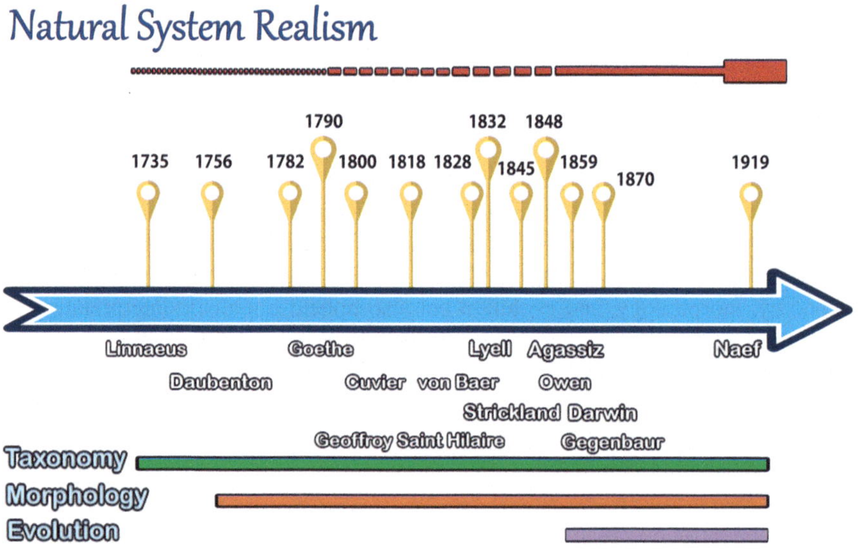

Fig. 2.2 *From a typological perspective, this timeline represents the contributions of taxonomy, morphology, and evolution to the formation of the Natural System.* The blue arrow indicates some dates and relevant naturalists that were mentioned in the writing. Below, the involved disciplines are depicted with colored lines representing the timing of their integration. Above, a red line represents the degree of increase in credibility regarding the realism of the Natural System. Thinner dotted lines symbolize low credibility, while solid and thicker lines symbolize higher credibility. Basically, the diagram shows how the type concept—through the synthesis of taxonomy, morphology, and evolution—constructed a realistic framework of the Natural System

2.2 Archetypes, Ancestors, and Homologues

According to general agreement, upon adopting an evolutionary view, the concept of the "Unity of Type"—viewed as an anti-evolutionary idea that refers to a preexisting mold in the mind of God—was replaced by that of the "common ancestor." This shift in perspective stands as a pivotal milestone in the field of biological classification. It led to the narrative of the "Darwinian revolution" and fostered dichotomies such as "typological thinking" versus "population thinking," or "old systematics" versus "new systematics." For example, in his seminal work *The Growth of Biological Thought* (1982), Ernst Mayr highlighted the challenges faced by "idealistic morphologists" who struggled to explain why organisms exhibited the "unity of plan." They did not understand why there was a static pattern of connections among organisms despite their structures being accommodated to different functions. However, Darwin untangled this puzzle by recasting these patterns into an evolutionary framework:

> The real explanation, says Darwin, is as simple as the egg of Columbus. All mammals, birds, or insects share the same morphological type, resulting in an extraordinary anatomical similarity, because they all have descended from a common ancestor from whom they have inherited this structural pattern. Natural selection will be constantly at work to modify the components of this pattern so as to make them most efficient for the functions they have to serve but this does not require destruction of the basic pattern. **Darwin thus replaced the archetype of idealistic morphology by the common ancestor**. (Mayr 1982, pp. 464–465; emphasis added)

Mayr further elucidates that this shift also influenced our modern understanding of the concept of homology. For example, while Richard Owen initially coined the term "homology" in the context of typology, lacking a strictly evolutionary explanation, Darwin prompted a redefinition of homologous structures: "Attributes of two organisms are homologous when they are derived from an equivalent characteristic of the common ancestor." Interestingly, Mayr notes that "Darwin himself never clearly gave this definition but it is implied in his discussions." In addition, he argues that this redefinition of homologues "is applicable not only to structural elements but to any other properties, including behavioral ones, which might have been derived by inheritance from a common ancestor" (p. 465).

However, this narrative of the term "type" being replaced by that of "common ancestor" is nothing but one of the historical interpretations crafted by actors of the Modern Synthesis, driven by their aversion to typology—a sentiment evident throughout the development of this conceptual framework (see Mayr 1959; Simpson 1959; Witteveen 2015, 2016a). Nonetheless, this historical narrative not only undermined the evolutionary type concept but also hindered the contemporary understanding of the phenomenon of structural constraints, homology, and the establishment of the Natural System. In order to promote a more favorable perspective on the type concept and homology, this section will explore some historical insights into the use of two distinct notions of type in an evolutionary context: First, the general type, or Archetype, served as a concept for identifying homologous structures within a higher group of organisms (e.g., either animals or plants); second, the specific type, or *Bauplan*, denotes the morphological stability of characters perceived hierarchically in both ontogeny and phylogeny.

In this regard, the Archetype (general type) was entirely independent of the common ancestor; that is, it was not directly comparable to it and served as a methodological model to identify homologous structures. Similarly, the *Bauplan* (specific type) was not entirely comparable to the common ancestor either. For example, within the framework of evolution, the type concept shed its metaphysical connotations, transitioning instead to recognizing that these patterns of constraints (stability) had an evolutionary explanation. While the type here sometimes refers to an ancestor, it can also denote a specific developmental stage. In other words, through the type concept, both phylogeny (ancestors) and ontogeny (embryos) present genealogical connections and structural constraints.

Considering this, the idea of evolution neither significantly altered the morphological program nor replaced the type concept. Put differently, typology and evolution are not opposing doctrines but compatible ones. A brief analysis of the morphological ideas of Richard Owen and Charles Darwin will demonstrate this.

2.2.1 Richard Owen and the Evolutionary Archetype

Richard Owen, one of the most prominent scientific figures of the Victorian era, earned the moniker of British Cuvier by his peers due to his mastery of morphology. However, despite this epithet, Owen's philosophical leanings aligned more closely with the ideas of transcendental morphology advocated by Oken, Goethe, and Carus as well as Geoffroy Saint Hilaire's philosophical anatomy. His goals mirrored theirs, namely, the exploration of anatomical relationships based on a common body plan that underlies all morphological structures in their most basic form and its subsequent modifications that account for diversity. This body plan, known as the "Unity of Type," was expressly referred to by Owen as the Archetype, with particular emphasis on the Vertebrate Archetype (Fig. 2.3).[12] According to Rupke (1993, p. 231): "THE VERTEBRATE ARCHETYPE… is one of the most fascinating constructs of what has been called the 'morphological period' in the history of biology… It represented the fullest expression of a belief in the fundamental relatedness, if not of all organisms, at least of all animals with endoskeletons."

[12] As mentioned earlier, naturalists of the time held the belief that there were at least four distinct types of animals with no relation to each other, a notion primarily championed by Georges Cuvier. Others, such as Geoffroy Saint-Hilaire, argued that all animals could be reduced to a single type. Indeed, this difference of opinion constituted one of the most interesting contentious issues regarding the type during that period. It should be noted that Owen supported Cuvier's view for a long time. Yet, in his later works, he began to advocate for the idea of a single type encompassing all animals (see Ochoa and Barahona 2014, p. 83).

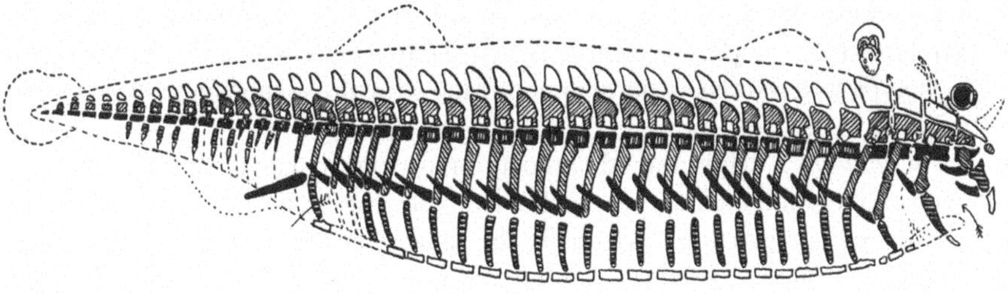

Fig. 2.3 *The Vertebrate Archetype, according to Richard Owen* (1849)

Admittedly, Owen presented the Archetype as a Platonic idea and a preexisting mold in the mind of God. According to this view, the Creator designed a primordial and common framework for organisms, unfolding into different forms for each group to fulfill specific functions according to their natural purpose. However, as some scholars rightly point out (e.g., Desmond 1982; Rupke 1994; Camardi 2001; Sloan 2003), the Vertebrate Archetype was more closely associated with the scientific methodologies and practices of its time rather than being solely a metaphysical belief: "It is no mere transcendental dream, but a true knowledge and legitimate fruit of inductive research, that clear insight into the essential nature of the organ…" (Owen 1849, p. 70). As Sloan (2003, p. 59) asserts: "[T]he archetype both is discovered inductively by generalization from anatomical research, summarizing the results of the comparative study of a wide range of vertebrate skeletons, and is a genuine Idea with an internal source that is also grounded in the divine mind."

The main point is that Owen (1848), basing his reasoning on the idea of the Archetype, began to explain the biological phenomenon through "natural laws," which govern the elaboration and modification of organisms over geological time. For example, Owen delineates two types of forces driving organismal development: the "polarizing force" and the "adaptive force" or "specific organizing principle." The polarizing force operates during embryonic development, maintaining the unity of the organization and the equality of forms. Conversely, using the same underlying structures, the adaptive force adjusts organisms to different lifestyles. By considering these forces, Owen concluded that the same structures constrain vertebrate organisms via the polarizing force, while variations in these structures arise from the adaptive force.

Considering this, Owen (1848) introduced his research program on homologous structures, categorizing them into three different types:

1. *Special homology*: "[It is] the determination of which homology indicates that such animals are constructed on a common type" (p. 7). For example, if the bone structures of a whale's fin, a bat's wing, and a human's arm are compared, they can be recognized as homologous because they are built from the same basic elements.
2. *General homology*: "A higher relation of homology… in which a part or series of parts stands to the fundamental or general type, and its enunciation involves

and implies a knowledge of the type on which a natural group of animals, the vertebrate for example, is constructed" (p. 7). In simple terms, by employing the Archetype as a reference, the source of each structure composing the vertebrate body can be traced.

3. *Serial homology*: "[It is a] kind of repetition or representative relation in the segments of the same skeleton" (p. 8). For example, the Archetype is represented by a set of segments that are essentially the same. Hence, the bodies of vertebrate organisms comprise a sequence of repetitions of a fundamental structure (ideal vertebra).

Given this context, it could be argued that, for Owen, the Vertebrate Archetype served as a tool enabling him to identify homologues and, thereby, formulate hypotheses about the transformation of these structures to suit their various functions. Therefore, the debate begins here, considering that this dynamic process was neither based on a static view of species nor the independent creation of species. Some researchers have pointed out that, as early as the 1840s, Owen was already considering that species had a natural biological origin (Richards 1987; Rupke 1994, 2024; Camardi 2001; Gould 2002; Amundson 2005; Ochoa and Barahona 2014). Still, doubts persist regarding his evolutionary stance during that period (Caponi 2013). After the publication of Darwin's *On the Origin of Species*, Owen publicly admitted his acceptance of evolution. Here, it is noteworthy that for any naturalist embracing evolution as a fact, the concept of the "common ancestor" holds important significance. Now, if homologues correspond to the Archetype, how did Owen maintain the conception of the Archetype considering that homologues serve as evidence of ancestral relationships? Did Owen change his concept of the Archetype to align with the idea of common ancestry once he adopted evolution?

To answer these questions, one must first grasp Owen's evolutionary perspective. According to Camardi (2001), Owen's embrace of evolution stemmed, not directly from Darwin, but from the implication of von Baer's law. As discussed in the previous section, von Baer proposed that all vertebrates share basic characteristics during early embryonic development. Subsequently, as their development progresses, specific characteristics diverge sequentially, beginning with those of classes followed by those of orders, those of families, then those of genera, and finally those of species. Provided that developmental divergence aligns in parallel with the sequence of historical transitions observed in the fossil record, Owen interpreted this developmental divergence as a reflection of events occurring within lineages over time (Ospovat 1981, pp. 136–137; Camardi 2001, pp. 493–496).

Desmond (1989) asserts that Owen had gleaned from von Baer's law that both recapitulation and Lamarck's progressive transformism were difficult to sustain. But this was not a blanket critique of evolutionism. Owen's evolutionary progress needed to occur from the general to the special; organisms obtained their basic structures from the beginning and subsequently developed specific differentiations over time: "progressive departure from a general to a special type" (Owen 1863, p. 66). Regarding this, Camardi (2001) concludes that, for Owen, evolution occurs only within the confines of the Archetype. Amundson (2005, 2007) arrives at a similar conclusion suggesting that species are simple variations of the Archetype. In contrast, for Owen, the Archetype did not represent the earliest ancestor of vertebrates: an organism that lived in the past and transmitted its traits to its offspring.[13] Instead, the Archetype served as the model whereby the same structures are represented, with which all vertebrates begin. Therefore, species represent the modifications of those structures over time "under every variety of form and function."

In other words, the Archetype was a necessary device in Owen's research program methodology, facilitating the recognition of homologues. However, it could never equate to an ancestor, as ancestral species were governed by the same law of constraint as present-day species.[14] The Archetype represents the foundational blueprint of vertebrate structure, applicable to both past and present species. Nonetheless, Owen's view did not negate the notion of common ancestry within a specific group of organisms. Ancestors are real, whereas the Archetype is merely an "artifice." In his evolutionary phase, in *Monograph on the Aye-Aye*—the first work in which he openly declared his belief in evolution—Owen (1863 p. 66; emphasis added) wrote: "That this law [of adaptive force] works by derivation of one species from a previous species, of a new from an old species, is made probable by demonstrated *unity of plan.*" Similarly, in volume three of his *On the Anatomy of Vertebrates*, Owen maintained that:

> To demonstrate the evidence of the community of organisation, I found that the *artifice* of an *archetype vertebrate* animal was as essential as that of the *archetype plant* had been to Goethe in expressing analogous ideas; and as the like reference to an "ideal type" must be to all who undertake to make intelligible the "unity in variety" pervading any group of organisms. (Owen1868, pp. 788–789; emphasis added)

Owen's references to the Unity of Type can be seen as an artifice, even considering his alignment with evolutionary principles at the time. If this is the case, it becomes plausible to argue that the Archetype is entirely compatible with evolution. Of course, Owen still maintains the idea of the invariability of structures present in a higher group of organisms. In contrast to the adaptive force, the polarizing force acts as a mechanism for stability, even extending to the most fundamental piece that can be abstracted. While organisms undergo modifications, they always preserve their basic structures. This underscores the essence of homologues, irrespective of the phylogeny.

An essential aspect of this discussion is that the Archetype, as a device for recognizing the same structures, is independent of the concept of the "specific type." While the Archetype aids in recognizing homologues, specific types represent abstract models that help perceive the stability parameters within the organization of specific related groups at different scales. A point of clarification is needed before proceeding. In order to differentiate the general type from the concept of the specific

---

[13] This is indeed one of Darwin's interpretations, as will be discussed later on.

[14] As well as any species that could exist anywhere in the universe, according to Jenner's (2022) interpretation.

type, the former can be referred to as the Archetype, while the latter might be called *Bauplan*. So, in Owen's analysis of aye-aye evolution, attention should be paid to his language in delineating and associating the *Bauplan* with the "common ancestor":

> … the Aye-aye has been derived, in common with other existing *Lemuridae*, from some *pre-existent animal* [this is the common ancestor] of a more generalized *Lemurine type of organization* [this is the *Bauplan*], in departing from which it has gained a character, e.g. the dental one, very like that which prevails in the *Rodentia*, without losing the more *numerous and essential characters of its inherited Lemurine organization*. The terms in which the anatomist would express the sum of his observations on the structural resemblances traceable from the Aye-aye throughout the *Lemuridae* would be, that the principle of "*unity of organization*" prevailed through such group. And such term would have a more intelligible meaning on the hypothesis that these singularly diversified Lemurs were genetically related by descent from *a common ancestral form*. (Owen 1863, p. 67; emphasis added)

Keep in mind that when Owen (1868, p. 807; emphasis added) states: "I deem an innate tendency to deviate from *parental type*… whereby species have been derived one from the other," he employs the notion of Bauplan to emphasize a genealogical relationship. Note that this "parental type" does not correspond to a common ancestor. Still, this association may seem odd for researchers committed to the idea of types as eternal and metaphysical entities. Simply stated, the parental type denotes the ancestral *Bauplan* from which the descendants were derived. However, as far as the research has shown, the use of types aligns well with the evolutionary approach. In short, typology and evolution are complementary doctrines.

2.2.2 Charles Darwin and the Typological Common Ancestor

When Charles Darwin embarked on developing his theory of evolution in the 1840s, his objective was to translate the "laws" and biological phenomena described by his contemporaries through the lens of "descent with modification." Darwin aimed to unify all branches of natural history—paleontology, biogeography, taxonomy, morphology, embryology, among others—into a coherent argument. When addressing morphology, Darwin encountered two fundamental laws: the "conditions of existence" (relating to the functional purposes that adapt organisms to different lifestyles) and the "Unity of Type." According to Darwin's theory, the law of the conditions of existence "is fully embraced by the principle of natural selection," while the Unity of Type "is explained by unity of descent" (Darwin 1859, p. 206). Although much historical discussion has been devoted to Darwin's view on the preference for these laws—known as the conflict between formalism and functionalism (e.g., Russell 1916; Asma 1996; Gould 2002, Amundson 2005; Ochoa and Barahona 2014)—here, the specific focus will be on Darwin's interpretation of the type phenomenon and its relationship with the notion of common ancestry.

As mentioned in the previous section, the idea of type did not stem from a particular school of thought. Instead, this concept emerged from observing structural constraints underlying organisms: "We have seen that the members of the same class, independently of their habits of life, resemble each other in the general plan or their organisation" (Darwin 1859, p. 434). For Darwin, the constancy of these traits was evidence that evolution occurs, i.e., this pattern was proof of genealogical relationships between groups. While Darwin refers to the Unity of Type as the "unity of descent," it is important to understand that his writings consistently recognize the phenomenon of structural constraints. In fact, the sole distinction between Owen's type and Darwin's type is that Darwin removed the metaphysical connotation of the type, which was associated with its origin from God's plan. Darwin materialized the type by portraying it either as the earliest embryonic state or as the earliest ancestor that lived in ancient times within the group in question. In a letter to Thomas Henry Huxley, Darwin remarked: "I should have thought that the archetype in imagination was always in some degree embryonic" (Darwin in Darwin and Seward 1903, p. 73). Meanwhile, on the back of his copy of Owen's *On the Nature of Limbs*, he wrote: "I look at Owen's Archetypes as more than ideal, as a real representation as far as the most consummate skill & loftiest generalization can represent the parent form of the Vertebrata" (Darwin in Desmond 1982, p. 50).

However, Darwin's conception of type is not very different from that conceived by his morphological predecessors. Asma (1996, p. 116), for example, suggests that Darwin's materialistic view of the Unity of Type closely resembles that of Étienne Geoffroy Saint Hilaire. From the perspective of both naturalists, the projection of the type preexists due to the laws of structural constraints. Darwin interprets this law as "the strong principle of inheritance" (or the less altered steady state of form), which manifests during the earliest stage of embryonic development and in the earliest adult ancestor of the group under consideration (see below).

Nevertheless, the issue of constraints became more intricate within Darwin's evolutionary framework because, behind this array of phenomena, natural selection had to play a primary role in evolutionary change. If evolution occurs gradually, for example, how does this process contribute to organisms departing from the type? In theory, Darwin (1859) posited that the "whole organisation seems to have become plastic, and tends to depart in some small degree from that of the parental type" (p. 12). Despite this, morphological evidence suggested the opposite: "In changes of this nature [gradualism], there will be little or no tendency to modify the original pattern" (p. 435). Interestingly, Darwin upheld Owen's definitions of homology and maintained the core argument of homologues as structures that represent a common archetypical organization: "What can be more curious than that the hand of a man, formed for grasping, that of a mole for digging, the leg of the horse, the paddle of the porpoise, and the wing of the bat, should all be constructed on the same pattern, and should include the same bones, in the same relative positions?" (pp. 434–435). Therefore, despite vertebrate limbs acquiring different adaptations, "in all this great amount of modification there will be no tendency to alter the framework of bones or the relative connexion of the several parts" (p. 435).

In Darwin's view, the departure of descendant organisms from the parental type occurs through gradual change. This process involved small heritable variations that accumulated over geological time, gradually contributed to the formation of distinct basic frameworks. However, "as the whole amount of modification will have been effected by slight successive steps, we need not wonder at discovering in such parts or organs, a certain degree of fundamental resemblance, *retained by the strong principle of inheritance*" (p. 438, emphasis added).

Another important point to discuss is the hierarchy of types. Unlike Owen's perspective, where the Archetype served as a methodological tool or an abstract model to identify homologues, Darwin's view is that the Archetype represented a depiction of any common ancestor.

> If we suppose that the *ancient progenitor, the archetype as it may be called, of all mammals*, had its limbs constructed on the existing general pattern, for whatever purpose they served, we can at once perceive the plain signification of the homologous construction of the limbs throughout the whole class. (Darwin 1859, p. 435; emphasis added)

In other words, the Archetype can be depicted as the common ancestor of vertebrates, but it can also be interpreted as the common ancestor of mammals, even though their homologous characters may be recognizable. This suggests that Linnaean categories may denote hierarchical representations of different archetypical levels: "[M]any species descended from a single progenitor grouped into genera; and the genera are included in, or subordinate to, sub-families, families, and orders, all united into one class" (p. 413). This view of the hierarchical structure, however, was not easy to interpret within Darwin's scheme either. Mostly because this was primarily an issue related to embryology.

Darwin's primary challenge stemmed from von Baer's law. Initially, von Baer's embryological conclusions likely contributed to Darwin's formulation of the principle of divergence of the Natural System (Fig. 2.4). For example, in 1837, Martin Barry illustrated a diagram resembling a genealogical tree to depict von Baer's embryological theory of type divergence (Fig. 2.4a). Richards (1992, p. 111) demonstrates a striking resemblance between this diagram and the famous phylogenetic tree that Darwin drew the same year in his Notebook B (the initial sketch where Darwin conceived evolution). Interestingly, Darwin wrote there: "Thus genera would be formed—bearing relation to ancient types" (Fig. 2.4b). According to Richards' interpretation, "the nodes representing the generic patterns would also represent the morphological types of the ancestors."

Nonetheless, von Baer's law implied two problematic aspects of Darwin's mature theory. First, in their early stages, the embryos of a closely related group exhibited unchanged structures, which differentiated in later stages of development. Although this embryological similarity was evidence of ancestry for Darwin, these patterns also revealed structural constraints. So, how did natural selection influence the change in the form? Since embryos exhibited few modifications in their early stages, natural selection seemed to operate only on variations emerging after birth. If this was the case, how were new variations introduced in the ontogeny of the descendants? Darwin aimed to adequately address this dynamic of processes of constraint and embryological change through his theory (Ospovat 1981), but in my view, he fell short.

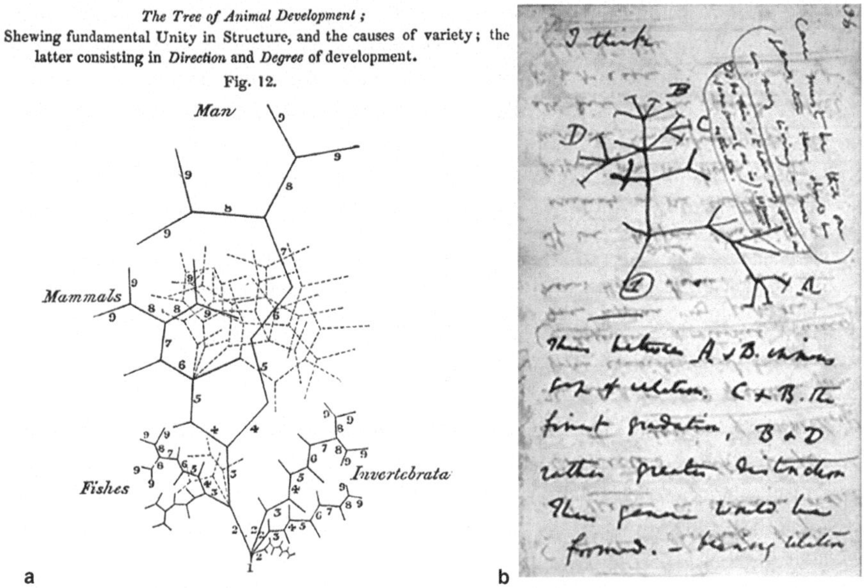

Fig. 2.4 *A possible source of Darwin's principle of divergence from von Baer's law.* (**a**) Martin Barry's diagram depicting divergence during development according to von Baer's law (From Barry 1837). (**b**) The famous phylogenetic tree drawn by Darwin in his Notebook **b**. According to Richards (1992), the similarity between both trees is evident. If this is the case, Darwin might have conceived the principle of evolutionary divergence through the implications of von Baer's law. Interestingly, this perspective holds a typological context; see the text for more details

Second, as mentioned earlier, von Baer's law suggests a diverging tree that progresses from the general to the specific. In other words, the characters of the general type (phylum) arise first, followed by those of specific types (class, order, family, genus, and species). For Darwin, embryos in their early stages provide a glimpse into the past, revealing the form that the earliest ancestor of the group likely possessed, which would have had less differentiated structures: "[W]e may readily believe that the unknown progenitor of the vertebrata possessed many vertebrae; the unknown progenitor of the articulata, many segments; and the unknown progenitor of flowering plants, many spiral whorls of leaves" (p. 437). Thus far, Darwin's observations seem to align perfectly with von Baer's findings and the apparent parallelism in organisms' evolution over geological time. However, Darwin's theory demands two requirements. The first is his commitment to gradualism, and the second is that the line of evolutionary succession must occur from the specific (species) to the general (genera, families, classes, orders, etc.). Accepting von Baer's law implies that general characters constrain evolutionary changes from top to bottom. For instance, according to Owen, evolutionary changes occur at the confines of the Vertebrate Archetype. If Darwin accepted this, specific changes would be expected to manifest at the limit of the earliest vertebrate ancestor. But how did the most remote vertebrate ancestor come into being? Darwin (1872, p. 384) maintained that:

"On the theory of natural selection, we can, to a certain extent, answer these questions. We need not here consider how the bodies of some animals first became divided into a series of segments... for such questions are almost beyond investigation."[15] In addition, von Baer's law could lead to the endorsement of saltational evolution, as some naturalists of Darwin's time interpreted it. Thus, as Richards (1992) suggests, Darwin resolved these challenges by embracing the theory of recapitulation, which was a necessary deduction for his theory of evolution:

> The archetype is the ancestor. But if you add to von Baer's law Darwin's theory of adaptation (in which new traits are joined to the terminal, that is, more adult phases of development) and situate both in the framework of descent (in which the archetype is the ancestor), then it follows that ideally *the embryo must recapitulate* the ancient forms of its adult progenitors. (Richards 1992, p. 166; his emphasis)

In contrast, Gould (1977, p. 73) proposes that Darwin embraced von Baer's law over recapitulation and suggests that Darwin teetered on the brink of accepting recapitulation when he stated that "[a]s the embryonic state of each species and group of species partially shows us the structure of their less modified ancient progenitors, we can clearly see why ancient and extinct forms of life should resemble the embryos of their descendants,—our existing species" (Darwin, 1859, p. 449). Despite this passage, Darwin refrained from outright endorsement, leaving the door ajar for further evidence to validate the hypothesis: "I am bound to confess that I only hope to see the law hereafter proved true" (p. 449). This series of inconsistencies in Darwin's writings suggests that his theory—which focused primarily on small changes and natural selection—encountered many difficulties when facing the phenomenon of structural constraints. Thus, it would be possible to assert that the synthesis between taxonomy, morphology, and evolution remained incomplete from the Darwinian perspective. Although Darwin made some contributions, particularly by materializing the type within the framework of evolution, he could not fully integrate his central arguments on gradualism and natural selection.

Another aspect to contemplate is how historians later interpreted the narrative surrounding the common ancestor as a replacement for the type (hereafter called the "replacement paradigm"). Jenner (2022), for example, argues that the idea of common ancestry, in the modern evolutionary sense, emerged due to a metaphysical revolution wherein the conception of the Archetype (as a universal law) was replaced by the recognition of a particular (and, therefore, contingent) ancestral species visualized in deep time.[16] However, what remains notable from Darwin's thinking is that despite equating the Archetype with the earliest ancestor, he also associated it with

---

[15] Interestingly, in the first edition of the *Origin of Species*, Darwin (p. 437; emphasis added) wrote: "On the theory of natural selection, we can *satisfactorily* answer these questions." Surely, upon reconsideration, he changed his mind, replacing the word "satisfactorily" with the more cautionary phrase "to a certain extent."

[16] However, Delisle (2019, p. 59) categorizes ancestors metaphysically as "shadows" since they are not entirely tangible and are mentioned from abstractions projected in geological times, even though they may be considered real.

the earliest state of the embryo. For example, although Darwin (1872) occasionally alluded to homologues as ancestral structures, he primarily relied on embryology to explain their metaphysical nature: "[r]elation between parts which results from their development from corresponding embryonic parts" (p. 434). Consequently, it becomes apparent that evolution did not replace the type with the common ancestor, as both the earliest ancestor and the earliest embryonic state served as explanation devices for Darwin to address the phenomenon of structural constraints. Notably, whether depicted as ancestors or embryonic states, types are portrayed as the least differentiated structural states. This encapsulates the precise question that Darwin attempted to explain: Why do organisms maintain an unchanging pattern despite their successive modifications? Jenner (2022, p. 36) points out that "[t]he mysterious 'force by which all the modifications of the vertebrate skeleton… are still subordinated to a common type'[17] ... is simply inheritance from common ancestors" in Darwin's explanation. But this assertion is not entirely accurate, as the underlying cause that allows either the ancestors or embryonic states to maintain the least differentiated states—structural constraints—differs from the straightforward explanation that the descendant characters were inherited from a common ancestor.[18] Thus, "the strong principle of inheritance" remained mysterious even within Darwin's writings.

For this reason, with the advent of evolution, it becomes evident that there were no significant changes, neither in the morphological program (see Nyhart 1995) nor in the use of the type concept (see Di Gregorio 1995; Camardi 2001; Amundson 2005). For example, Thomas Henry Huxley (1853) appropriated the term Archetype to describe the body plan of mollusks. Huxley clarified that his conception of the Archetype did not pertain "to any real or imaginary 'ideas' upon which animal forms are modeled"; rather, it served akin to "the diagram to a geometrical theorem, and like it at once imaginary and true" (p. 50). In short, Huxley employed the Archetype as a device to identify homologous structures in mollusks, akin to what Owen did with the Vertebrate Archetype. Through this concept, for example, he could conclude that: "the Cephalopoda and Gasteropoda are morphologically one, are modifications of the same archetypal molluscous form" (p. 62). Interestingly, in corresponding with Huxley, Darwin wrote: "The discovery of the type or 'idea' (in your sense, for I detest the word as used by Owen, Agassiz & Co) of each great class, I cannot doubt, is one of the very highest ends of Natural History" (Darwin in Darwin and Seward 1903, p. 73).

Another interesting case to consider is that of Carl Gegenbaur, a prominent figure in German morphology. In an early study, probably working under the replacement paradigm, Coleman (1976) argued that, during the 1850s, Gegenbaur adhered to the morphological program of types and homologies, characterized by the school of Goethe, Oken, and Owen. However, after reading Darwin's *On the Origin of Species*, Gegenbaur relinquished the idea of type, instead adopting the notion of

---

[17] Owen (1848, p. 171) as cited by Jenner.

[18] This discussion is akin to the current causal explanation of the homologues; see Chap. 8, Sect. 8.2.

common ancestry: “Gegenbaur abandoned the zoologist’s long-familiar and useful type concept and gave full allegiance to the idea that patterns of relationship among organisms were due exclusively to genealogical connection. His change of mind rested upon acceptance of the descent theory” (p. 150). Nevertheless, upon revisiting Gegenbaur’s works from this new angle, Coleman observed that certain elements of the typological approach remained integrated into Gegenbaur’s thought. Notably, Coleman concludes his analysis by arguing that: “Reconciliation of the two interpretations [typology and evolution] was possible, but it could only be effected by the typologists at the cost of ignoring the central process of evolutionary change [natural selection]” (p. 174).

Some years later, Di Gregorio (1995) embarked on a similar study of Gegenbaur, considering Haeckel’s perspective as well. But unlike Coleman, Di Gregorio steered his analysis away from the replacement paradigm. According to Di Gregorio, Gegenbaur indeed integrated Darwin’s common descent doctrine into the morphological program of homologies. In so doing, he gave historical connotation to both the recognition of homologies and the patterns of structural arrangement in organisms. Still, Di Gregorio noticed that there was no substantial departure in Gegenbaur’s methodology concerning homologies, as it continued to echo Owen’s work. Di Gregorio posits that acceptance of Darwin’s view on common ancestry was only possible because Darwin explained types as evidence of genealogical connections without neglecting the morphological and embryological aspects of structural constraint: “This, then, is ‘evolutionary typology’: a particular marriage of descent and development, which preserves a focal role for the morphological type, and thus for comparative anatomy and the study of homology” (p. 258). Nevertheless, Di Gregorio ultimately concludes that this scheme is not “Darwinian” after all because morphologists such as Gegenbaur were not really committed to Darwin’s main argument. Put differently, their approach resembled Owen’s focus on homology studies rather than Darwin’s theory of natural selection and individual variation: “the old wolf had survived in sheep’s clothing.” However, the overarching conclusion of this study undoubtedly indicates that the idea of the type not only survived the advent of evolutionary theory but also gained strength by assuming an evolutionary significance. Such types were no longer mere metaphysical “ideas”; rather, they were conceived as real entities based on years of extensive empirical research:

> For such as Gegenbaur, Huxley, or Haeckel, the best term might be “empirical realist,” allowing for ontological commitment to constructs such as types—that is, to entities beyond the concrete physical organism; these commitments are strictly constrained by empirical research and thus not perniciously “ideal.” (Di Gregorio 1995, p. 278)

In conclusion, there are two ways in which the type concept can be conceived within an evolutionary framework. First, the Archetype served as a methodological model to recognize homologues within a higher group of organisms. In this context, the type is not equivalent to the common ancestor, as each concept entails distinct ontological implications. This interpretation is consistent with how Owen and Huxley employed the type concept. Second, the *Bauplan* facilitated the recognition of stable patterns of characters, perceived hierarchically in both phylogeny and ontogeny. This interpretation is consistent with how Owen, Darwin, and Gegenbaur employed the type concept. In this context, for example, while Darwin associated types with ancestors, they also manifested through different states during embryonic development.

When comparing Owen's and Darwin's interpretations and uses of the type concept, it should be noted that, despite their differences, there is no discernible shift in the perception of structural constraint—except for Darwin's reinforcement of the type through a materialistic explanation (see Table 2.1). Given this, it can be demonstrated that the type concept was not replaced by that of the common ancestor. Rather, the type continued to prevail and gained greater support since it became a

Table 2.1 *A comparison of Charles Darwin's and Richard Owen's interpretations and applications of the type concept*

	Owen	Darwin
General type	Archetype	Earliest ancestor of a lineage Earliest embryological steady state of an individual
Specific type (*Bauplan*)	Type (unity of organization at different scales)	Specific ancestors of a lineage Later specific embryological steady states of an individual
Ultimate causation of the Archetype/ ancestor	Divine plan	Evidence of a parental relationship
Structural constraints explanation	Law of polarizing force	The strong principle of inheritance
Homologues definition	The same structures, shown in the Archetype, under every variety of form and function	Relationship between parts, which results from their development from corresponding embryonic parts
Can an Archetype be used as a device?	Yes	No, but the general type as an "idea" could be used for methodological purposes.[a]
Can an Archetype be equivalent to an ancestor?	No	Yes, the earliest ancestor but also the earliest embryonic steady state

It should be noted that neither naturalist replaces the type with the common ancestor. In the case of Owen, despite the fact that he refers to the Archetype as a divine plan, he uses the term for the identification of homologue structures. In addition, Owen uses the specific type to denote steady states in the morphological organization. In contrast, Darwin did not replace the Archetype, but rather he materialized it; instead of representing it as an "idea," he offered the Archetype a real expression in either the earliest common ancestor or the earliest embryonical steady state. Darwin explains the specific type similarly to the Archetype, but this involves a hierarchical view based on ontogeny and phylogeny. In this regard, Darwin's definition of homologue structures has an embryological context, and its explanation lies in parental relationships. Furthermore, although Darwin materialized the type, he did not oppose the view of the Archetype as an "idea" since the general type might have methodological purposes in the program of morphology. Finally, Owen and Darwin use different elements to describe types, and it follows that they recognize the phenomenon of structural constraints. Owen accounted for this phenomenon with the law of polarizing force, while Darwin explained it as the strong principle of inheritance

[a]Conclusion based on Darwin's letter to Huxley (see Darwin and Seward 1903, p. 73)

complementary concept for morphological explanations in light of evolutionary theory.

Unfortunately, this scientific progress was hindered and replaced during the mid-twentieth century by the success of the Modern Synthesis and its competitive alternative known as population thinking. In order to promote his version of the Natural System, Ernst Mayr wrote:

> What was the new systematics? It was not a specific technique and is, perhaps, best described as a viewpoint, an attitude, or a general philosophy. It started primarily as a rebellion against the nominalistic, typological, and thoroughly nonbiological approach of certain (alas, all too many) taxonomists of the preceding period. The new systematist appreciates that all organisms occur in nature as members of populations. He studies the biological properties of organisms rather than the static characters of dead specimens. He utilizes the greatest number of kinds of characters—physiological, biochemical, and behavioral characters as well as morphological ones... (Mayr 1982, pp. 276–277)
>
> A true understanding of natural selection, speciation, and adaptation was not possible until population thinking had displaced typological thinking. (Mayr 1982, p. 561)

2.3 Conclusions

Undoubtedly, the essentialism story has tarnished the reputation of the type concept, leading to distorted narratives that fuel inadequate debates and confusing interpretations about what the units of evolution are. In other words, the traditional narrative developed from the Modern Synthesis framework has impeded our understanding of typology's significance in the development of the Natural System. For example, some philosophers of biology argue that, because the Linnaean hierarchy was initially steeped in a creationist and essentialist context, its relevance is now challenged in light of contemporary evolutionary theory (Ereshefsky 2001; Kunz 2012). They argue, for instance, that categories lack any ontological meaning; thus, a hierarchical structure, as traditionally conceived for representing the Natural System, is no longer necessary (e.g., de Queiroz and Gauthier 1994; Ereshefsky 2001).

However, the evidence presented here suggests quite the contrary. The type does not stem from Platonic idealism, nor is it an ancient concept rooted in species creationism. Instead, it is a concept born out of the observable structural pattern in closely related groups of organisms, suggesting the presence of hierarchically operating structural constraints within their internal properties. Typology has facilitated a seamless integration between taxonomic practices, morphological empiricism, and its congruence with evolutionary theory. The type concept has provided a coherent explanation for applying the Linnaean system, which has proven instrumental in representing the hierarchical order observed in evolving biological entities. If this narrative holds true, it is possible to conclude that recognizing the type through structural constraints has been a necessary phenomenon in our comprehension of the Natural System. Notably, the emergence of the H-Evo-Devo Theory reinstates the relevance of typology by emphasizing structural constraints, thereby endeavoring to comprehend the structure of the Natural System through a contemporary lens. Typology is back, and it is here to stay.

References

Agassiz L (1859) An essay on classification. Longman, Brown, Green, Longmans, and Roberts, London

Amundson R (2005) The changing role of the embryo in evolutionary thought: roots of evo-devo. Cambridge University Press, Cambridge

Amundson R (2007) Richard Owen and animal form. In: Owen R (first published 1849) On the nature of limbs: a discourse. Facsimile edition. The University of Chicago Press, Chicago, pp XV–LI

Asma ST (1996) Following form and function: a philosophical archeology of life science. Northwestern University Press, Illinois

Barry M (1837) Further observations on the unity of structure in the animal kingdom, and on congenital anomalies, including "hermaphrodites"; with some remarks on embryology. Edinb New Philos J 22:345–364

Camardi G (2001) Richard Owen, morphology and evolution. J Hist Biol 34:481–515

Caponi G (2013) Entre el Dios de Paley y el Dios de Bonnet: el parco evolucionismo teísta de Richard Owen. Principia 17:71–101

Chung (2003) On the origin of the typological/population distinction in Ernst Mayr's changing views of species, 1942–1959. Stud Hist Phil Biol and Biomed Sci 34:277–296

Coleman W (1976) Morphology between type concept and descent theory. J Hist Med Allied Sci 31:149–175

Darwin C (1859) On the origin of species by means of natural selection, or preservation of favored races in the struggle for life. John Murray, London

Darwin C (1872) The origin of species by means of natural selection, or preservation of favored races in the struggle for life, 6th edn. John Murray, London

Darwin F, Seward AC (eds) (1903) More letters of Charles Darwin: a record of his work in a series of hitherto unpublished letters. Volume 1. John Murray, London

de Queiroz K, Gauthier J (1994) Toward a phylogenetic system of biological nomenclature. Trends Ecol Evol 9:27–31

Delisle RG (2019) Charles Darwin's incomplete revolution: the origin of species and the static worldview. Springer, Cham

Desmond A (1982) Archetypes and ancestors: palaeontology in Victorian London 1850–1875. Blond and Briggs, London

Desmond A (1989) The politics of evolution: morphology, medicine, and reform in radical London. The University of Chicago Press, Chicago/London

Di Gregorio MA (1995) A wolf in sheep's clothing: Carl Gegenbaur, Ernst Haeckel, the vertebral theory of the skull, and the survival of Richard Owen. J Hist Biol 28:247–280

Ereshefsky M (2001) The poverty of the Linnaean hierarchy: a philosophical study of biological taxonomy. Cambridge University Press, Cambridge

Farber PL (1975) Buffon and Daubenton: divergent traditions within the histoire naturelle. Isis 66:63–74

Farber PL (1976) The type-concept in zoology during the first half of the nineteenth century. J Hist Biol 9:93–119

Gould SJ (1977) Ontogeny and phylogeny. The Belknap Press of Harvard University Press, Cambridge, MA

Gould SJ (2002) The structure of evolutionary theory. The Belknap Press of Harvard University Press, Cambridge, MA

Huxley TH (1853) On the morphology of the Cephalous Mollusca, as illustrated by the anatomy of certain Heteropoda and Pteropoda collected during the voyage of H.M.S. "Rattlesnake" in 1846–50. Philos Trans R Soc Lond 143:29–66

Jenner RA (2022) Ancestors in evolutionary biology: linear thinking about branching trees. Cambridge University Press, Cambridge

Kunz W (2012) Do species exist?: principles of taxonomic classification. Wiley-Blackwell, Weinheim
Linnaeus C (1753) Species plantarum, exhibentes plantas rite cognitas, ad genera relatas, cum differentiis specificis, nominibus trivialibus, synonymis selectis, locis natalibus, secundum systema sexuale digestas. Tome I. Laurentius Salvius, Stockholm
Lyell C (1832) Principles of geology: being an attempt to explain the former changes of the earth's surface, by reference to causes now in operation. Volume 2. John Murray, London
Mayr E (1959) Darwin and the evolutionary theory in biology. In: Meggers BJ (ed) Evolution and anthropology: a centennial appraisal. The Anthropological Society of Washington, Washington, DC, pp 1–10
Mayr E (1982) The growth of biological thought: diversity, evolution, and inheritance. The Belknap Press of Harvard University Press, Cambridge, MA
Müller-Wille S (2011) Making sense of essentialism. Crit Q 53:61–77
Naef A (1919) Idealistische Morphologie und Phylogenetik: Zur Methodik der systematischen Morphologie. Verlag von Gustav Fischer, Jena
Nyhart L (1995) Biology takes form: animal morphology and the German universities, 1800–1900. The University of Chicago Press, Chicago and London
Ochoa C, Barahona A (2014) El Jano de la morfología: de la homología a la homoplasia, historia, debates y evoluci6n. UNAM; Centro de Estudios Filosóficos, Políticos y Sociales Vicente Lombardo Toledano, México
Ospovat D (1981) The development of Darwin's theory: natural history, natural theology, and natural selection, 1838–1859. Cambridge University Press, Cambridge
Owen R (1848) On the archetype and homologies of the vertebrate skeleton. John van Voorst, London
Owen R (1849) On the nature of limbs: a discourse. John van Voorst, London
Owen R (1863) Monograph on the aye-aye (*Chiromys madagascariensis*, Cuvier). Taylor and Francis, London
Owen R (1868) On the anatomy of vertebrates. Volume 3, Mammals. Longmans, Green and Co, London
Pavlinov IY (2023) The species problem: a conceptual history. CRC Press, Boca Raton
Richards E (1987) A question of property rights: Richard Owen's evolutionism reassessed. Br J Hist Sci 20:129–171
Richards RJ (1992) The meaning of evolution: the morphological construction and ideological reconstruction of Darwin's theory. The University of Chicago Press, Chicago
Richards RA (2010) The species problem: a philosophical analysis. Cambridge University Press, New York
Rieppel O, Williams DM, Ebach MC (2013) Adolf Naef (1883–1949): on foundational concepts and principles of systematic morphology. J Hist Biol 46:445–510
Rupke NA (1993) Richard Owen's vertebrate archetype. Isis 84:231–251
Rupke NA (1994) Richard Owen: Victorian naturalist. Yale University Press, New Haven/London
Rupke NA (2024) Richard Owen's deconstruction of Darwinian natural selection. In: Delisle RG, Esposito M, Ceccarelli D (eds) Unity and disunity in evolutionary biology: deconstructing Darwinism. Springer, Cham, pp 123–140
Russell ES (1916) Form and function: a contribution to the history of animal morphology. John Murray, London
Schindewolf OH (1993) Basic questions in paleontology: geologic time, organic evolution, and biological systematics. Translated by Judith Schaefer. Original (1950). University of Chicago Press, Chicago
Simpson GG (1959) Anatomy and morphology: classification and evolution: 1859 and 1959. Proc Am Philos Soc 103:286–306
Sloan PR (2003) Whewell's philosophy of discovery and the archetype of the vertebrate skeleton: the role of German philosophy of science in Richard Owen's biology. Ann Sci 60:39–61
Sokal RR (1962) Typology and empiricism in taxonomy. J Theoret Biol 3:230–267

Strickland H (1845) Report on the recent progress and present state of ornithology. Report of the fourteenth meeting of the British Association for the Advancement of Science. John Murray, Albemarle Street, pp 170–221

van der Hammen L (1981) Type-concept, higher classification and evolution. Acta Biotheor 30:3–48

Wagner MJ (2024) Typology/population distinction and its role in the marginalization of nineteenth-century non-Darwinian theories in modern historiography. In: Delisle RG, Esposito M, Ceccarelli D (eds) Unity and disunity in evolutionary biology: deconstructing Darwinism. Springer, Cham, pp 277–300

Whewell W (1847) The philosophy of the inductive sciences, founded upon their history. John W Parker, London

Wilkins JS (2009a) Defining species: a sourcebook from antiquity to today. Peter Lang, New York

Wilkins JS (2009b) Species: a history of the idea. University of California Press, Berkeley/Los Angeles/London

Wilkins JS (2013) Essentialism in biology. In: Kampourakis K (ed) The philosophy of biology: a companion for educators, history, philosophy and theory of the life sciences. Springer, Dordrecht, pp 395–419

Winsor MP (2003) Non-essentialist methods in pre-Darwinian taxonomy. Biol Philos 18:387–400

Winsor MP (2006) The creation of the essentialism story: an exercise in metahistory. Hist Phil Life Sci 28:149–174

Witteveen J (2015) "A temporary oversimplification": Mayr, Simpson, Dobzhansky, and the origins of the typology/population dichotomy (part 1 of 2). Stud Hist Phil Biol Biomed Sci 54:20–33

Witteveen J (2016a) "A temporary oversimplification": Mayr, Simpson, Dobzhansky, and the origins of the typology/population dichotomy (part 2 of 2). Stud Hist Phil Biol Biomed Sci 57:96–105

Witteveen J (2016b) Suppressing synonymy with a homonym: the emergence of the nomenclatural type concept in nineteenth century natural history. J Hist Biol 49:135–189

Witteveen J (2020) Linnaeus, the essentialism story, and the question of types. Taxon 69:1141–1149. https://doi.org/10.1002/tax.12346

Hierarchical Evolutionary Synthesis 3

Abstract

This chapter explores the history of the H-Evo-Devo Theory that covers the late nineteenth century to mid-twentieth century, a crucial period intertwined with the emergence and establishment of the Modern Synthesis. First, it challenges the narrative of the "synthesis" between Darwinism and Mendelism, shedding light on the significant role played by anti-Darwinian theories such as neo-Lamarckism, orthogenesis, and saltationism. Notably, it argues that these theories are rooted in an organism-centered perspective, emphasizing the organism's interaction with its environment and developmental processes to explain the complexities of variation phenomena. In so doing, it concludes that population geneticists did not resolve the issues of Darwinism, as their interpretations were confined to a gene-centered perspective. However, adopting an organism-centered perspective is needed in order to gain insight into the behavior of variation as well as the phenomena of inheritance. Second, it illustrates that Goldschmidt and Schindewolf (representing the Unmodern Synthesis) along with Dobzhansky, Mayr, and Simpson (who represent the Modern Synthesis) formulated their evolutionary ideas under a hierarchical view. Despite their perspectives stemming from different evolutionary frameworks, the following common points are recognized for the emergence of a potential hierarchy theory of evolution: (1) It represents a hierarchy of types, (2) it presents two decoupled evolutionary levels, and (3) it portrays a phylogenetic tree constructed from the top down, rather than bottom up.

C. Ochoa, *Hierarchical Evolutionary-Developmental Theory*, Evolutionary Biology – New Perspectives on Its Development 9,
https://doi.org/10.1007/978-3-031-88723-9_3

Keywords

H-Evo-Devo Theory · Modern synthesis · Unmodern synthesis · Neo-Lamarckism · Orthogenesis · Saltationism · Darwinism · Organism-centered perspective · Gene-centered perspective · The eclipse of Darwinism · Mendelism · Population genetics · Evo-Devo · Typology · Goldschmidt · Schindewolf · Dobzhansky · Mayr · Simpson

This chapter covers the late nineteenth century to the mid-twentieth century. In traditional historiography, this period is used to highlight the achievements of the Modern Synthesis. However, the narrative presented here primarily focuses on the historical foundation that erects the H-Evo-Devo Theory, which revitalizes the role of the "organism" in evolutionary biology and reinterprets the units of evolution through a new taxonomic-hierarchical perspective. Under this approach, I delve into two main aspects: First, I examine the transition from an organism-centered perspective to a gene-centered perspective in early twentieth-century evolutionary discussions. Second, I explore the persistent consideration of a hierarchical structure in any synthetic evolutionary model.

Considering the above, Sect. 3.1 challenges the history of the "synthesis" between Darwinism and Mendelism. It will become clear that anti-Darwinian theories—such as neo-Lamarckism, orthogenesis, and saltationism—emerged from an organism-centered perspective, which focuses on the organism's interaction with the environment and developmental processes to explain the complexities of variation phenomena. Through this inquiry, it will be observed that the problems of variation and inheritance within Darwinism persisted despite the emergence of population genetics. While population geneticists seemingly settled the debate, they did so by overlooking the phenomena of variation highlighted by the anti-Darwinians, phenomena that could not be adequately explained from a gene-centered perspective. Section 3.2 examines the ideas of Goldschmidt and Schindewolf (representing the Unmodern Synthesis) alongside those of Dobzhansky, Mayr, and Simpson (representing the Modern Synthesis) regarding an evolutionary hierarchical view. Despite arising from different theoretical frameworks, their proposals converge on some key aspects to construct a potential hierarchy theory of evolution that (1) represents a hierarchy of types, (2) presents two decoupled evolutionary levels, and (3) depicts a phylogenetic tree constructed from the top down rather than bottom up.

3.1 The Eclipse of the Organism

The term "Modern Synthesis" refers to an evolutionary framework developed between the 1930s and 1940s, which involves the integration of different disciplines to support the claim that population genetics mechanisms are sufficient to account for evolution at all levels (Ochoa 2017). Needless to say, the Modern Synthesis framework has significantly influenced most evolutionary studies since the 1950s, e.g., educational programs and contemporary evolutionary models are centered around this framework. According to the traditional view, the Modern Synthesis was introduced as a widespread agreement in which (a) Darwin's original idea was supported, (b) previously incompatible research traditions were synthesized, and (c) anti-Darwinian evolutionary theories were excluded, which hindered interdisciplinary agreement (Mayr 1980). Moreover, the development of the Modern Synthesis occurred in two stages: (1) the fusion of Mendelian inheritance theory with Darwin's theory, leading to the emergence of population genetics and (2) the consensus and integration of traditional biological disciplines (e.g., experimental genetics, systematics, paleontology, etc.) to endorse the assertion of the first stage (Gould 2002).

A central period that underpins this historical narrative is the interval between the publication of Darwin's *On the Origin of Species* in 1859 and the emergence of population genetics in the 1930s. This timeframe is commonly referred to as the "eclipse of Darwinism." Coined by Julian Huxley (1942)—who remarked that Darwinism was going through a dark period of acceptance—the term gained widespread traction in modern historiography, notably following Peter Bowler's (1983) book, which stressed the significance of theories that did not align with Darwin's point of view.

The narrative of the "eclipse of Darwinism" is told as follows: Despite the relative success of Darwin's ideas, by the late nineteenth and early twentieth centuries, the theory of natural selection faced significant skepticism among most evolutionists. Three alternative theories challenging Darwin's approach emerged during this period: neo-Lamarckism (or the inheritance of acquired characters), orthogenesis, and saltationism. However, despite their initial dominance, Darwinism subsequently gained more acceptance due to its integration with Mendelian theory, which effectively explained mechanisms of inheritance and variation. Consequently, anti-Darwinian theories vanished, and the Modern Synthesis emerged as the prevailing model, reinforcing Darwin's central argument.

In this regard, some authors have raised concerns about the conceptual implications of the term "eclipse of Darwinism" when linked to the historical process leading to the Modern Synthesis (see Largent 2009; Delisle 2017, pp. 159–160; Meulendijks 2021; Baedke et al. 2024; Ceccarelli 2024; Wagner 2024). As Wagner (2024) has demonstrated in a more detailed analysis, the early interpretations of the "eclipse" period were grounded in erroneous historiographical methods that blamed anti-Darwinian theories for being irrational or based on outdated philosophical assumptions. For example, although Bowler (1983) originally intended to highlight the contributions of anti-Darwinian research, he regarded the reconciliation between Darwinism and Mendelism as inevitable.[1] Likewise, according to him, theories such

[1] Traditionally, given that Mendelism was rooted in saltationism, historians have argued that the synthesis between Mendelism and gradual evolution became feasible as early researchers recognized their compatibility. However, a closer examination reveals a challenge to this narrative: The early Mendelians were also deeply immersed in embryology; hence, a synthesis between Mendelism and gradual evolution was practically impossible (see below; also see Ochoa 2017, 2021b).

as neo-Lamarckism and orthogenesis simply collapsed due to their reliance on outdated and teleological perspectives based on embryonic development, which were incompatible with modern interpretations of inheritance. In light of this, Amundson (2005, pp. 107–108) points out that this period might be more significant than Bowler initially considered, as these anti-Darwinian theories endeavored to address issues that have reemerged in contemporary debates: "I consider those same [anti-Darwinian] theories to be legitimate and productive attempts to solve problems that have been marginalized during most of the twentieth century, problems that are once again receiving scientific attention." In fact, much later, with the advancement of Evo-Devo, Bowler ended up challenging his initial argument:

> In recent decades, our interpretation of this episode has been transformed by the emergence of evolutionary developmental biology. This has reopened issues once marginalized by genetics and the modern Darwinian synthesis. Some enthusiasts see 'evo-devo' as reintroducing a role for non-selectionist factors such as Lamarckism, while even those sceptical of this view acknowledge that the older theories were not as wide of the mark as was once claimed. (Bowler 2017, p. 196)

Although there is a consensus that the term "eclipse of Darwinism" is associated with historiography that favors the achievements of the Modern Synthesis, questions persist regarding the value of this period in understanding early evolutionary debates as well as their connection to contemporary knowledge.[2] As presented below, the Modern Synthesis framework was not a comprehensive solution to Darwinian issues surrounding variation. The origin of phenotypic variation is more complex than merely invoking genetic mechanisms to explain it.[3] Considering this, even though the mechanisms postulated by the anti-Darwinians to explain observed patterns of variation were not entirely clear, their arguments warrant serious consideration and should not be overlooked.

Below, it is revealed that, while the issues surrounding variation appeared to have been resolved within the framework of the Modern Synthesis, it was achieved at the cost of disregarding an organism-centered perspective. Instead, it attempted to explain all phenomena of inheritance and variation through a gene-centered perspective. In other words, population geneticists prioritized genes as the main actors in evolution, perceiving them as the hereditary medium through which lineages persist over time, and their alteration simply accounted for the origin of new variations. From this point of view, the organism and its internal properties were deemed irrelevant to the study of evolution. In this context, it becomes clear that early population geneticists lacked the tools to explain the patterns of variation highlighted by the anti-Darwinians. Instead, these patterns might have found a solution through embryological aspects, such as the complex interactions occurring during the developmental process, structural constraints, the organism's responses to environmental stimuli, etc.

---

[2] See Wagner (2024) for an alternative perspective that explores new avenues for interpreting this period.

[3] See, for example, Chaps. 11 and 12.

Thus, the persistent belief among scientists and historians that the synthesis between Darwinism and Mendelism unraveled the problems of evolution leads to misconceptions in contemporary interpretations of evolutionary theory (see Stoltzfus 2017; Ochoa 2024). Indeed, Amundson and Bowler are correct in suggesting that anti-Darwinian theories might hold relevance today, particularly in a reevaluation of embryology. The Modern Synthesis did not provide any definitive solution to the challenges Darwinism faced regarding the facts of variation. Reinterpreting these early debates is impossible without bringing this "eclipse of the organism" to light (Walsh 2015), as neglecting it distorts not only the history of evolutionary thought but also contemporary debates on the theory of evolution.

3.1.1 Organism-Centered Perspective

The organism-centered perspective entails viewing the organism as the basic unit of biological systems. The organism is a complex biological entity composed of hierarchical internal parts that directly interact with the environment, shaping its development and evolutionary trajectory. This perspective has its roots in "organicism," a philosophical school in biology that emerged between the 1910s and 1930s, championed by select groups of specialized biologists from various fields and countries (Nicholson and Gawne 2015; Esposito 2017; Baedke 2019; Baedke and Fábregas-Tejeda 2023).

Initially, organicism arose as a sort of synthesis between mechanism (which aimed to reduce biological phenomena to physicochemical processes) and vitalism (which sought to find some exclusive element or force in living beings not reducible to physicochemical laws). These scientists concluded that living beings are not machines and that they do not possess special elements that characterize them either; thus, they opted for a "third way." In other words, organicism took its holistic perspective of the organization of living beings from vitalism, recognizing that it cannot be reduced to simple inorganic processes alone, and, from mechanism, it retained that the processes of living beings are subject to material explanations (Etxeberria and Umerez 2006; Nicholson and Gawne 2015; Peterson 2016; Esposito 2017).

However, in their later years, organicists devoted considerable attention to challenging the reductionist view of population genetics, which posited that the phenotypic outcomes were exclusively determined by hereditary genetic material (Nicholson and Gawne 2015; Esposito 2017). Notable organicists involved in these discussions included Ludwig von Bartalanffy, John Scott Haldane, Lawrence Joseph Henderson, Oscar Hertwig, Conway Lloyd Morgan, Joseph Needham, William Emerson Ritter, Edward Stuart Russell, Julius Schaxel, Conrad Hal Waddington, Joseph Henry Woodger, Charles Otis Whitman, among others. In any case, according to Baedke and Fábregas-Tejeda (2023), this organicist movement was characterized by attributing three fundamental roles to organisms:

1. *Contextualizing parts in development.* The organism must be considered as a unified whole constructed by itself through a complex dynamic interaction among its constituent parts.
2. *Framing organism-environment causal reciprocity.* The environment directly influences the organism, which can actively respond in positive ways during its development. Concurrently, the organism influences the environment in which it lives.
3. *Understanding the role of agential processes in evolution.* The organism is an active entity that shapes its own evolution. It maintains a stable organization despite rearrangements of its parts, can react to a new environment favorably, chooses the environment it can live in, and influences its subsequent evolution through its behavior.

Nevertheless, this does not imply that, before this conceptual framework, the "organism," as a fundamental biological entity, was excluded from evolutionary discussions. Late nineteenth-century naturalists examined the internal properties of the organisms, including their interactions with the environment, and set forth hypotheses about the origin of variation based on these observations. For example, although embryonic development is an important property within the organism-centered perspective, certain scientists supported developmental models based on either mechanism (exemplified by Wilhelm Roux) or vitalism (as seen in the work of Hans Driesch). The distinction lies in the following points regarding organicism:

1. It existed as a movement aiming to maintain the concept of "organism" as a central entity within the study of biology (Baedke and Fábregas-Tejeda 2023).
2. It made it possible to conceptualize a hierarchical view of developmental processes, highlighting that the organism cannot be understood by reducing it to one of its parts (Nicholson and Gawne 2015).
3. It emphasized the intimate relationship between the organism and its environment (Baedke and Fábregas-Tejeda 2023).
4. It motivated some researchers to establish a definitive synthesis between development and evolution (Esposito 2017).

Below, it will become clear that anti-Darwinian theories—such as neo-Lamarckism, orthogenesis, and saltationism—adopted an organism-centered perspective. In other words, these anti-Darwinian theories aligned with the criteria established by organicists, recognizing the organism as the basic unit of biology and evolution.[4]

[4]This does not imply that only anti-Darwinian theories adhered to an organism-centered perspective and that Darwinism would not find support within this conceptual framework. As Walsh (2015, p. 63) rightly points out, Darwin also considered the organism as the fundamental unit of organization, unlike the perspective of the Modern Synthesis, which focused on the gene.

3.1.2 Anti-Darwinian Theories as Organism-Centered Frameworks

According to historical records, late nineteenth- and early twentieth-century Darwinism was defined in terms of the nature of variation, emphasizing the creative role of natural selection (Kellogg 1907; Gayon 1998; Gould 2002; Beatty 2016; Ochoa 2024). In order to consider natural selection as a creative force, two primary conditions must be met. First, selection must primarily act on continuous variation. Over geological time, this process would explain the origin of macroevolutionary biological units (e.g., new structures, species, higher taxa). Second, variation must be random (isotopic); under such circumstances, selection would be able to choose the most favorable variations without limitation. Thus, anti-Darwinian theories can be defined by models that fail to meet these criteria for variation (e.g., Kellogg 1907; Bowler 1983; Gould 2002; Ochoa 2017). Saltationism, for example, challenges the relevance of continuity in variation, while Lamarckism and orthogenesis disagree on the randomness of variation. Nonetheless, what becomes evident is that the studies in this period pointed toward questioning the traditional view of Darwinism and grew alongside an organism-centered perspective of biology. This perspective, however, was overshadowed by a gene-centered perspective of evolution.

Neo-Lamarckism[5] This is the idea that organisms acquire new variations due to their direct interaction with the environment, which stimulates them to change favorably during their lifetime. Subsequently, after a few generations, offspring inherit these new traits from birth without requiring environmental stimulus. Associating neo-Lamarckism with organicism is far from new. In fact, such a framework grasps the idea that organisms are active agents, not passive ones. In other words, organisms have the capacity to self-adjust to new circumstances by either interacting with the environment or adopting new behavior. Initially, Bowler (1983) noted that the organicists "were expounding a theme that had once lain at the heart of Lamarckism—yet their supporters made no reference to the inheritance of acquired characters" (p. 105), with some exceptions. More recently, Esposito (2017, p. 227) acknowledged that "[a]lthough [John Scott] Haldane did not mention Lamarck directly, his vision was much more akin to a Lamarckian understanding of

[5] Although the term "Lamarckism" is commonly associated with Jean-Baptiste Lamarck, this movement is not precisely a return to his original ideas. Late-nineteenth century naturalists began calling it such due to their recognition of certain phenomena that were reminiscent of Lamarck's idea of what is popularly known as the "inheritance of acquired characters." In fact, historians have identified different evolutionary schools of thought under the same term. Bowler (2017) argues that this presents challenges for those seeking a unified movement. For instance, Loison (2011) states that, in France, at least two types of Lamarckism gained prominence at different periods. One was materialistic, focusing on adaptation, while the other was vitalistic and progressive. The description in this paragraph refers to the former, covering the period between 1880 and 1950, which is based solely on experimental aspects and whose explanations are rooted in natural processes. See also Wagner (2024, p. 22), who provides other categorizations.

the evolutionary process than it was to the blind mechanism of Darwinian selection." But, here, the issue is not precisely whether the organicists recognized their views within the category of "inheritance of acquired characters." Rather, it deals with delineating which biological phenomena support the interpretation that organisms are active agents that develop and evolve on their own. This statement fundamentally embodies "the heart of Lamarckism" from the late nineteenth and early twentieth centuries.

For instance, Edward Drinker Cope (1896) categorized the phenomenon of the "inheritance of acquired characters" into physiogenesis and kinetogenesis. Physiogenesis involves the influence of physical factors of the environment that lead to structural modification in response to new circumstances. Instances of physiogenesis include the growth or reduction of mollusk shells when exposed to different environments, changes in animal pigmentation in response to light availability (e.g., pigment decrease in the absence of light), alteration in animal coloration due to either temperature or diet, cave-dwelling fish exhibiting reduced eyes, among others. Moreover, kinetogenesis concerns how use and disuse affect body structure from a mechanical standpoint. Examples of this phenomenon include the adjustment of muscles and bones to different circumstances. When these structures are exercised, they strengthen, while lack of use can lead to a decrease in size or strength. For example, in bipedal animals that leap with their hind legs or in animals that climb using their front legs, these limbs elongate accordingly. Similarly, the development of calluses in animals or the formation of human palms and soles due to repeated friction with external surface exemplifies this process. In all these instances, variation initially emerges as an organism's response to environmental stimuli; yet subsequently, such variation occurs without external influences.

The challenges to neo-Lamarckism stemmed from theories of inheritance endorsed by the Darwinian perspective, such as August Weismann's theory of germ-plasm continuity (Gould 2002, p. 201; Loison and Herring 2017)[6] and the genetics advocated by proponents of the Modern Synthesis, which in one way or another supported Weismann's ideas. According to Bowler (1983), neo-Lamarckism was not refuted because of a lack of experimental evidence; rather, scientists began to lose interest once Mendelism started to be defined as the correct conceptual framework for inheritance. Another explanation is that neo-Lamarckists attempted to adjust the theory within the limitations imposed by Mendelian inheritance, but they were unsuccessful in doing so. Bowler ensured that the only viable option for Lamarckism's survival was for its proponents to formulate their own mechanistic theory of inheritance as a worthy rival to Mendelism, which never came to pass.

However, an examination of historical records reveals that none of these inheritance frameworks impeded the organicist interpretations associated with the

[6]According to Weismann, the germ-plasm (formed by germ cells) is isolated from the somatic plasma (formed by cells from the rest of the body). Any environmental influence affecting the somatic plasma is deemed irrelevant to the individual's inheritance, which depends entirely on the germ-plasm.

phenomena stressed by Lamarckists. For example, contrary to Weismann's view, in 1892, Oscar Hertwig argued that an organism's development relies not on a single cell but rather on the interaction among cells during its development, which is also mediated by the environment. Building upon this, Hertwig presented empirical examples of inheritance of acquired characters (Bowler 1983, pp. 91–92). Similarly, in the early twentieth century, while paleontologist Henry Fairfield Osborn did express doubts about the strict inheritance of acquired characters, he still believed that organisms would undergo adjustments during embryonic development to meet environmental demands. Specifically, he proposed that vertebrate bodies solve various mechanical problems during their development by adapting to factors like gravity, water resistance, and air resistance (Ochoa and Barahona 2014, p. 184). In fact, early geneticists such as William Bateson, Wilhelm Johannsen, and Richard Goldschmidt supported the idea that phenotype was not solely determined by gene action but rather that an organism's body responded to environmental stimuli, even though these changes were not exclusively inherited (see further details below).

Moreover, following Osborn's and C. Lloyd Morgan's considerations, psychologist James Mark Baldwin (1897) proposed an alternative to consider both hereditary perspectives. According to him, any organism acquires favorable characteristics through environmental stimuli during its lifetime; if those characteristics persist in the same environment over generations, the organism will eventually manifest spontaneous variations in the same adaptive direction. These variations will gradually accumulate through natural selection, resulting in the inheritance of these characteristics.[7] In a similar vein, albeit more sophisticated, a comparable explanation emerged in the early genetics era from Conrad Hal Waddington, a prominent figure in the organicist movement. In a 1942 paper, Waddington argued that the inheritance of acquired characters could be feasible through a developmental phenomenon termed "canalization."[8] He proposed that environmental stimuli could trigger a specific developmental pathway that is subject to selection. As some genes could induce the same variation along the same embryological pathway, a character might first be acquired due to environmental influence and then become "captured" by gene action. Using the example of ostrich callosities—developed in response to continuous friction during an individual's lifetime and observed to be inherited—he argued that: "[T]he callosities were formed as responses to external friction, but that during the course of evolution the environmental stimulus has been superseded by an internal genetical factor" (p. 563) (see Fig. 3.1a). Moreover, Waddington sought to test this hypothesis in a series of experiments with *Drosophila melanogaster* in the mid-twentieth century. In one experiment (1953b), he applied heat shock to pupae; in another (1956), he exposed young fly embryos to ether. Both actions simulated environmental stimuli. As a result of these inductions, he produced "crossveinless" and "bithorax" phenotypes, respectively (see Fig. 3.1b and c).

[7] Baldwin found this framework particularly helpful in explaining some instances of inherited behavior. Hence, this process has been labeled as "organic selection" or the "Baldwin effect."

[8] According to Waddington, developmental canalization refers to the stabilization of an organism's developmental pathways, despite the minor genetic or environmental changes it may undergo.

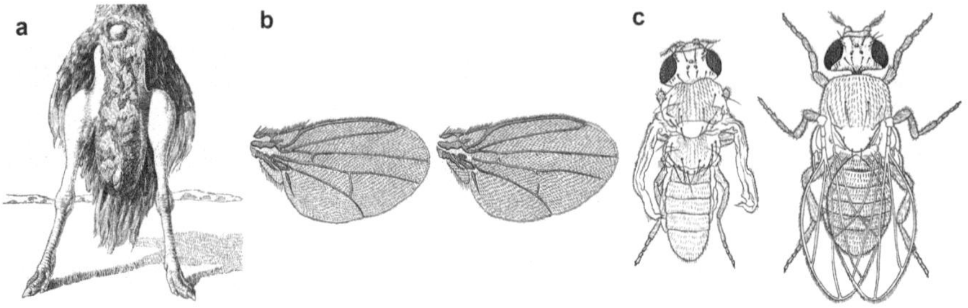

Fig. 3.1 *Evidence of the inheritance of acquired characters according to Waddington.* (**a**) Ostrich calluses. Ostriches develop these calluses, as depicted in the image, due to skin friction when they rest their weight while lying down. These calluses appear at pressure points on the breast and near the tail. Interestingly, similar growths also appear during embryonic development. (**b**) *Drosophila* wings. Typically, fruit flies have two cross veins (as illustrated on the left wing). However, exposure to temperature fluctuations during the pupal stage results in adult wings lacking cross veins (as seen on the right wing). (**c**) Modification in thoracic segments of *Drosophila*. On the left, a fruit fly exhibiting a thoracic segment modification induced by an ether treatment. On the right is a normal fruit fly. In cases like these, modifications begin with an environmental stimulus (friction, temperature, ether), and, subsequently, they persist to appear after several generations even without the environmental stimulus. (From Waddington 1953a)

Interestingly, Waddington found that, although these variations initially arose due to environmental stimulus, the flies subsequently developed these phenotypes over a few generations, even without the environmental trigger. He coined this phenomenon "genetic assimilation": "A character which had originally been an 'acquired' one might then be said to have become genetically assimilated" (Waddington 1956, p. 1).

In this regard, it is evident that the dismissal of the inheritance of acquired characters theory is more likely attributable to rhetorical issues rather than the trend of Mendelism, as initially proposed by Bowler (1983). However, as far as research shows, this theory undoubtedly had compelling experimental evidence, even during the era of genetics.

Orthogenesis This theory suggests that certain characters among related groups follow similar evolutionary trends, which are primarily attributed to processes occurring within organisms. For example, species of different genera within the same family can independently develop the same characters even if they do not necessarily confer adaptive advantage. Orthogeneticists asserted that an internal factor guided the trajectory of variation, and while natural selection could also produce these transformations, it was confined to acting with few alternatives of change (Ochoa and Barahona 2014; Ochoa 2017, 2021a).

Unfortunately, orthogenesis has been tarnished by its association with metaphysical, teleological, or divine notions. Yet, in recent years, some researchers have claimed that the theory was grounded in well-founded empirical and epistemological bases (e.g., Grehan and Ainsworth 1985; Gould 2002; Levit and Olsson 2006; Popov 2009, 2018; De Renzi 2014; Ochoa and Barahona 2014; Ulett 2014, 2016;

Ochoa 2017, 2021a). For example, in its descriptive sense, orthogenesis highlighted the independent origin of characters emerging in different groups of closely related organisms (a phenomenon currently associated with the term "parallelism"). Researchers endeavored to uncover the underlying causes of these trends, positing them as distinct from those influenced by external factors (e.g., either the inheritance of acquired characters or natural selection). In the late nineteenth century, some naturalists—including Theodor Eimer, E. D. Cope, and Alpheus Hyatt—employed the theory of recapitulation to support the idea that closely related organisms underwent internal channeling imposed by development that inevitably guided them toward the same pathway of variation. Similarly, paleontologist William Berryman Scott attributed the explanation to structural constraints. He suggested that variation was significantly restricted by the structure of an ancestral group, consequently limiting the range of options available for natural selection. By contrast, Osborn proposed that these tendencies originated in the germ-plasm through some yet-to-be-discovered "biomechanism" (see Ochoa and Barahona 2014).

In any case, in the context of the "eclipse of Darwinism," Bowler (1983) argued that when orthogeneticists attempted to fit their observations into the theory of recapitulation, they failed to provide a coherent explanation for these patterns of variation that was consistent with the Mendelian principle of inheritance. Bowler argued that this discrepancy contributed to the decline of orthogenesis. However, a closer examination of the perspectives of certain orthogeneticists reveals that they were not necessarily threatened by the burgeoning field of genetics. For example, the American organicist Charles Otis Whitman, in his work *Orthogenetic Evolution in Pigeons* (1919), criticized Hugo de Vries' Mendelism, particularly because this theory suggested that phenotypic characters are inherited entirely from one generation to another via specific hereditary factors, thus neglecting to recognize the organism's development as a unified process:

> The germ, as I believe and have long maintained, stands for an organized whole [...] [A]ll the features that arise in course of development are within the sphere of the individual unity and integral parts of it, and whatever specificity they possess is completely determined and not of independent origin... I am familiar with them [alleles], and deeply interested; but I am unable to see in them the sum total of all we know about heredity [...] Mendelism [...] neglects the natural history of the characters it experiments with and is not primarily concerned to know how characters have originated and multiplied. (Whitman 1919, p. 6)

Whitman's goal was to illustrate certain developmental laws underlying the color patterns of the pigeons' feathers. He stressed that the organism is constructed through a dynamic and complex interaction among its components rather than by relying solely on simple "chemical substitution" as suggested by Mendelism. Thus, according to Whitman, orthogenesis relied on a materialistic explanation concerning developmental constraints that operate along defined pathways in the channeling of variation without invoking any teleological process of divinity, as some scientists erroneously misinterpreted:

> If a designer sets limits to variation in order to reach a definite end, the direction of events is teleological; but if organization and the laws of development exclude some lines of variation and favor others, there is certainly nothing supernatural in this, and nothing which is incompatible with natural selection. (Whitman 1919, p. 11)

Even a few years later, as a prominent figure in the field of genetics, Richard Goldschmidt mentioned that orthogenetic tendencies could be attributed to developmental constraints. He proposed that, during the organism's development, only certain pathways allow for its normal progression. Mutations may disrupt numerous pathways, leading to an unviable outcome, but other pathways will lead to a proper and well-coordinated ontogeny:

> [I]f there are only a few avenues free for the action of mutational changes without knocking out of order the whole properly balanced system of reactions, the probability is exceedingly high that repeated mutations will go in the same direction, [that they] will be orthogenetic. (Goldschmidt 1933, p. 544)

Therefore, orthogenesis, as a phenomenon indicating that variation is channeled, found its explanation in developmental constraints, which remained in harmony with the progress in the field of genetics. Hence, a gene-centered perspective, which led to a misinterpretation of this theory, appears to be the primary reason for its decline, rather than the suggestion by some historians that it was due to the implications of Mendelism.

Saltationism This is primarily the idea that higher-ranking taxonomic units (e.g., species and higher taxa) emerge through sudden changes in the phenotype of descendants, occurring in one or a few generations. For example, proponents of this theory argue that the gradual changes hypothesized by Darwin (gradualism) are insufficient to account for the origin of characters of taxonomic significance.

Despite emerging as an alternative to Darwinism during Darwin's era, saltationism has historically been associated with early Mendelism. The historiography of the "eclipse of Darwinism" has recounted that early Darwinians held misconceptions about inheritance (e.g., blended inheritance), even though they had a correct understanding of evolution. Conversely, while early Mendelians understood the right principles of inheritance, they advocated for flawed saltational evolution. However, researchers had not yet realized that Mendelism served as the "missing piece" of Darwin's theory, which could solve issues related to inheritance and variation. Consequently, the reconciliation between these two schools became an inevitable outcome of scientific progress, paving way for the development of population genetics (see Provine 1971). Nevertheless, some contemporary researchers have begun challenging the "missing piece" narrative by realizing that a misinterpretation of early Mendelians has skewed history (e.g., Stoltzfus and Cable 2014; Ochoa 2017, 2021b). Notably, early Mendelians included embryonic developmental processes in their notion of inheritance. This incorporation reflects certain attributes of the "organism" highlighted by classical organicists. Thus, it is safe to assert that their viewpoint was organism-centered rather than gene-centered.

For example, Darden (1977, 1991) points out that early Mendelians did not believe genes were static material entities that were simply transmitted through germ cells across generations. Genes alone were insufficient to explain the organized dynamic changes observed during development. For example, William Bateson, regarded as the primary proponent of Mendelism in the early twentieth century, established that hereditary factors triggered complex embryological pathways. He even drew analogies from physics to account for different characters, such as animal color patterns and the repetition of morphological structures. Although his explanation was inspired by Wilhelm Roux's mechanism, Bateson considered a dynamic interaction of processes among the parts of organisms rather than merely invoking the simple action of genes. Moreover, while not advocating for the inheritance of acquired characters, Bateson accepted that the environment influences organisms' variation. In this regard, he proposed that many variations induced by the environment were continuous and—being nonheritable—were not subject to selection. This was one of the arguments he used to challenge the typical gradualism of Darwinism (Ochoa 2021b). Similarly, Wilhelm Johannsen, drawing inspiration from Oscar Hertwig's embryology, argued that an organism's phenotype resulted partly from interactions among developing parts and partly from environmental influence, notwithstanding the genetic constraints present in the species (i.e., the genotype):

> The agencies of normally varying ambient conditions and the interactions of specialized parts in the developing individual may exercise their strong influence upon the whole phenotypical state of the resulting particular individual. But these factors will as a rule not change or shift the fundamental genotypical constitution of the biotype in question. (Johannsen 1911, p. 154)

In Johannsen's argument, continuous variations were not extensions of particular genes but rather side effects of either environmental or developmental factors. Consequently, selecting these variations would have no impact on subsequent inheritance. So, in order to give rise to a new species, a significant genetic change that completely modifies the common type was necessary (see Ochoa 2017).

Another example is the German holistic morphological tradition of the early twentieth century (Levit and Hoßfeld 2017), which paved way for German organicism, chiefly championed by embryologist Julius Schaxel (Nicholson and Gawne 2015). In his work *Das biologische Individuum* (1931), Schaxel challenged the traditional view of genetics, stating that organisms are built through a series of intricate processes rather than solely through hereditary expression. For him, the key lies in the influence genes exert on these processes. To illustrate this, he referred to Richard Goldschmidt's works. Goldschmidt's argument about the "hopeful monster,"[9] for example, highlights the importance of understanding evolution within the context of "the nature of the developmental system of the organism." This scheme describes the dynamic interaction among the organism's parts during its development and the developing organism's response to its environment.

To summarize, while some historians and scientists have posited that anti-Darwinian theories were discarded because of the inevitable "Mendelian revolution," they might not have foreseen that Mendelian genetics was also compatible

[9] For further details on Goldschmidt's theory of the hopeful monster, see below Sect. 3.2.1.

with such theories through an organism-centered perspective. This viewpoint, however, was overshadowed by the gene-centered discourse framed by actors of population genetics, as I explore below.

3.1.3 Gene-Centered Perspective

The scholars who particularly address twentieth-century organicism agree that the main reason for the decline of the organism-centered perspective was the emergence and subsequent popularity of population genetics of the Modern Synthesis, which was reinforced by the prevalence of a gene-centered perspective (Etxeberria and Umerez 2006; Nicholson 2014; Walsh 2015; Baedke 2019; Baedke and Fábregas-Tejeda 2023). Within the Modern Synthesis framework, the gene is viewed as a discrete entity that controls the production of phenotypes, hence leading to the belief that the organism is entirely determined by its genetic expression mechanisms. As Walsh (2015, p. 21) remarks: "One consequence of prioritising genes in this way is that it fosters the impressions that what happens within an organism… has little if any direct influence on evolution." Consequently, the characteristics attributed to the organism by organicists—such as the interaction between its developing parts, its interaction with the environment, and its agential activities—faded away as the gene-centered perspective began to dominate during the mid-twentieth century (Baedke and Fábregas-Tejeda 2023).

In fact, it is no coincidence that anti-Darwinian theories were marginalized alongside the organicist movement of the first third of the twentieth century because, as highlighted, these theories were rooted in an organism-centered perspective. In other words, the "eclipse of anti-Darwinism" (Ochoa 2017) coincides with the "eclipse of the organism" (Walsh 2015). If this is the case, then the period identified by Gould (2002) as the "restriction" of anti-Darwinian theories was actually a period of the restriction of the organism-centered perspective. However, such a restriction or eclipse did not arise from a conclusive resolution provided by Darwinism regarding the issues of the nature of variation. Instead, it occurred because early population geneticists found themselves unequipped when confronted with anti-Darwinian theories, as they lacked an organism-centered perspective.

Regarding neo-Lamarckism, for example, Ronald A. Fisher (1930, pp. 12–13) argued that, based on genetic behavior, neither the organism's response to motor activity nor direct environmental action seemed to affect the direction of genetic mutations. For Sewall G. Wright (1931), that possibility was also dismissed because "[u]nfortunately, the results of experimental study have given no support to this view" (p. 142). Genetic mutations consistently seem unfavorable and despite factors such as X-rays, radioactive components, and, to a lesser extent, temperature influencing the mutation rate, these agents "merely speed up the rate of random mutation" (p. 142). Similarly, John Burdon S. Haldane (1932) found no support for this theory because "whenever sufficiently careful experiments have been done, this has been shown not to occur" (p. 20).

In this regard, it is clear that for the pioneers of population genetics, the rejection of the inheritance of acquired characters stemmed from their inability to conceive changes occurring outside the "hereditary" parameter. In other words, what happens in an individual's development is deemed irrelevant since it does not alter the hereditary information. In this regard, Walsh (2015, pp. 73–76) points out that, for population geneticists, inheritance and development are separate processes, and adaptation is merely contemplated as genetic variation filtered by natural selection. In contrast, the neo-Lamarckian viewpoint posits that "[t]he processes of individual adaptive change, development and inheritance are all intimately intertwined" (p. 75).

With respect to orthogenesis, Fisher (1930, pp. 19–20) contended that the available experimental evidence provided no signal for the existence of mutations that tend to generate variations in the same direction; rather, these trends occur randomly and are mostly lethal. Even if there were a slight bias in mutations, the mutation rate would not be high enough to alter genetic frequencies across the entire population. Interestingly, in contrast, Wright (1931, 1932) and Haldane (1932) held favorable views toward orthogenetic trends, particularly in light of paleontological evidence. But they attempted to explain them within a gene-centered context. For example, Wright (1932, p. 365) introduced his renowned "shifting balance theory," arguing that phenomena such as "intergroup selection" and "crossbreeding" offered the most plausible explanations within the framework of Mendelian theory. In contrast, Haldane (1932) proposed that orthogenesis found an explanation "not only with the accumulation of a number of genes having a similar action, but with the very slow modification of single genes, each changing in turn into a series of multiple allelomorphs" (pp. 108–109).

In this regard, it can be asserted that, adhering to a gene-centered perspective, these geneticists believed that orthogenesis implied that genetic mutations were channeled in the same direction by the genes themselves. Even so, within this conceptual framework, randomness in mutations was still expected. In light of compelling evidence from paleontology (which indicated non-random trends), they looked for alternative explanations within a gene-centered perspective.[10] Yet, lacking an organism-centered perspective, they failed to recognize that development imposed constraints on the structure, which would eventually be the most plausible explanation for such trends.

Considering saltationism, Fisher (1930) proposed that small variations offer greater advantages compared to discontinuous variations, which generally result in nonadaptive malformations; hence, "in any highly adapted organism the probability of advantage through any considerable evolutionary step (saltation) rapidly becomes infinitesimal as the step is increased in magnitude" (p. 114). In Wright's case (1931), although he did not dismiss this possibility outright, he simply stated: "apparent continuity as the rule, discontinuity the rare exception" (p. 153). Finally, Haldane (1932, pp. 104–105), while more inclined towards saltational evolution than Fisher

[10] Even George Gaylord Simpson, widely regarded as one of the leading proponents of the Modern Synthesis, recognized these trends through the concept of "parallelism" and had to resort to genetic mechanisms to account for these patterns. See Ochoa 2021a.

and Wright, believed this mode was only possible under specific conditions: reduced selection pressures, decreased inter-individual competition, and greater genetic diversity among individuals. In other cases, it could be explained by the phenomenon of hybridization.

Regardless, for early population geneticists, considering saltational evolution becomes unquestionably complicated when setting aside the idea that the organism is built by complex processes of ontogeny. Given that Mendelian mutations also produce continuous phenotypes, as Fisher (1930) sets forth, there would be no reason "to suppose that the discontinuity so produced is in any way connected with the discontinuity of the genetic particles in Mendelian inheritance" (p. 112).

In conclusion, the idea that population genetics resolved the issues of inheritance and variation and thereby restricted anti-Darwinian theories is mistaken. If the argument posits that anti-Darwinian theories either mistook conceptions of inheritance (Mayr 1980) or had no place in the Mendelian world (Bowler 1983), then it follows that there was no such "Darwinian revolution." Indeed, under the gene-centered perspective, explaining the patterns of variation recognized by anti-Darwinists proved exceedingly challenging, a task that only found a solution through an organism-centered perspective (see Table 3.1). In short, ignoring or failing to adequately address anti-Darwinian challenges definitely did not provide any solution to the problems of Darwinism. Therefore, the historical narrative surrounding the synthesis between Darwinism and Mendelism is wrong.

3.2 Unmodern and Modern Synthesis: In Looking for a Hierarchical View of Life

Although the Modern Synthesis is often hailed as the predominant evolutionary theoretical framework, it is worth mentioning that it was not the sole synthetic theory to emerge in the first half of the twentieth century (see Esposito 2017; Ochoa 2017; Baedke et al. 2024). Gilbert et al. (1996), for example, highlight that while there was a synthesis between genetics and evolution during that period (the Modern Synthesis), there also must have been a synthesis between embryonic development and evolution, a perspective that was largely marginalized. They termed this viewpoint the "Unmodern Synthesis." This clarification is important, as recognizing two competing theoretical frameworks allows us to better grasp the evolutionary debates during that period (Grene 1958; Ochoa 2017). Nevertheless, my intention here is not to delve into the reasons for the incompatibility of these two frameworks; rather, I aim to highlight their similarities—specifically, that both sought to develop a hierarchy theory of evolution.[11] Despite being based on apparently different ideas, they share some common points, for example: (1) the development of a hierarchy of

[11] Traditional interpretations often assume a clear division between these ways of thinking. However, as Delisle (2017) points out, the plurality of ideas proposed by these authors leads to the conclusion that there is no "single dominant intellectual thread." Consequently, labels such as "Darwinism" or "the hardening of the Evolutionary Synthesis" are ultimately imprecise.

Table 3.1 *A comparison between the organism-centered perspective and the gene-centered perspective for explaining anti-Darwinian theories*

	Definition	Phenomena	Explanation from an organism-centered perspective	The theory from a gene-centered perspective
Neo-Lamarckism	It refers to the idea that organisms acquire new variations due to their direct interaction with the environment, which stimulates them to change favorably during their lifetime. Subsequently, after a few generations, offspring inherit these new traits from birth without requiring environmental stimulus	Physiogenesis: Changes in size when exposed to different environments; reduction in pigment in animals inhabiting light-absent environments; the effects of temperature or diet on changes in animal coloration, and so forth Kinetogenesis: The adjustment of muscles and bones to different circumstances; the development of calluses in animals in response to the skin's repeated friction with external surfaces. All these traits, initially triggered by environmental stimuli, can manifest in subsequent generations even in the absence of the environmental stimulus	The environmental stimulus can trigger a specific developmental pathway subject to selection. Since some genes can also trigger the same embryological pathway, a trait might initially be acquired due to the environmental stimulus and subsequently assimilated by the action of genes	While the environment's influence on the phenotype is widely acknowledged, the theory of inheritance of acquired characters is impossible because there is no evidence suggesting that direct interaction with the environment induces genetic mutations in a favorable adaptive direction. Genetic mutations are random
Orthogenesis	It indicates that certain characters among related groups follow similar evolutionary trends, which are primarily attributed to internal factors within organisms	Closely related groups tend to develop the same traits independently (parallelism)	Developmental constraints (bias). During embryonic development, genetic mutations can affect many pathways, potentially leading to unfavorable outcomes. However, among these possibilities, some may lead to proper development. Thus, variation tends to follow certain pathways more easily than others	Since genetic mutations occur randomly, there is no reason to believe they tend to follow the same direction

(continued)

Table 3.1 (continued)

	Definition	Phenomena	Explanation from an organism-centered perspective	The theory from a gene-centered perspective
Saltationism	It involves the idea that higher taxonomic units (e.g., species and higher taxa) arise from abrupt changes occurring in one or a few generations	Taxonomic groups are distinguished by different levels of morphological discontinuities Transitional forms are absent in the fossil record	Developmental systems are discrete entities. Interactions during embryonic development produce stable phenotypes. However, any genetic or environmental perturbation affecting these interactions results in discontinuous outcomes within the system	Genetic mutations with significant effects on the phenotype are less likely to prevail than those with continuous phenotypic outcomes

Note that these theories have a fairly coherent explanation when viewed through the organism-centered perspective, whereas the gene-centered perspective shows clear interpretational limitations

types, (2) the recognition of two decoupled evolutionary levels, and (3) the establishment of a phylogenetic tree constructed from the top down rather than bottom up.

Considering this, I will examine the ideas surrounding the hierarchy theory of evolution during the first half of the twentieth century. On the one hand, proponents of the "Unmodern Synthesis" include Richard Goldschmidt in developmental genetics and Otto Schindewolf in paleontology. On the other hand, the leading figures of the "Modern Synthesis" include Theodosius Dobzhansky in classical genetics, Ernst Mayr in systematics, and George Gaylord Simpson in paleontology.

3.2.1 Unmodern Synthesis: Goldschmidt and Schindewolf

Richard Benedict Goldschmidt, a renowned German geneticist credited with introducing the concept of the "hopeful monster" (see Gould 1982; Dietrich 2011), and Otto Heinrich Schindewolf, a prominent paleontologist known for his theory of typostrophism (see Seilacher 1972; Reif 1993), have been considered the most influential scientists to take on an anti-Darwinian viewpoint during the first half of the twentieth century. While it seems that these scientists formulated their evolutionary framework independently, the following aspects suggest a synthesis between their hierarchical perspectives:

1. *Goldschmidt and Schindewolf challenged the orthodox view of the Modern Synthesis.* Goldschmidt focused primarily on experimental embryology, while Schindewolf concentrated on patterns in the fossil record.
2. *They envisage a hierarchy of types*, where taxonomic categories represent a series of types ordered hierarchically from greater to lesser inclusivity.[12]
3. *They emphasized the hierarchical nature of their evolutionary system, arguing against the extrapolationist premise.* They mention that at least two levels of evolution are entirely distinguishable from each other. Hence, they point out that events on a larger scale (macroevolution) cannot be simply reduced to lower-level processes (microevolution).[13]
4. *Embracing a hierarchy of types, they assert that the taxonomic structure is built from the top down*, i.e., a higher taxonomic category emerges first, followed by subsequent lower taxonomic categories.[14]

[12] In systematics and morphology, the type concept suggests that closely related groups of organisms are abstractly recognized by a highly integrated common pattern, both morphologically and embryologically; see Chap. 8. For a historical analysis of the Natural System based on a hierarchy of types, see Chap. 2.

[13] As discussed in more detail in Chap. 10, Sect. 10.2, the H-Evo-Devo Theory qualitatively distinguished at least two levels of evolution.

[14] The inverted direction of the taxonomic structure is a consequence of the hierarchy of types. The H-Evo-Devo Theory supports this implication; see Chap. 12, Sect. 12.1.1.

Below, I delve slightly deeper into their proposals, exploring Goldschmidt's stance on developmental genetics and Schindewolf's approach to paleontology.

Goldschmidt and the Hopeful Monster Theory Goldschmidt portrayed his hierarchy theory of evolution primarily through two publications: a brief paper in 1933 and a further developed exposition in his 1940 book titled *The Material Basis of Evolution*. According to his theory, Goldschmidt claimed that there were two entirely decoupled levels of evolution: microevolution and macroevolution. Microevolution constitutes the first level of evolution, encompassing processes occurring within the bounds of a species, while macroevolution corresponds to the evolution of species and the other higher taxonomic categories.[15] Regarding this, Goldschmidt argued that population geneticists supported the view that macroevolution was explained through the accumulation of microevolutionary processes. However, they were limited to such explanations because, in matters of macroevolution, "the paleontologist, the comparative anatomist, and the embryologists are supreme" (Goldschmidt 1940, p. 5). Notably, Goldschmidt emerged as a pioneer in developmental genetics, attempting to unravel the link between development and evolution. Under this view, he proposed a new challenge to the Modern Synthesis, concluding that the processes of microevolution are distinct from those responsible for macroevolution: "I may challenge the adherents of the strictly Darwinian view… to try to explain the evolution of [complex organs] … by accumulation and selection of small mutants" (Goldschmidt 1940, p. 6).

For Goldschmidt, microevolution pertains to the evolution of subspecies. From this standpoint, subspecies are not models that explain the origin of species (incipient species),[16] but rather their role is to adapt species to different environments without exceeding the species' boundaries. A characteristic of subspecies is that they do not form a continuous line (cline) but rather resemble a chessboard arrangement. In other words, each variety of the species occupies a distinct and well-defined geographical area, as each one is specialized to the environmental conditions within that region. To account for their adaptations, he argues that some of such characters respond physiologically to the environment rather than being directly inherited (a phenomenon now referred to as phenotypic plasticity). In other cases, hereditary adaptations may not have originated where the organisms currently reside; rather, they may have emerged as pre-adaptive forms, i.e., structures that fortunately found a suitable environment to exploit. Goldschmidt stressed the significance of natural selection in such cases; however, he emphasizes that it only acts as the decisive step in determining the acceptance or rejection of these forms in newly colonized environments. He concluded that the distinctions between species are well marked, and

[15] For a review of the historical and contemporary debate on the terms of microevolution, macroevolution, and megaevolution, see Chap. 10, Sect. 10.2.1.

[16] According to the Modern Synthesis framework, some subspecies have the potential to turn into new species as long as they are geographically isolated from the ancestral population. See Chap. 10, Sect. 10.1.

the variations within the same species do not provide significant characteristics that extend beyond species' boundaries. In other words, microevolution can produce changes within the structural limits of the species but not beyond them:

> Below this level, microevolution has even less significance for evolution… *Subspecies are actually, therefore, neither incipient species nor models for the origin of species. They are more or less diversified blind alleys within the species. The decisive step in evolution, the first step toward macroevolution, the step from one species to another, requires another evolutionary method than that of sheer accumulation of micromutations.* (Goldschmidt 1940, p. 183; his emphasis)

Therefore, for Goldschmidt, macroevolution—which corresponds to the origin of species or higher taxa—requires explanations distinct from those governing simple microevolution. Hence, he resorted to aspects of embryology. According to him, the processes of individual development are so intricately coordinated that only one associated process between each part can be conceived as resulting in the same outcome. Yet certain genetic changes may occasionally perturb the rates of one or many reactions, causing a disturbance in embryonic development processes and resulting in ontogenetically nonviable organisms. Although most mutations cause lethal disruptions in the complex order of interactions taking place during an individual's development, Goldschmidt believed that these sudden and discontinuous changes could hold the key to explaining "macrotaxonomy." This was because a genetic change capable of controlling new possible transformation pathways could also produce a change with macroevolutionary effects in a single step. Goldschmidt (1933) set forth that "in most cases such a shift of a partial process would lead to the production of monstrosities… But we must not forget that what appears to-day as a monster will be to-morrow the origin of a line of special adaptations" (p. 544). Here, Goldschmidt alludes to his renowned concept of the *hopeful monster*. A mutation might affect many pathways during an organism's embryonic development, resulting in a nonviable outcome. However, other pathways will also lead to proper and well-coordinated development. Once these individuals are developmentally established, some of these monstrosities might also facilitate the occupation of a new habitat, leading to the origin of a new type. Undoubtedly, natural selection plays a crucial role in this process, but it only acts on the new system as a whole, excluding the nonviable kinds of monstrosities from an external point of view.

Goldschmidt's theory posits two decoupled evolutionary levels: microevolution corresponds to subspecies, whose role is to adapt species to different environments, and macroevolution pertains to species or higher taxa, representing the evolution of types. It is worth noting that although Goldschmidt does not explicitly mention species as types, his discussion about "true species" implied such an association (see Goldschmidt 1940, pp. 139–155). Additionally, he evidently considered taxonomic categories as types, thus supporting the perspective of the Natural System as a hierarchy of types. Nevertheless, one consequence of such an evolutionary stance lies in the taxonomic structure, as it is represented upside down compared to the traditional approach. Put simply, the direction is not from bottom to top, where subspecies gradually build higher taxonomic categories over time during microevolutionary differentiation; instead, the higher taxonomic categories develop first, followed by the lower-level ones. In a 1952 article, Goldschmidt wrote:

> This means that a phylum consists of a number of classes all of which are basically recognizable as belonging to the phylum but, in addition, are different from each other. The same principle is repeated at each taxonomic level. All the genera of a family have in common the traits which characterize the family; e.g., all genera of penguins are penguins. But among themselves they differ from genus to genus. So it goes on down to the level of the species. Can this mean anything but that the type of the phylum was evolved first and later separated into the types of the classes, then into orders, and so on down the line? This natural, naive interpretation of the existing hierarchy of forms actually agrees with the historical facts furnished by paleontology… Thus logic as well as historical fact tells us that the big categories existed first, and that in time they split in the form of the genealogical tree into lower and still lower categories. (Goldschmidt 1952, pp. 91–92)

Schindewolf and typostrophism In his seminal books, *Paläontologie, Entwicklungslehre und Genetik* [Paleontology, Evolution, and Genetics] (1936) and *Grundfragen der Paläontologie* [Basic Questions in Paleontology] (1950), Schindewolf argued that higher taxa, represented as types, exhibited well-defined patterns of evolution: *typogenesis* manifests a period of their abrupt origin, followed by *typostasis*, characterized by the stability of their basic characters, and *typolysis*, defined by their extinction. According to Schindewolf, these three phases of evolutionary periodicity—collectively called "typostrophism"—can be analogous to an individual organism's life cycle: birth, stability, and death (see Fig. 3.2).

During the typogenesis phase, Schindewolf proposed that geological strata reveal the sudden emergence of new groups; this observation suggests that a type underwent an abrupt change in its basic form without intermediate stages, giving rise to a new type. He emphasized that these discontinuities in the fossil record could not be explained solely by the lack of intermediate fossils, as this absence often appeared as a pervasive pattern rather than an exception. But what processes underlie these abrupt transformations? Schindewolf stated that every significant evolutionary change in the type complex stems from changes in the embryonic development of organisms. Disruptions at the early developmental stage would undermine the constraints imposed by an established ancestral type. In this regard, Schindewolf advocated for a genetic perspective focused on embryonic development, asserting its potential to explain observed patterns in the fossil record:

> The question of the nature and mode of operation of macromutations lies beyond the capacities of paleontology and must be left to experimental genetics to answer. This branch of research will have to discover what is responsible for far-reaching, discontinuous evolutionary transformations—whether they are due simply to a genetic mutation with pleiotropic, complex consequences, to a "repatterning" of a chromosome or of the entire genome in the sense of Goldschmidt, or to some other mechanism. (Schindewolf 1993, pp. 351–352)[17]

[17] Interestingly, in the footnote citing this quote, Schindewolf (1993, p. 352, n. 22) articulated his perspective on Goldschmidt's works as follows: "[They] have had considerable influence on my thinking or have strengthened it, but in essence, the concepts described here grew out of my own

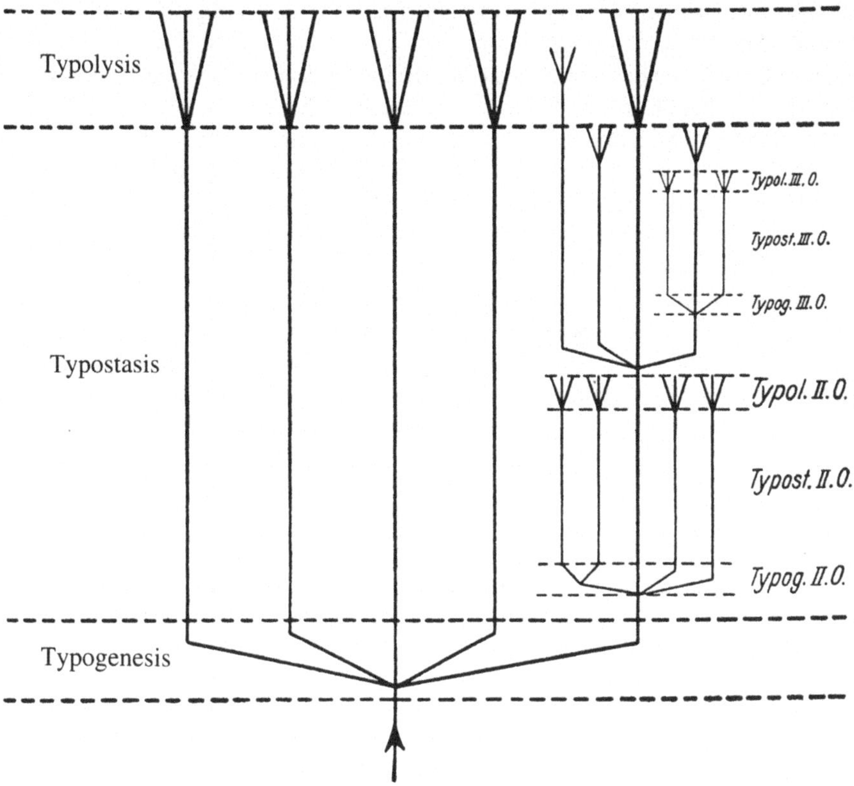

Fig. 3.2 *An illustration depicting the theory of typostrophism.* Schindewolf proposed that the evolution of types consists of three fundamental stages. The first stage is typogenesis, which marks the origin of types. The second stage is typostasis, referring to the period when types become spatiotemporally stable. The third stage is typolysis, or the extinction of types. Note that this theory presents a hierarchy of types, i.e., types are arranged inclusively at different scales. Furthermore, according to this view, a higher taxon emerges before the lower one. See text for more details. (From Schindewolf 1993; with permission from the University of Chicago Press)

With respect to the typostasis phase, according to Schindewolf, once a higher taxonomic group emerges, it stabilizes and maintains nearly unchanged for a long geological time. However, it undergoes gradual and linear changes while maintaining its basic structure and never exceeds the boundaries of the type. Schindewolf presented some clear cases of remarkable structural constancy, notably evident in what is commonly termed "living fossils." Although Schindewolf did not explicitly delineate the mechanism underlying this stability, he likely had in mind what is now referred to as "developmental constraints" since he argued in favor of orthogenesis

analysis of paleontological material… I regard this convergence of views arising from extremely different premises as a welcome sign that we are on the right track."

at this stage, which is primarily explained by the "narrowing of morphological potential" (p. 268).

Finally, the last pattern observed is the decline or death of the groups is known as typolysis. During this phase, Schindewolf described that when the types persist over an extended geological period, the adaptation that initially supported their survival in a particular environment might ultimately lead them to extinction. Nonetheless, in 1963, in a paper titled "Neokatastrophismus?" he changed his mind, suggesting that mass extinctions could be attributed to global catastrophic events, such as radiation generated by a supernova.

Regardless, what is noteworthy about Schindewolf's hierarchy theory is its distinction between two distinct levels of evolution: one responsible for producing the types and the other for producing type adaptations. For example, regarding natural selection and small variations, he argues that these mechanisms only apply to a lower level and are not involved in producing type characters:

> *[S]election concerns and affects only the most superficial layer of characters… It is not increasing divergence of racial and specific characters* that produces the supraordinate categories and structural designs; rather, these are added on to the structural characters of the higher typal units, in a certain sense, only as capstones; they [natural selection and micromutations] introduce the last variations and arabesques—indeed the most important futures for specialized fitness for life—into the existing framework of the structural design. (Schindewolf 1993, p. 341; his emphasis)

In addition, he asserted that this peculiarity of such a hierarchical view implies that higher taxonomic categories do not emerge "as the result of continuous modification in a long chain of species, but rather by means of a *sudden, discontinuous direct refashioning of the type complex*" (Schindewolf 1993, p. 214; his emphasis). In other words, the taxonomic structure of a hierarchy of types entails that the origin of more inclusive taxonomic categories occurs first, followed by less inclusive ones:

> Since, further, the organizational traits of a higher category are common to all the categories subordinate to it, we must conclude *that the type features of the higher units arose historically-temporally before those of the lower ones, that the general features arose before the progressively more specialized sets of characters.* (Schindewolf 1993, p. 415; his emphasis)

3.2.2 Modern Synthesis: Dobzhansky, Mayr, and Simpson

Theodosius Dobzhansky, Ernst Walter Mayr, and George Gaylord Simpson are credited with establishing the foundations of the so-called Modern Synthesis. Although some researchers have noted a lack of a common framework among these "architects of the Modern Synthesis" (Delisle 2011; Ochoa 2021a; Smocovitis 2023), their perspectives converge on certain fundamental aspects:

1. *They maintain a gene-centered perspective.* They believe that evolution focuses only on genes, often overlooking the internal properties of organisms.[18]

---

[18] For further details on this discussion, see Sect. 3.1 of this chapter.

2. *They are Darwinian adaptationists*. For them, most structures of the organism are adaptations resulting from natural selection acting upon small genetic variations.
3. *They advocate for the extrapolationist premise*. Microevolutionary mechanisms are sufficient to explain macroevolutionary events.

However, despite these differences, their proposals for a hierarchy theory of evolution share commonalities with those of Goldschmidt and Schindewolf, as I will explore. For example, although they expressed their aversion to the type concept in their writings (see Mayr 1959; Simpson 1959; Witteveen 2015, 2016), their interpretations of the hierarchical structure suggest an apparent hierarchy of types. Similarly, concerning Mayr and Simpson, despite upholding the conceptual framework of the Modern Synthesis that favored the extrapolationist argument, they seem to invoke two decoupled levels of evolution. Likewise, in Simpson's case on the evolution of higher taxa, occasionally, he envisioned the phylogenetic tree as apparently structured from the top down. I examine these points in greater detail below.

Dobzhansky and the Hierarchy of Adaptive Types Dobzhansky's commitment to the extrapolationist premise is well known (see Adams 2021). For example, the main argument developed in *Genetics and the Origin of Species* is that, although the origin of species cannot be reproduced in the laboratory, mere observations of changes occurring within the span of a human lifetime (evolutionary uniformitarianism) make it possible to infer these processes. Moreover, by advocating for a shift in perspective regarding the meaning of species—now viewed as genetic units occupying specific geographic zones—he argued against the traditional taxonomic approach of his time, which referred to the recognition of the type: "Such a description is more adequate than the usual method of finding the abstract average phenotypes of 'races'... The geography of the genes, not of the average phenotypes, must be studied" (Dobzhansky 1937, p. 63). However, Dobzhansky devised a hierarchy theory of apparent adaptive types within his general evolutionary framework. Although he consistently upheld the extrapolationist argument, it is important to consider that an evolutionary hierarchy theory implies, to some extent, separate evolutionary levels.

In his argument about the Natural System, Dobzhansky emphasized morphological discontinuities as a necessary phenomenon for identifying natural groups. He claimed that biological groups are organized by a discontinuous arrangement, i.e., lacking intermediate forms, and that each arrangement represents a grouping of individuals sharing common characters; hence, they can be defined by a "modal point in their variations." Thus, the Natural System consists of a series of discontinuities that are hierarchically organized:

> Small clusters are grouped together into larger secondary ones, these into still larger ones, and so on in an hierarchical order. (p. 4)

> [T]he variation is hierarchical, since the small discrete arrays are grouped into larger ones, these into still larger ones, etc. The discontinuity is preserved throughout the hierarchy of arrays. (p. 303)

But how "natural" is the Natural System? According to Dobzhansky, classification is natural as long as the morphological separations between groups are inherent aspects of nature: "[T]he same classification is natural so far as it reflects the objectively ascertainable discontinuity of variation, and the dividing lines between species, genera, and other categories are made to correspond to the gaps between the discrete clusters of living forms" (p. 4). However, classification is also artificial because there are no specific natural categories that distinguish each level of discontinuity. For example, there is no specific level of discontinuity that differentiates the category "family" from the category "genus," or the "family" from the "order." Accordingly, for Dobzhansky, the only real category is the species, as it represents the first recognized level of discontinuity that is applicable to any lineage (see p. 306).

Of course, recognizing a hierarchy of discontinuities does not invalidate his Darwinian conceptual framework. Discontinuities arise from a gradual process—intermediate forms are erased by natural selection, preserving the extremes—and hierarchical levels of discontinuity are attained over time as organisms become reproductively isolated from other groups. Although Dobzhansky argued that character averages—or types in the language of morphology—were merely an observer's illusion about the arrangement of variation, he recognized that taxonomy should be structured by inclusive phenotypes. Since genetic changes are the cause of new variations, groups would be arranged based on different combinations of genes on a sort of "topographic map," and the resulting phenotypes from those combinations would represent adaptations to different ecological niches. Herein, he resorts to the concept of an "adaptive peak," influenced by Sewall Wright's view.[19] Thus, Dobzhansky conceives of groups as "adaptive types," so to speak, arranged hierarchically. In the third edition of his book, he wrote:

> Thus, the ecological niche occupied by the species "lion" is relatively much closer to those occupied by tiger, puma, and leopard than to those occupied by wolf, coyote, and jackal. The feline adaptive peaks form a group different from the group of the canine "peaks." But the feline, canine, ursine, musteline, and certain other groups of peaks form together the adaptive "range" of carnivores, which is separated by deep adaptive valleys from the "ranges" of rodents, bats, ungulates, primates, and others. In turn, these "ranges" are again members of the adaptive system of mammals, which are ecologically and biologically segregated, as a group, from the adaptive systems of birds, reptiles, etc. The hierarchic nature of the biological classification reflects the objectively ascertainable discontinuity of adaptive niches, in other words the discontinuity of ways and means by which organisms that inhabit the world derive their livelihood from the environment. (Dobzhansky 1951, p. 10)

---

[19] Wright (1932) uses the metaphor of geographical landscapes—including mountains, mountain ranges, valleys, and others—to analogize an abstract adaptive landscape reflecting potential genetic combinations within populations. The landscape (taxonomic structure) comprises numerous peaks (taxonomic groups), with each peak's height symbolizing the population's degree of adaptability. These peaks (mountains) are separated by valleys representing lower levels of adaptability.

Mayr as a Typological Thinker Mayr's original argument in *Systematics and the Origin of Species* (1942) is similar to that of Dobzhansky in the sense that he supported the extrapolationist premise: "There is only a difference of degree, not one of kind, between the two classes of phenomena [micro and macroevolution]" (p. 291). For example, when Mayr emphasized the difference between the "old systematics" and the "new systematics,"[20] he indicated that the former "is characterized by the central position of the species," while in the latter, "[the] importance of the species as such is reduced, since most of the actual work is done with subdivisions of the species, such as subspecies and populations" (pp. 6–7). In order to support this "new systematics," Mayr claimed that phenotypic discontinuities stem from mechanisms of geographic isolation. In other words, initially, populations exhibit slight continuous differences, forming a gradual chain of forms (cline). Subsequently, populations become isolated by a geographical barrier. But, because gene interruption fosters evolutionary divergence, a minimum of geographic isolation is sufficient to generate a certain degree of discontinuity. Accordingly, for Mayr, the entire taxonomic structure (species, genera, families, etc.) is built upon processes observed in the lowest taxonomic category of the system, namely, the subspecies: "All the processes and phenomena of macroevolution and of the origin of the higher categories can be traced back to intraspecific variation, even though the first steps of such processes are usually very minute" (p. 298).

During this period, Mayr maintained the view that the basic unit of evolution was the subspecies rather than the species. However, in later years, he changed his mind. In publications after *Systematics*, Mayr asserted that species are real entities, distinct from subspecies, aligning closely with Goldschmidt's perspective on the differences between microevolution and macroevolution. Moreover, he advocated a typological perspective on species.[21] For example, in 1954, Mayr outlined two levels of variation in populations: "ecotypic variation" and "typostrophic variation." Ecotypic variation pertains to population properties that adapt species to different environments: "This variation adapts to the local environment populations which are members of a continuous series of contiguous populations," that is, "adaptive within the established 'type'" (Mayr 1954, pp. 159–160). On the other hand, typostrophic variation represents "a type of variation so different from that of contiguous populations that we may be dealing with something entirely new." In other words, "they may have morphological or ecological features which deviate quite strikingly and unexpectedly from the 'parental' pattern" (p. 160). This reference to "typostrophic variation" is not mere chance; clearly, the inspiration comes from Schindewolf's theory of typostrophism. In fact, Mayr linked the perception of the origin of the type with "the origin of a real evolutionary novelty," i.e., "a new type."

---

[20] Julian Huxley's terminology.

[21] Of course, this may seem surprising, given that such a statement contradicts his later position, especially since he has been considered one of the main critics of typological thinking. See Chap. 2; Witteveen 2016; Ochoa 2017.

Surprisingly, Mayr conceptualized species as types, similar to Goldschmidt, arguing that intraspecific variation, which is adaptive, does not generate any significant change capable of surpassing the boundaries of its type: "Even though a species may have many local races (ecotypes), it arrives sooner or later at a geographical line, its species border, beyond which it cannot expand" (p. 166). The difference from Goldschmidt is that, for Mayr, the level of ecotypic variation is depicted as a clinal variation, yet he concurs with Goldschmidt that speciation constitutes a special kind of divergence that is not reducible to microevolutionary explanations. Hence, he agrees that an identified special and distinctive process is necessary to account for this level of evolution. Accordingly, Mayr explained that allopatric peripheral populations were the only candidates that could potentially give rise to true species. Furthermore, he claimed that these populations were capable of producing higher taxonomic categories as well: "[I]t is quite evident that they [peripheral populations] are not only incipient species but in many instances also incipient genera and higher categories" (p. 176).

Although in his book *Animal Species and Evolution* (1963), Mayr refrained from making as many references to the type—probably because he had already launched a campaign against typological thinking—he consistently advocated for the species as an autonomous evolutionary unit. For him, species possessed a sort of genetic cohesion that bound together all their parts (subspecies) as a whole. Now, Mayr stated that speciation was not exclusively dependent on identifying allopatric populations; hence, he invoked the founder effect as the most likely cause for speciation. For example, the founder effect would facilitate divergence by breaking the homeostatic forces of the ancestral population, leading the new population to deviate from the "common type" of the species. Mayr referred to this process as the "genetic revolution."[22] If Mayr leaned toward typology in this sense, it is no surprise that he might have also formulated a hierarchy of types, thereby potentially endorsing two apparently decoupled evolutionary levels.

Simpson and the Higher Taxa Evolution Undoubtedly, Simpson envisioned a hierarchy theory of evolution in its entirety. The main message of Simpson in *Tempo and Mode in Evolution* (1944) is that evolution at the higher hierarchical level could not be solely studied through inference from microevolutionary genetics since the observation of patterns in the fossil record contributes significantly to this inquiry as well. Hence, he proposed different rates and modes of evolution for each hierarchical level and discussed them in detail. In fact, while Simpson accepted that macroevolution (at the species level) could be extrapolated as the sum of microevolutionary processes—explained by geneticists—he recognized a detached higher level corresponding to the evolution of higher taxa, exclusively studied by paleontology, namely, megaevolution. For example, in order to explain the absence of transitional sequences in the emergence of higher taxonomic groups in the fossil record, Simpson asserted that a mechanism other than natural selection was needed. He

[22] See this model with more detail in Chap. 10, Sect. 10.1.

referred to this evolutionary mode as "quantum evolution," which describes a rapid (tachytelic) change within a lineage, attributing its cause to genetic drift.

However, in *The Major Features of Evolution* (1953), Simpson apparently changed his mind, presenting a discourse more in line with the traditional extrapolationist premise. First, Simpson claimed that megaevolution is not such a special level. For example, he reasoned that if the distinction between macroevolution and microevolution is merely one of degree, there is no reason to think that the distinction between megaevolution and macroevolution is one of type (see p. 339). Second, Simpson stated that higher taxa arise from traditional processes of speciation rather than from special processes of the higher taxa evolution:

> [T]he higher categories did not arise as such, that there was no "archetype," no "generalized" ancestor with the characters-in-common of the order... but that there was simply normal speciation among forms at first closely related and closely similar in all respects... The family and all its distinctive characters [for example] developed gradually as time went on. (Simpson 1953, p. 345)

Third, he diminished the importance of quantum evolution and, by extension, genetic drift as the cause of the instantaneous origin of higher taxa: "Genetic drift is certainly not involved in all or in most origins of higher categories, even of very high categories such as classes or phyla" (p. 355). Therefore, he asserts that "[t]here is no reason to believe that any different factors are involved than those seen in lower categories or in 'microevolution.' On the contrary, those factors are fully consistent with what we know of higher category evolution and quite capable of explaining it" (p. 376).

But this does not mean that Simpson sacrificed his initial observations to surrender to the extrapolationist premise. Despite making an argument that attempts to empathize with microevolutionary studies, in *The Major Features of Evolution*, Simpson continued to consider that laboratory studies have little to say about the nature of higher taxa. In fact, unlike the many contemporary positions maintaining that higher taxa have no evolutionary value,[23] Simpson believed that higher taxa were legitimate units of evolution and, therefore, worthy of investigation:

> The study of the evolution of higher categories, say for present purposes from families upward, does involve some of the most important problems of evolution. For one thing, evolution on this scale cannot be directly studied by experimental method... The main themes of this book relate to evolution on a broad scale, hence largely to categories above the species. The time has now come to undertake explicit consideration of what higher categories are and how they arise. (Simpson 1953, pp. 339–340)

Although Simpson consistently showed antipathy toward the type concept, in a sense, he also devised a hierarchy of types. For example, he viewed the type as representing common abstract characters or "archetypes." Still, according to Simpson, referring to higher taxa through this concept did not make sense, as that version

[23] See Chap. 8, Sect. 8.1.1

implies that "idealized" characters would have nothing to do with adaptation. Remember that Simpson is a Darwinian adaptationist; therefore, he considered that the characters that exemplify a higher taxon must be adaptive. Likewise, he believed that diagnosing a higher taxon through shared characters should rely on a materialistic and realistic explanation—rather than an "idealistic" one—that underlies the genealogical relationships. Nonetheless, it is clear that his view on higher taxa is not so different from his interpretation of types in Schindewolf's sense, as he recognized an underlying "type" behind diversity: "Indeed the diagnosis of a higher category is not a description of any ancestor or of a type of organization ever embodied in or definitive of an actual organism or population. It is *an abstraction* of the broader adaptive features present, *usually with variations, in members of the category*" (p. 175; my emphasis). Therefore, it is no coincidence that Simpson refers to higher taxa as adaptive types.

Furthermore, it is important to note that Simpson's deductions about the evolution of higher taxa led him to conclusions similar to those envisioned by Goldschmidt and Schindewolf regarding taxonomic structure. For instance, according to Simpson, higher taxa cumulatively emerge from speciation through an adaptive process: one or more adaptive lineage characters enable entry into a new "adaptive zone" (i.e., an ecological niche). If a character immediately benefits the population that possesses it, the group expands and diversifies in numbers within the adaptive zone, making the character common. Interestingly, Simpson believed that when a group attempts to expand within an adaptive zone that is already occupied by another dominant group, it undergoes such narrow evolutionary progress that it does not lead to evolutionary radiation. However, it possesses such distinctive adaptive characteristics that some systematists might consider that the groups belong to a higher taxon despite the lack of diversification. Once a group occupies an adaptive zone, it tends to undergo rapid evolutionary changes and broad diversification. Thus, according to Simpson, in "these usual cases it is true that occupation of the zone, which in retrospect is the origin of the higher category, precedes the origin of numerous genera, species, and other units that come to comprise the higher category" (p. 350). In other words, Simpson also proposed that, occasionally, the taxonomic structure of the phylogenetic tree may be constructed from the top down:

> In *this sense*, and this only, we can agree with Wright[24]... that "there seems to be a large measure of truth in the contention of Willis and Goldschmidt [also Schindewolf, G.G.S.] that evolution works down from the higher categories to the lower rather than the reverse." (Simpson 1953, p. 350; his emphasis and brackets)

In conclusion, Dobzhansky, Mayr, and Simpson demonstrated the endeavor to formulate a hierarchy theory of evolution. This analysis leads us to deduce that such a hierarchical structure possesses the following characteristics:

[24] Wright 1949, p. 387.

1. *A hierarchy of types.* While the architects of the Modern Synthesis rejected the structuralist version of the type, they recognized an underlying type whose boundaries were explained either by adaptation (Dobzhansky and Simpson) or by genetic constraint (Mayr).
2. *A hierarchy theory that recognizes at least two separate levels of evolution.* One level corresponds to the evolution of the type (e.g., species or higher taxa), while the other level involves evolution within the type (e.g., subspecies). Although the architects of the Modern Synthesis endorsed the extrapolationist premise, their insights from the hierarchy theory incited a reevaluation of their own assumptions.
3. *The taxonomic structure is built in reverse*, i.e., the higher taxonomic category emerges first, followed consecutively by the lower taxonomic category. Even though the interpretation of taxonomy in the Modern Synthesis does not imply this sense (i.e., the phylogenetic tree arises from subspeciation), Simpson's investigation into the study of higher taxa eventually landed on this conclusion (for an abstract, see Table 3.2).

3.3 Conclusions

It is worth emphasizing that some contemporary evolutionary models are built upon the Modern Synthesis framework. This is probably because many researchers consider this framework as part of scientific progress, marking a chapter that addressed the inheritance and variation issues faced by Darwinism during the mid-twentieth century. Therefore, the next step was to complement evolutionary theory with what was left aside in the Modern Synthesis. However, as concluded in Sect. 3.1, this scientific progress was somewhat limited because it attempted to address the challenges of Darwinism primarily from a gene-centered perspective. In other words, the debate was solved but at the expense of ignoring an organism-centered perspective, which could have provided insight into the issues of the variation's behavior highlighted by anti-Darwinians.

Considering this, could revisiting anti-Darwinian theories from a fresh perspective be worthwhile? In my opinion, there is no need to resurrect these theories in order to counterpose them against an organism-centered Darwinism because, in recent years, research has shown that, for example, (1) the organism is an active entity that can generate new variations through its direct interaction with the environment, (2) variations are channeled by embryonic development (thus, they are not random), and (3) variations arise discontinuously because they are the options produced by the developmental systems, etc. Therefore, if a new evolutionary theory is to emerge, it must be structured through an organism-centered perspective. This means departing from the essential core of the Modern Synthesis, which relies strictly on a gene-centered perspective.

Turning to another aspect, as seen in Sect. 3.2, evolutionists associated with both the Unmodern Synthesis and the Modern Synthesis demonstrated a commitment to formulating a hierarchy theory of evolution. This perspective leads to the following conclusions about a prospective new theory of evolution: (1) it establishes a

Table 3.2 *Characteristics of a potential hierarchy theory of evolution (1, 2, and 3) derived from the common points observed in the hierarchical framework developed by the proponents of the Unmodern Synthesis and Modern Synthesis*

	(1) A hierarchy of types	(2) Discrete separation into two levels of evolution	(3) The taxonomic structure is built from the general to the specific
Goldschmidt	His theory entails a hierarchy of types	He distinguished two entirely decoupled levels: Microevolution corresponds to subspecies evolution, whereas macroevolution encompasses both the evolution of species and the other higher taxonomic categories. He argued that subspecies cannot give rise to a species, and their function is only to adapt species to different environments. Therefore, species and higher taxa require another explanation. Hence, he proposed the hopeful monster theory as an alternative to macroevolution	He strongly supported this view
Schindewolf	His theory supports a hierarchy of types	He distinguished two entirely decoupled levels Macroevolution corresponds to the evolution of type, while microevolution is the level that accounts for adaptations within the type. For example, according to him, the mechanism of natural selection working on small variations only produces adaptations within the type but is never involved in the origin of complex structures	He strongly supported this view

(continued)

Table 3.2 (continued)

	(1) A hierarchy of types	(2) Discrete separation into two levels of evolution	(3) The taxonomic structure is built from the general to the specific
Dobzhansky	He advocated for the importance of phenotypic discontinuities that separate taxonomic groups with a hierarchical arrangement. While he did not support the idea of "types" as archetypes, he recognized an underlying type in such groups based on adaptation. In other words, he argued in favor of a hierarchy of "adaptive types"	He did not support this view. However, his view of hierarchy could imply separate levels	He did not support this view. However, a hierarchy of types implies such taxonomic structure
Mayr	His evolutionary view could entail a hierarchy of types. He stated that species have a sort of genetic cohesion that maintains their parts together (species as types). Then, he believed that the processes occurring within allopatric populations were the key to understanding how to break free from the species boundaries. However, he stated that these populations should not only be considered incipient species but also incipient genera and higher taxa.	It seems that he made a distinction between two separate levels. In 1954, Mayr claimed that populations forming clines possess variations that adapt the species to different environments within an established type. This kind of variation differs from type variation, which corresponds to characteristics defining the species. Additionally, according to Mayr, type variations arise through a special and distinctive process compared to adaptive clinal variations	He did not support this view. However, a hierarchy of type implies such a taxonomic structure

(continued)

Table 3.2 (continued)

	(1) A hierarchy of types	(2) Discrete separation into two levels of evolution	(3) The taxonomic structure is built from the general to the specific
Simpson	He recognized an underlying type within higher taxa, characterized by common adaptive characters. While he admitted a hierarchy of types, he interpreted these types as adaptive	He apparently distinguished two separate levels. Microevolution is related to the level below species, and macroevolution pertains to the species level. Despite maintaining that these two levels are distinguished by degree, he introduced an additional and exclusive level for the study of higher taxa, namely, megaevolution. In fact, in *Tempo and Mode in Evolution*, he proposed that genetic drift, rather than natural selection, was the mechanism involved at this level. Then, it could be said that he separated two levels related to the mode of origin: micro-macroevolution and megaevolution. Although, in later publications, he argued that the difference between macroevolution and megaevolution was also a matter of degree, he always supported the idea that studies at the higher taxa level could not be reduced to the microevolutionary explanations of genetics	Although he apparently supported the traditional theory that taxonomic structure is built from the conventional processes of speciation, his study on the nature of higher taxa led him to consider that higher taxonomic categories might originate before lower categories

See text for more details

hierarchy of types, (2) it recognizes at least two decoupled evolutionary levels, and (3) it delineates a phylogenetic tree constructed from the top down rather than the bottom up. While I find Goldschmidt's and Schindewolf's hierarchy theories more compelling, this does not mean the architects of the Modern Synthesis should be discredited for their attempts to establish a hierarchical view of life. The challenge lies in the fact that they articulated this hierarchy through approaches that are now considered outdated, such as (1) a gene-centered perspective, (2) an exclusively

Darwinian adaptive framework, (3) a commitment to the extrapolationist premise, etc.

Julian Huxley was right in envisioning the development of a synthetic evolutionary theory, but the purpose of synthesis extended beyond merely unifying diverse disciplines. Rather, it aimed to construct a purely hierarchical model by incorporating various perspectives addressing the topic of evolution as well. Indeed, the model presented in this book not only seeks to bring the "organism" back to the center of investigation but also aims to structure a hierarchy theory of evolution akin to that envisioned by the architects of both syntheses. The history of the H-Evo-Devo Theory is rooted in these considerations, echoing Huxley's sentiments when he introduced his synthetic movement in his book *Evolution: The Modern Synthesis* (1942, p. 8): "If this book contributes to such a synthetic point of view, I shall be well content."

References

Adams MB (2021) Little evolution, BIG evolution: rethinking the history of Darwinism, population genetics, and the "synthesis". In: Delisle RG (ed) Natural selection: revisiting its explanatory role in evolutionary biology. Springer, Cham, pp 195–229

Amundson R (2005) The changing role of the embryo in evolutionary thought: roots of evo-devo. Cambridge University Press, Cambridge

Baedke J (2019) O organism, where art thou? Old and new challenges for organism-centered biology. J Hist Biol 52:293–324

Baedke J, Fábregas-Tejeda A (2023) The organism in evolutionary explanation: from early twentieth century to the extended evolutionary synthesis. In: Dickins TE, Dickins BJ (eds) Evolutionary biology: contemporary and historical reflections upon core theory. Springer, Cham, pp 121–150

Baedke J, Böhm A, Reiners-Selbach S, Straetmanns V (2024) A synthesis without Darwin: unification attempts in early theoretical biology. In: Delisle RG, Esposito M, Ceccarelli D (eds) Unity and disunity in evolutionary biology: deconstructing Darwinism. Springer, Cham, pp 333–356

Baldwin JM (1897) Organic selection. Nature 55:558

Beatty J (2016) The creativity of natural selection? Part I: Darwin, Darwinism, and the Mutationists. J Hist Biol 49:659–684

Bowler PJ (1983) The eclipse of Darwinism: anti-Darwinian evolutionary theories in the decades around 1900. The Johns Hopkins University Press, Baltimore

Bowler PJ (2017) Alternatives to Darwinism in the early twentieth century. In: Delisle RG (ed) The Darwinian tradition in context: research programs in evolutionary biology. Springer, Cham, pp 195–217

Ceccarelli D (2024) Redrawing the boundaries of Darwinism: addressing Darwin's endorsement of the inheritance of acquired characteristics in Darwin's celebrations, 1909–1959–2009. In: Delisle RG, Esposito M, Ceccarelli D (eds) Unity and disunity in evolutionary biology: deconstructing Darwinism. Springer, Cham, pp 509–537

Cope ED (1896) The primary factors of organic evolution. The Open Court Publishing Co, Chicago

Darden L (1977) William Bateson and the promise of Mendelism. J Hist Biol 10:87–106

Darden L (1991) Theory change in science: strategies from Mendelian genetics. Oxford University Press, New York

De Renzi M (2014) Evolutionary trends promoted by internal causes versus orthogenesis-a discussion. In: Royo-Torres R, Verdú FJ, Alcalá L (eds) XXX Jornadas de Paleontología de la Sociedad Española de Paleontología. ¡Fundamental! vol 24, pp 57–60

Delisle RG (2011) What was really synthesized during the evolutionary synthesis? A historiographic proposal. Stud Hist Phil Biol Biomed Sci 42:50–59
Delisle RG (2017) From Charles Darwin to the evolutionary synthesis: weak and diffused connections only. In: Delisle R (ed) The Darwinian tradition in context. Springer, Cham, pp 133–167
Dietrich MR (2011) Reinventing Richard Goldschmidt: reputation, memory, and biography. J Hist Biol 44:693–712
Dobzhansky T (1937) Genetics and the origin of species. Columbia University Press, New York
Dobzhansky T (1951) Genetics and the origin of species, 3rd edn. Columbia University Press, New York
Esposito M (2017) The organismal synthesis: holistic science and developmental evolution in the English-speaking world, 1915–1954. In: Delisle R (ed) The Darwinian tradition in context. Springer, Cham, pp 219–241
Etxeberria A, Umerez J (2006) Organización y organismo en la biología teórica: ≚vuelta al organicismo? Ludus Vitalis 14:3–38
Fisher RA (1930) The genetical theory of natural selection. Oxford University Press, Oxford
Gayon J (1998) Darwinism's struggle for survival: heredity and the hypothesis of natural selection. Cambridge University Press, Cambridge
Gilbert SF, Opitz JM, Raff RA (1996) Resynthesizing evolutionary and developmental biology. Dev Biol 173:357–372
Goldschmidt RB (1933) Some aspects of evolution. Science 78:539–547
Goldschmidt RB (1940) The material basis of evolution. Yale University Press, New Haven
Goldschmidt RB (1952) Evolution, as viewed by one geneticist. Am Sci 40:84–98
Gould SJ (1982) The uses of heresy: an introduction to Richard Goldschmidt's the material basis of evolution. Yale University Press, New Haven, pp xiii–xlii
Gould SJ (2002) The structure of evolutionary theory. The Belknap Press of Harvard University Press, Cambridge, MA
Grehan JR, Ainsworth R (1985) Orthogenesis and evolution. Sys Zool 34:174–192
Grene M (1958) Two evolutionary theories (I). Br J Philos Sci 9:110–127
Haldane JBS (1932) The causes of evolution. Longmans, Green and Co, London
Huxley J (1942) Evolution: the modern synthesis. Harper and Brother, London
Johannsen W (1911) The genotype conception of heredity. Am Nat 45:129–159
Kellogg VL (1907) Darwinism to-day: a discussion of present-day scientific criticism of the Darwinian selection theories, together with a brief account of the principal other proposed auxiliary and alternative theories of species-forming. G. Bell and Sun, London
Largent MA (2009) The so-called eclipse of Darwinism. Trans Am Philos Soc 99:3–21
Levit GS, Hoßfeld U (2017) Major research traditions in twentieth-century evolutionary biology: the relations of Germany's Darwinism with them. In: Delisle R (ed) The Darwinian tradition in context. Springer, Cham, pp 169–193
Levit GS, Olsson L (2006) "Evolution on rails": mechanisms and levels of orthogenesis. Ann Hist Philos Biol 11:99–138
Loison L (2011) French roots of French neo-Lamarckisms, 1879–1985. J Hist Biol 44:713–744
Loison L, Herring E (2017) Lamarckian research programs in French biology (1900–1970). In: Delisle R (ed) The Darwinian tradition in context. Springer, Cham, pp 243–269
Mayr E (1942) Systematics and the origin of species. Columbia University Press, New York
Mayr E (1954) Change of genetic environment and evolution. In: Huxley J, Hardy AC, Ford EB (eds) Evolution as a process. Allen and Unwin, London, pp 157–180
Mayr E (1959) Darwin and the evolutionary theory in biology. In: Meggers BJ (ed) Evolution and anthropology: a centennial appraisal. The Anthropological Society of Washington, Washington, DC, pp 1–10
Mayr E (1963) Animals species and evolution. Harvard University Press, Cambridge
Mayr E (1980) Prologue: some thoughts on the history of the evolutionary synthesis. In: Mayr E, Provine WB (eds) The evolutionary synthesis: perspectives on the unification of biology. Harvard University Press, Cambridge, pp 1–48

Meulendijks M (2021) Eclipsing the eclipse?: a neo-Darwinian historiography revisited. J Hist Biol 54:403–443
Nicholson DJ (2014) The return of the organism as a fundamental explanatory concept in biology. Philos Compass 9:347–359
Nicholson DJ, Gawne R (2015) Neither logical empiricism nor vitalism, but organicism: what the philosophy of biology was. HPLS 37:345–381
Ochoa C (2017) El eclipse del antidarwinismo: la historia detrás de la síntesis moderna. Centro de Estudios Filosóficos, Políticos y Sociales Vicente Lombardo Toledano, México
Ochoa C (2021a) Inertia, trend, and momentum reconsidered: G. G. Simpson—an orthogeneticist? In: Delisle RG (ed) Natural selection: revisiting its explanatory role in evolutionary biology. Springer, Cham, pp 261–290
Ochoa C (2021b) The origins of theoretical developmental genetics: reinterpreting William Bateson's role in the history of evolutionary thought. In: Delisle RG (ed) Natural selection: revisiting its explanatory role in evolutionary biology. Springer, Cham, pp 137–170
Ochoa C (2024) Deconstructing the extended evolutionary synthesis: do we need a new theory of evolution? In: Delisle RG, Esposito M, Ceccarelli D (eds) Unity and disunity in evolutionary biology: deconstructing Darwinism. Springer, Cham, pp 403–439
Ochoa C, Barahona A (2014) El Jano de la morfología: de la homología a la homoplasia, historia, debates y evolución. UNAM; Centro de Estudios Filosóficos, Políticos y Sociales Vicente Lombardo Toledano, México
Peterson EL (2016) The life organic: the theoretical biology club and the roots of epigenetics. University of Pittsburgh Press, Pittsburgh
Popov I (2009) The problem of constraints on variation, from Darwin to the present. Ludus Vitalis 17:201–220
Popov I (2018) Orthogenesis versus Darwinism, Translated by Natalia Lentsman. Springer, Cham
Provine WB (1971) The origins of theoretical population genetics. University of Chicago Press, Chicago
Reif W-E (1993) Afterword of OH Schindewolf, basic questions in paleontology. University of Chicago Press, Chicago, pp 435–453
Schaxel J (1931) Das biologische Individuum. Erkenntnis 1:467–492
Schindewolf OH (1963) Neocatastrophism? Z Dtsch Geol Ges 114:430–445
Schindewolf OH (1993) Basic questions in paleontology: geologic time, organic evolution, and biological systematics. Translated by Judith Schaefer. Original (1950). University of Chicago Press, Chicago
Seilacher A (1972) Otto H. Schindewolf, 7 Juni 1896–10 Juni 1971. N Jb Geol Paläont 2:69–71
Simpson GG (1944) Tempo and mode in evolution. Columbia University Press, New York
Simpson GG (1953) The major features of evolution. Columbia University Press, New York
Simpson GG (1959) Anatomy and morphology: classification and evolution: 1859 and 1959. Proc Am Philos Soc 103:286–306
Smocovitis VB (2023) Every evolutionist their own historian: the importance of history, context, and the extended evolutionary synthesis. In: Dickins TE, Dickins JA (eds) Evolutionary biology: contemporary and historical reflections upon core theory. Springer, Cham, pp 25–54
Stoltzfus A (2017) Why we don't want another "synthesis". Biol Direct 12. https://doi.org/10.1186/s13062-017-0194-1
Stoltzfus A, Cable K (2014) Mendelian-mutationism: the forgotten evolutionary synthesis. J Hist Biol 47:501–546
Ulett MA (2014) Making the case for orthogenesis: the popularization of definitely directed evolution (1890–1926). Stud Hist Phil Biol Biomed Sci 45:124–132
Ulett MA (2016) Directed evolution, history of. In: Kliman RM (ed) The encyclopedia of evolutionary biology. Academic, Oxford, pp 441–443
Waddington CH (1942) Canalization of development and the inheritance of acquired characters. Nature 150:563–565
Waddington CH (1953a) Experiments in acquired characteristics. Sci Am 189:92–99
Waddington CH (1953b) Genetic assimilation of an acquired character. Evolution 7:118–126

Waddington CH (1956) Genetic assimilation of the bithorax phenotype. Evolution 10:1–13
Wagner MJ (2024) Revisiting the eclipse of Darwinism: a historiographical and philosophical analysis. Springer, Cham
Walsh DM (2015) Organisms, agency, and evolution. Cambridge University Press, Cambridge
Whitman CO (1919) Orthogenetic evolution in pigeons. In: Riddle O (ed) Posthumous works of Charles Otis Whitman. Volume 1. Carnegie Institution of Washington, Washington, DC
Witteveen J (2015) "A temporary oversimplification": Mayr, Simpson, Dobzhansky, and the origins of the typology/population dichotomy (part 1 of 2). Stud Hist Phil Biol Biomed Sci 54:20–33
Witteveen J (2016) "A temporary oversimplification": Mayr, Simpson, Dobzhansky, and the origins of the typology/population dichotomy (part 2 of 2). Stud Hist Phil Biol Biomed Sci 57:96–105
Wright S (1931) Evolution in Mendelian populations. Genetics 16:97–159
Wright S (1932) The roles of mutation, inbreeding, crossbreeding and selection in evolution. Proc VI Int Congr Genetics 1:356–366
Wright S (1949) Adaptation and selection. In: Jepsen GL, Mayr E, Simpson GG (eds) Genetics, paleontology, and evolution. Princeton University Press, Princeton, pp 365–389

Part II

Hierarchies and Evolution

Challenging Traditional Hierarchies of Evolution

4

Abstract

This chapter challenges the traditional hierarchical frameworks within evolutionary theory. It begins by offering a concise overview of hierarchy theory in its general sense, tracing its early applications in an evolutionary context during the 1960s and 1980s. This chapter mostly presents theoretical and philosophical arguments to challenge the two prevailing models of the hierarchy theory of evolution, the Hierarchical Theory of Selection and the Hierarchy Theory of Evolution (sensu Eldredge and Salthe), with a particular emphasis on criticizing the latter model. For example, the Hierarchical Theory of Selection is mainly considered by species selection; however, selection does not seem to operate at that level, and no concrete examples exist to validate it. Meanwhile, the Hierarchy Theory of Evolution is flawed because its structure of two hierarchies (genealogical and ecological) is based on the outdated replicator-interactor framework for identifying evolving entities. Overall, this chapter argues that such models lack proper integration of Evo-Devo insights.

Keywords

H-Evo-Devo Theory · Hierarchy theory · Hierarchical Theory of Selection · Hierarchy Theory of Evolution · Species selection · Genealogical hierarchy · Ecological hierarchy · Replicator · Interactor · Criteria for individuality · Linnaean hierarchy · Evo-Devo · Units of selection · Units of evolution

C. Ochoa, *Hierarchical Evolutionary-Developmental Theory*, Evolutionary Biology – New Perspectives on Its Development 9,
https://doi.org/10.1007/978-3-031-88723-9_4

Today, when biologists address a hierarchy theory of evolution, they generally have two models in mind: the *Hierarchical Theory of Selection* (hereafter H-TS) and the *Hierarchy Theory of Evolution* (hereafter H-TE). H-TS proposes that the mechanism of natural selection operates on different levels of organization (e.g., genes, organisms, and species), while H-TE claims that evolutionary processes result from interactions between genealogical and ecological entities at different scales. This chapter will examine the framework of these traditional models to question their validity. Specifically, it will scrutinize H-TE in more detail since it stands as the most accepted proposal of a hierarchical evolutionary view in this century.

Section 4.1 examines the development of hierarchy theories of evolution. To begin, it explores hierarchy theory as a model that studies complex systems within a context of scale. Then, it traces its application to evolutionary approaches during the 1960s and 1980s. Specifically, it focuses on the origin of the Hull/Dawkins interactor-replicator framework, which has profoundly influenced the hierarchical evolutionary models that are currently accepted. Subsequently, this section critiques H-TS, particularly in relation to species selection. For example, it challenges the notion of species as units of selection despite their recognition as units of evolution. Section 4.2 presents arguments against H-TE, particularly examining its structure of two hierarchies (genealogical and ecological), which relies on the outdated replicator-interactor framework to identify units of evolution. Generally, it stresses that those models lack a proper integration of Evo-Devo notions.

4.1 Hierarchy Models in Evolution

Given the intricate complexity of the processes interacting across different biological levels, numerous biologists have advocated adopting a hierarchical approach in evolutionary theory for decades. As Gontier (2021, p. 315) points out: "The study of any kind of evolution involves a search for units, levels, and mechanisms or processes, and allocating these into ontological hierarchies." Unfortunately, biologists have inadequately employed hierarchy theory in the evolutionary context, largely because they have formulated these ideas within the framework of the Modern Synthesis. However, as discussed in Chap. 3, this model relegated the organism and its structural properties to the evolutionary scope. This section will examine the key aspects of hierarchical theory and explore its integration into the traditional evolutionary perspective.

4.1.1 The Use of Hierarchies to Study Complex Systems

Hierarchy theory is a theoretical model that makes it possible to study complex systems within the context of scale. Researchers dissect a system into multiple levels of organization, thus gaining insights into its components. According to Wimsatt (2007), these organizational levels are divided based on hierarchy to form an integrated whole. They represent nonarbitrary divisions that ontologically structure natural systems.

Hierarchy theory encompasses a broad range of phenomena and avoids reductionist approaches by acknowledging that the whole is more than the sum of its parts. This approach enables the reconciliation of seemingly contradictory theories, demonstrating their applicability at different scales within a holistic framework (Simon 1962; O'Neill 1989; Ahl and Allen 1996; Allen 2009; Wu 2013). Hierarchical systems consist of entities arranged into levels, each embodying ontological duality. Koestler (1967) coined the term "holon" to illustrate that an entity is both a whole and simultaneously a component of a larger whole. In this regard, entities at the same level are made up of parts at a lower level, while these, in turn, make up entities at the higher level (Wimsatt 2007). Put simply, entities arrange themselves hierarchically based on greater or lesser inclusiveness. Such a hierarchical organization is known as "a nested compositional hierarchy" (Tëmkin and Serrelli 2016), and it is commonly observed in various biological systems. According to some authors, different types of hierarchies are used for biological systems:

1. *Compositional hierarchy.* This hierarchy denotes the part-whole relationships between entities (Wimsatt 2007), which can manifest as either "a" or "b."
   (a) *Aggregative hierarchy.* The nested entities are inclusively aggregated, one by one, at different levels to form the whole (Valentine and May 1996).
   (b) *Constitutive hierarchy.* The nested entities are physical components of the whole (Valentine and May 1996).
2. *Procedural hierarchy.* This hierarchy refers to interconnected processes at different levels occurring sequentially in a specific order for a particular outcome (Love 2006; see Chap. 5).

Another crucial property of biological systems is that their parts are "nearly decomposable" (Simon 1962). This implies that complex systems decompose into subparts whose interactions are more pronounced among themselves than among other parts of the whole. Given that these parts cannot be entirely separated from the whole system, they are related to each other in a complex and asymmetric manner. Higher-level entities are larger in scale and accompanied by processes occurring at lower frequencies. Conversely, lower-level entities tend to be smaller with processes occurring at higher frequencies. Moreover, entities of similar sizes and processes are typically ranked at the same level (Wimsatt 2007). In this regard, the relationship between entities at the same level is referred to as a "horizontal distinction." In contrast, the relationship between entities above or below levels is labeled as a "vertical distinction" (Allen 2009; Wu 2013).

To describe the arrangement of the hierarchical system, Salthe (1985) introduced the "basic triadic structure," which consists of recognizing three contiguous levels within the hierarchy (Fig. 4.1). *The focal level* is situated at the center of the triadic hierarchy, representing the primary entity or process of research interest. Adjacent to the focal levels, the *lower level* is the least inclusive, while the *higher level* is the most inclusive. According to Salthe, although the observer is responsible for

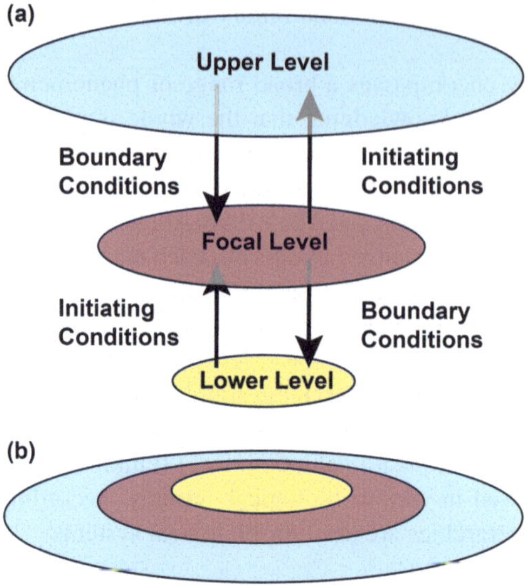

Fig. 4.1 *Triadic hierarchical structure*. (**a**) The focal level is located at the center of the triadic hierarchy, representing the primary entity or process under study. The lower level is positioned below, which indicates the less inclusive level than the focal level. Conversely, the higher level is situated above, corresponding to the more inclusive level than the focal level. The arrows in the diagram indicate the interrelations between levels. The lower level establishes *initiating conditions* for the focal level, while the higher level sets *boundary conditions* for the focal level. Similarly, the focal level exerts *boundary conditions* on the lower level or *initiating conditions* on the higher level. Although the levels are depicted separately in diagram (**a**), these levels are nested into a whole (**b**). (Modified from Tëmkin and Eldredge 2015)

proposing the criteria to delineate between levels, the degree of discontinuity among entities is often so conspicuous that it typically allows for identifying boundaries. This constraint upon the adjacent levels yields different effects. When a lower level exerts constraints or its interactions affect the focal level, it is termed "initiating conditions." On the contrary, when higher levels impose constraints or determine the processes of the focal level, this is termed "boundary conditions." According to Tëmkin and Eldredge (2015, p. 189), the dynamic of interactions between two adjacent levels can also be termed "upward causation," when the lower level affects the higher level, and "downward causation," when the higher level affects the lower level. As stated by these authors, although the effects among levels are bidirectional—i.e., both levels affect each other simultaneously—they can also be unidirectional when one level influences the other without reciprocal impact. In addition, Tëmkin and Eldredge (2015, p. 186) mention two specific attributes of entities at any given level. On the one hand, higher-level entities might exhibit "emergent properties," in which their processes cannot be reduced by the properties of lower-level entities. On the other hand, higher-level entities emerge through "aggregate properties" resulting from the accumulation of properties exhibited by lower-level entities.

Another aspect to consider surrounding hierarchical approaches revolves around the issue of ontology, especially within the context of physical and biological systems. Even though the existence of hierarchical systems in nature is taken for granted by many scholars, several aspects still require consideration. Ahl and Allen (1996) argue, for example, that entities and levels are contingent upon the observer's judgments, i.e., they rely upon the interest driving a given investigation. They claim that complexity and its parts are not "things" that exist independently from observers; instead, their structuring is arbitrary and established based on the observer's decision. On the other hand, Wu (2013) comments that although the observer plays a very important role in constructing hierarchies, these roles reflect a certain reality of complex systems: "The degree of 'realness' of hierarchy not only depends on the nature of the real system but also [on] the observer's abilities, including the theoretical framework, methods, and data used to discover the hierarchy" (p. 290). Indeed, Salthe (1985, p. 30) states that the recognition of boundaries—which makes it possible to delimit levels—arises precisely from the interaction between the observer and nature. In a similar vein, Allen and Starr (1982, p. 86) raise the question of boundaries in relation to whether these entities are continuous or discontinuous. According to Salthe (1985, p. 49), the degree of discontinuity is an important issue for identifying entities in nature. However, this recognition relies on the observer's criteria as well. For Wimsatt (2007), while organizational levels are an ontological reality, as long as connections between levels increase (particularly at higher levels), identifying the levels becomes more diffuse and complex. The division can result in either "perspectives" (divisions based on the observer's criteria) or "causal thickets" (causal relationships that are so complex that their organization is nearly impossible to capture). In fact, the challenge of organizing nature into levels is one of the most recurring themes in hierarchy theory, emphasizing either the inadequacy of the notion of a "level" (Potochnik and McGill 2012; Potochnik 2021) or the consideration of various criteria to define what constitutes a "level" (Craver and Bechtel 2007; DiFrisco 2017; Brooks 2021).

In summary, hierarchy theory is essential for studying complex systems since it aims to analyze asymmetrically related processes within the context of the scale. Employing hierarchies makes it possible to delve into the system part by part and provide a more detailed examination of its components, serving as an approach that improves the understanding of the sequential causes and mechanisms involved in the entire system. Hierarchy theory is a valuable instrument for addressing specific research issues. For example, within the field of biology, this perspective has found widespread application in ecological models (e.g., Allen and Starr 1982; Allen 2009) as well as in evolutionary frameworks (e.g., Jablonski 2007; Tëmkin and Eldredge 2015; Pievani 2016; Tëmkin and Serrelli 2016; Umerez 2016; Gontier 2021; DiFrisco 2021; Spiridonov and Eldredge 2024; Duran-Nebreda et al. 2024). My focus lies particularly within the latter field, which will guide the subsequent sections of this study. Below, I will examine the application of hierarchy theory within the traditional evolution framework.

4.1.2 Hierarchy Theory Applied to Evolutionary Approaches

Hierarchy theory in the evolutionary context is far from new. For example, as discussed in Chap. 2, early naturalists who linked taxonomy, morphology, and evolution maintained that the phylogenetic tree was depicted through a hierarchy of types. Regrettably, the "new systematics" of the Modern Synthesis replaced such a hierarchical perspective. However, as seen in Chap. 3, some proponents of the Modern Synthesis also addressed a hierarchy theory of evolution despite their apparent inclination toward a reductionist perspective of genetics populations. In fact, the Modern Synthesis framework sparked the interest of a subsequent generation of evolutionary biologists who began developing a hierarchy theory of evolution as its theoretical extension, building upon aspects that it supposedly left aside.[1]

Several evolutionary hierarchical approaches emerged between the 1960s and 1980s. Evolutionary models, such as those developed by Valentine (1968) and Salthe (1975) focused primarily on an ecological point of view. Even the theory of punctuated equilibrium proposed by Eldredge and Gould (1972) incorporated a hierarchical perspective, suggesting that macroevolutionary phenomena were not entirely reducible to microevolutionary processes.[2] In addition, MacMahon et al. (1978) proposed an evolutionary model consisting of four interconnected biological hierarchies, all anchored to the "organism" as the central entity. Subsequently, Gould (1980) argued that a new evolutionary theory should adopt a hierarchical structure that recognizes a wide range of phenomena across all levels. Also, such a view would reintroduce the concept of the "organism" as the focal point in discussions within evolutionary biology. Similarly, Arnold and Fristrup (1982) supported a new hierarchical structure in evolutionary theory: "An exploration of the implications of such an approach leads us to the assertion that the development of a unified theory of evolution demands the recognition and incorporation of hierarchical structure as a conceptual foundation" (p. 113).

Still, one of the most important debates regarding a hierarchical perspective stems from the assertion that natural selection could operate at different levels of organization. Emerson (1960) was among the first to envision a hierarchical framework of natural selection. For example, in his essay titled *The Evolution of Adaptation in Population Systems*, he wrote: "Natural selection operates at each level of integration from the gene and complex polygenic characters within the individual, to the whole individual, and to various levels of intraspecific population systems and interspecific interadapted community systems and ecosystems" (p. 340). Subsequently, Lewontin (1970) supported such a hierarchical view,

---

[1] For example, when H-TE was initially introduced, Eldredge (1985) argued that the Modern Synthesis was not incorrect but rather incomplete. In this regard, Eldredge puts forward that the main objective is to extend the theory by incorporating a perspective that involves two evolutionary hierarchies, see Sect. 4.2.

[2] However, as will be presented in Chap. 9, Sect. 9.1.1, the punctuated equilibrium supports the macroevolutionary pattern, but it does not constitute a real macroevolutionary theory since it explains those patterns through microevolutionary mechanisms.

suggesting that natural selection could potentially operate at the levels of molecules, organelles, cells, gametes, individuals, populations, species, and communities. Additionally, Stanley (1975) introduced the concept of species selection, "which is analogous to natural selection but acts upon species within higher taxa rather than upon individuals within populations" (p. 646).[3]

Nevertheless, the publication of *Animal Dispersion in Relation to Social Behaviour* by Wynne-Edwards (1962) was a crucial step in this debate. In this book, Wynne-Edwards argued that group selection might account for the regulation of population growth through behavior. In response, Maynard Smith (1964) introduced the term "kin selection" as a sort of true group selection, specifying that this type of selection occurs exclusively within closely related groups that are partially isolated from others. Yet, Williams (1966), a prominent critic of group selection, claimed that neither groups nor individuals could efficiently act as units of selection. He maintained that these entities have a brief existence and do not transmit faithful copies to the next generation. According to Williams, genes were the true agents of selection. Dawkins (1976, 1982) later strongly supported this stance, leading to a fundamental division that has persisted in hierarchical debates to the present day. He coined the term "replicator" to name entities that faithfully transmit their structure across generations, e.g., genes. In contrast, he referred to "vehicles" as transient entities that contain replicators, e.g., the bodies of organisms. Based on this dichotomy, Dawkins supported the perspective that genes, rather than organisms, are true units of selection because organisms cannot faithfully transmit their structure to the next generation. However, Hull (1980) maintained this distinction and formulated his own interpretation of both concepts. He defined a *replicator* as "an entity that passes on its structure directly in replication" and an *interactor* as "an entity that directly interacts as a cohesive whole with its environment in such a way that replication is differential" (p. 318). In contrast to Dawkins, Hull argued that both notions were necessary to explain how natural selection works at any level. Interestingly, Hull established two types of hierarchies: one for replicators and another for interactors. By asserting that the units of selection must be both interactors and replicators, he sought to identify the appropriate level of organization at which natural selection could operate.

In any case, these assumptions gave rise to two current models that adopt a hierarchical view of evolution. One model is the *Hierarchical Theory of Selection* (H-TS), also known as "multilevel selection," which has primarily focused on debating the levels at which natural selection operates (see Gould 2002; Okasha 2007; Goodnight 2018; Hertler et al. 2020) and particularly emphasizes the species level (see Lieberman and Vrba 2005; Jablonski 2008). Another model is the *Hierarchy Theory of Evolution* (H-TE), which has been a central focus in the major evolutionary debates of this century (see Eldredge 2008; Miller III 2008; Pievani and Serrelli 2013; Tëmkin and Eldredge 2015; Eldredge et al. 2016; Pievani 2013,

---

[3] For a critique of the species selection hypothesis, see below.

2016). Below, I explore criticisms of both models, but considering the prominence of H-TE as the prevailing hierarchical model in recent years, I will look into that analysis in a separate section.

4.1.3 Reconsidering the Hierarchical Theory of Selection

Broadly speaking, H-TS postulates that natural selection can operate hierarchically at different levels of organization (Gould 2002; Okasha 2007; Goodnight 2018). In this regard, some biologists recognize units of evolution as units of selection (e.g., Gould 2002). However, this theoretical framework—also known as "multilevel selection"—has been sporadically discussed in this century, with ongoing debates about its validity (e.g., Gould 2002; Okasha 2005, 2007; Jablonski 2008, 2017; Jeler 2018; Goodnight 2018; Hertler et al. 2020; DiFrisco 2021; Eronen and Ramsey 2022; Simpson et al. 2024). The issue with H-TS as a general stance lies in two main points. First, as highlighted by DiFrisco (2021), there is an overvaluation of the mechanism of natural selection, as the identification of evolutionary entities is only examined under a single criterion, which overlooks structural aspects based on embryonic development:

> Levels decomposed under the process of selection do not capture the sources and structures of variation, which are developmental and genetic, but only the *outcomes* of development… [L]evels of selection alone are too limited to serve as the hierarchical framework for explaining things like major transitions in individuality, which involve changes in development and inheritance. (DiFrisco 2021, p. 129; his emphasis)

Second, the H-TS may not be readily identifiable due to a lack of clear delineations of the organizational levels where it could potentially operate. Eronen and Ramsey (2022), for example, argue that the H-TS framework is inadequately framed considering that the traditional levels of biological organization (gene, cell, organism, population, etc.) have been used as a guide to identify the entities that are subject to selection, but this association has only caused a lack of clarity in the debate.[4]

Nevertheless, it is worth noting that in recent years, the only level that has received significant attention in this context is the species level, a phenomenon known as species selection. Although Steven Stanley postulated species selection as early as 1975, it gained prominence during the 1980s and 1990s, with involvement from prominent paleontologists such as Niles Eldredge, Stephen Jay Gould, Elisabeth Lloyd, Elizabeth Vrba, and philosopher Todd Grantham (Vrba 1984; Vrba

[4] In order to reformulate this debate, Eronen and Ramsey (2022) suggest a deflationary account, claiming that all that is needed is to identify individuals at a lower level who could be potential targets of selection and, simultaneously, form a collective at a higher level that is also subject to selection. Considering this, "multilevel selection" arguably holds conceptual validity in other contexts. For instance, the H-Evo-Devo Theory states that natural selection operates differently with simple traits (represented as subspecies) than with complex traits (represented as species or higher taxa). If this is the case, the H-Evo-Devo Theory would support a sort of "multilevel" selection phenomenon. See Chap. 11, Sect. 11.2.1 and Fig. 11.4.

and Eldredge 1984; Gould and Vrba 1986; Lloyd 1988; Gould and Lloyd 1999; Grantham 1995). However, occasional support for and references to this concept can still be found in the work of paleontologist David Jablonski (2007, 2008, 2017). In any case, the most extensive discussion surrounding this topic focuses on whether selected characters result from the sum of their parts at a lower level (aggregate characters approach) or whether such characters are proper at the species level and, hence, irreducible at lower levels (emergent characters approach). According to Vrba (1984), true species selection occurs only in the case of emergent characters. In contrast, Lloyd (1988) argues that species selection also operates on aggregate characters. Subsequently, some authors have suggested that both ideas could be reconciled. For example, while certain traits emerge from the accumulation of lower-level components, their interaction may occur at the species level, making them irreducible to the lower level (Lloyd and Gould 1993; Grantham 1995; Gould 2002).

Regardless, as I will present, the issue of species selection does not solely revolve around whether it generally operates through emergent characters or aggregate characters, but rather, whether it truly occurs in nature. While a hierarchy theory is essential for clarifying the vast complexity of biological systems, the pertinent question remains: Does the hypothesis of species selection still hold validity within the contemporary evolutionary framework? Below, I will provide two fundamental criticisms of species selection as the most accepted notion supporting H-TS.

(1) *Given that units of evolution are not equivalent to units of selection, species do not seem to be units of selection*

One of the issues with H-TS is its tendency to create confusion in the identification of units of evolution. As discussed in Chap. 6, units of evolution are entities characterized by exhibiting specific patterns during their life history, such as definable birth, cohesion, and punctual death. Meanwhile, units of selection are merely entities that interact with the environment, causing differential survival and reproductive success. Importantly, these two types of entities are not ontologically similar (Shanahan 1990). This oversight arises from conflating the evolutionary processes observed in organisms with those of other entities within the hierarchy. For instance, in a traditional context, Lewontin (1970) suggested three central principles in Darwinian theory: variation, differential fitness, and inheritance. In this regard, the question is whether these principles also extend to entities other than individual organisms (see above). However, what do units of evolution mean according to the Darwinian framework? Interestingly, in his opus magnum, Gould (2002) employs the criterion of individuality to consider species as evolutionary individuals. In doing so, he attempts to expand the role of species by applying analogous Darwinian processes at the level of the individual organisms (i.e., individuals of a population).[5]

[5] These criteria of individuality differ from the framework I propose in Chap. 6, Sect. 6.2. In order to distinguish my proposal from Gould's, I refer to this traditional view of individuality as "Darwinian individuals."

Nevertheless, individual organisms and species should not be considered the same kinds of entities. Individual organisms are units of selection but not units of evolution, whereas species are units of evolution but not necessarily units of selection. In addition, treating entities as Darwinian individuals prevents the recognition of other evolutionary units that do not require Darwinian criteria for individuality. For example, when considering the criterion of variation, a single individual organism does not exhibit variation unless its phenotypic differences at distinct phases of its life cycle are contemplated. In contrast, a species exhibits variation particularly when considering the differences between its constituent parts (subspecies) that make up its body plan. Thus, variation is an emergent property of species, while individual organisms lack this criterion.[6] For instance, there is general agreement that a single individual organism does not evolve; instead, what evolves is the population and it does so through its constituent individual organisms. However, equating individual organisms with species could lead to the erroneous conclusion that a species does not evolve; instead, what evolves is the higher taxon, and it does so through its constituent species. But this is not the case. Moreover, under the Darwinian framework, species selection only means the differential survival and reproductive success of species through their characters at that level. Thus, although species interact with the environment, from this perspective, viewing them as units of evolution is problematic.

For example, while higher taxa could potentially be considered as units of evolution,[7] most biologists assume they are not because they do not function as interactors. More specifically, given that species within a higher taxon are independent entities living in different ecological areas, no selective mechanisms operate at the level of the higher taxon as a whole (Gould 2002, p. 712). With this in mind, describing species selection in terms of functionality, Damuth wrote:

> [N]atural selection at a given level… requires that the units of selection involved share an environment, a set of external conditions that allows the assignment of fitness values to the phenotypes of the population… The local population is an evolutionary entity, united by its common ecological context and, in sexual species, by gene flow among its members. It follows that for simple selection the evolving entity must be restricted to a specific geographical area that is more or less homogeneous relative to the units. Consider a set of organisms, all of which are ultimately the descendants of a single arbitrarily chosen organism, including all those that may have emigrated from the region occupied by the ancestor and are now dispersed throughout the species' range in various habitats. We do not consider this set of lineage members a biological "population," because at any given time its members are not necessarily functioning as economic interactants in a common milieu. For simple selection alone to occur within it, the set must be an ecological, localized entity. (Damuth 1985, p. 1133)

---

[6] According to H-Evo-Devo Theory, variation (as an emergent property) is a criterion of individuality that also applies to other entities such as subspecies, higher taxa, and homologues; see Chap. 6, Sect. 6.2.

[7] Indeed, from the H-Evo-Devo Theory point of view, higher taxa are units of evolution. See Chap. 8, Sect. 8.1 and Chap. 12, Sect. 12.1.

Considering this, it becomes clear that the species encounter the same issue as higher taxa since species consist of subspecies that can live in different environments. This raises doubts about the existence of genuine species selection because the constituent parts of the species might not interact with the environment as a whole. From this perspective, both higher taxa and species might fail to qualify as units of selection, although they could be units of evolution under different criteria. If species were units of selection, they would likely be small populations isolated from other groups in an ecological context (e.g., Damuth's avatars), but this does not precisely correspond with the conception of species.[8] In a nutshell, given the criteria for identifying units of selection, species do not seem to be ideal entities that satisfy that criterion.

(2) *Examples of selection in emergent characters do not constitute true species selection*

In the Lloyd-Vrba debate, although most scholars have reached a consensus by accepting the aggregative characters approach as an unusual case of species selection, the truth is that many of them do not realize that true species selection is only possible through the emergent characters approach. By emergent characters, they mean properties that exist exclusively at the species level and cannot be reduced to lower levels of the hierarchy. The issue is that no such emergent properties satisfy the traditional selection criteria. Furthermore, these characters are not exclusive to species in a taxonomic sense; they are simple demographic characteristics that ignore the organisms' variational and intrinsic properties. Within the context of evolution, this leads to significant neglect of hierarchical considerations in morphology.

For instance, some examples of emergent characters at the species level include geographic range size, genetic population structure, and sex ratio (Vrba 1984; Jablonski 2007, 2008). Now consider that, for these to be considered as selection properties, they must require Lewontin's criteria of variation, fitness, and heritability. Yet, as Cracraft (1985) noted early in this debate, heritability is lacking in some cases of emergent characters, e.g., geographic range. More recently, Jablonski (2008) argued that "geographic range appears to be heritable at the species level, in the sense that closely related species tend to attain range sizes that are more similar than expected by chance" (p. 509). Notably, geographic range—the factor presented as the paradigmatic element of strict species selection—is not an exclusive property of species but rather a property of populations in any Linnaean category (subspecies, species, genus, family, etc.). Furthermore, the "heritability" of geographic range in closely related populations is neither a replicable property nor the type of "replicator" typically referenced in biological systems. The key point in this discussion, however, is that neither the geographic range nor the genetic structure of a population is a property of the species as a taxonomic category.

Setting aside the issue of the species concept, the truth is that advocates of species selection completely ignore morphological aspects within the context of

[8] Regarding the species problem, see Chap. 6, Sect. 6.1 and Chap. 7, Sect. 7.2.

evolution and hierarchy. None of the emergent characters are morphological characters that define groups as species. Instead, species are considered only in terms of their demographic aspects. Even if this were the case, the only way to accept species selection, as Dietrich (2010) rightly claimed, is through the emergent characters approach. But this implies that species selection is an extremely rare phenomenon in nature if it exists at all.

In conclusion, the H-TS model has been one of the most commonly used frameworks for considering the hierarchical structure of evolutionary theory. Yet with the progress seen in the field of Evo-Devo knowledge, its foundations have lost support and credibility (see DiFrisco 2021). Species selection is the only proposal that has survived this century; however, its structure lacks strength: (1) units of selection are not equivalent to units of evolution and (2) there is no true species selection when considering examples of emergent characters. Finally, it is crucial to highlight Gontier's (2010) contention that most scholars take natural selection for granted as the sole mechanism for identifying evolving entities at different levels. Still, a hierarchical evolutionary framework should not confine itself solely to this mechanism. While natural selection holds significance, other mechanisms also play an important role in the evolutionary process: "[W]e need to try and move beyond the metaphysical issues as well as allow for the inclusion of evolutionary theories other than natural selection in regard to the identification of levels and units" (p. 171). Therefore, it seems the H-TS does not fully encapsulate the complexities of evolutionary processes within the context of scale. Clearly, the exploration of alternative directions is necessary in order to propose a more comprehensive hierarchy theory of evolution. But what other alternatives should be considered?

4.2 Hierarchy Theory of Evolution Under Fire

The so-called *Hierarchy Theory of Evolution* (H-TE) represents this century's predominant and widely accepted hierarchical perspective of evolution. It was initially formulated by Eldredge and Salthe (1984; Eldredge 1985, 1989; Salthe 1985), and subsequently bolstered by contributions from various scholars (Vrba and Eldredge 1984; Grene 1987; Eldredge and Grene 1992). Currently, this model has been championed and popularized, particularly by paleontologists and philosophers of biology who identify themselves as the *hierarchy group*. The members of this group include Niles Eldredge, Telmo Pievani, Emanuele Serrelli, Ilya Tëmkin, among others.

Broadly speaking, H-TE stipulates that evolutionary biological systems are nested and connected by two types of hierarchies reflecting their attributes in terms of reproduction and survival: the genealogical hierarchy and the ecological or economic hierarchy. The genealogical hierarchy relates to the transmission of heritable information over time, i.e., it concerns the ability of systems to transfer their structure through replication. This continuous transmission of information from generation to generation reflects its historical nature. Entities identifiable within the genealogical hierarchy include genes, organisms, demes, species, and monophyletic taxa. In contrast, the ecological hierarchy corresponds to the exchange of matter and energy among entities, resulting from their interaction with the environment. Recognizable entities within the ecological hierarchy comprise proteins, organisms, populations, communities, ecosystems, and the biosphere. Therefore, advocates of the H-TE affirm that the outcome of the interactions between entities within the ecological hierarchy results in the differential replication of the entities within the genealogical hierarchy. This dynamic of interactions and processes occurring between the two hierarchies explains all the patterns and events seen in the evolving entities (Eldredge and Salthe 1984; Eldredge 1985, 1989; Salthe 1985; Tëmkin and Eldredge 2015).

But why is there dissatisfaction with H-TE? Pievani (2016) presented H-TE as another version of "extended synthesis," "a multilevel extended synthesis" that is, however, "still strongly Darwinian in its core" (p. 359). His criticism of the Extended Evolutionary Synthesis (EES)[9] is based mainly on its lack of a hierarchical perspective of evolution: "[T]he extended synthesis is not yet a multilevel explanation of the evolutionary phenomena because it lacks a theoretical frame for the interconnections between different levels of change (ecological and genealogical)" (p. 358).[10] Despite this effort to present an alternative evolutionary theory with a hierarchical and pluralistic structure at its core, the truth is that H-TE still lacks an adequate integration of Evo-Devo insights. While Pievani (2016) has attempted to move in this direction, the structure of H-TE falls short of achieving this goal. Below, I will examine some issues associated with H-TE. See Table 4.1 for a summary.

(1) *Biological entities within the two hierarchies do not correspond to each other*

During the development of H-TE, at least five types of biological hierarchies have been identified: (1) the genealogical hierarchy, (2) the ecological hierarchy, (3) the somatic hierarchy, (4) the hierarchy of homology, and (5) the Linnaean hierarchy. In an early study, Vrba and Eldredge (1984) assumed that the hierarchy of homology and the Linnaean hierarchy should be included within the genealogical hierarchy, as they solely represent the historical record of the flow of heritable information. Given that the organism—which represents the somatic hierarchy (organelles–cells–tissues–organs)—was simply viewed as a vehicle containing genetic information, they thought this hierarchy should also be implicitly included in the genealogical hierarchy. Thus, only the ecological hierarchy was excluded. But how could the genealogical and ecological hierarchy be integrated into a single framework?

[9] Amid the challenges faced within the Modern Synthesis (MS) framework, certain evolutionary biologists have introduced an explanatory model that is positioned as an extension of it: the so-called "Extended Evolutionary Synthesis" (EES). This model strives to incorporate certain topics purportedly overlooked by the MS, including inclusive inheritance, phenotypic plasticity, niche construction, and more (see Laland et al. 2015; Lala et al. 2024). Yet, in another work, I challenge the EES framework, contending that it demands more thorough refinement (see Ochoa 2024).

[10] Interestingly, Fábregas-Tejada and Vergara-Silva (2017) have recently analyzed the possible unification of H-TE and EES.

Table 4.1 *Summary of the shortcomings of Hierarchy Theory of Evolution (H-TE)*

Difficulties	Summary
1. Biological entities within the two hierarchies do not correspond to each other	Given that entities of the two hierarchies do not simultaneously function as both interactor and replicator—except the organism—they can hardly be called units of evolution
2. Biological entities do not correspond among the levels within the same hierarchy	There is no consensus on defining the kinds of biological entities that occupy the nested sequence within their two hierarchies (genealogical and ecological). Additionally, the complexity of biological systems makes it challenging to establish discrete and universal entities across different levels of organization. More importantly, the entities occupying different levels within a nested hierarchy are not of the same kind; for example, in the genealogical hierarchy, a cell is not merely a collective of genes
3. The problem of replicators in the structure of the genealogical hierarchy	Most entities within the genealogical hierarchy are not replicators since they do not produce faithful copies of the same kind for the next generation. Moreover, the replicator scheme fails because it does not account for multiple types of replicators at lower levels. Additionally, the complete replicator-interactor framework is flawed because it ignores the processes involved in the organism's development
4. The problem of considering the entities of the genealogical hierarchy as packages of genetic information	Neither organisms nor successive genealogical entities are just packages of genetic information. H-TE does not give any evolutionary significance to the organism and its structural properties. This model does not take into account the vast variety of factors that contribute to the development of organisms either
5. The entities within the genealogical hierarchy are not evolutionary individuals but historical entities; also, they do not exhibit hierarchy in their structure	The genealogical hierarchy comprises entities that transmit hereditary information over time, reflecting their historical nature. Consequently, species and monophyletic taxa are not evolutionary individuals; instead, they are historical entities represented in a phylogenetic tree that does not exhibit a true hierarchy
6. The entities within the ecological hierarchy are not units of evolution; rather, they are systems that foster the interaction of true evolving entities	The entities of the ecological hierarchy are neither interactors nor evolutionary individuals; rather, they are systems that influence the evolution of the true evolving entities

According to his narrative on the origin of the double hierarchy, Niles Eldredge (2015) commented that his initial intention was to formulate a single hierarchy that would encompass both genealogical and ecological elements: "I attempted in 1982 to cram the Theodosius Dobzhansky–style hierarchy of genes–organisms–populations–species–monophyletic higher taxa, together with regional biotas (larger-scale ecosystems), and, at the highest rung of the ladder of biological system scale, the entire 'biosphere'" (p. 301). Yet upon reviewing one of Stanley Salthe's early studies (1975), Eldredge realized that the ecological aspect of evolution was crucial, notwithstanding that little attention had been devoted to this topic. Subsequently, after meeting Stanley, they collaborated on deciphering how to connect the genealogical and ecological hierarchies, despite their marked differences. They concluded that the solution was not to integrate both hierarchies into one but rather to recognize that each harbors distinct entities that interrelate and mutually affect each other, causing common evolutionary outcomes:

> This observation led us for years to pursue, in vain, an attempt to find other exact correspondences between individuals within both the ecological and genealogical hierarchies—obscuring the main point … that the relationship between the two hierarchies consists of interactive effects between individuals found within the two different hierarchies. Individual elements within the ecological hierarchy are largely responsible for the shuffling of properties of individuals within the genealogical hierarchy—while the reproducing individuals within the genealogical hierarchy determine the very components, the actual interactive biotic players in the game, within the ecological hierarchy. (Eldredge and Salthe 1984, p. 191)

However, they identified a common entity between both hierarchies: the organism. According to these authors, the organism is a replicator of the package of genetic information, and it also plays an important ecological role as a matter-energy transfer machine. The issue with this view arises when realizing that no other entities correspond between the two hierarchies besides the organism: "We… find this situation just a trifle unsettling, if only because it is only organisms that are so easily agreed upon by all to be individuals. Our failure to find other individuals common to both systems may reflect more the difficulty to recognizing other sorts of individuals in the biotic realm" (Eldredge and Salthe 1984, p. 191). This issue becomes even more problematic within the hierarchy when considering the philosophical source that inspired them.

As becomes evident, the structure of the double hierarchy (genealogical/ecological) is rooted in Dawkins/Hull terminology, particularly the distinction between replicator and interactor. In Dawkins' original version (1976), the replicator solely refers to the gene, while the interactor (vehicle) refers to the organism. As discussed earlier, Dawkins used this classification to recognize the gene as a unit of selection. In Hull's proposal (1980), units of selection are distinguished only when an entity simultaneously possesses the properties of both a replicator and an interactor. Lineages, for example, evolve as a result of this duality, and the double hierarchy seen in Hull's study is an integral aspect of this analysis. Similarly, this perspective finds ample representation in Brandon's study (1988), which established that units of selection are essentially units of interactors. Brandon's primary goal was to identify the hierarchy of interactors in order to assess whether they also function as replicators: "[T]he replicator hierarchy is derivative from the interactor hierarchy in the sense that we need to determine the level of interaction in order to determine the level of replication but not vice versa" (pp. 64–65). Conversely, Gould (2002) argues that the units of selection are exclusively interactors, suggesting that being a replicator is merely a property of some interactors, but not an exclusive one.[11]

So, according to Hull's view, to be a unit of evolution, an entity must function as a unit of selection;[12] and to be a unit of selection, an entity must be both a replicator and an interactor. Since the entities (individuals) of the double hierarchy do not correspond ontologically within the replicator–interactor framework, except for the organism, it becomes challenging to identify them as units of evolution. For instance, within the genealogical hierarchy, entities such as genes, demes, species, and monophyletic taxa lack direct interaction with the environment. Similarly, within the ecological hierarchy, entities such as communities, ecosystems, and biotic regions do not replicate faithful copies of themselves, or even worse, they do not make more entities of the same kind. If anything, Damuth's avatar[13] could aspire to such recognition, but the hierarchy would remain stagnant at that level. Moreover, problems arise when advocates of H-TE admit that species are not entities belonging to the ecological hierarchy; however, species function as interactors for other researchers (e.g., species selection).

Finally, recently Spiridonov and Eldredge (2024) have proposed a new hierarchy that seeks to find the "suture zone" between the genealogical and ecological hierarchies. In analyzing this interaction, they identify two types of entities: "holobionts" and "geobiomes." It should be noted that while holobionts represent organic symbiotic entities, geobiomes encompass non-living components. Thus, if identifying units of selection between the two hierarchies is problematic under the traditional argument, this becomes even more challenging when considering the complex series of interrelationships required to identify units within this new hierarchy.

(2) *Biological entities do not correspond among the levels within the same hierarchy*

One issue with H-TE is the lack of a precise consensus on defining the kinds of biological entities that occupy the nested sequence within either of the two hierarchies (see Fig. 4.2). For example, the studies that discuss the genealogical hierarchy mention entities such as codons, chromosomes, cells, and even the entire biosphere, though the latter is typically located within the ecological hierarchy according to other contributions (e.g., Salthe 1985). In contrast, within the ecological hierarchy, entities such as enzymes, cells, ecosystems (and even the entire surface of the Earth!) are suggested

[11] As a challenge to the stance of Williams and Dawkins, who considered replicators as units of selection, Gould (2002) claimed that interactors, not replicators, were the real causal agents of selection. Accordingly, evolutionary entities do not require a replication that is faithful to their hereditary transfer, but rather a relative plurifying of such hereditary contribution (i.e., producing more individuals of the same kind than the competitor). This perspective led him to advocate for a single hierarchy of Darwinian individuals, excluding unnecessary models of double hierarchies.

[12] This is true only when strictly adhering to the replicator–interactor framework; however, as I have argued previously, units of selection are not equivalent to units of evolution. See Sect. 4.1.3.

[13] Damuth (1985) coins "avatars" to refer to those populations within species that share a common environment and hence respond to similar selective pressures.

Eldredge and Salthe's (1984) Original Proposal

Genealogical Hierarchy	Ecological Hierarchy
Codons	Enzymes
Genes	Cells
Organisms	Organisms
Demes	Populations
Species	Local ecosystems
Monophyletic taxa	Biotic regions
(Special case: all life)	Entire biosphere

Eldredge's (1985) proposal. See that he introduces an alternative ecological hierarchy.

Genealogical	Ecological (I)	Ecological (II)
Monophyletic taxa	Regional biotas	Biosphere
Species	Communities	Ecosystems
Demes	Populations	Populations
Organisms	Organisms	Organisms
Chromosomes	Cells	Cells
Genes	Molecules	Molecules

Salthe's (1985) proposal. *He introduced an alternative level for addressing cases of coevolution. **According to Salthe, the term "organism" should be reserved only as an entity within the ecological hierarchy. For him, the counterpart to the organism in the genealogical hierarchy should be termed the "integrated genotype-phenotype."

Ecological Hierarchy	Genealogical Hierarchy
The Level of the Surface of the Earth	The Level of the Total Biosphere
	The Level of the Historical Biota*
The Biogeographic Regional Level	The Level of the Monophyletic Lineage
The Ecosystem Level	The Level of the Species
The Population Level	The Level of the Deme
The Organismic Level	The Level of the Integrated Genotype-Phenotype**
The Molecular Level	The Level of the Gene

Eldredge's (2016) proposal. In this version, Eldredge distinguishes the two hierarchies by considering the differences between soma and germline.

Economic Hierarchy	Level	Genealogical Hierarchy
Biosphere	Planetary Level	Tree of life
Biocenosis	Biotic Level(s)	Monophyletic taxon Species
Avatar	Population Level	Deme
Organism (soma)	Organismal Level	Organism (germline)
Cell (cytoplasm)	Cellular Level	Cell (nucleoplasm)
Proteome	Molecular Level	Genome

Fig. 4.2 *Comparison of the tables describing the nested sequences of the entities within the genealogical and ecological/economic hierarchies.* Note the lack of consensus in defining the entities (or levels) that occupy the nested sequences within either of the two hierarchies. For further details, see the text

The changes within the nested sequences are notably depicted in Eldredge's tables across different periods. For example, in 1985, Eldredge presented an alternative interpretation of ecological hierarchy whereby the distinction between the two versions involved incorporating the abiotic realm into one of the two hierarchies.

Meanwhile, in his latest version, Eldredge (2016) split the two hierarchies based on soma and germline, following August Weismann's nineteenth-century terminology. Conversely, for Salthe (1985), the equivalent of the "organism" within the ecological hierarchy is the "integrated genotype-phenotype" within the genealogical hierarchy. He even introduced an additional level to address the phenomenon of "coevolution" within the genealogical hierarchy. Of course, defining the stratigraphic sequences of entities is not an easy task. Some philosophers have observed that systems are too complex, making it challenging to establish discrete and universal entities at different levels of organization, particularly in biological systems that tend to be more heterogeneous (see Potochnik and McGill 2012; DiFrisco 2017; Eronen 2015, 2021; Potochnik 2021).

Another issue is that some entities occupying different levels within a single hierarchy do not possess the same ontological nature throughout the entire sequence. A more realistic perspective would consider distinct hierarchical views. For example, regarding the "gene-cell-organism" nested sequence within the genealogical hierarchy, while the organism can be described as "a collective of cells," the cell itself is not a collective of genes. A more appropriate perspective would recognize two types of hierarchies: a genetic hierarchy, comprising the "gene–genetic network–genome" sequence, and a somatic hierarchy, comprising the "cell–tissue–organ–system–organism" sequence. Valentine and May (1996) comment, for example, that including the enzyme and the cell in Eldredge and Salthe's ecological hierarchy is misplaced: "[E]nzymes are only one of the constituents of cells at the molecular level and do not comprise a rank themselves; their collective is not a cell, and processes within a cell that are not exclusive to enzymes are unfortunately excluded from this hierarchy" (p. 31). Similarly, Hull (1988) asserts that the "gene–organism–species" sequence within any evolutionary hierarchy (e.g., genealogical hierarchy) is not natural. In other words, arbitrarily arranging entities into a hierarchy leads to flawed ideas about the patterns and regularities of evolutionary change: "One reason why evolutionary biologists have been unable to discover universal regularities in the evolutionary process is that they are not comparing like with like. They are dividing up the organizational hierarchy inappropriately" (p. 26).

In brief, H-TE faces the issue of defining the kinds of entities comprising the sequence in the hierarchy. Likewise, these entities lack natural nesting because, among the levels, none of the entities are the same kind. For example, an individual organism is not simply a collective of genes, nor is a community merely a collective of populations (in addition to the fact that a community also includes abiotic factors). An appropriate evolutionary hierarchy should incorporate the same kinds of entities at all levels. Alternatively, distinct types of hierarchies could be employed to delineate various complex phenomena; for example, genes and proteins within a procedural hierarchy can represent entities operating at different levels in developmental processes, while species and higher taxa may indicate distinct levels of body plans, etc.[14]

[14] The H-Evo-Devo Theory aims to leverage hierarchical frameworks in this context. See Chap. 5.

(3) *The problem of replicators in the structure of the genealogical hierarchy*

As previously mentioned, the genealogical hierarchy considers that nested entities should be formed from the transfer of hereditary information over time. According to this view, such a transmission of information underlies the ability of "evolutionary individuals" to reproduce by themselves. In such terms, the genealogical hierarchy is a hierarchy of replicators, or, in Eldredge's words, it is a hierarchy of "moremaking" entities. For example, genes can replicate into more genes, organisms replicate into more organisms, demes into more demes, and species into more species. For Eldredge, the only exception is monophyletic taxa because they cannot produce more individuals of the same kind; rather, they only reflect the genealogical connections between closely related species. Nevertheless, the primary concern with this mode is that the entities within the genealogical hierarchy are not true replicators.

In the original terminology of Dawkins/Hull, a replicator is "an entity that passes on its structure directly in replication"; this implies that entities in replication must transmit their structure intact to the next generation as faithfully as possible. The issue here is that not all levels consist of faithful replicas of their kind. According to Dawkins (1976), the gene is the only entity that satisfies this criterion. For Hull (1980), replication is not only observed in genes but also asexual organisms, some colonies, and, in extremely rare cases, at higher levels. Of course, even supporters of H-TE acknowledge that strict replication is generally not feasible for most entities in the sequence. Pievani (2013) claims, for example: "In Eldredge's model no faithful inheritance is required and the levels are wider units of evolutionary change (ecological and genealogical)" (p. 45). Similarly, Tëmkin and Eldredge (2015) assert: "*Simple replicators* that produce descendants identical to ancestors cannot evolve… whereas *hereditary replicators* that generate progeny that might differ from its ancestral template can" (p. 199; their emphasis).

But if entities are not replicators in the strict sense, then what kind of entities are they? This question would seem less contentious in the case that the hierarchical group was open to the possibility that these entities could also function as interactors, but this is not the case. For example, Hull's concept of an "interactor" differs from Dawkins' concept of a "vehicle." According to Hull, an interactor is an entity that interacts as a cohesive whole within the environment, resulting in differential replication. Whereas for Dawkins, the vehicle is simply a carrier of genes. With this in mind, it becomes clear that Eldredge's argument is similar to Dawkins' argument in this regard. According to Dawkins (1976), the organism is the vehicle bearing the genes. Similarly, for Eldredge (2015), not only the organism but also the "genealogical systems are *all packages of genetic information* that are composed of next-lower-level entities that 'reproduce'" (p. 305; my emphasis).

Needless to say, this perspective is completely flawed, as the organism is not solely developed by genes alone. That is, the organism is not simply an extension of genetic information over time. Genes cannot perform alone without the immense complexity of the processes involved in an individual's development (Lewontin

1983; Oyama 1985; Wimsatt 2001; Danchin et al. 2011).[15] Interestingly, for some time now, advocates of Developmental Systems Theory (DST) have stressed that organisms are constructed from diverse ontogenetic resources and not solely from genes (Griffiths and Gray 2001, 2005; Oyama et al. 2001; Griffiths and Stotz 2018; Stotz 2019). For instance, many studies have demonstrated that inheritance goes beyond the simple transmission of genes. In addition to DNA, examples of inheritance factors include RNAs, methyl groups, proteins, cellular elements, glandular secretions, hormones, nutrients, pathogens, organelles, symbionts, ecological niches, behaviors, and so on. According to some authors, all these elements influence the evolutionary process in one way or another (Jablonka and Lamb 2005; Bonduriansky and Day 2018; Lala et al. 2024). In light of this, genes are not the only replicators at this level. To this point, Griffiths and Gray (2001) comment: "[T]he idea that developmental systems can be reduced to a collection of independent replicators is either inadequate or has to recognize relationships to persistent features of the environment as an addition to the cast of replicators" (p. 197). Likewise, Bonduriansky and Day (2018) confirm this perspective: "The existence of nongenetic inheritance means that genes… are not the only types of replicators—we could also speak of epigenetic replicators, structural replicators, behavioral replicators, and so on—and extended heredity can be seen as the combined outcome of all these different types of replicators" (p. 19).

Indeed, Griesemer (2000b) emphasized that the replicator-interactor framework relies on an inadequate view wherein genes operate autonomously from the embryonic process. Following this terminology, the phenomenon of ontogeny, with all its processes, goes unnoticed[16]:

> Dawkins-Hull replicators are things that pass on their structure intact, invariant in the developmental processes that construct new vehicles each generation that are exposed to the buffeting of selection… interactors are characterized in terms of selective interaction, though development is clearly the process that gives rise to interactors. However, development does not figure in the analysis of either replicators or interactors. (Griesemer 2000b, p. 260)

Similarly, in recent comments on the replicator-interactor framework, DiFrisco (2021) asserted that:

> [It] includes only two narrow subprocesses out of the wider process of developmental evolution—namely, inheritance and selection. This leads to a severely impoverished picture of evolution. By excluding development and its subprocesses, we lose any chance of capturing the causal structures that link genetic inheritance and selection. Moreover, by excluding development, the framework *cannot* incorporate insights about the hierarchical structure of

[15] See point 4 below.

[16] As an alternative to this, Griesemer (2000a, b) introduces the term "reproducer" to identify units of evolution. In his view, reproducers are entities (developmental systems) that reproduce other entities with the same capacity to reproduce. Reproducers do not transmit but rather they transfer "pieces of development" to their offspring, conferring upon them the ability to reproduce other entities of the same kind.

> variation, including features like network-level causation, variational independence above the genetic level, and shifting causality horizons in evolution. (DiFrisco 2021, p. 130; his emphasis)

In conclusion, the model of a hierarchy of replicators as the basis for the genealogical hierarchy is problematic since not all entities within the hierarchy produce faithful copies of the same kind. Strict adherence to the original term of the replicator would exclude many levels of replicators. Furthermore, the view that different nested entities behave as packages of genetic information closely resembles Dawkins' notion of vehicles. However, from the contemporary point of view, it is difficult to conceive that genealogical entities are simply carriers of genes, considering that organisms are built from diverse ontogenetic resources and not from genes alone. Given this fact, the replicator framework fails at its most fundamental premise because there would be not just one type of replicator at the lower level but many others as well. Even the complete replicator-interactor framework is inadequate since it ignores the processes involved in the organism's development.

(4) *The problem of considering the entities of the genealogical hierarchy as packages of genetic information*

Undoubtedly, a notable error in H-TE lies in considering that nested entities only make sense within the context of transmitting hereditary information over time. In other words, H-TE supports the gene-centered perspective inherited from the Modern Synthesis framework.[17] Under this view, the organism is depicted as emerging from a "genetic blueprint" or a "genetic information program" without any causal reference to its embryonic development. Heredity is narrowly defined as the "transmission of genetic information," while variation is viewed merely as "altered genetic information," disregarding the organism's structural properties and their significance in the evolutionary scope (Wagner 2015). In this respect, Caianiello (2016) argues that, in its early stages, H-TE "tended to downplay the evolutionary relevance of levels and entities below the organism and above the gene," in this regard, "the so-called somatic or phenotypic hierarchy… could be deprived of evolutionary significance." In addition, the hierarchy of homology was relegated "as a merely descriptive record of evolutionary history" (p. 153).

The idea that an organism merely transmits "genetic information" is a vestige of the traditional view of population genetics, which suggests that evolution is solely the change in allele frequencies within a population (Beurton 2000). Additionally, this was partly due to an inadequate analogy drawn from computing (Oyama 1985; Keller 2000; Lewontin 2000). In any case, Wagner (2015) asserts that one of the key contributions of integrating developmental biology with evolution is to restore the importance of the organism in the evolutionary scope: "[T]he variational properties of organisms have theoretical significance—they are more than a vehicle for genes

[17] Regarding the foundation of the Modern Synthesis framework through a gene-centered perspective, see Chap. 3, Sect. 3.1.

or an epiphenomenon of gene frequency changes" (p. 333). Therefore, from a developmental point of view, it is impossible to recognize the function of genes without considering the great complexity of ontogenetic interactions that take place at other levels (Gawne et al. 2017). Furthermore, not only does the environment play a crucial role in organism differentiation (West-Eberhard 2003; Sultan 2015; Gilbert 2016), but external physical forces also contribute significantly to this process (Newman and Müller 2005; Newman and Bhat 2009; Newman 2021).

In summary, genealogical entities are not merely packages of genetic information. H-TE has relegated the organism and its structural properties, diminishing its evolutionary significance. Its view of inheritance is confined to the flow of genetic information; thus, this theory overlooks the vast array of factors contributing to organismal development. Considering this, H-TE still lacks proper integration of Evo-Devo, as it supports the traditional gene-centered perspective inherited from the Modern Synthesis framework.

(5) *The entities within the genealogical hierarchy are not evolutionary individuals but historical entities; also, they do not exhibit hierarchy in their structure*

As I previously described, during the development of H-TE, Vrba and Eldredge (1984) included two historical hierarchies within the genealogical hierarchy: the hierarchy of homology and the Linnaean hierarchy. So, from the point of view of H-TE, homologous characters are "resemblances" or "attributes" found in different organisms due to common ancestry (synapomorphies).[18] If homologues are explained by common ancestry, then they can be traced back through a continuous chain of phenotypic changes in ancestors and descendants: "[H]omologous characters are simply the various known conditions that a phenotypic feature has assumed in the course of evolutionary history" (Eldredge 1985, p. 192). If homologues function as resemblances serving to identify nested monophyletic groups, then these groups could be recognized within the Linnaean hierarchy as species, genera, families, and so forth. Nevertheless, Eldredge (1985, p. 145; 1989, pp. 158–160) claimed that the genealogical hierarchy does not correspond with the Linnaean hierarchy. For example, although the term "species" is used in both hierarchies, in the Linnaean hierarchy it is related to a category, whereas in the genealogical hierarchy it is linked to "evolutionary individuals."[19] Moreover, Eldredge (1993) puts forward that species are real entities, but only within the context of being "reproductive communities" (i.e., members of species have the potential to interbreed and possess a mate-recognition system). According to the thesis of individuality (sensu Ghiselin

[18] As will be explored in Chap. 8, Sect. 8.2, this point of view is typical of the phylogenetic systematics approach, which emphasizes identifying monophyletic groups. It contrasts, the morphology approach focuses on the causes of constancy of modules (highly integrated parts of an organism in developmental terms) over time. This latter view is supported by the H-Evo-Devo Theory.

[19] Eldredge and Salthe (1984) adopted Ghiselin's and Hull's proposal that species function more effectively as individuals (historical entities) rather than as natural kinds (categories). However, the new perspective on individuality contradicts the view that entities are solely historical. For further details, see below and Chap. 6, Sect. 6.2.

and Hull), species are "spatiotemporally bounded" due to such reproductive connections. Therefore, for Eldredge, the morphological aspect of species—where the Linnaean hierarchy applies—appears vague, while the reality of species lies in their potential for interbreeding, which is associated with the genealogical hierarchy.

With respect to higher levels, Eldredge (1985, p. 145) states that, unlike the Linnaean hierarchy, monophyletic taxa in the genealogical hierarchy are "unranked." The main reason is that monophyletic taxa are not reproductive entities, i.e., they do not produce more entities of the same kind; instead, they are simple groups of species that share genealogical connections (Eldredge 1989, pp. 158–159). From this point of view, there is no nested hierarchy between categories as seen in the Linnaean hierarchy: "The reason why the Linnaean hierarchy is not simply tacked onto the top of the genealogical hierarchy is that there is no formal difference between genera and kingdoms. All monophyletic taxa are defined as 'all species (one or more) descended from a single ancestral species'" (Eldredge 1985, p. 161). In other words, according to Eldredge, monophyletic taxa represent the terminal entities of the genealogical hierarchy, i.e., they indicate only the genealogical relationships between species without any hierarchical structure as such. Rather, they are "just the branches of the genealogical 'tree of life,' and not each and every bifurcation need be, or ever is, named" (Eldredge 1985, p. 192).[20]

The issue with this model is that Eldredge conceptualizes each entity within the genealogical hierarchy as an evolutionary individual: "Genes, organisms, demes, species, and monophyletic taxa form one nested hierarchical system of individuals that is concerned with the development, retention, and modification of *information* ensconced, at base, in the genome" (Eldredge 1985, p. 7; his emphasis). However, Ereshefsky (2001, p. 112) points out that some scholars who argue for species as evolutionary individuals do so considering them instead as historical entities; hence, the meaning of individuality becomes highly ambiguous. In this regard, it is worth emphasizing that when the advocates of H-TE refer to their genealogical entities as evolutionary individuals, they actually mean that these entities are historical. Nevertheless, as Ereshefsky correctly notes (1991, 2001), historical entities are not evolutionary individuals. In fact, the entire discourse of the genealogical hierarchy seems to go in that direction: their entities are merely historical, as they only represent the flow of genetic information over time (see point 4).

In addition, it should be noted that this sequence lacks any hierarchical structure. For example, several scholars have highlighted significant differences between the Linnaean hierarchy and a phylogenetic hierarchy (Williams 1992; Valentine and May 1996; Valentine 2004; Vasilyeva and Stephenson 2008; Brower 2023). The former implies a true hierarchy (inclusive hierarchy) used to depict morphological states (disparity) at different scales. In contrast, the latter is a phylogenetic hierarchy (generative hierarchy) that employs Linnaean ranks metaphorically to signify the separation of two or more taxa within the phylogenetic tree. Valentine and May

[20] Nevertheless, as will be explored throughout this book, within the framework of H-Evo-Devo Theory, higher taxa not only appear real but also play a significant role in evolution. See Chap. 8, Sect. 8.1 and Chap. 12, Sect. 12.1.

(1996, pp. 29–30) assert that such a phylogenetic hierarchy lacks the essential characteristics of any hierarchy.[21] Instead, it manifests a positional structure, where all entities exist at the same level and whose arrangement in "nests" does not involve hierarchical ranks.[22] For example, consider the genealogical tree of a family line of women illustrated in Fig. 4.3. At the lowest level (1), the grandmother represents the most inclusive ancestor. Moving up to the next level (2), you find the mother and aunt as the split from the last common ancestor (the grandmother); thus, they represent the less inclusive ancestors. At the subsequent higher level (3), you find the daughters and cousins as the split from the mother and aunt, respectively; thus, they represent the descendants. Although this tree appears to have a hierarchical structure, it does not exhibit the characteristics of a true hierarchy. For example, daughters neither aggregate their mother nor are they physical components of their mother. Neither the lower levels nor the higher levels are united by collectives because they all exist at the same level. While Eldredge (1985) recognizes this issue for

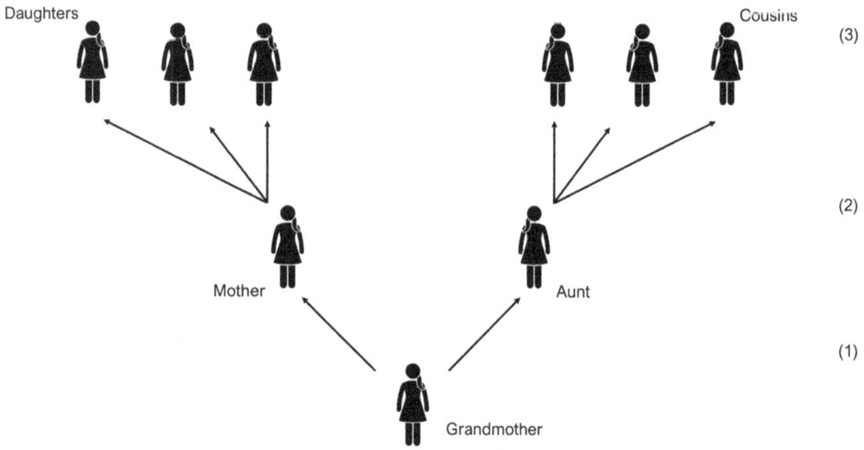

Fig. 4.3 *A genealogical tree outlining a family line of women.* At level 1, the grandmother represents the more inclusive ancestor. At level 2, the mother and the aunt split from the grandmother; thus, they represent the less inclusive ancestors. At level 3, the daughters and cousins split from the mother and aunt, respectively, and thus, they represent the descendants. Superficially, this type of tree resembles a hierarchy, but it is not one; instead, it has a positional structure. Daughters neither aggregate inclusively to make their mother nor are they physical components of their mother. This arrangement lacks discernible ranks or distinct scales; all entities (women) are placed at the same level

[21] Valentine and May (1996) identify two types of hierarchies in complex systems: the aggregative hierarchy and the constitutive hierarchy. The aggregative hierarchy involves the aggregation of parts at various scales to form a whole (e.g., the Linnaean hierarchy). In contrast, the constitutive hierarchy entails parts at different levels being the physical components of the whole (e.g., the somatic hierarchy).

[22] Regarding the distinction between a Linnaean hierarchy and a phylogenetic hierarchy, see also Chap. 5, Sect. 5.1.3 and Chap. 8, Sect. 8.1.1.

monophyletic taxa, he does not extend this observation to species since, according to him, species generate entities of the same kind. However, Ereshefsky (1991, 1999) clearly asserts that species and higher taxa share the same ontological situation under this view; namely, they are historical entities.[23]

In short, the advocates of H-TE affirm that a genealogical hierarchy comprises entities responsible for transmitting hereditary information over time, thus reflecting their historical nature. Interestingly, they also claim these entities are evolutionary individuals. Nevertheless, upon analyzing the structure of the genealogical hierarchy, it becomes apparent that neither species nor monophyletic taxa are truly evolutionary individuals; instead, they are historical entities represented in a phylogenetic tree that lacks a true hierarchical structure.

(6) *The entities within the ecological hierarchy are not units of evolution; rather, they are systems that foster the interaction of true evolving entities*

For proponents of H-TS, interpreting the ecological hierarchy as a hierarchy of interactors follows the traditional replicator-interactor framework. However, as mentioned above, some scholars state that true units of evolution must be both replicators and interactors simultaneously. The issue is that the entities within the two hierarchies do not strictly follow this framework. For example, most entities within the ecological hierarchy are supposedly interactors rather than replicators. But are they truly interactors? Given that units of selection must be both interactors and replicators, the entities within the ecological hierarchy are not units of selection. First, this is because they are not "moremaking" entities: "Even though the term evolution is used to describe a change in ecological entities (such as community or ecosystem evolution), ecological individuals do not reproduce" (Tëmkin and Eldredge 2015, p. 196). Second, and more importantly, they are not interactors either because, according to Hull (1980), an interactor is "an entity that directly interacts as a cohesive whole with its environment in such a way that replication is differential." However, some entities within the ecological hierarchy (e.g., the community, the ecosystem, the biosphere) do not interact as a cohesive whole with their environment because, in fact, such systems are the environment itself. Indeed, I doubt that a protein, a community, an ecosystem, or the entire biosphere[24] could act as agents of selection. This notion lacks coherence, even for the advocates of H-TE. For example, Eldredge (1985) states: "If my arguments... [regarding H-TE]

[23] Nevertheless, according to the H-Evo-Devo Theory, species and higher taxa are individuals as well as natural kinds; see Chap. 6.

[24] Spiridonov and Eldredge (2024) recently referred to the entire biosphere as "Gaia," the most inclusive geobiomic unit in the Bretskyan hierarchy, which includes all the planet's physical and biological components. Interestingly, they argue that this entity possesses characteristics of interactors, albeit with distinctive properties compared to lower levels: "The Gaian unit is modulating fitness of all lower-level units... and also possesses its own fitness, defined here as resilience to its ultimate extinction... Contrary to popular conception and models that Gaia increases its chances of survival by means of regulation of environment... current evidence shows that such regulation is absent... and life resists this unchecked change at the longest time scales" (p. 14).

are correct, and accepting the validity of Hull's twin necessary conditions for selection [replicator-interactor], no entity above the organism level can be selected (at least in a sense consistent with use of the term *natural selection*)" (p. 9; his emphasis)

In discussing the individuality thesis, cohesion is a fundamental criterion for identifying evolutionary individuals. According to H-TE, the cohesion of higher-level entities within the ecological hierarchy results from interactions among entities at lower levels. For example, interactions among individuals "glue" a population (avatar) together, while interactions among populations "glue" a community together, and so forth. Nonetheless, "[b]oundaries in ecological hierarchies above individual levels are complex and difficult to identify precisely in practice" (Tëmkin and Eldredge 2015, p. 196). If accepting evolutionary individuals within the genealogical hierarchy presents issues (see point 5), the situation becomes even more complicated for entities within the ecological hierarchy. This has led some H-TE proponents to relinquish the ontological aspect of individuality regarding this aspect: "Despite the historical role of Eldredge and colleagues in exploring the individuality thesis in evolution… there doesn't seem to be a necessary commitment to the individuality thesis in Hierarchy Theory for any particular level" (Pievani and Serrelli 2013, p. 399). But if entities within the ecological hierarchy are neither interactors nor individuals, then what kind of entities are they?

Considering such a double hierarchy, Caponi (2012, 2016, 2023) claims there are two kinds of entities: lineages and systems. Lineages are those entities whose parts are examples of themselves. In other words, a lineage might exhibit varying levels of inclusiveness, where one of its parts can represent the whole, e.g., an individual of *Panthera leo persica* represents the subspecies *persica*, the species *leo*, the genus *Panthera*, and so forth. Conversely, systems are entities whose parts are not examples but rather their constituents, i.e., unlike lineages, a subsystem cannot represent the entirety of the system. For instance, a population (avatar) represents neither the community nor the ecosystem. Considering this, Caponi states that it is erroneous to equate systems as if they were lineages: "Lineages evolve; systems develop, grow, and age" (Caponi 2016, p. 49). But if entities within the ecological hierarchy are systems rather than lineages, then these kinds of entities should not be considered units of evolution. Hence, true evolving entities must be exclusively lineages.[25]

This does not mean ignoring the ecological aspects of evolution. Nonetheless, as Lewontin (1983, 2000) highlighted, interactions with the environment must be appropriately incorporated into the context of how it affects the organism during its development. Indeed, research has demonstrated that the environment plays an instructive role in development, either by facilitating the normal development of an organism or by disrupting its stability (Sultan 2007, 2015, 2017; Gilbert 2016; Schwab and Moczek 2017). In this context, it is well known that the environment can directly contribute to evolutionary change, i.e., the environment is a source of variation akin to genes (West-Eberhard 2003, 2005; Crispo 2007; Dayan et al. 2019).

[25] While Caponi identifies these types of entities within the genealogical hierarchy, they align better with the structure of the Linnaean hierarchy since, as mentioned in the previous point, the genealogical hierarchy lacks a real hierarchical structure.

However, the notion of the environment within an evolutionary scope is so broad that it is necessary to identify it by using different mechanisms. For example, phenotypic plasticity is an important phenomenon concerning the organism and its interaction with the environment. This phenomenon indicates that organisms have the capacity to alter their development in response to signals coming from outside of them (West-Eberhard 2003; Sultan 2015; Levis and Pfennig 2016; Dayan et al. 2019; Pfennig 2021). Niche construction represents another crucial mechanism in relation to the environment. This concept emphasizes that organisms shape their environment in a way that influences their own evolution, e.g., either by modifying the selective pressure on offspring or by introducing new instructive signals during their development (Laland et al. 1999, 2019; Odling-Smee et al. 2003; Sultan 2015). Likewise, behavioral choice represents another kind of process within organism–environment interaction. It can be defined as contingent behavior—whether conscious or unconscious—that organisms exhibit in response to external stimuli (Diogo 2017). Importantly, behavioral choice not only influences the niches construction but also the selective pressures exerted on other organisms (Duckworth 2009; Diogo 2017). Therefore, the claim is that the interaction between an organism and the environment should not solely focus on the survival and differential reproductive success of the organism through the environment (natural selection). It should also consider how the environment actively influences the development of an organism and, conversely, how the organism itself contributes to the construction of the environment.[26]

This is precisely the type of interaction expected in evolutionary theory. Admittedly, the organism and the environment are not independent; rather, they are interdependent (Lewontin 1983; Sultan 2015). Nevertheless, this does not mean that the organism should be interpreted as a parallel entity existing between two hierarchies. The organism exhibits different morphological states of inclusiveness (as lineages, in Caponi's sense), while its evolution is profoundly influenced (although not exclusively) by ecological systems through what is termed "interaction." In brief, entities within the ecological hierarchy are neither interactors (they are not units of selection) nor evolutionary individuals (because they lack well-defined boundaries). Instead, they represent systems that sway the evolution of true evolving entities. For as much as H-TE advocates might attribute equal importance to the ecological hierarchy and the genealogical hierarchy (e.g., Eldredge 1989, pp. 187–190; Miller III 2008; Tëmkin and Serrelli 2016), the truth is that the ecological hierarchy lacks the structure necessary for its entities to be recognized as units of evolution.

[26] This organism-centered perspective lies at the heart of the H-Evo-Devo Theory. For instance, in order to explain adaptations in subspecies, it considers phenotypic plasticity, niche construction, behavioral choice, and natural selection as the key factors in organism-environment interaction. See Chap. 11, Sect. 11.1.1.

4.3 Conclusions

Undoubtedly, evolutionary theory requires a hierarchical structure. However, based on contemporary evolutionary knowledge, neither H-TS nor H-TE represents adequate models for achieving that goal. These proposals are rooted in an outdated framework lacking the integration of evolutionary developmental biology. For example, the proponents of H-TS, via species selection, consider nonmorphological characters as emergent properties (e.g., geographic range). Meanwhile, H-TE advocates maintain that evolution is nothing more than the transmission of genetic information over time, diminishing the role of the organism and its structural properties to mere vehicles of genetic information. With this in mind, I do not intend to discourage an evolutionary hierarchical approach. On the contrary, throughout this book, I introduce a novel hierarchical model that effectively integrates embryology with evolution. A hierarchy theory of evolution is indeed necessary, but it must be structured in appropriate terms. Definitely, the H-Evo-Devo Theory is a better choice for this job.

References

Ahl V, Allen TFH (1996) Hierarchy theory: a vision, vocabulary, and epistemology. Columbia University Press, New York

Allen TFH (2009) Hierarchy theory in ecology. In: Jørgensen SE (ed) Ecosystem ecology. Elsevier, Amsterdam, pp 114–120

Allen TFH, Starr TB (1982) Hierarchy: perspectives for ecological complexity. University of Chicago Press, Chicago/London

Arnold AJ, Fristrup K (1982) The theory of evolution by natural selection: a hierarchical expansion. Paleobiology 8:113–129

Beurton PJ (2000) A unified view of the gene, or how to overcome reductionism. In: Beurton PJ, Falk R, Rheinberger H-J (eds) The concept of the gene in development and evolution: historical and epistemological perspectives. Cambridge University Press, Cambridge, pp 286–314

Bonduriansky R, Day T (2018) Extended heredity: a new understanding of inheritance and evolution. Princeton University Press, Princeton/Oxford

Brandon RN (1988) The levels of selection: a hierarchy of interactors. In: Plotkin HC (ed) The role of behavior in evolution. MIT Press, Cambridge, MA/London, pp 51–71

Brooks DS (2021) Levels of organization as tool and doctrine in biology. In: Brooks DS, DiFrisco J, Wimsatt WC (eds) Levels of organization in the biological sciences. The MIT Press, Cambridge, MA, pp 39–59

Brower AVZ (2023) Hierarchies, classifications, cladograms and phylogeny. Cladistics 39:229–239

Caianiello S (2016) Revisiting the phenotypic hierarchy in hierarchy theory. In: Eldredge N, Pievani T, Serrelli E, Tëmkin I (eds) Evolutionary theory: a hierarchical perspective. University of Chicago Press, Chicago, pp 151–173

Caponi G (2012) Linajes y sistemas: dos tipos de individuos biológicos. Scientiae Studia 10:243–268

Caponi G (2016) Lineages and systems: a conceptual discontinuity in biological hierarchies. In: Eldredge N, Pievani T, Serrelli E, Tëmkin I (eds) Evolutionary theory: a hierarchical perspective. University of Chicago Press, Chicago, pp 47–62

Caponi G (2023) Linajes: esas cosas que evolucionan. Universidad El Bosque, Colombia

Cracraft J (1985) Species selection, macroevolutionary analysis, and the "hierarchical theory" of evolution. Syst Zool 34:222–229

Craver CF, Bechtel W (2007) Top-down causation without top-down causes. Biol Philos 22:547–563
Crispo E (2007) The Baldwin effect and genetic assimilation: revisiting two mechanisms of evolutionary change mediated by phenotypic plasticity. Evolution 61:2469–2479
Damuth J (1985) Selection among "species": a formulation in terms of natural functional units. Evolution 39:1132–1146
Danchin E, Charmantier A, Champagne FA, Mesoudi A, Pujol B, Blanchet S (2011) Beyond DNA: integrating inclusive inheritance into an extended theory of evolution. Nat Rev Genet 2:475–486
Dawkins R (1976) The selfish gene. Oxford University Press, New York
Dawkins R (1982) The extended phenotype. Oxford University Press, New York
Dayan DI, Graham MA, Baker JA, Foster SA (2019) Incorporating the environmentally sensitive phenotype into evolutionary thinking. In: Uller T, Laland KN (eds) Evolutionary causation biological and philosophical reflections. MIT Press, Cambridge, MA/London, pp 81–107
Dietrich M (2010) Microevolution and macroevolution are governed by the same processes. In: Ayala FJ, Arp R (eds) Contemporary debates in philosophy of biology. Wiley-Blackwell, Malden, pp 169–179
DiFrisco J (2017) Time scales and levels of organization. Erkenn 82:795–818
DiFrisco J (2021) Integrating composition and process in levels of developmental evolution. In: Brooks DS, DiFrisco J, Wimsatt WC (eds) Levels of organization in the biological sciences. The MIT Press, Cambridge, MA, pp 111–134
Diogo R (2017) Evolution driven by organismal behavior: a unifying view of life, function, form, mismatches, and trends. Springer, Cham
Duckworth RA (2009) The role of behavior in evolution: a search for mechanism. Evol Ecol 23:513–531
Duran-Nebreda S, Bentley RA, Vidiella B, Spiridonov A, Eldredge N, O'Brien MJ, Valverde S (2024) On the multiscale dynamics of punctuated evolution. Trends Ecol Evol 39:734–744
Eldredge N (1985) Unfinished synthesis: biological hierarchies and modern evolutionary thought. Oxford University Press, Nueva York
Eldredge N (1989) Macroevolutionary dynamics: species, niches, and adaptive peaks. McGraw-Hill, New York
Eldredge N (1993) What, if anything, is a species? In: Kimbel WH, Martin LB (eds) Species, species concepts, and primate evolution. Plenum, New York, pp 3–20
Eldredge N (2008) Hierarchies and the sloshing bucket: toward the unification of evolutionary biology. Evo Edu Outreach 1:10–15
Eldredge N (2015) Eternal ephemera: adaptation and the origin of species, from the nineteenth century, through punctuated equilibria and beyond. Columbia University Press, New York
Eldredge N (2016) The checkered career of hierarchical thinking in evolutionary biology. In: Eldredge N, Pievani T, Serrelli E, Tëmkin I (eds) Evolutionary theory: a hierarchical perspective. University of Chicago Press, Chicago, pp 1–16
Eldredge N, Gould SJ (1972) Punctuated equilibria: an alternative to phyletic gradualism. In: Schopf TJM (ed) Models in paleobiology. Freeman, Cooper and Co, San Francisco, pp 82–115
Eldredge N, Grene MJ (1992) Interactions: the biological context of social systems. Columbia University Press, New York
Eldredge N, Salthe SN (1984) Hierarchy and evolution. In: Dawkins R, Ridley M (eds) Oxford surveys in evolutionary biology, vol I. Oxford University Press, Oxford, pp 182–206
Eldredge N, Pievani T, Serrelli E, Tëmkin I (eds) (2016) Evolutionary theory: a hierarchical perspective. University of Chicago Press, Chicago
Emerson AE (1960) The evolution of adaptation in population systems. In: Tax S (ed) Evolution after Darwin, The evolution of life, vol I. University of Chicago Press, Chicago, pp 307–348
Ereshefsky M (1991) Species, higher taxa, and the units of evolution. Philos Sci 58:84–101
Ereshefsky M (1999) Species and the Linnaean hierarchy. In: Wilson RA (ed) Species: new interdisciplinary essays. MIT Press, Cambridge, MA, pp 285–305

Ereshefsky M (2001) The poverty of the Linnaean hierarchy: a philosophical study of biological taxonomy. Cambridge University Press, Cambridge
Eronen MI (2015) Levels of organization: a deflationary account. Biol Philos 30:39–58
Eronen MI (2021) Levels, nests, and branches: compositional organization and downward causation in biology. In: Brooks DS, DiFrisco J, Wimsatt WC (eds) Levels of organization in the biological sciences. The MIT Press, Cambridge, MA, pp 77–87
Eronen MI, Ramsey G (2022) What are the 'levels' in levels of selection? Br J Philos Sci. https://doi.org/10.1086/718990
Fábregas-Tejada A, Vergara-Silva F (2017) Hierarchy theory of evolution and the extended evolutionary synthesis: some epistemic bridges, some conceptual rifts. Evol Biol 45:127–139
Gawne R, McKenna KZ, Nijhout F (2017) Unmodern synthesis: developmental hierarchies and the origin of phenotypes. Bio Essays. https://doi.org/10.1002/bies.201600265
Gilbert S (2016) Developmental plasticity and developmental symbiosis: the return of eco-devo. Curr Top Dev Biol 116:415–433
Gontier N (2010) Evolutionary epistemology as a scientific method: a new look upon the units and levels of evolution debate. Theory Biosci 129:167–182
Gontier N (2021) Hierarchies, networks, and causality: the applied evolutionary epistemological approach. J Gen Philos Sci 52:313–334
Goodnight C (2018) The theory of multilevel selection. In: Scheiner SM, Mindell DP (eds) The theory of evolution: principles, concepts, and assumptions. University of Chicago Press, Chicago, pp 194–210
Gould SJ (1980) Is a new and general theory of evolution emerging? Paleobiology 6:119–130
Gould SJ (2002) The structure of evolutionary theory. The Belknap Press of Harvard University Press, Cambridge, MA
Gould SJ, Lloyd EA (1999) Individuality and adaptation across levels of selection: how shall we name and generalize the unit of Darwinism? Proc Natl Acad Sci USA 96:11904–11909
Gould SJ, Vrba ES (1986) The hierarchical expansion of sorting and selection: sorting and selection cannot be equated. Paleobiology 12:217–228
Grantham TA (1995) Hierarchical approaches to macroevolution: recent work on species selection and the "effect hypothesis". Annu Rev Ecol Syst 26:301–321
Grene M (1987) Hierarchies in biology. Am Sci 75:504–510
Griesemer J (2000a) The units of evolutionary transition. Selection 1:67–80
Griesemer J (2000b) Reproduction and the reduction of genetics. In: Beurton PJ, Falk R, Rheinberger H-J (eds) The concept of the gene in development and evolution: historical and epistemological perspectives. Cambridge University Press, Cambridge, pp 240–285
Griffiths PE, Gray RD (2001) Darwinism and developmental systems. In: Oyama S, Griffiths PE, Gray RD (eds) Cycles of contingency developmental systems and evolution. MIT Press, Cambridge, MA/London, pp 195–218
Griffiths PE, Gray RD (2005) Discussion: three ways to misunderstand developmental systems theory. Biol Philos 20:417–425
Griffiths PE, Stotz K (2018) Developmental systems theory as a process theory. In: Nicholson DJ, Dupré J (eds) Everything flows: towards a processual philosophy of biology. Oxford University Press, Oxford, pp 225–245
Hertler SC, Figueredo AJ, Peñaherrera-Aguirre M (eds) (2020) Multilevel selection: theoretical foundations, historical examples, and empirical evidence. Palgrave Macmillan/Springer, Cham
Hull DL (1980) Individuality and selection. Ann Rev Ecol Syst 11:311–332
Hull DL (1988) Interactors versus vehicles. In: Plotkin HC (ed) The role of behavior in evolution. MIT Press, Cambridge, MA/London, pp 51–71
Jablonka E, Lamb MJ (2005) Evolution in four dimensions: genetic, epigenetic, behavioral, and symbolic variation in the history of life. MIT Press, Cambridge, MA
Jablonski D (2007) Scale and hierarchy in macroevolution. Palaeontology 50:87–109
Jablonski D (2008) Species selection: theory and data. Annu Rev Ecol Evol Syst 39:501–524
Jablonski D (2017) Approaches to macroevolution: 2. Sorting of variation, some overarching issues, and general conclusions. Evol Biol 44:451–475

Jeler C (ed) (2018) Multilevel selection and the theory of evolution: historical and conceptual issues. Palgrave Macmillan/Springer, Cham

Keller EF (2000) Decoding the genetic program: or, some circular logic in the logic of circularity. In: Beurton PJ, Falk R, Rheinberger H-J (eds) The concept of the gene in development and evolution: historical and epistemological perspectives. Cambridge University Press, Cambridge, pp 159–177

Koestler A (1967) The ghost in the machine. Random House, New York

Lala KN, Uller T, Feiner N, Feldman MW, Gilbert SF (2024) Evolution evolving: the developmental origins of adaptation and biodiversity. Princeton University Press, Princeton/Oxford

Laland KN, Odling-Smee FJ, Feldman MW (1999) Evolutionary consequences of niche construction and their implications for ecology. Proc Natl Acad Sci USA 96:10242–10247

Laland KN, Uller T, Feldman M, Sterelny K, Müller GB, Moczek A, Jablonka E, Odling-Smee J (2015) The extended evolutionary synthesis: its structure, assumptions and predictions. Proc R Soc B 282:20151019. https://doi.org/10.1098/rspb.2015.1019

Laland KN, Odling-Smee FJ, Feldman MW (2019) Understanding niche construction as an evolutionary process. In: Uller T, Laland KN (eds) Evolutionary causation biological and philosophical reflections. MIT Press, Cambridge, MA/London, pp 127–152

Levis NA, Pfennig DW (2016) Evaluating 'plasticity-first' evolution in nature: key criteria and empirical approaches. Trends Ecol Evol 31:563–574

Lewontin RC (1970) The units of selection. Annu Rev Ecol Syst 1:1–18

Lewontin RC (1983) Gene, organism, and environment. In: Bendall DS (ed) Evolution: from molecules to men. Cambridge University Press, Cambridge, pp 273–285

Lewontin RC (2000) The triple helix: gene, organism, and environment. Harvard University Press, Cambridge, MA/London

Lieberman BS, Vrba ES (2005) Stephen Jay Gould on species selection: 30 years of insight. Paleobiology 31:113–121

Lloyd EA (1988) The structure and confirmation of evolutionary theory. Greenwood, New York

Lloyd EA, Gould SJ (1993) Species selection on variability. Proc Natl Acad Sci USA 90:595–599

Love AC (2006) Evolutionary morphology and evo-devo: hierarchy and novelty. Theory Biosci 124:317–333

MacMahon JA, Phillips DL, Robinson JV, Schimpf DJ (1978) Levels of biological organization: an organism-centered approach. Bioscience 28:700–704

Maynard Smith J (1964) Group selection and kin selection. Nature 201:1145–1147

Miller W III (2008) The hierarchical structure of ecosystems: connections to evolution. Evo Edu Outreach 1:16–24

Newman SA (2021) Inherency. In: Nuño de la Rosa L, Müller GB (eds) Evolutionary developmental biology. Springer, Cham, pp 121–131

Newman SA, Bhat R (2009) Dynamical patterning modules: a "pattern language" for development and evolution of multicellular form. Int J Dev Biol 53:693–705

Newman SA, Müller GB (2005) Origination and innovation in the vertebrate limb skeleton: an epigenetic perspective. J Exp Zool (Mol Dev Evol) 304B:593–609

O'Neill RV (1989) Perspectives in hierarchy and scale. In: Roughgarden J, May RM, Levin SA (eds) Perspectives in ecological theory. Princeton University Press, Princeton, pp 140–156

Ochoa C (2024) Deconstructing the extended evolutionary synthesis: do we need a new theory of evolution? In: Delisle RG, Esposito M, Ceccarelli D (eds) Unity and disunity in evolutionary biology: deconstructing Darwinism. Springer, Cham, pp 403–439

Odling-Smee FJ, Laland KN, Feldman MW (2003) Niche construction: the neglected process in evolution. Princeton University Press, Princeton

Okasha S (2005) Multilevel selection and the major transitions in evolution. Philos Sci 72:1013–1025

Okasha S (2007) The units and levels of selection. In: Sarkar S, Plutynski A (eds) A companion to the philosophy of biology. Blackwell Publishing, Malden, pp 138–156

Oyama S (1985) The ontogeny of information: developmental systems and evolution. Cambridge University Press, Cambridge

Oyama S, Griffiths PE, Gray RD (eds) (2001) Cycles of contingency developmental systems and evolution. MIT Press, Cambridge, MA/London
Pfennig DW (2021) Key questions about phenotypic plasticity. In: Pfennig DW, West-Eberhard MJ (eds) Phenotypic plasticity and evolution: causes, consequences, controversies. CRC Press, Boca Raton, pp 55–88
Pievani T (2013) Kinds of pluralism: Stephen J. Gould and the Future of evolutionary theory. In: Danieli GA, Minelli A, Pievani T (eds) Stephen J. Gould: the scientific legacy. Springer, Verlag Italia, pp 37–50
Pievani T (2016) Hierarchy theory and the extended synthesis debate. In: Eldredge N, Pievani T, Serrelli E, Tëmkin I (eds) Evolutionary theory: a hierarchical perspective. University of Chicago Press, Chicago, pp 351–364
Pievani T, Serrelli E (2013) Bucket thinking: the future framework for evolutionary explanation. Revista Internacional de Filosofía: Suplemento 18:389–405
Potochnik A (2021) Our world isn't organized into levels. In: Brooks DS, DiFrisco J, Wimsatt WC (eds) Levels of organization in the biological sciences. The MIT Press, Cambridge, MA, pp 61–76
Potochnik A, McGill B (2012) The limitations of hierarchical organization. Philos Sci 79:120–140
Salthe SN (1975) Problems of macroevolution (molecular evolution, phenotype definition, and canalization) as seen from a hierarchical viewpoint. Amer Zool 15:295–314
Salthe SN (1985) Evolving hierarchical systems: their structure and representation. Columbia University Press, New York
Schwab DB, Moczek AP (2017) Evo-devo and niche construction. In: Nuño de la Rosa L, Müller GB (eds) . Evolutionary developmental biology. Springer, Cham, pp 1179–1192
Shanahan T (1990) Evolution, phenotypic selection, and the units of selection. Philos Sci 57:210–225
Simon HA (1962) The architecture of complexity. Proc Am Philos Soc 106:467–482
Simpson C, Halling A, Leventhal S (2024) Levels of selection and macroevolution in organisms, colonies, and species. Paleobiology:1–9. https://doi.org/10.1017/pab.2024.12
Spiridonov A, Eldredge N (2024) The Bretskyan hierarchy, multiscale allopatry, and geobiomes—on the nature of evolutionary things. Paleobiology 50:194–213
Stanley S (1975) A theory of evolution above the species level. Proc Natl Acad Sci USA 72:646–650
Stotz K (2019) Biological information in developmental and evolutionary systems. In: Uller T, Laland KN (eds) Evolutionary causation biological and philosophical reflections. MIT Press, Cambridge, MA/London, pp 323–344
Sultan SE (2007) Development in context: the timely emergence of eco-devo. Trends Ecol Evol 22:575–582
Sultan SE (2015) Organism and environment: ecological development, niche construction, and adaptation. Oxford University Press, New York
Sultan SE (2017) Eco-evo-devo. In: Nuño de la Rosa L, Müller GB (eds) Evolutionary developmental biology. Springer, Cham, pp 1165–1177
Tëmkin I, Eldredge N (2015) Networks and hierarchies: approaching complexity in evolutionary theory. In: Serrelli E, Gontier N (eds) Macroevolution: explanation, interpretation and evidence. Springer, pp 183–226
Tëmkin I, Serrelli E (2016) General principles of biological hierarchical systems. In: Eldredge N, Pievani T, Serrelli E, Tëmkin I (eds) Evolutionary theory: a hierarchical perspective. University of Chicago Press, Chicago, pp 19–25
Umerez J (2016) Biological organization from a hierarchical perspective: articulation of concepts and interlevel relation. In: Eldredge N, Pievani T, Serrelli E, Tëmkin I (eds) Evolutionary theory: a hierarchical perspective. University of Chicago Press, Chicago, pp 63–85
Valentine JW (1968) The evolution of ecological units above the population level. J Paleontol 42:253–267
Valentine JW (2004) On the origin of phyla. University of Chicago Press, Chicago/London
Valentine JW, May CL (1996) Hierarchies in biology and paleontology. Paleobiology 22:23–33

Vasilyeva LN, Stephenson SL (2008) The Linnaean hierarchy and "extensional thinking". Open Evol J 2:55–65

Vrba ES (1984) What is species selection? Syst Zool 33:318–328

Vrba ES, Eldredge N (1984) Individuals, hierarchies and processes: towards a more complete evolutionary theory. Paleobiology 10:146–171

Wagner GP (2015) Reinventing the organism: evolvability and homology in post-Dahlem evolutionary biology. In: Love AC (ed) Conceptual change in biology: scientific and philosophical perspectives on evolution and development. Springer, Dordrecht, pp 327–342

West-Eberhard MJ (2003) Developmental plasticity and evolution. Oxford University Press, New York

West-Eberhard MJ (2005) Developmental plasticity and the origin of species differences. Proc Natl Acad Sci USA 102:6543–6549

Williams GC (1966) Adaptation and natural selection: a critique of some current evolutionary thought. Princeton University Press, Princeton and Oxford

Williams PA (1992) Confusion in cladism. Synthese 91:135–152

Wimsatt WC (2001) Generative entrenchment and the developmental systems approach to evolutionary processes. In: Oyama S, Griffiths PE, Gray RD (eds) Cycles of contingency developmental systems and evolution. MIT Press, Cambridge, MA/London, pp 219–237

Wimsatt WC (2007) Re-engineering philosophy for limited beings: piecewise approximations to reality. Harvard University Press, Cambridge, MA

Wu J (2013) Hierarchy theory: an overview. In: Rozzi R, Pickett STA, Palmer C, Armesto JJ, Callicott JB (eds) Linking ecology and ethics for a changing world: values, philosophy, and action. Springer, New York, pp 281–301

Wynne-Edwards VC (1962) Animal dispersion in relation to social behaviour. Oliver and Boyd, Edinburgh

The Structure of the H-Evo-Devo Theory

5

Abstract

In this book, I introduce a new theory of evolution within a hierarchical framework called *Hierarchical Evolutionary-Developmental Theory* or *H-Evo-Devo Theory*. This chapter provides the first description of this model. By incorporating hierarchies into the Evo-Devo context, this model emphasizes the concept of developmental constraints as the primary mechanism. Developmental constraints are established from a procedural hierarchy, exemplifying the interactions within the developmental process across different levels. However, the phenotypic outcomes of this process are examined by a compositional hierarchy, allowing for the identification of evolving entities in a phylogenetic context. In this regard, developmental constraints operate hierarchically, both upward and downward, producing distinct states of morphological stability that result in different evolutionary entities. The homologue represents the highest entity in the hierarchy, and the character states of homologues within the whole organism generate a series of steady states at various scales, which can be interpreted through the Linnaean hierarchy as subspecies, species, and higher taxa. In short, developmental constraints draw a single compositional hierarchy that enables the identification of homologues, higher taxa, species, and subspecies as units of evolution.

Keywords

H-Evo-Devo Theory · Developmental constraints · New theory of evolution · Pere Alberch · Procedural hierarchy · Compositional hierarchy · GRN hierarchy · Modular hierarchy · Linnaean hierarchy · Natural system · Hierarchy theory of evolution · Evo-Devo · Units of evolution · Levels of evolution

C. Ochoa, *Hierarchical Evolutionary-Developmental Theory*, Evolutionary Biology – New Perspectives on Its Development 9,
https://doi.org/10.1007/978-3-031-88723-9_5

In this book, I introduce a new model that integrates Evo-Devo insights into a hierarchy theory of evolution called *Hierarchical Evolutionary-Developmental Theory* or *H-Evo-Devo Theory*. This model proposes that evolution is not reducible to a single level but occurs at multiple levels, each operating under its own causes and effects, resulting in a complete perspective of the phenomenon of biological evolution. This theory focuses on the organism—rather than solely on genes—as the primary subject of study, i.e., it is based on an organism-centered perspective. It posits that the internal properties of the organism reflect entities that are highly integrated in developmental terms that evolve hierarchically. Central to this theory is the concept of "developmental constraints," understood as a mechanism that limits morphological variation by maintaining it at a specific position within the morphospace. These developmental constraints operate hierarchically, both upward and downward, producing distinct states of morphological stability, which can be recognized as subspecies, species, higher taxa, and homologues, all representing units of evolution.

This chapter explores pivotal elements that form the structure of this model. Section 5.1 discusses the integration of Evo-Devo into a hierarchy theory of evolution. It argues for the use of different types of hierarchies to provide an understanding of the processes responsible for generating and maintaining the evolving entities. In doing so, the concept of developmental constraints is proposed as the fundamental mechanism for establishing a hierarchical structure. Furthermore, it advocates for the significance of the Linnaean hierarchy as a relevant element within this model. Section 5.2 provides an overview of the H-Evo-Devo Theory, suggesting that developmental constraints are established from a procedural hierarchy. These processes lead to phenotypic outcomes that delineate a single compositional hierarchy, allowing the identification of homologues, higher taxa, species, and subspecies as units of evolution.

5.1 Hierarchies, Constraints, and Taxonomy

Before providing a detailed description of the H-Evo-Devo Theory, it is essential to analyze some fundamental aspects of its structure: (1) the application of hierarchies within the context of evolutionary developmental biology (Evo-Devo); (2) the proposal that developmental constraints are the primary evolutionary mechanism of this model, which makes it possible to identify units of evolution; and 3) the demonstration that ranks in the Linnaean hierarchy serve as a taxonomic basis reflecting different steady states of morphological disparity.

5.1.1 The Use of Hierarchies Through the Evo-Devo Approach

As discussed in Chap. 4, ontological hierarchies are valuable tools for understanding the complex systems present in nature, and applying this hierarchical approach is essential for evolutionary theory. Interestingly, for years, many evolutionary biologists have relegated morphological and embryological aspects to the background of their hierarchical structures, despite acknowledging that the developmental process is structured across multiple levels of organization (Hall 1999; Newman 2003; Love 2006; Wake 2015; Caianiello 2016; Eronen 2020; DiFrisco 2021). As Caianiello (2016) notes when criticizing the traditional Hierarchy Theory of Evolution: "Ontogeny in itself cannot be adequately conceived as a hierarchic process in a scalar sense… although differences in the timing of developmental processes at different scales are considerable, as well as mechanistically crucial in the temporal ordering of morphogenesis." Undoubtedly, embryonic development—and its relationship with evolution—necessitates hierarchical analysis (Love 2006; Wake 2015; Eronen 2020; DiFrisco 2021). But how can developmental biology be synthesized within a hierarchical perspective of evolution?

According to Love (2006), there are at least two types of hierarchies within the Evo-Devo context. These hierarchies are described either in developmental time within a single generation (ontogeny) or in evolutionary time across generations (phylogeny):

1. *Compositional (or scalar) hierarchy*, which pertains to part-whole relationships.
2. *Procedural (control or organizational) hierarchy*, which involves interconnected processes at different levels occurring sequentially in a specific order for a particular outcome.

Compositional Hierarchy According to its most basic definition, compositional hierarchy involves entities at the same level that are made up of entities at a lower level, while these, in turn, make up entities at a higher level (Wimsatt 2007). This type of hierarchy encompasses at least two descriptions (Valentine and May 1996):

(a) *Aggregative or inclusive hierarchy*, which nests entities inclusively, one by one, at different levels to form the whole.
(b) *Constitutive hierarchy*, whose entities are nested as the physical components of the whole.

According to Wimsatt (2007), the choice of level in a compositional hierarchy is typically determined by size. For example, entities at the same organizational level tend to exhibit similar sizes, thereby possessing the same mechanisms, properties, regularities, etc. Regarding spatial scale, as noted by DiFrisco (2017), levels can also be identified by their discontinuities, providing a more reliable criterion for delineating spatial boundaries between entities.

As an illustration, consider that, in vertebrates, the forelimb is a homologue made up of other homologues such as the stylopod, zeugopod, and autopod. In other words, these three elements are physical components within a more inclusive structure—the forelimb. Thus, in this instance, the hierarchy of nested homologues represents a constitutive hierarchy. Another hierarchy, however, could focus on the variational modalities of these homologues, referred to as homogenies (Ochoa and

Rasskin-Gutman 2015). Such homogenies are organized in a nested sequence of greater or lesser inclusiveness within a homologue. In this case, the hierarchy of nested homogenies represents an aggregative hierarchy. For example, when considering the forelimbs of a bird (a homologue), one character state represents the wing shape, while a more inclusive state represents a hummingbird's wing shape, and an even more inclusive state represents the bee hummingbird's wing shape (*Mellisuga helenae*), and so on (see Chap. 8, Fig. 8.4). Similarly, when examining all homologous components within an organism, their collective homogenies exhibit discrete states of variation at different scales. In order to represent such steady states, the ranks of the Linnaean hierarchy are useful. Notably, the Linnaean hierarchy is an aggregative hierarchy historically established as a hierarchy of types.[1] Another important aspect is distinguishing the Linnaean hierarchy from one that focuses on homologues. Therefore, I suggest the term "modular hierarchy" to indicate different scales of both the character identities (homologues) and their nested character states (homogenies).[2]

Procedural Hierarchy This type of hierarchy denotes interconnected processes across different levels that occur sequentially in a specific order to achieve a particular outcome. In philosophical debates on levels of organization, some scholars have argued that the compositional hierarchy is often represented by static structures, notwithstanding that the developmental systems dynamically change through processes (Baedke 2021; Griesemer 2021). In this regard, the levels in a procedural hierarchy could be tiers of "mechanisms" in Craver's sense (2007). So, these processes are represented in sequential stages that are causally related. For example, entities at lower levels are organized to produce entities at higher levels. However, lower and higher levels of such processes maintain a relationship of mutual interactions. A similar idea is Baedke and Mc Manus's (2018) proposal of a "dynamic hierarchy," which describes how ontogenetic processes emerge, stabilize, and persist in a temporal sequence.

In brief, a procedural hierarchy involves a series of dynamic processes across different levels, each entailing specific interconnected steps in a defined order to achieve a specific outcome. Unlike a compositional hierarchy, which is entirely structural, a procedural hierarchy has functional properties with a focus on process dependence (Love 2006). For example, the developmental process itself can reflect a dynamic hierarchy. One level highlights the causal interactions occurring within the gene regulatory network, another denotes the protein functions within the system, and another represents the cellular interactions that give rise to resulting phenotypes. Similarly, by understanding the processes occurring at the level of gene regulatory network interactions, a procedural hierarchy that indicates the relational functions of the genetic system in question is appropriate (e.g., the GNR hierarchy).

[1] See Chap. 2.

[2] For more detail on this concern, see Chap. 8, Sect. 8.2.

Defining the Levels of Organization To my knowledge, DiFrisco's study (2021) is the only one that has attempted to pinpoint the levels of organization in the Evo-Devo context. His approach begins by highlighting that many ambiguities arise in delineating hierarchies because structural (compositional) aspects are decoupled from functional (causal-dynamic) ones. Addressing structure without reference to function, or vice versa, results in a lack of generalization about levels of organization. According to DiFrisco, incorporating both structural and functional aspects is essential for identifying the properties that distinguish different levels. He suggests that the most effective approach is to decompose the system through primary causal processes. This criterion minimizes ambiguity in distinguishing levels and facilitates organizing the system by identifying entities and processes at higher or lower levels.

For example, considering that evolution is driven by changes in embryonic development, DiFrisco argues that entities involved in the process of evolution (populations) are composed of those participating in the process of development (organisms). Evolution, as the constituted process, occupies a higher position, while development, as the constitutive process, holds a lower position. Consequently, the population, as the unit of evolution, represents the higher-level entity; meanwhile, the organism, as the unit of development, is positioned at the lower level. The next step involves decomposing the organism under the criterion of partial constitution, i.e., the parts comprising the organism and its processes are hierarchically arranged according to this criterion. Nevertheless, as stated by DiFrisco (2021), for entities to participate in the primary process (evolution), they must be inheritable. In addition, these entities must vary "quasi-independently" from other entities within the system, a property he terms "variational independence." According to him, this condition captures the levels by unifying the relationship between the entities and the processes that identify them (see Table 5.1).

While I agree with DiFrisco (2021) on his approach to identifying levels, I differ slightly in terms of "perspective." Although I advocate for integrating structural and functional aspects, I recognize that these relationships can also be asymmetrical. Building on Baedke's (2021) insights, a "dynamic" hierarchy can delve into the different processes that generate the system, while a "static" hierarchy can be employed to identify specific parts of that system. For instance, the procedural hierarchy

Table 5.1 *The levels of organization in the Evo-Devo context, according to DiFrisco*

Entity level	Process level
Population	Evolution
Organism	Development, ontogenesis
Character	Development
Cells/tissues	Morphogenesis
Gene network	Regulatory interactions, signaling, transcriptional state specification
Gene	Gene expression and regulation

The shaded area in the left column emphasizes the unit of evolution. Modified from DiFrisco (2021)

explains how character identities and character states[3] emerge and are maintained in a dynamic sense. Meanwhile, a compositional hierarchy focuses on integrating the patterns of such outcomes at the phenotypic level as character identities (homologues) and character states (nested in a sequence of greater or lesser inclusiveness). The former hierarchy represents the "process"—requiring a comprehensive understanding of numerous functions and interactions across different levels during organism development—whereas the latter hierarchy represents the "pattern," which can be studied through comparative morphology (see Love 2006).

According to DiFrisco's proposal, in Evo-Devo, levels of organization are arranged within a single hierarchy that aligns entities parallel to processes (Table 5.1). In contrast, the framework of the H-Evo-Devo Theory integrates causal interactions across different hierarchies to account for the entities that are subject to evolution (see Table 5.2). Gontier (2021) refers to these types of interactions as "reticulate causation," indicating how different hierarchies can interact causally within entities and levels that make up a particular hierarchy: "Hierarchies need to be conceptualized as multiple, and the challenge is to examine how many hierarchies can be distinguished, and how *reticulate causal interactions* occur between *distinct* hierarchical systems" (p. 327; her emphasis).

In this regard, I do not advocate for integrating different hierarchies as proposed by the Hierarchy Theory of Evolution (H-TE). Instead, I stress the use of diverse hierarchical models to illustrate processes and patterns within the Evo-Devo approach. Yet to determine the type of evolving entities, it is advisable to employ a single hierarchy (see below) but with the necessary precautions. Note that the choice

Table 5.2 *The levels of organization and the use of hierarchies according to the H-Evo-Devo Theory*

Units or levels	Hierarchy involved	Hierarchy type
Higher taxa	Linnaean hierarchy	Compositional hierarchy
Species		
Subspecies		
Homologue	Modular hierarchy	
Homogeny		
Cells and tissues	Developmental-process hierarchy	Procedural hierarchy
Proteins		
Gene regulatory network (GRN)		
Positional information signals	GRN hierarchy	
Character identity network (ChIN)		
Realizer genes		

The left column specifies the units or levels under consideration. The middle column indicates the specific hierarchy involved. The right column denotes whether it pertains to the procedural or compositional hierarchy. The shaded area in the left column represents the units of evolution

[3] By "character states," I mean the attributes or variational modalities through which a "character identity" is expressed. For further details on these concepts, see Chap 8, Sect. 8.2.2.

of hierarchical models depends on the specific research goals. However, attempting to consolidate all phenomena into a single hierarchical framework can be misleading. For example, H-TE advocates argue that the genealogical hierarchy represents a nested compositional hierarchy. According to this view, evolving entities at the lowest level are genes, followed sequentially by cells, organisms, demes, species, and, ultimately, monophyletic groups at the highest level. Nevertheless, as mentioned in Chap. 4, the cell is not a collective of genes, nor are organisms merely aggregations of cells. Indeed, some entities function as units of processes (e.g., genes), while others exemplify units of patterns (e.g., species). Thus, depicting the genealogical hierarchy solely as a compositional hierarchy in this context is inadequate.[4]

Therefore, the issue is not simply about identifying entities within a single hierarchy that indiscriminately conflates "units of processes" with "units of patterns." Instead, it is about understanding how hierarchies should be employed to represent processes and patterns in evolutionary research. Unlike the perspective that prioritizes the gene as the primary unit of inheritance and variation, the H-Evo-Devo Theory posits that the genome is just one element contributing to the generation of phenotypic variation. As mentioned above, a procedural hierarchy is employed to analyze these processes, while the outcomes as phenotypic variations (patterns) can be traced through a compositional hierarchy. In conclusion, one type of hierarchy delineates sequential process interactions across different levels of organization, while the other depicts phenotypic patterns at different scales. It is important to note that units of evolution should be represented within a single hierarchy because patterns arise from the complexities of developmental processes.

In DiFrisco's scheme (Table 5.1), the developmental process specifies the type of entities sequentially generated based on the criterion of partial constitution. All entities that participate in the evolutionary process must possess the property of "variational independence." Nonetheless, the units of evolution are mere "populations," while the units of development are "organisms." In contrast, within the structure of the H-Evo-Devo Theory (Table 5.2), units of evolution are not restricted to being "populations," nor are units of development confined to being "organisms." Although there is a general agreement that populations evolve, it is important to recognize that different steady states manifest in a scalar manner. Thus, various types of "populations" should be considered, and the application of Linnaean ranks can assist in categorizing these distinctions. Furthermore, the concept of the "organism" was removed as a level within the hierarchy because, as I will emphasize later, the organism is not a level but an entity that encompasses all the hierarchical relationships necessary for the study of evolution as a whole.[5]

---

[4] See Chap. 4, Sect. 4.2.

[5] The confusion may have arisen because researchers have conflated the notion of "organism" with "individual." While a population comprises individuals, the concept of "organism" has a broader scope, referring to the biological unit applicable in all cases, typically denoting a multicellular entity.

Regarding the structure of the H-Evo-Devo Theory, the developmental process is outlined through a procedural hierarchy where the sequence of levels is established based on the functional order of processes leading to a specific outcome. This developmental process is represented by a causal mechanism known as "developmental constraints," whose result manifests as the pattern's stability at different scales. The unit produced at the highest level is the homologue, while at the lower level it is the homogeny (character state). Both are represented in a compositional hierarchy through a modular hierarchy. However, since the organism consists of homologous structures and these homologues collectively produce character states, these patterns are observed across different scales as steady states within the organism (body plans). They are also represented through a compositional hierarchy via the Linnaean hierarchy. Of course, the criterion for positioning levels adheres to the rule of constitution processes, albeit now centered on developmental constraints as the causal process underlying the hierarchy's entire main system, as detailed in the following section.

5.1.2 Developmental Constraints as the Main Mechanism of the H-Evo-Devo Theory

At the core of this hierarchy theory of evolution lies a crucial mechanism that not only integrates functional and structural aspects between different levels of organization but also enables the recognition of the units of evolution and their respective levels. This mechanism is known as *developmental constraints*. Although the concept of developmental constraints has been interpreted in various ways (e.g., Amundson 1994; Raff 1996, pp. 292–320; Salazar-Ciudad 2006; Brigandt 2015; Rasskin-Gutman and Esteve-Atava 2018; Nuño de la Rosa and Müller 2024), it fundamentally refers to the limitations that the embryonic development process imposes on evolutionary change. Nevertheless, my use of this concept lies directly in Pere Alberch's (1980, 1982) proposal.

According to Alberch (1982), developmental interactions "drastically constrain processes of morphological change during evolution." These constraints define the distribution of morphological variation within morphospace (i.e., the universe of abstractly possible organic forms), resulting in "discontinuities and directionality in morphological transformations" (p. 313). Although Alberch's original proposal involves two types of constraints—which Gould (2002, pp. 1025–1061) categorizes as negative constraints (stasis) and positive constraints (channeling change in certain directions)—I focus here on the negative sense.[6] Developmental constraints, then, reflect the stability of the developmental system's organization, emphasizing that they "fall into a discrete number of stable states, i.e., we should expect a

[6]Throughout my explanation of the processes of divergence and stability, I clearly differentiate between two types of constraints. I refer to the constraints in the negative sense as "developmental constraints," and in the positive sense, I use the term "developmental bias." See Chap. 11, Sects. 11.1.2 and 11.2.2; Chap. 12, Sect. 12.1.2.

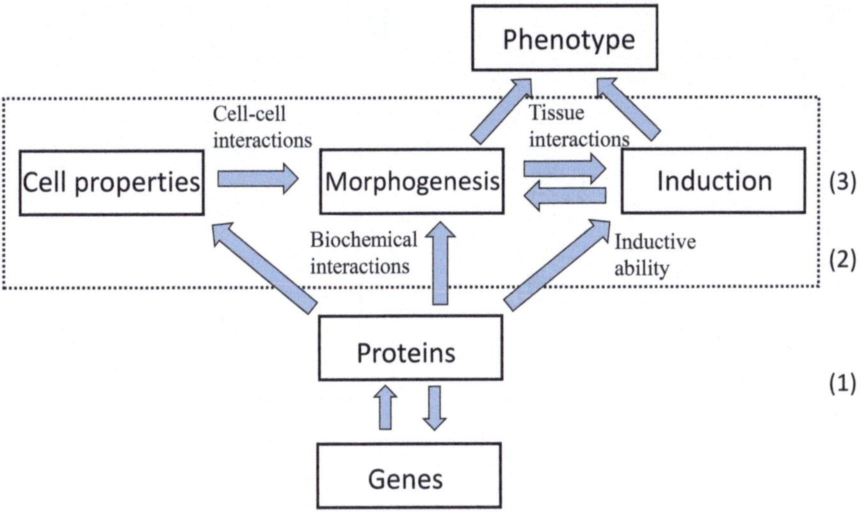

Fig. 5.1 *Alberch's original diagram depicts the pattern of interactions within developmental processes*. According to Alberch, developmental constraints manifest through these hierarchical interactions. (Modified from Alberch 1982) Compare this diagram with my adapted version in Fig. 5.4

discrete and bounded distribution of phenotypes." (Alberch 1982, p. 328). In short, developmental constraints produce forms that are separated by morphological discontinuities, each maintaining a specific steady state within the morphospace.

Interestingly, Alberch proposes that this development dynamic implies inherent hierarchical organization, which can be divided into three levels: The first level corresponds to interactions within the genome. The second level relates to interactions between proteins and enzymes. The third level involves interactions in cells and tissues (see Fig. 5.1). According to Alberch, while the organization at the first level imposes limitations on potential phenotypic changes (see Davidson and Erwin 2006; Gerhart and Kirschner 2007; Hu et al. 2017), the second and third levels ultimately determine the type of products to be assessed during phylogeny:

> At these levels one needs to define macroscopic variables… that, although ultimately under genetic control, are necessary to allow us to understand the generation of morphology from a mechanistic basis. One can define morphogenetic processes as the result of a set of "assembly" rules. An understanding of these… rules can provide valuable insights into the apparent complexity of many developmental processes and, furthermore, allows us to determine the relationships among different phenotypes, since the set of possible phenotypic transformations will be constrained by the generative potentialities of the morphogenetic rules involved in the process. According to this scheme, evolution can be viewed as the process of phenotypic transformations resulting from the genetically mediated perturbations of these basic developmental parameters through phylogeny. (Alberch 1982, pp. 321–322)

In this context, while a hierarchical structure following organizational rules delineates the distribution and constraints of the morphological variation, the crucial

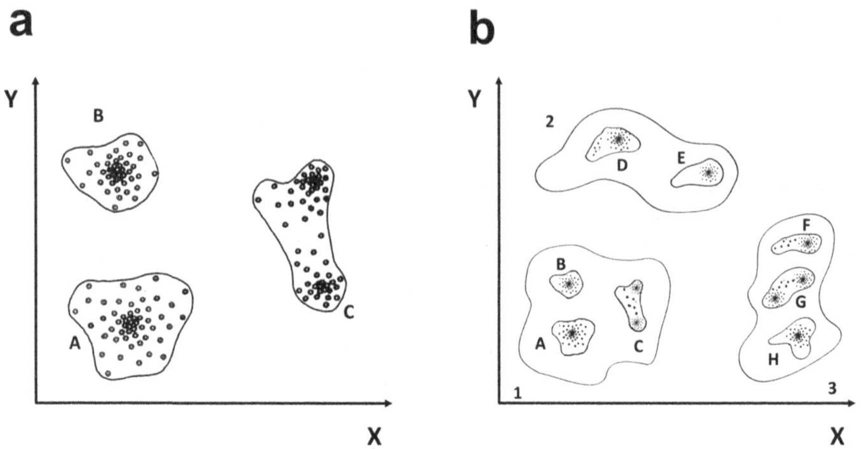

Fig. 5.2 *The distribution of morphological variation in measurements "x" and "y" within the morphospace*. According to Alberch, this distribution is discontinuous. Each entity is in a steady state, maintaining a specific position within the morphospace. From a developmental point of view, empty spaces within the morphospace are impossible to fill. (**a**) The original diagram depicts three clusters—A, B, and C (Modified from Alberch 1982). (**b**) The modified diagram illustrates the hierarchical view of developmental constraints. Clusters 1, 2, and 3 denote higher-ranking phenotypic distribution. Note that clusters A, B, and C, from "diagram a" belong to "higher cluster 1" in "diagram b." Here, this cluster can be interpreted as a "genus 1" with three species (A, B, C). This scheme suggests that developmental constraints operate at different levels, contributing to the hierarchical structure of the Natural System (see also Fig. 5.3). For more details, see the text

parameter for a phylogenetic study is the resulting phenotypic byproduct of these constraints. This outcome facilitates an assessment of the origin and stability of developmental systems over time.[7] In light of this, the following statements should be taken into consideration (see Fig. 5.2a):

1. A developmental system is in a steady state (A), maintaining a specific position within the morphospace as much as possible, despite any genetic or environmental perturbations it undergoes: "Small perturbations will be regulated back to the system's original domain" (Alberch 1980, p. 655).
2. Developmental systems are separated by discontinuities (gaps), each exhibiting fixed morphological states (A, B, C). This mode of change differs from gradual evolution; thus, transitioning from one state (A) to another (either B or C) necessitates a discontinuous change. Once the change occurs, the new developmental system takes on a different position than that of the ancestral system: "Large perturbations, however, may place the system into another domain of attraction

[7] Here, I have employed the term "developmental system" in Waddington's sense: "An animal is, in fact, a developmental system, and it is these systems, not the mere adult forms which we conventionally take as typical of the species, which becomes modified during the course of evolution." (Waddington 1952, p. 155). This notion of the "developmental system" emphasizes that the entity is generated by numerous processes and interactions across different levels.

where it will be pushed into a 'steady state' different from the original one" (Alberch 1980, p. 655).
3. Once the new developmental system relocates within the morphospace, it remains in a steady state, making it (either B or C) identifiable through time: "Thresholds in developmental processes imply discontinuities in morphological transformations and, consequently, constraints in phyletic processes" (Alberch 1982, p. 323).
4. Within a hierarchical context, systems consist of subsystems while simultaneously being subsystems of a higher-level system. These nested systems create distinct levels of morphological steady states. Thus, developmental systems can be perceived as lineages that form a nested compositional hierarchy. Since the distribution of morphological variation at different scales is caused by developmental constraints operating at various levels, these hierarchical steady states reflect the Natural System (see Fig. 5.2b).[8] In this regard, Alberch (1989, p. 23) asserted that: "These discontinuities in phenotypic space account for the order observed in nature and they are the reason why a classification of forms is possible. Forms can only be described and arranged in a hierarchical classification (order, family, genus, species…) if they are discrete."

Accordingly, the hierarchical approach to developmental constraints can be described as follows: developmental constraints are established within a procedural hierarchy, operating at different levels, drawing a single compositional hierarchy. By acting at the highest level, developmental constraints are reflected through a set of factors named "Character Identity Mechanisms" (ChIMs) by DiFrisco et al. (2020). These processes culminate in the identification of a homologue as a module. However, this homologue exhibits many character states or variational modalities, manifesting as steady states at different scales. Such stabilities arise due to developmental constraints working at a lower level, which are reflected in a set of factors I refer to as "Character State Mechanisms" (ChSMs). Similarly, homologues and their variational modalities collectively generate a sequence of phenotypic steady states across different scales in the entire organism. These differences in states can be interpreted as the outcomes of developmental constraints operating at different levels and can be recognized as categories within the Linnaean hierarchy, as will be discussed in the next subsection.

In summary, developmental constraints yield a discrete and constrained distribution of morphological variation. They maintain the forms in specific positions within the morphospace as much as possible. Developmental constraints are established from a procedural hierarchy, exhibiting functional interactions among different levels of biological organization. These developmental interactions not only contribute to the generation of structural properties in multicellular organisms but also explain why entities display a highly integrated common pattern (type or body plan). Importantly, I emphasize that the phenotypic outcomes of the developmental process need to be analyzed through a compositional hierarchy as parameters for

[8] See Sect. 5.1.3.

phylogenetic studies. Hence, developmental constraints represent the primary mechanism in this model, which illustrates the hierarchical structure observed in organisms' internal properties.

Of course, recognizing developmental constraints as the central mechanism of the H-Evo-Devo Theory does not entail dismissing natural selection. Nevertheless, as Pere Alberch highlighted, the role of natural selection is basically to prune non-functional morphologies, not in terms of internal organization but rather in terms of adaptive interaction. Put differently, the development process determines the types of morphologies that can be generated while natural selection determines whether these forms persist or are eliminated based on their functional role.[9] In this regard, it follows that the mechanism of natural selection operates hierarchically. Yet, this hierarchical approach differs from the traditional concept of the so-called Hierarchical Theory of Selection (e.g., species selection). In any case, this view should be employed to address questions about the fitness of evolving entities at different levels considering their interaction with the environment.[10]

Finally, in the quest to identify units of evolution, this approach diverges from the traditional replicator–interactor framework. As Gontier (2010) comments, even though the replicator–interactor framework has been used to decide which units qualify as units of evolution, they "do not necessarily have the intrinsic properties demanded of replicators, interactors, reproducers, etc." (p. 171). For example, developmental systems are not replicators, as they do not faithfully transfer their exact structure to the next generation. Admittedly, developmental systems are steady despite genetic and ontogenetic disruptions, yet they do not replicate as perfect copies of their ancestors. Similarly, these entities do not function as interactors either since, as mentioned earlier, the value of an entity lies not in its selection capacity but in its generation. In addition, as emphasized throughout this book, the interaction of an entity involves various factors (such as phenotypic plasticity, niche construction, and behavioral choice) and is not solely through natural selection. Moreover, the replicator–interactor framework excludes embryonic development, which is necessary for interpreting the hierarchical structure of variation (DiFrisco 2021). If these entities are anything, they should be "reproducers" according to Griesemer's view (2000). However, this debate seems outdated and unnecessary since, as detailed in Chap. 6, from a philosophical point of view, it is more consistent to identify units of evolution as either evolutionary individuals, natural kinds, or both simultaneously.

---

[9] As an analogy with a soccer game, Pere Alberch said that natural selection can determine which team wins, but developmental biology determines which teams play. See Chap.11, Sects. 11.1.2 and 11.2.2, Fig. 11.7.

[10] For more details on the H-Evo-Devo Theory interpretation of the multilevel selection model, see Chap. 11, Sect. 11.2.1, Fig. 11.4.

5.1.3 The Linnaean Hierarchy as the Taxonomic Structure of the H-Evo-Devo Theory

The Linnaean hierarchy is a biological classification system that organizes organisms into nested ranks or categories—species, genera, families, orders, and so on—based on their morphological traits. Although the Linnaean system was founded in the eighteenth century, the names of the ranks remain used in modern taxonomy. Despite this, the Linnaean hierarchy has faced significant criticism from contemporary systematists and philosophers of biology. Critics argue that this classification is outdated because its theoretical foundation is rooted in a creationist and essentialist framework, rather than in the modern theory of evolution. They claim that phylogenetic analysis should focus on assessing genealogical relationships, making the exclusive reliance on morphological characters unnecessary. In addition, they advocate for developing phylogenetic nomenclature that eliminates the Linnaean system's hierarchical implications (Ereshefsky 2001; Zachos 2011; Cantino and de Queiroz 2020; Mishler 2022). In other words, they propose excluding the use of ranks, arguing that they lack ontological meaning from an evolutionary perspective (see Griffiths 1976; de Queiroz 1992, 1994; de Queiroz and Gauthier 1992, 1994; Ereshefsky 1994, 1997, 2001, 2002; Moore 1998; Mishler 1999, 2022).[11]

On the other hand, as discussed in Chap. 2, the Linnaean hierarchy—originally rooted in typology—played a key role in developing the Natural System. Importantly, typology did not represent an outdated biological approach that viewed species as unchangeable Platonic ideals. Instead, it focused on observing natural morphological patterns (types) without committing to particular ontological causes. The main issue is that evolutionary biology, as framed by the Modern Synthesis framework, rendered the typological perspective incompatible with evolution. Scholars influenced by this framework established new criteria for defining species and higher taxa. For instance, they argued that species exist based on their relational properties (such as interbreeding), rather than their intrinsic properties (like morphological traits). Likewise, they rejected categories above the species level as nonexistent, contending that these categories merely reflect the genealogical relationships among species.[12]

Nonetheless, several researchers distinguish two types of hierarchies used in taxonomic analysis, each with significant differences: the Linnaean hierarchy and the phylogenetic hierarchy (Williams 1992; Valentine and May 1996; Valentine 2004; Vasilyeva and Stephenson 2008; Brower 2023). The Linnaean hierarchy, as previously noted, involves an aggregative hierarchy based on morphological characters. From a typological point of view, the Linnaean hierarchy represents a hierarchy of types, arranged from greater to lesser inclusiveness. For example, the phylum stands as the most inclusive type, followed by the class, which is a less inclusive type than the phylum type, then the order, which is a less inclusive type than the

[11] For a well-argued rebuttal to these criticisms, see Schuh (2003).

[12] For a more detailed discussion of the traditional views on species and higher taxa, see Chap. 7, Sect. 7.2.1 and Chap. 8, Sect. 8.1.1, respectively.

class type, and so on. However, while criticized for supposedly lacking an evolutionary context, the Linnaean hierarchy—as a hierarchy of types—also reflects genealogical relationships, making it compatible with an evolutionary perspective (Valentine 2004; see Chap. 2).

In contrast, the phylogenetic hierarchy, also known as the generative hierarchy, represents a phylogenetic tree where the position of taxa is determined solely by the sequence of ancestors and descendants, irrespective of morphological characters. Systematists use various characteristics—physiological, chemical, behavioral, and even the absence of traits—to infer relationships between taxa. In addition, the phylogenetic hierarchy incorporates the Linnaean system's rank names to indicate the timing of species divergence (de Queiroz and Gauthier 1992). The earlier the divergence, the more inclusive the rank assigned to the compared taxa. Initially, within the framework of phylogenetic systematics, a category above the species level was delimited by the position of the ancestral species in time and the inclusion of all its descendant species (Williams 1992).[13] Nevertheless, a problem with the phylogenetic hierarchy is its lack of a true hierarchical structure, as it exhibits a positional structure instead (Valentine and May 1996).[14]

Advocates of the Linnaean hierarchy argue that such a system does not contradict the representation of genealogical connections; the conflict arises when compared with the phylogenetic hierarchy (Schuh 2003; Vasilyeva and Stephenson 2008). For instance, Williams (1992) points out the considerable confusion in taxonomic representation within phylogenetic systematics due to the elements that its hierarchy adopted from the Linnaean system. In fact, opponents of the Linnaean hierarchy criticize it precisely because of this incompatibility (e.g., Ereshefsky 1994, 2001). Likewise, proponents such as Vasilyeva and Stephenson (2008) contend that the Linnaean hierarchy—and its association with typology—reflects the structure of natural order, hence providing ontological support for its ranks. They argue that the hierarchy shows the sequential appearance of morphological traits over time, correlating the emergence of new characters with increased diversity. Similarly, Valentine (2004) asserts that the advantage of the Linnaean hierarchy lies not only in depicting a phylogenetic tree but also in possessing a structured hierarchy of inclusiveness, unlike the phylogenetic hierarchy, which lacks a genuine hierarchical structure.

Another important point is that many traditional evolutionary biologists depict the evolution of lineages in the phylogenetic tree as a gradual process. Similarly, critics of typology—and thus of the Linnaean hierarchy—take for granted that acknowledging the stability of characters over time implies fixed taxa with no capacity for change.[15] Within this context, the gaps in the fossil record and among extant taxa are explained as nothing more than the outcome of a smooth and gradual

[13] See Chap. 8, Sect. 8.1.1.

[14] See Chap. 4, Sect. 4.2 and Fig. 4.3.

[15] However, traditional evolutionary biologists may not fully recognize the patterns observed in evolving entities. As discussed in this book, units of evolution emerge abruptly, then enter a phase of stasis, and finally undergo a punctual extinction. See Chap. 9.

process where intermediate forms were progressively erased over time. This assumption of gradualism has also influenced advocates of the Linnaean system within an ontological context. For example, while Vasilyeva and Stephenson (2013) support the Linnaean hierarchy, they claim that these gaps should not unduly influence rank designation; instead, they argue that ranks should reflect the weighting of the characters based on their phylogenetic relationship: "Unfortunately, using the 'amount' of difference as the delimitative basis for classifying taxa strongly reminds one of a 'gap' and thus has no meaning in taxonomy. It is not the 'amount' of difference but rather the weight assigned to the differences that matters" (p. 154). However, units of evolution are not committed to gradual evolutionary change. In fact, as explored throughout this book, there are compelling reasons to support a theory of saltational evolution.[16] Bearing this in mind, I will now examine how the properties of evolutionary developmental systems connect (discussed in Sect. 5.1.2) with the Linnaean hierarchy (refer to Fig. 5.2a).

1. *A developmental system is in a steady state (A), maintaining a specific position within the morphospace as much as possible.* This principle is congruent with the typological framework employed in taxonomy. These steady states "account for the order observed in nature and they are the reason why a classification of forms is possible" (Alberch 1989, p. 23).
2. *A developmental system undergoes a discontinuous change, giving rise to another system occupying a different position within the morphospace.* The three groups (A, B, and C) are recognized ontologically because they are separated by morphological gaps. There is no ambiguity in the taxonomic recognition of these entities due to their sudden origin and subsequent stabilization. Likewise, the category is assigned based on the scale of discontinuity. In this case, the three groups belong to the category of "species" because, according to this model, species represent the least inclusive body plan within a hierarchy of body plans.
3. *A developmental system remains in a steady state, making it recognizable over time.* This assertion also corroborates the typological basis of taxonomy. For example, because developmental systems persist in a steady state—i.e., they maintain their basic characters as much as possible—they can be identifiable over time within a phylogenetic context.
4. *A developmental system can generate a nested compositional hierarchy.* This statement forms the ontological basis for the ranks within the Linnaean hierarchy that characterize the Natural System (see Fig. 5.2b). For example, an ancestral group (1A) can give rise to descendant groups either at the same level (2 or 3) or at a lower level (1B or 1C). Here, the different degrees of discreteness in the character states are evident. For example, from a hierarchical perspective, there are three taxa at a higher level (1, 2, 3), ranked as genera, and seven taxa (A, B, C, D, E, F, G, H) at a lower level, ranked as species: genus 1 comprises three species (A, B, C), genus 2 comprises two species (D, E), and genus 3 comprises three species (F, G, H).

[16] See Chap. 9, Sect. 9.2; Chap. 11, Sect. 11.2.1; Chap. 12, Sect. 12.1.1.

From this point of view, the Linnaean ranks represent the degree of morphological discreteness among character states at different scales, each maintaining stability over time. For example, when a taxon achieves a significant level of discontinuity (disparity), it may subsequently experience a lesser degree of discontinuity in its form, giving rise to lower taxa (diversity). Commonly, a greater degree of discontinuity appears early in the lineage's history. Importantly, within this framework, the construction of the phylogenetic tree does not follow the conventional bottom-up approach but rather adheres to the top-down approach. This means that body plans at higher taxonomic levels emerge first, followed by those at lower levels.[17] This arrangement not only reveals genealogical connections between groups but also provides ontological validation for the Linnaean ranks. The fact that developmental constraints operate at different scales provides an explanation for the stability of basic forms within a hierarchy. This process facilitates the identification and classification of groups of organisms into categories during taxonomic exercise.

As mentioned earlier, the Linnaean hierarchy represents a hierarchy of body plans (types). Therefore, according to the interpretation of the H-Evo-Devo Theory, species are the least inclusive body plan, while higher taxa represent the body plans that are more inclusive than the species body plan. It is important to stress that although subspecies do not characterize a type—as this level corresponds to the evolution of simple characters such as color, size, proportion, etc.—they are included in the hierarchy because they are character states of homologues as well. Thus, when integrating all the units of evolution into the Linnaean hierarchy, the scheme may be illustrated according to Pere Alberch's hierarchical model of developmental constraints, depicted in a triadic structure (see Fig. 5.3). In conclusion, the Linnaean hierarchy constitutes a fundamental component of the H-Evo-Devo Theory's framework, as it represents the units of evolution as character states of homologues.

5.2 Hierarchical Evolutionary-Developmental Theory

The *Hierarchical Evolutionary-Developmental Theory*—or the *H-Evo-Devo Theory*—is centered on developmental constraints as the primary mechanism of evolution.[18] As detailed in Sect. 5.1.2, these developmental constraints are established from a procedural hierarchy that delineates the functional interactions between different levels of organization. This hierarchical process generates the internal (phenotypic) properties observed in multicellular organisms. Thus, to

[17] See Chap. 12, Sect. 12.1.1.

[18] I want to clarify that highlighting the significance of developmental constraints does not diminish the importance of other evolutionary mechanisms. Instead, developmental constraints hold significance because they enable the recognition of the units of evolution. This model advocates for the integration of processes rather than their exclusion. In this regard, any processes, concepts, or mechanisms that shed light on the immense complexity demanded by the study of biological systems are welcome (see Wake 1990, 2008).

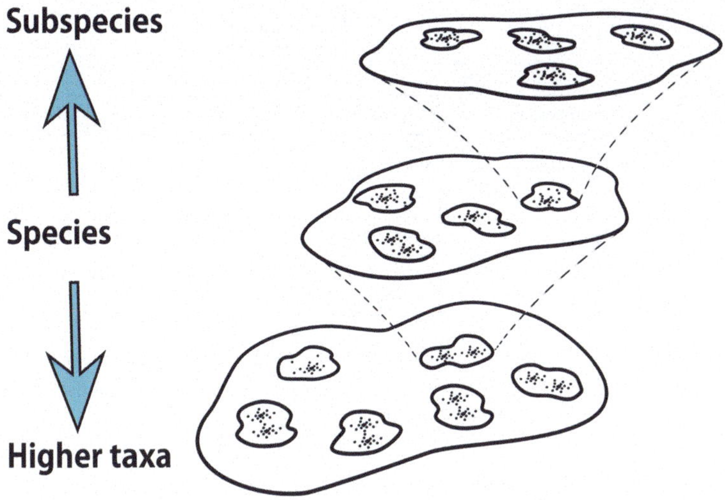

Fig. 5.3 *Triadic structure diagram of developmental constraints representing the Linnaean hierarchy*. According to the H-Evo-Devo Theory, each form represents a morphological steady state at different scales established by developmental constraints (see Fig. 5.2b). (Diagram courtesy of Julio Velasco Olmos)

address the question "What kind of entities evolve?" or, more precisely, "What are the units of evolution?"[19] the steady states of these phenotypic properties must be considered as an assessment parameter. In this regard, I suggest a single compositional hierarchy to identify homologues, higher taxa, species, and subspecies as units of evolution.

Importantly, the H-Evo-Devo Theory regards the "organism" as the focal point of study, i.e., it is based on an organism-centered perspective. According to this view, the organism is not merely a hierarchical level but a fundamental unit that encapsulates all the hierarchical relationships vital for evolutionary inquiry. Put differently, the organism stands as an entity that exhibits internal (functional interactions and structural properties) and external (interactions with the environment) relationships at different scales. In this regard, my focus lies in describing this intricate organizational structure, first, by using a procedural hierarchy that accentuates the main processes of its generation and stabilization and second, by establishing a compositional hierarchy based on its phenotypic properties, allowing for the identification of the kinds of entities that evolve.

Fig. 5.4 depicts the hierarchical pattern of interactions in the developmental process. Level 1 represents a dual system depicted by two boxes: the lower box indicates the "Gene Regulatory Network" (GRN), which is responsible for regulating

[19] This question pertains to what Gontier (2010, 2021) terms Applied Evolutionary Epistemology, which aims to study evolution from a philosophical standpoint, focusing on identifying units and levels of evolution within an ontological hierarchy delineated by specific mechanisms and processes.

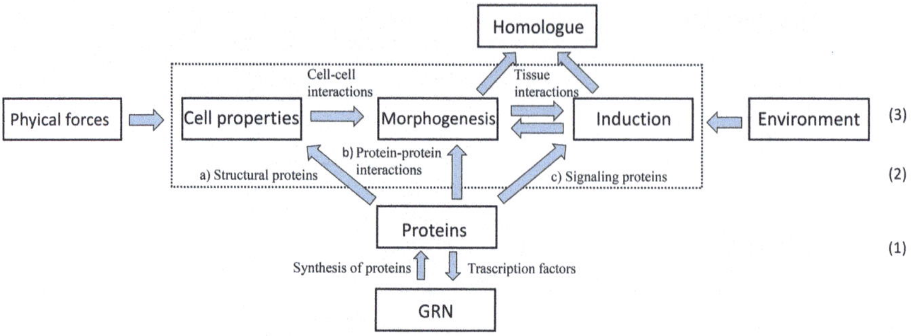

Fig. 5.4 *A diagram depicting the pattern of interactions within developmental processes.* (Adapted from Alberch 1982; see Fig. 5.1). Developmental constraints are established from these hierarchical interactions. See the text for details

the synthesis of proteins crucial for differentiating specific cell lineages that form a module. The higher box indicates "Proteins," which interact with the GRN as transcription factors to control cellular and developmental functions. Notably, this tier not only encompasses the protein-coding region and the regulatory sequences of DNA (e.g., promoters and enhancers) but also includes epigenetic regulatory elements such as histones, modifying enzymes, transcription factors, and splicing factors. Level 2 corresponds to the three types of protein functions: (a) *structural proteins* that contribute to the physical properties of the cell (such as adhesion, motility, and contractility); (b) *protein-protein interactions* that involve enzymatic pathways and interactions with the products of other cells, etc.; and (c) *signaling proteins* that facilitate cell communication and signal transduction. Finally, level 3 is represented by three boxes: *cell properties*, *morphogenesis*, and *induction*, i.e., cell and tissue interactions that control morphogenesis. This level integrates two inputs: on the left side, there are the *physical forces* that contribute to cell properties (Newman and Bhat 2009; Newman 2017), and on the right side, there is the *environment* that influences the determination of cell fate and specification (West-Eberhard 2003; Gilbert and Epel 2015; Sultan 2015). This process culminates in a module referred to here as a homologue.[20]

It is important to note that the GRN also exhibits a hierarchical structure, which is crucial for understanding distinctions at the phenotypic level. According to Wagner (2007, 2014, 2015), this system comprises three levels (Fig. 5.5). The first level corresponds to *positional information signals*. These genes use inductive signals to instruct cells on both their location and specific fate during differentiation. The second level is the *Character Identity Network* (ChIN), which is considered the GRN's most conserved part. The role of these genes is to interpret positional information signals to determine specific cell activities. Wagner regards this "core network" as the genetic basis that specifies character identity in phenotypes. Finally,

[20] According to modularity theory in morphology, I equate a module with a homologue. See Chap. 8, Sect. 8.2.2.

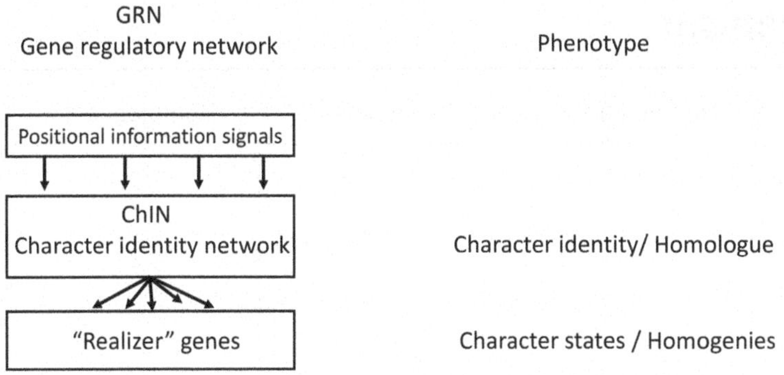

Fig. 5.5 *The GRN hierarchy*. According to Wagner (2007, 2014, 2015), a gene regulatory network exhibits a hierarchical structure. The top box represents the "positional information signals" responsible for indicating the location and fate of cells during morphogenesis. In the middle box, the "character identity network" (ChIN) comprises genes that interpret the positional information signals, thus determining specific cellular activities. This is the most highly conserved part of the system, and it is crucially linked to character identities in phenotypes (homologues). The bottom box represents the "realizer genes," which are responsible for generating character states at the phenotypic level (homogenies)

the third level corresponds to *realizer genes* whose expression exhibits significant variability. These genes are responsible for producing the specific character state at the phenotypic level.[21]

The structure of the completed model is depicted in Fig. 5.6. The organism comprises the entire system under study, with its genome represented at the bottom. As the genome does not simultaneously generate all cell types, the figure illustrates the division of the genome into several GRNs. Differentially expressed genes (level 1) produce different types of proteins (level 2), leading to the differentiation of various cell types and tissues (level 3). Consequently, each GRN generates distinct modules, represented in the diagram as homologues. The organism is composed of morphological units or homologues, and the modifications in each homologue are indicated as changes in character states. However, when considering the entire system (the organism), all character state changes in homologues collectively result in phenotypic steady states at different levels. These steady states, projected as types, reflect the Natural System, which can be represented through the Linnaean hierarchy.

[21] In an earlier model, Davidson and Erwin (2006) argued that the development of the body plan in animals was controlled by GRNs, which consist of various components. According to them, within GRN, there are highly conserved parts, referred to as "kernels." These parts are closely associated with the stability of the body plan. Also, they identified other components, e.g., "differentiation gene batteries," which are responsible for executing the final instructions of development and are therefore linked to specific changes in later stages. Wagner, on the other hand, proposed a different perspective on the composition of GRN, emphasizing a modular framework. According to this approach, it is possible to distinguish between "character identity" and "character state," establishing a direct relationship between genotype and phenotype. However, more recently, he and other authors have expanded this argument to include additional levels of organization (see DiFrisco et al. 2020 and below).

Organism

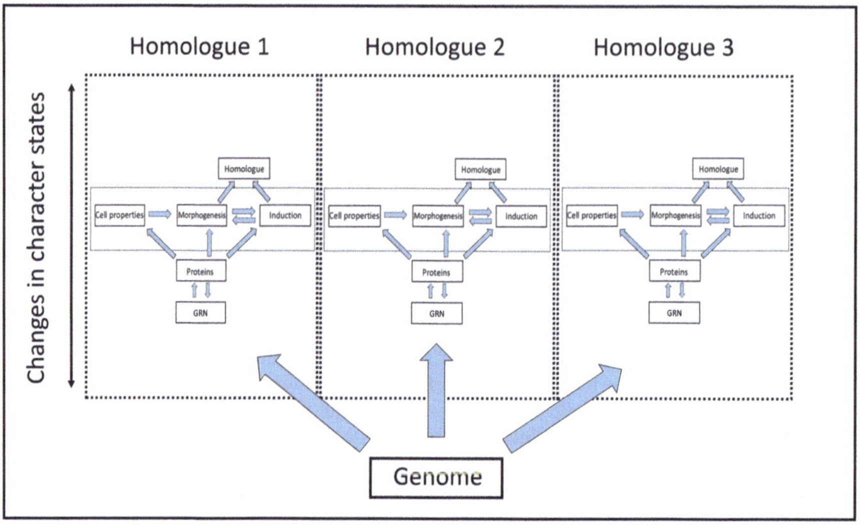

Fig. 5.6 *Depiction of the organism's complete internal structure*. The largest rectangle represents the entire organism. Within the organism, three small squares depict three homologues. Each homologue includes an illustration outlining a pattern of interactions, starting with a box at the bottom indicating the genome. Changes in character states are indicated by opposing arrows on the left side of the homologues. According to this framework, homologues can undergo variational modalities, collectively forming discrete states of variation within the organism. These steady states are reflected by taxonomic categories in the Linnaean hierarchy

As mentioned earlier, generative potentialities are defined by the patterns of interactions in developmental dynamics at different levels. While it should theoretically be possible to differentiate the developmental constraints operating across various scales within a compositional hierarchy, understanding the meaning of developmental constraints in relation to the procedural hierarchy remains an area of ongoing research. For instance, DiFrisco et al. (2020) introduce the concept of "Character Identity Mechanisms" (ChIMs) to mechanistically explain the phenomenon of homology. This model makes it possible to formulate hypotheses about the mechanisms underlying the most conserved parts of the system, thereby facilitating the identification of traceable characters. In this regard, developmental constraints are reflected at the highest level as ChIMs. Similarly, there are variational modalities in organs whose stability is achieved through other identifiable mechanisms, interpreted as developmental constraints operating at a lower level. I use the term "Character State Mechanisms" (ChSMs) to refer to these mechanisms, which underlie the causes of homogenies. In short, developmental constraints work at different levels, both upward and downward, reflecting different kinds of biological entities. Below, I describe the evolving entities identified by this model and their levels, briefly referring to the processes or mechanisms involved.

Units of Evolution According to Gontier (2021), identifying the units of evolution requires specifying their position within the hierarchy and the mechanisms that justify them. As demonstrated, developmental constraints are the evolutionary mechanism that enables this distinction. Therefore, under this interpretation, the homologue represents the highest evolving entity within the hierarchy. *By homologue I mean those quasi-autonomous parts of the organism that are highly integrated in developmental terms and maintained phylogenetically over a long time.*[22] These homologues are discernable because developmental constraints act in this system at the highest level (Ochoa and Rasskin-Gutman 2015). Yet, when considering all homologues and their character states collectively—i.e., assessing the entire organism—a set of phenotypic steady states is observed at different scales. These steady states can be categorized through an aggregative hierarchy, employing the Linnaean hierarchy for their identification.

The Natural System must reflect the evolutionary processes acting upon the "organism" that gives rise to different biological entities. Following the Linnaean hierarchy, as discussed in Sect. 5.1.3, ranks are determined by the degree of morphological discreteness among character states at different scales, each maintaining a certain position within the morphospace. Of course, these steady states are caused by developmental constraints operating at different levels. Thus, a ranking within a triadic structure representation can be established to represent this hierarchy (see Fig. 5.3). The focal taxonomic category is the *species*, representing groups of organisms that exhibit the first step of significant morphological gaps. In this regard, *"species" are the least inclusive body plan within a hierarchy of body plans.* Consequently, *subspecies* are positioned beneath the species category. *"Subspecies" are groups that share the least inclusive body plan (species) but are separated from each other by small morphological gaps in simple characters—such as color, pattern, size, proportion,* etc.*—as well as by the fact that they live in different environments*. Notably, subspecies do not represent body plans; rather, they manifest as variations within the species body plan. The *genus* is posited above the species category that encompasses groups of species. Above the genus is the family, which includes groups of genera, and so on. While there are distinct ranks within the Linnaean hierarchy that categorize different steady states at a higher level, I employ the term *higher taxa* to refer to all the higher categories above the genus category (e.g., family, order, class, and phylum).[23] So, according to this model, higher taxa are groups distinguished by great distances within the morphospace. Specifically, *they are the more inclusive body plans than the species body plan within a hierarchy of body plans*.

[22] This definition of homologue corresponds to the notion of the morphological module. See Chap. 8, Sect. 8.2.

[23] I distinguish higher taxa above the genus level because, as Chap. 7 will explore, species and genera are very similar ontologically.

Levels of Evolution According to Gontier (2021, p. 317), "levels refer to the locus or place where the evolution of units occurs." In other words, once the units of evolution are identified, they can be positioned within distinct ontological layers of the hierarchy. This approach enables the recognition of specific patterns, mechanisms, and processes pertinent to each entity. As mentioned in this book's introduction (Chap. 1), a precise definition of evolution should emphasize not only the origin of evolving entities but also their stability over time as well as their eventual extinction. Put differently, units of evolution exhibit three distinct evolutionary patterns: a discrete origin, stability, and a well-defined end.[24] Therefore, the challenge lies in recognizing the mechanisms underlying each of these patterns to provide a better understanding of the hierarchical landscape of evolution.

Importantly, the H-Evo-Devo Theory distinguishes at least two levels of evolution: microevolution and macroevolution. This distinction arises from the ability to differentiate between two types of variation: simple and complex characters. Furthermore, different factors operate between the two levels, at least in the patterns of origin and extinction. Additionally, macroevolution also comprises two distinct levels: macroevolution *per se* and megaevolution. Although the causes explaining macroevolution and megaevolution are the same, their separation is justified due to the degree of discontinuity between them (more pronounced in higher taxa than in species). Finally, the H-Evo-Devo Theory incorporates an additional level called "modular-evolution" that focuses on studying the patterns and processes of homologues. To clarify this framework, I suggest the following definitions:

- *Modular-evolution* is related to processes tied to the origin and stability of homologues.
- *Megaevolution* comprises processes related to the origin, stability, and extinction of higher taxa (such as families, orders, classes, and phyla).
- *Macroevolution* corresponds to processes related to the origin, stability, and extinction of species and genera.
- *Microevolution* involves processes related to the origin, stability, and extinction of subspecies.

However, more specific terminology may be necessary to indicate the patterns of each unit of evolution. For this purpose, I suggest the following terms as modes:

- *Modular-evolution*: morphological novelty and morphological stability.
- *Megaevolution*: supraspeciation, suprastasis, and supraextinction.
- *Macroevolution*: speciation, specific stasis, and specific extinction.
- *Microevolution*: subspeciation, substasis, and subextinction.

Thus, with respect to origin, *morphological novelty* refers to a character that arises from the co-option of developmental mechanisms from a preexisting

[24] See Chap. 6, Sect. 6.2 and Chap. 9.

character. Once a new module emerges, it tends to exhibit variation across different modalities. When all homologues are considered, their collective character states are represented by distinct taxonomic units. How do these entities emerge? *Supraspeciation* and *speciation* are explained through *SupraDarwinian mechanisms*, i.e., the prospecies model (hopeful monster model), while *subspeciation* is attributed to *Darwinian mechanisms*, i.e., the dispersal model.

Considering stability, *morphological stability* is explained by developmental constraints acting at the highest part of the hierarchy, or ChIMs (sensu DiFrisco et al. 2020). *Suprastasis* and *specific stasis* are caused by developmental constraints operating on body plan characters, while *substasis* results from developmental constraints acting on simple characters. Nevertheless, the causes leading to the disappearance of units of evolution entail a hierarchical viewpoint. For instance, *supraextinction* and *specific extinction* are due to the loss of parts and mass extinction events, whereas *subextinction* results from factors altering their local habitat and their population structure.[25]

In summary, the framework of the H-Evo-Devo Theory is outlined in Table 5.2. The organism is not a level but an entity that comprises all the necessary hierarchical relationships for the study of evolution. To understand the causes of its generation and stability, a procedural hierarchy should be applied. The developmental process involves interactions across different levels of organization, beginning at the level responsible for gene expression mechanisms and culminating in cell and tissue interactions. Moreover, to delve deeper into some of these levels, additional hierarchies can be employed. For example, within the GRN hierarchy, at least three levels are discernible concerning gene functions and interactions among genes: positional information signals, ChIN, and realizer genes. The developmental process results in distinct and highly integrated phenotypes that could be studied using a compositional hierarchy, which potentially involves different hierarchies. For example, to study homologues and their variational modalities—called homogenies—a modular hierarchy is helpful. To assess the steady states of variation throughout the entire organism—resulting from a discontinuous change in character states as a whole—the Linnaean hierarchy is appropriate.

Given that homologues, higher taxa, species, and subspecies can be recognized as units of evolution due to developmental constraints operating at different levels, it should be noted that these units can be represented through a single compositional hierarchy. However, it is important to clarify that this hierarchy is not a nested compositional kind. For instance, while a set of subspecies is nested within a species and a set of species is nested within a higher taxon, higher taxa are not nested within a homologue. Nonetheless, this hierarchy is compositional because the organism consists of homologues that generate various character states. These character states collectively undergo discontinuous changes, resulting in distinct Linnaean ranks. Within this framework, homogenies are not included as units of evolution, as they

[25] For more details regarding the patterns and processes across different levels, see Chaps. 11 and 12.

are already encompassed within the entire organism's overall set of character states through the Linnaean hierarchy. Therefore, according to the H-Evo-Devo Theory, homologues, higher taxa, species, and subspecies are units of evolution.

5.3 Conclusion

This book introduces a new evolutionary model called *Hierarchical Evolutionary-Developmental Theory* or *H-Evo-Devo* Theory. This theory suggests that the internal properties of the organism reflect highly integrated entities that evolve hierarchically. Recognizing the need for a hierarchical structure in evolutionary theory, my proposal aims to integrate embryology, taxonomy, and evolution effectively through a hierarchical approach. This model encompasses both functional (process) and structural (pattern) aspects across different levels of organization. In this regard, developmental constraints are the main mechanism that shapes the structure of the H-Evo-Devo Theory: homologues are the most prominent entities that represent the work of developmental constraints operating at the highest level. By collectively examining all homologues and their variational modalities within the entire organism, phenotypic steady states are observed at various scales. The Linnaean hierarchy's ranks or categories are nothing but a reflection of these phenotypic steady states (types) caused by developmental constraints operating both upward and downward in a nested sequence of greater or lesser inclusiveness. In a nutshell, developmental constraints draw a single compositional hierarchy, making it possible to identify homologues, higher taxa, species, and subspecies as units of evolution. Ultimately, the H-Evo-Devo Theory fosters a comprehensive evolutionary approach considering the different levels at which evolution can be studied. While the Hierarchy Theory of Evolution (*sensu* Eldredge and Salthe) currently attempts to achieve this goal, the H-Evo-Devo Theory competes with it by suggesting alternative evolutionary modes, aiming to convincingly decouple macroevolution from microevolution.

References

Alberch A (1980) Ontogenesis and morphological diversification. Am Zool 20:653–667

Alberch P (1982) Developmental constraints in evolutionary processes. In: Bonner JT (ed) Evolution and development. Springer, New York, pp 313–332

Alberch P (1989) The logic of monsters: evidence for internal constraint in development and evolution. Geobios 22:21–57

Amundson R (1994) Two concepts of constraint: adaptationism and the challenge from developmental biology. Philos Sci 61:556–578

Baedke J (2021) Mechanisms in evo-devo. In: Nuño de la Rosa L, Müller GB (eds) Evolutionary developmental biology. Springer, Cham, pp 383–395

Baedke J, Mc Manus SF (2018) From seconds to eons: time scales, hierarchies, and processes in evo-devo. Stud Hist Phil Biol Biomed Sci 72:38–48

Brigandt I (2015) From developmental constraint to evolvability: how concepts figure in explanation and disciplinary identity. In: Love AC (ed) Conceptual change in biology: scientific and philosophical perspectives on evolution and development. Springer, Dordrecht, pp 305–325
Brower AVZ (2023) Hierarchies, classifications, cladograms and phylogeny. Cladistics 39:229–239
Caianiello S (2016) Revisiting the phenotypic hierarchy in hierarchy theory. In: Eldredge N, Pievani T, Serrelli E, Tëmkin I (eds) Evolutionary theory: a hierarchical perspective. The University of Chicago Press, Chicago, pp 151–173
Cantino PD, de Queiroz K (2020) International code of phylogenetic nomenclature (PhyloCode). CRC Press, Boca Raton
Craver CF (2007) Explaining the brain mechanisms and the mosaic unity of neuroscience. Oxford University Press, Oxford
Davidson EH, Erwin DH (2006) Gene regulatory networks and the evolution of animal body plans. Science 11:796–800
de Queiroz K (1992) Phylogenetic definitions and taxonomic philosophy. Biol Philos 7:295–313
de Queiroz K (1994) Replacement of an essentialistic perspective on taxonomic definitions as exemplified by the definition of "Mammalia". Syst Biol 43:497–510
de Queiroz K, Gauthier J (1992) Phylogenetic taxonomy. Annu Rev Ecol Syst 23:449–480
de Queiroz K, Gauthier J (1994) Toward a phylogenetic system of biological nomenclature. Trends Ecol Evol 9:27–31
DiFrisco J (2017) Time scales and levels of organization. Erkenn 82:795–818
DiFrisco J (2021) Integrating composition and process in levels of developmental evolution. In: Brooks DS, DiFrisco J, Wimsatt WC (eds) Levels of organization in the biological sciences. The MIT Press, Cambridge, MA, pp 111–134
DiFrisco J, Love AC, Wagner GP (2020) Character identity mechanisms: a conceptual model for comparative-mechanistic biology. Biol Philos. https://doi.org/10.1007/s10539-020-09762-2
Ereshefsky M (1994) Some problems with the Linnaean hierarchy. Philos Sci 61:186–205
Ereshefsky M (1997) The evolution of the Linnaean hierarchy. Biol Philos 12:493–519
Ereshefsky M (2001) The poverty of the Linnaean hierarchy: a philosophical study of biological taxonomy. Cambridge University Press, Cambridge
Ereshefsky M (2002) Linnaean ranks: vestiges of a bygone era. Philos Sci 69:S305–S315
Eronen MI (2020) Levels of organization in evo-devo. In: Nuño de la Rosa L, Müller GB (eds) Evolutionary developmental biology. Springer, Cham, pp 447–455
Gerhart J, Kirschner M (2007) The theory of facilitated variation. Proc Natl Acad Sci USA 104:8582–8589
Gilbert S, Epel D (2015) Ecological developmental biology: the environmental regulation of development, health, and evolution. Sinauer Associates, Sunderland
Gontier N (2010) Evolutionary epistemology as a scientific method: a new look upon the units and levels of evolution debate. Theory Biosci 129:167–182
Gontier N (2021) Hierarchies, networks, and causality: the applied evolutionary epistemological approach. J Gen Philos Sci 52:313–334
Gould SJ (2002) The structure of evolutionary theory. The Belknap Press of Harvard University Press, Cambridge, MA
Griesemer J (2000) The units of evolutionary transition. Selection 1:67–80
Griesemer J (2021) Levels, perspectives, and thickets: toward an ontology of complex scaffolded living systems. In: Brooks DS, DiFrisco J, Wimsatt WC (eds) Levels of organization in the biological sciences. The MIT Press, Cambridge, MA, pp 89–109
Griffiths GCD (1976) The future of Linnaean nomenclature. Syst Zool 25:168–173
Hall BK (1999) Evolutionary developmental biology. Kluwer, Dordrecht
Hu H, Uesaka M, Guo S, Shimai K, Lu T-M, Li F, Fujimoto S, Ishikawa M, Liu S, Sasagawa Y, Zhang G, Kuratani S, Yu J-K, Kusakabe TG, Khaitovich P, Irie N, the EXPANDE Consortium (2017) Constrained vertebrate evolution by pleiotropic genes. Nat Ecol Evol 1:1722. https://doi.org/10.1038/s41559-017-0318-0
Love AC (2006) Evolutionary morphology and Evo-devo: hierarchy and novelty. Theory Biosci 124:317–333

Mishler BD (1999) Getting rid of species? In: Wilson RA (ed) Species: new interdisciplinary essays. MIT Press, Cambridge, MA, pp 306–315
Mishler BD (2022) Ecology, evolution, and systematics in a post-species world. In: Wilkins J, Zachos FE, Pavlinov IY (eds) Species problems and beyond: contemporary issues in philosophy and practice. CRC Press, pp 177–190
Moore G (1998) A comparison of traditional and phylogenetic nomenclature. Taxon 47:561–579
Newman SA (2003) Hierarchy. In: Hall BK, Olson WM (eds) Keywords and concepts in evolutionary developmental biology. Harvard University Press, Cambridge, MA, pp 169–174
Newman SA (2017) Inherency. In: Nuño de la Rosa L, Müller GB (eds) Evolutionary developmental biology. Springer, Cham, pp 121–131
Newman SA, Bhat R (2009) Dynamical patterning modules: a "pattern language" for development and evolution of multicellular form. Int J Dev Biol 53:693–705
Nuño de la Rosa L, Müller GB (2024) The legacy and evolvability of Pere Alberch's ideas. Interface Focus 14:20240011. https://doi.org/10.1098/rsfs.2024.0011
Ochoa C, Rasskin-Gutman D (2015) Evo–devo mechanisms underlying the continuum between homology and homoplasy. J Exp Zool B Mol Dev Evol 324:91–103
Raff RA (1996) The shape of life: genes, development, and the evolution of animal form. The University of Chicago Press, Chicago and London
Rasskin-Gutman D, Esteve-Altava B (2018) Concept of burden in evo-devo. In: Nuño de la Rosa L, Müller GB (eds) Evolutionary developmental biology. Springer, Cham, pp 39–49
Salazar-Ciudad I (2006) Developmental constraints vs. variational properties: how pattern formation can help to understand evolution and development. J Exp Zool (Mol Dev Evol) 306B:107–125
Schuh RT (2003) The Linnaean system and its 250-year persistence. Bot Rev 69:59–78
Sultan SE (2015) Organism and environment: ecological development, niche construction, and adaptation. Oxford University Press, New York
Valentine JW (2004) On the origin of phyla. The University of Chicago Press, Chicago/London
Valentine JW, May CL (1996) Hierarchies in biology and paleontology. Paleobiology 22:23–33
Vasilyeva LN, Stephenson SL (2008) The Linnaean hierarchy and "extensional thinking". Open Evol J 2:55–65
Vasilyeva LN, Stephenson SL (2013) An essentialistic view of the species problem. In: Pavlinov I (ed) The species problem: ongoing issues. IntechOpen. https://doi.org/10.5772/52625
Waddington CH (1952) The evolution of developmental systems. In: Herbert DA (ed) Proceedings of the twenty-eighth meeting of the Australian and New Zealand Association for the Advancement of Science. A. H. Tucker Government Printer, Brisbane, pp 155–159
Wagner GP (2007) The developmental genetics of homology. Nat Rev Genet 8:473–479
Wagner GP (2014) Homology, genes, and evolutionary innovation. Princeton University Press, Princeton
Wagner GP (2015) Reinventing the organism: evolvability and homology in post-Dahlem evolutionary biology. In: Love AC (ed) Conceptual change in biology: scientific and philosophical perspectives on evolution and development. Springer, Dordrecht, pp 327–342
Wake MH (1990) The evolution of integration of biological systems: an evolutionary perspective through studies on cells, tissues, and organs. Am Zool 30:897–906
Wake MH (2008) Integrative biology: science for the 21st century. Bioscience 58:349–353
Wake MH (2015) Hierarchies and integration in evolution and development. In: Love AC (ed) Conceptual change in biology: scientific and philosophical perspectives on evolution and development. Springer, Dordrecht, pp 405–420
West-Eberhard MJ (2003) Developmental plasticity and evolution. Oxford University Press, New York
Williams PA (1992) Confusion in cladism. Synthese 91:135–152
Wimsatt WC (2007) Re-engineering philosophy for limited beings: piecewise approximations to reality. Harvard University Press, Cambridge, MA
Zachos FE (2011) Linnean ranks, temporal banding, and time-clipping: why not slaughter the sacred cow? Biol J Linn Soc 103:732–734

Part III

The Units of Evolution: Philosophical Criteria

Criteria for Recognizing Units of Evolution

6

Abstract

According to the H-Evo-Devo Theory, homologues, higher taxa, species, and subspecies are units of evolution. To support this perspective, this chapter employs a metaphysical argument for their identification. Initially, it explores the philosophical debate on whether species should be categorized as natural kinds or individuals. In this regard, it upholds the traditional principle of natural kinds, arguing that the essentialism story is misleading and that the fundamental attributes of natural kinds can be sustained within the theoretical framework of the H-Evo-Devo Theory. However, it supports the idea that natural kinds and individuals are not opposing categories but rather metaphysically compatible. This assertion is based on the premise that developmental constraints act as the causal mechanism that simultaneously configures both categories. Finally, it develops metaphysical criteria concerning the individuality thesis, which not only serves to distinguish the evolving entities spatiotemporally but also to identify specific evolutionary patterns and modes within these units.

Keywords

H-Evo-Devo Theory · Units of evolution · Natural kinds · Individuality · Species-as-individuals · HPC kinds · Species realism · Species problem · Essentialism · Typology · Bauplan · Criteria for individuality · Homologues · Higher taxa · Species · Subspecies

Following the H-Evo-Devo Theory, this chapter employs a metaphysical argument to support the recognition of homologues, higher taxa, species, and subspecies as units of evolution. The contemporary controversy in the philosophy of biology aims to determine whether evolving entities are natural kinds or individuals. However, I

C. Ochoa, *Hierarchical Evolutionary-Developmental Theory*, Evolutionary Biology – New Perspectives on Its Development 9,
https://doi.org/10.1007/978-3-031-88723-9_6

argue here that they can simultaneously be both natural kinds and individuals. Since these units of evolution encompass both ontological categories, it follows that they truly exist in nature.

Considering the above, Sect. 6.1 examines the debate on whether species are natural kinds or individuals. In this regard, it provides support for the traditional principle of natural kinds, arguing that the essentialism story is misleading and that the characteristics of natural kinds remain valid from the perspective of the H-Evo-Devo Theory. Furthermore, it holds the view that natural kinds are metaphysically compatible with the individuality thesis, considering that developmental constraints are the same causal mechanisms that produce both ontological configurations. To complement the metaphysical identification of the units of evolution, Sect. 6.2 proposes some criteria for individuality: birth, cohesion, death, reproduction, variation, and phylogeny. These criteria not only serve to distinguish these entities spatiotemporally but also to identify specific evolutionary patterns and modes within these units.

6.1 Natural Kinds and Individuals as Metaphysical Criteria

A contentious matter in the philosophy of biology is whether species are natural kinds or individuals. Debates on this topic are associated with epistemological and metaphysical issues regarding species. As Zachos (2016, p. 45) points out: "[T]he species problem is in large parts a philosophical rather than a biological issue. For no other part of the problem is this as true as for the question of the ontological status of species, i.e. the question 'what kind of things' species taxa are." For example, if species are fundamental units of evolution, why are there so many concepts to define them? Is the category species real, or is it merely a category that satisfies practical needs?

These philosophical issues were similarly extended to other biological entities, e.g., homologues, higher taxa, and subspecies. Thus, the main discussion involves the ontological status of these evolving entities alongside species. What does it mean for units of evolution to be considered natural kinds or individuals? Below, I explore the debates surrounding this distinction, which will support a metaphysical criterion for identifying units of evolution.

6.1.1 Natural Kinds or Individuals?

Natural kinds are ontological categories or natural divisions characterized by a specific set of properties shared collectively by all their members. In other words, members of a particular kind possess a sufficient number of properties that define their essence. Those lacking any of these necessary properties are excluded from belonging to that kind. These properties are typically considered intrinsic, ahistorical, and derived from generalizing objects governed by universal laws. The kinds are "natural" in the sense that they represent groups of things that reflect true divisions in the world, i.e., they exist independently of human interests.

Since the early days of Linnaean taxonomy, species have often been considered paradigmatic examples of natural kinds. Nevertheless, from a Darwinian standpoint, viewing species as natural kinds is not the most appropriate approach. According to most philosophers and biologists, the idea that species are natural kinds stems from essentialism in biology—an outdated notion asserting that species cannot change.[1] For Darwin and his followers, on the other hand, species were considered more pragmatic than real (i.e., they were nominalists). After the Modern Synthesis, it became widely accepted that species were not defined by a fixed set of abstract traits. Instead, they were seen as groups of individuals capable of interbreeding, sharing the same ecological niche, or having a genealogical connection. While many neo-Darwinians acknowledge that species boundaries can be fuzzy, they are not completely opposed to the idea of species realism.

For example, in a series of works, Ghiselin (1966, 1974) and Hull (1976, 1978) initially suggested that addressing the ontological status of species could be resolved by considering them as individuals rather than natural kinds. Viewing species as individuals entails recognizing them as cohesive wholes, i.e., concrete spatiotemporally localized entities whose parts (individuals) exhibit continuity and a genealogical nexus. According to Ghiselin (1997, p. 15), within this metaphysical context, it becomes possible to contemplate the reality of species, albeit with some degree of ambiguity. In this regard, Richards (2010, p. 159) comments that, by treating species as individuals, Ghiselin killed two birds with one stone, since he rejected the essentialist notion of species—which is inherently realistic but incompatible with evolutionary perspectives—and simultaneously eliminated Darwin's radical nominalist stance on species. In doing so, Ghiselin proposed a sort of "moderate nominalism."

Much ink has been spilled debating the ontological status of species in the philosophy of biology. The divide is between those who consider species as natural kinds and those who assert that species are individuals. For example, proponents of the individuality thesis argue the following:

1. *Species possess well-defined spatiotemporal boundaries; if they were natural kinds, their members would be present anywhere and at any time in the universe.* Natural kinds are spatiotemporally unrestricted categories of objects defined abstractly by their intrinsic properties based on universal laws. That is, members of a kind can theoretically exist at any point in space and at any time. However, applying this view to species appears implausible (Zachos 2016, p. 46; Kunz 2012, p. 51). Species are considered individuals because they can be traced historically. They have discrete beginnings and endings (Hull 1976; Mishler and Brandon 1987). Species are tangible entities rather than abstract things (Ghiselin 1997, p. 42; Burbrink et al. 2022). One key distinction between natural kinds and

[1] Nevertheless, as noted in Chap. 2, the essentialism story is misleading. See also: Sect. 6.1.2 below.

individuals is that natural kinds consist of members or instances, whereas individuals are composed of parts (Richards 2010, p. 160).

2. *Species evolve; natural kinds do not; organisms of the same species exhibit evolutionary continuity, while members of a kind are identified solely by a set of common fixed properties.* Natural kinds remain unchanged because their members maintain a set of fixed and invariable properties. But, if species evolve, then they must be considered individuals (Rosenberg 1985, p. 205; Zachos 2016, p. 46). Moreover, species lack essential traits and should not be solely defined by their phenotypic similarities. Instead, they are identified by their parts sharing a common evolutionary history, which makes them historical entities (Hull 1978; Sober 2000, p. 151; Reydon 2022).
3. *Species are particulars; they do not follow laws of nature, whereas members of a kind are always subject to laws that are universally valid at all places and times.* Natural kinds are constrained by universal laws, i.e., all members of a kind are governed by the same principles. Yet no such law should be established to define a particular species (Hull 1976; Ghiselin 1988, 1997; Kunz 2012, p. 53).
4. *Species have proper names; systematists name species as individuals, not as natural kinds.* Proper names imply naming individuals. When systematists name species, they allude to particular things, not categories (Rosenberg 1985, p. 204; Ghiselin 1997, p. 46; Coleman and Wiley 2001). For example, the name *Homo sapiens* denotes a particular species in terms of its spatiotemporal location rather than the properties it possesses (Hull 1976).

However, in response to the objections against the individuality stance, advocates of species as natural kinds argue, for example, that:

1. Since natural kinds can be historically delimited (Boyd 1999, p. 154–156; Griffiths 1999; Millikan 1999), species are examples of historical kinds (Kitcher 1984, p. 314; LaPorte 2004, p. 11; Magnus 2011; Godman and Papineau 2020; Godman 2021).
2. Natural kinds are compatible with the notion of evolution, i.e., members of a species can change their properties over time (Kitcher 1984, p. 311; LaPorte 2004, pp. 9–10; Maxwell et al. 2020; Devitt 2023).
3. Species, considered as particulars, can also be subject to natural laws (Kitcher 1984, p. 311–313; Griffiths 1999, pp. 222-223). Still, another consideration is that natural kinds in biology are not governed by laws (Boyd 1999, p. 152); thus, species can be lawless kinds (LaPorte 2004, pp. 13–14).
4. Proper names are also used to denote natural kinds; for instance, descriptions based on essential properties ensure the recognition that they belong to the same species and not another (Reydon 2003, pp. 51-54). Hence, species names do not exclusively refer to their genealogical connections (LaPorte 2004, p. 13).
5. Additionally, it is important to note that the theory of natural kinds allows for relational properties (e.g., interbreeding, ecological niche, phylogenetic relationships, etc.)—not only intrinsic properties (e.g., morphological characters, genetic similarities) (Griffiths 1999; Okasha 2002; LaPorte 2004; Brigandt 2009; Tahko

2024)—as well as a combination of both intrinsic and relational properties (Dumsday 2012; Devitt 2018).

Regarding this, it is evident that philosophers who endorse the individuality perspective strongly criticize the notion of units of evolution as natural kinds, mainly because these entities (1) have defined spatiotemporal boundaries, (2) evolve, (3) are not governed by natural laws, and (4) possess proper names. In contrast, advocates of natural kinds challenge these assumptions by asserting that these characteristics are also properties of natural kinds (see Table 6.1).

Alternatively, one of the most widely accepted approaches to the theory of natural kinds today is the proposal known as "Homeostatic Property Cluster Kinds" or simply the HPC account. This view was initially developed by Robert Boyd (1991, 1999, 2021) and has been strongly supported by other philosophers (e.g., Wilson 1999; Keller et al. 2003; Wilson et al. 2007; Craver 2009; Bird 2018). Unlike the traditional perspective of natural kinds, which uses a set of necessary and sufficient essential properties to identify members of a kind, the HPC kind thesis assumes that these properties are not random but result from causal mechanisms. For example, members of a kind (e.g., species) share homeostatic property clusters—such as common selective pressure, gene flow, reproductive isolation, developmental constraints, etc.—that collectively produce stable properties, facilitating the ability to perceive generalizations. In addition, the reality of natural kinds is grounded in the accommodation thesis, which asserts that reliable inductive practices align with causal factors in the world, thereby establishing the relevant properties of entities (see Boyd 1999). From this perspective, the kinds seem more real than arbitrary. As Wilson (1999, p. 200) puts it: "there are natural kinds in the world individuated by properties existing independent of us and that our schemes of categorization in science track these natural kinds."

Table 6.1 *Comparison of assumptions favoring either individuality or natural kinds*

In support of	Individuality	Natural kinds
Assumption 1	Individuals possess well-defined spatiotemporal boundaries, and natural kinds do not.	Natural kinds can be historically delimited.
Assumption 2	Individuals evolve, and natural kinds do not.	Natural kinds evolve.
Assumption 3	Individuals are not governed by laws as natural kinds are.	In biology, natural kinds are not subject to laws in the same way as in other disciplines.
Assumption 4	Individuals have proper names, while natural kinds do not.	Proper names are also used to describe natural kinds.

Note that assumptions about individuality challenge the view of natural kinds. Nevertheless, some philosophers who advocate for natural kinds contest this negative perspective. Furthermore, with deeper exploration, it can be concluded that both metaphysical categories are compatible

Many biologists, unfamiliar with the notion of "type,"[2] sought ontological alternatives for species. Both the individuality thesis and the HPC explanation, for example, have attempted to circumvent the implicit essentialism in the concept of natural kinds concerning species, arguing that this notion is rooted in a context often regarded as irrational. However, as discussed in Chap. 2, both the type concept and "species fixity" have played pivotal roles in the development of the Natural System and, hence, in the recognition of evolutionary entities. In light of this, I will briefly defend the classic notion of natural kinds in biology, considering that the essentialism story is mistaken and that the traditional characteristics of natural kinds remain valid within the framework of the H-Evo-Devo Theory. In addition, I will claim that the concept of natural kinds is metaphysically compatible with the thesis of individuality. Indeed, this endeavor aims to support both ontological approaches in identifying units of evolution through specific causal mechanisms.

6.1.2 Natural Kinds as Evolving Entities: The Lazarus Effect on Essentialism

Following the philosophical debates on the ontological status of species, essentialism asserts that groups of organisms are defined by their essential properties. These characteristics are immutable and eternal, derived from the abstraction of form. This implies that members can be added to or disappear within the kind, allowing entities to occupy these spaces at any time, as long as they possess those specific properties. According to some philosophers and historians of the late twentieth century, essentialism was typical of the pre-Darwinian era, where the notion of a species was developed through typological thinking that supported species fixity. Accordingly, most biologists took for granted that essentialism was an early biological tenet that was inconsistent with the current theory of evolution. As Sober (1980, p. 353) commented: "Essentialism about species is today a dead issue, not because there is no conceivable way to defend it, but because the way in which it was defended by biologists was thoroughly discredited."

Notably, essentialism is closely linked to the notion of natural kind (Tahko 2024). Traditionally, natural kinds are defined by their essential properties, considering that the essence of an object is what makes it one thing and not another. These essences are immutable and eternal; they are elements that all members of a kind must possess. This is Plato's and Aristotle's supposed metaphysical view concerning natural kinds. But, what does an object's essence hinge upon? In his distinction between nominal essence and real essence, John Locke suggested that the only access to an object's ontology was through the sensory recognition of the kind (nominal essence), e.g., through abstraction. However, he believed that such an object should possess some microstructure (real essence) that could account for this macroscopic perception. Although Locke was skeptical about discovering the real

[2] The "type" suggests that closely related groups of organisms are abstractly recognized by a highly integrated common pattern; see Chap. 8, Sect. 8.1.2.

essence, Kripke (1980) and Putnam (1975) asserted that these underlying microstructures can be discovered through scientific advances. For example, the nominal essence of gold is its being a "shiny, *malleable*, yellow metal," while its real essence—i.e., its underlying microscopic structure—is "having an atomic number of 79." Human senses enable the classification of objects of the same kind, but the cause of these perceptions lies in an internal element whose nature—and hence its reality—must be discovered through scientific methods, suggesting the idea behind Plato's famous phrase of "carving nature at its joints." This perspective on natural kinds in biology is generally applied to the notion of species. A species is conceived through the generalization of its morphological and physiological traits, which are assumed to be essential properties of the kind. Thus, to determine the ontology of species, it is necessary to address the following traditional challenges regarding species essentialism:

1. *Species should not be defined solely in terms of essential intrinsic properties.* According to Okasha (2002), Locke's theory of essentialism does not allow relational properties to define the real essence of objects. However, most modern species concepts are based on relational properties (e.g., interbreeding, ecological niches, phylogenetic relationships, etc.) rather than intrinsic properties (morphological traits). For example, under the Biological Species Concept, the property is "being able to interbreed with one group and no other"; under the Ecological Species Concept, the property is "occupying a particular ecological niche"; while under the Phylogenetic Species Concept, the property is "being a particular segment of the genealogical nexus," and so on (p. 201). Hence, Okasha argues that, according to the modern concepts of species, essential properties are relational rather than intrinsic. In fact, the HPC thesis is not entirely committed to strict essentialism, as it permits the properties of a kind to be relational as well (Boyd 1999; Brigandt 2009).
2. *Species are not defined by a set of similar traits since they are polymorphic.* According to Ereshefsky and Matthen (2005), defining natural kinds through morphological similarities overlooks the significant variability found within species. For example, sexual dimorphism—a widespread phenomenon—results in distinct morphological differences between males and females. Additionally, juvenile forms often differ greatly from adults throughout their life cycles. This variability is also evident in colonial species, such as ants and termites. Hull (1978) clearly noted this issue in his argument for supporting species as individuals: "Given the sort of variability characteristic of biological species, no one specimen could possibly be 'typical' in even a statistical sense… [T]hey are part of the same genealogical nexus, no matter how much they might differ phenotypically" (pp. 351–352). In this context, in addressing the issue of essentialism, the HPC account is open to accepting species polymorphisms because they are explained by specific underlying mechanisms (Wilson et al. 2007; Rieppel 2008).
3. *Species are not natural kinds because they have vague boundaries.* According to essentialism, natural kinds are discrete entities; nevertheless, if species evolve gradually, then species are not natural kinds. Considering this, Sober (1980)

states that biologists encounter issues with essentialism, as it cannot accurately explain the nature of variation. For example, from the essentialism point of view, variation is deemed unreal and represents a departure from the species' "natural state." In contrast, population genetics regards variation as real, resulting from processes that lead to statistical changes across the population. In order to salvage essentialism, Sober suggests that it can accept vague essences. Moreover, the HPC thesis offers a nuanced perspective on natural kinds, considering that speciation involves a continuous process (Rieppel 2007, 2009).

So, the challenge of recognizing species as natural kinds revolves around two main points. First, traditional biologists define species based on relational properties, not intrinsic properties (Okasha 2002). Second, they associate the philosophy of natural kinds with "essentialism" (Devitt 2023; Brigandt 2024). The issue at stake is that modern evolutionary theory is supposedly incompatible with the essentialism rooted in early taxonomic practices that identified taxa by their intrinsic properties (e.g., the identification of types). For example, the HPC account emerged as a response to traditional essentialism implicit in the philosophy of natural kinds. The title of Wilson et al.'s work (2007) illustrates that movement quite well: "When Traditional Essentialism Fails." For this reason, the HPC view embraces the relational properties of species. Now, while the emergence of a "new essentialism" (Pavlinov 2023) has been acknowledged, many proposals have steered clear of confining essential characteristics exclusively to intrinsic properties (e.g., Godman and Papineau 2020; Godman 2021). Likewise, some authors allude to moderate essentialism, which accepts the role of many relational properties in species identification even though it emphasizes the importance of intrinsic properties over relational ones (e.g., Dumsday 2012; Maxwell et al. 2020). Interestingly, other researchers have correctly connected developmental dynamics to the essential intrinsic properties of kinds (Boulter 2012); however, they have resorted to Aristotelian philosophy to reconcile essentialism with evolution (e.g., Walsh 2006; Austin 2019).

In any case, my defense of natural kinds aligns with that of Devitt (2008, 2010, 2021, 2023), who argues that species concepts based solely on relational properties fail to identify a species. Instead, it is the intrinsic properties that have proven effective in fulfilling this role. Regarding this, I do not aim to return to the early philosophical tradition of seeking real essences, as Reydon (2005, p. 77) states: "Although philosophical accounts commonly render the Kripke-Putnam theory as conceptualizing natural kinds in terms of microstructural essences, the explanatory core of the theory is formed by causal mechanisms rather than microstructural essences." In this regard, according to Reydon (2005), natural kinds can be interpreted as causal kinds, where generalizations among members of a kind are projected due to specific underlying causal mechanisms. In fact, this view is far from new (e.g., Sober 1980, p. 354). The HPC thesis, for example, identifies kinds through causal mechanisms. Considering this, the HPC account is metaphysically acceptable, as it connects inductive practices with the causal mechanisms responsible for these natural regularities (Boyd 1999).

The quest for an internal component enabling the identification of entities as natural kinds has been tackled from diverse perspectives. For instance, some philosophers have situated the underlying properties of a kind within a genetic context (e.g., Kitts and Kitts 1979; Devitt 2008, 2023), while others have placed them within a common evolutionary history, suggesting that members of a kind share the same genes, developmental pathways, morphological traits, and so on (Griffiths 1999; Wilson et al. 2007; Bird 2018; Godman and Papineau 2020; Godman 2021). According to Reydon (2005), these properties characterize historical kinds. Yet historical kinds alone do not explain the internal similarities among members. Therefore, a combination of both perspectives can provide insight into the causal processes that allow for their recognition (Devitt 2018) (see below).

Another aspect that requires clarification is the concept of "essentialism." It is important to note that this term is ambiguous and carries different connotations within the philosophical discourse of biology (Wilkins 2013a, b). Furthermore, the development of this concept, which forms the basis of the narrative known among historians as the "essentialism story," has notable inconsistencies (e.g., Winsor 2003, 2006; Amundson 2005; Wilkins 2009a, b, 2013b; Müller-Wille 2011; Witteveen 2020). If any line of thought represents a notion related to the ontological recognition of types within lineages (interpreted in philosophy as natural kinds), it should be termed "typologism" or "typology." However, as discussed in Chap. 2, typologism has positively influenced scientific progress in biology, aligning with the contemporary identification of units of evolution. This raises a significant challenge, as philosophers commonly associate the term "essentialism" with the theory of natural kind. For example, Brigandt (2024) argues that discussions about species as natural kinds have been influenced by essentialism, prompting questions about their metaphysical interpretation. Nonetheless, these ontological debates might unfold independently of the term "essentialism." As Devitt (2023, p. 3) points out: "It is the issue of in virtue of what an organism is a member of a certain Linnaean taxon; the issue of what makes an organism a member of that taxon; the issue of the very nature of the taxon."

It is therefore not coincidental that Brigandt (2021) has related typology and natural kinds as two concepts converging in the metaphysical discussion on species and higher taxa, even though these concepts appear to have emerged from different epistemological contexts. Both Brigandt (2021) and Rieppel (2006) are aware that these philosophical issues are somehow related to the notion of the "type." For example, under the essentialism story, philosophers interpreted the body plan as a natural kind characterized by essential and immutable properties. Conversely, Rieppel (2006) mentions that the type concept (*Bauplan*)—severely tarnished by the narrative of the Modern Synthesis—does not entail such irrational implications. Instead, its recognition relies on processes that maintain the internal organization of organisms, displaying a hierarchical structure:

> On that account, the "type," or "Bauplan" of comparative morphologists is not to be considered as an "*idealized, unchangeable abstraction... but... as a fundamental, structural, phylogenetic organization that is constantly being maintained and preserved because of the*

> *way that the ontogeny of organisms has become (is) organized*"[3]... The most extensive recent discussion of the hierarchy of "types," or "Baupläne," and its relevance for developmental biology, comparative morphology, paleontology, and phylogenetic systematics is that of Hall... All of these disciplines do not treat "types," or "Baupläne," as mere mental, ideal abstractions, but rather as an aspect of the hierarchically structured organismic diversity that requires, and deserves, a causal explanation. For that same reason, the hierarchy of "natural kinds" that is apparent in the hierarchy of life has explanatory force. (Rieppel 2006, p. 533; his emphasis)

Although the essentialism story should no longer be upheld—since it diminishes the significance of recognizing the type in evolutionary biology—it is possible to distinguish natural kinds as causal kinds in Reydon's sense, as stressed above. Thus, both homologues and species/higher taxa "become 'natural kinds' that have explanatory value in developmental biology as much as in phylogenetics" (Rieppel 2006, p. 533). Indeed, the reason why species/higher taxa can be identified as "types"—and hence as natural kinds—is based on an ontological causal mechanism; it "is the causal processes that govern the development of the tetrapod limb bud, which are constrained in such a way that the basic structural correspondences are maintained through all variations in shape and form" (p. 533). But what exactly is the causal (homeostatic) mechanism that allows for perceiving these projections regarding the theory of natural kinds? Of course, these mechanisms are the developmental constraints that—as described in the previous chapter—operate by producing a hierarchical structure in the multicellular organic kingdom.

Certainly, both Rieppel (2006) and Brigandt (2021) identify this mechanism as the potential cause for the recognition of the type. For example, Rieppel (2006) states: "The 'sameness' that marks out 'types' is one of structural correspondence (topology and connectivity), which is believed to be causally rooted in ontogeny (*developmental constraints*)." (p. 528; my emphasis). Meanwhile, Brigandt (2021) writes: "From an evo-devo perspective, however, homologous developmental processes and overall body plans are governed by *developmental constraints*, and thus have an important impact and shaping influence on future evolutionary trajectories" (p. 486; his emphasis).

Therefore, the aforementioned criteria for the issues surrounding the classic concept of natural kinds can now be challenged with the following assumptions (see Table 6.2 to compare the three metaphysical options regarding natural kinds):

(1) *Species should be defined solely in terms of their intrinsic properties*

Despite Devitt's (2008, 2010, 2021, 2023) accurate observation that no relational property contributes to identifying a species, he resorts to genetic properties as the intrinsic element that characterizes the members of a kind. In response, Godman and Papineau (2020) highlight the weakness of Devitt's argument by mentioning biological counterarguments on inheritance, which leads them to emphasize historical properties as more relevant than intrinsic properties.

[3] Hall 1992, p. 73, as cited by Rieppel.

Table 6.2 *Comparison of three metaphysical perspectives to determine whether species qualify as natural kinds*

Species are:	not natural kinds	HPC kinds	classic natural kinds
Intrinsic properties	Species are defined in terms of relational properties rather than intrinsic properties.	Species are defined in terms of both relational and intrinsic properties.	Species are defined solely in terms of intrinsic properties.
Developmental constraints that allow a wide range of variants	Species are not defined by a set of similar traits since they are polymorphic.	Species are polymorphic due to regularities of underlying mechanisms.	Species are defined by developmental constraints, which allow for a wide range of variants.
Clear-cut boundaries	Species have vague boundaries.	Species have vague boundaries.	Species have clear-cut boundaries.

Critics of the view of species as natural kinds argue that species are defined by (1) relational properties, (2) polymorphism, and (3) vague boundaries. They claim that classic natural kinds are presumed to hold those opposing assumptions. In this context, the HPC account attempts to soften the traditional perspective enough to allow species to be considered natural kinds despite the new criteria for defining them. However, such a traditional view of natural kinds does not conflict with Evo-Devo advancements, given that units of evolution are defined by (1) intrinsic properties, (2) developmental constraints that allow for a wide range of variants, and (3) clear-cut boundaries. Thus, the H-Evo-Devo Theory is consistent with the classical vision of natural kinds

However, more recently, by adhering to Aristotelian metaphysics, Austin (2019) supported the idea that intrinsic properties are the most important for determining the membership of the kind. Unlike Devitt, Austin is right, in my opinion, in identifying these properties as the capacities of developmental processes to shape a specific morphological profile that embryologically delimits organisms to recognize them as members of the same kind. Interestingly, Austin identified modules (homologous) as suitable examples of natural kinds but excluded species (or other taxonomic categories) from this classification. From an ontological standpoint, his rejection of taxonomy stems from the current taxonomy's reliance on identifying units through relational properties rather than intrinsic ones. Nevertheless, the H-Evo-Devo Theory proposes a different interpretation of taxonomy. According to this view, species (and higher taxa) are considered natural kinds whose intrinsic properties are defined by developmental constraints. This mechanism represents the causal process by which the "type" can be perceived as a natural entity. This perspective does not negate the importance of relational aspects, as evolutionary entities can also be recognized as historical kinds. Yet these relational properties gain significance upon acknowledging that closely related groups share similar developmental processes (Griffiths 1999).

(2) *Species are not defined by a set of similar traits; instead, they are defined by developmental constraints, which allow for a wide range of variants*

One of the criticisms leveled against polymorphic species is described by Magnus (2011) as "similarity fetishism": members of a kind do not necessarily have to be similar in all their properties; rather, the relevant factor is that they must be united by the same underlying causal mechanism. Drawing on Aristotelian metaphysical arguments, Walsh (2006) and Austin (2019) contend that the generative properties of the developmental process can lead to various phenotypic outcomes (triggered by various factors). Despite this morphological flexibility, these entities develop within certain specific limits.

Indeed, morphologists recognize that types exhibit a broad range of variations, including age differences, sexual dimorphism, and the diverse forms they present throughout their life cycles. According to the H-Evo-Devo Theory, species can be polymorphic, yet they maintain stability in their variations due to developmental constraints. In brief, what truly characterizes natural kinds is that they share developmental constraints, which ensures the preservation of their wide range of variants within the limits of their type.

(3) *Species are natural kinds since they have clear-cut boundaries*

As previously mentioned, to recognize entities as natural kinds, it is necessary to consider that the members of a kind must have clear boundaries and be distinctly separated from members of other kinds. The challenge arises because many biologists assume species evolve gradually. However, as Kitts and Kitts noted in their analysis of species as natural kinds:

> What is highly relevant… is not the historical continuity of diversity, but the development of diversity through the origin of discontinuity. If biologists can, even in principle, explain the origin and maintenance of discontinuous organic diversity in terms of the properties common to all the organisms of a species, then the origin of species must be seen as the origin of kinds. (Kitts and Kitts 1979, p. 621)

Similarly, Sober (1980, p. 358-359; 2000, pp. 150-151) asserts that theories of speciation involving discontinuity align with an "essentialist" perspective on kinds. Notably, the H-Evo-Devo Theory supports saltational evolution,[4] thus it is consistent with the classical perspective of natural kinds.

Finally, it is important to emphasize that the study of natural kinds has a respectable scientific foundation, as it allows for distinguishing natural categories through inductive and explanatory practices (Boyd 1999). In essence, while the concept of natural kinds is shaped by scientific theories and practices, it is also a step toward perceiving a natural distinction in the world (Keller et al. 2003; Bird 2018). In short, natural kinds are entities that exist in nature (Bird 2018), and it is in this metaphysical sense that units of evolution can be identified. But this does not imply that evolving entities are natural kinds instead of individuals. As will be discussed further, the perspective of entities as individuals is compatible with the notion of natural kinds.

[4] See Chap. 9, Sect. 9.2; Chap. 11, Sect. 11.2.1; and Chap. 12, Sect. 12.1.1.

6.1.3 Natural Kinds and Individuals as Metaphysical Compatible Entities

As mentioned earlier, the thesis of species individuality implies that species are discrete entities, cohesive wholes, and spatiotemporally localizable. Originally, the idea that species behave as individuals emerged as an alternative to viewing species as natural kinds. However, some philosophers argue that species can be both individuals and natural kinds simultaneously. This perspective has been considered even by those who support either the individuality thesis or the natural kinds' view. For example, David Hull (1976, p. 190), a prominent advocate of individuality, found this suggestion plausible: "Perhaps species might be viewed more profitably as belonging to some hybrid category such as a 'complex particular' or 'individualized class.'" Moreover, in the HPC account, Robert Boyd (1999, p. 141) maintains that: "those scientists who are convinced that species are individuals must conclude that they are natural kinds as well." But in what sense can these ontological categories be reconciled?

According to Brigandt (2009), both categories are metaphysically compatible and the pragmatic choice between one or the other depends on the epistemological context in which it is preferable. For example, concerning the metaphysics of species, when wishing to emphasize their origin, stability, and extinction, they function better as individuals. Conversely, when aiming to highlight the specific properties they have in common, they can be better interpreted as natural kinds. In addition, for Rieppel (2007), the traditional view of natural kinds ran afoul of the perspective of species as evolutionary individuals—i.e., natural kinds were seen as fixed entities, without historical restrictions and with clear-cut boundaries. Nevertheless, under the HPC thesis, such differences disappear because it is a weaker version of the natural kind hypothesis. As a result, reconciling both perspectives becomes possible. According to Rieppel (2009, 2013), species are spatiotemporally located complex wholes—i.e., individuals—whose parts are composed of intrinsic and relational properties, which together instantiate a homeostatic property cluster (HPC) natural kind.

The problem with these assumptions is that species have been defined very differently across various biological research programs. If this were the case, there would be distinct entities that could fall into different ontological interpretations (Reydon 2003, 2009). For instance, according to some authors, species function as individuals in certain contexts, while they behave as natural kinds in others (Dupré 1993, p. 43; Reydon 2003; LaPorte 2004, p. 16). Likewise, Baum (1998) and Reydon (2021) assert that different kinds of individuals can be recognized, e.g., entities may be individuals because they have the capacity to interbreed or because they share a genealogical nexus, etc. Additionally, individuals do not necessarily have to be concrete entities but processes (Dupré 2022). If this is so, in what kind of individuality does an entity fit as a natural kind? In this regard, Reydon (2009) asserts that philosophers seeking to unify both categories are not truly committed to a metaphysical synthesis; rather, they suggest that such compatibility occurs only in an epistemological sense.

Interestingly, Casetta and Vecchi (2021) argue that the individuality perspective and the natural kind view are compatible in a metaphysical sense. They propose that such compatibility depends on identifying specific biological processes that operate on organisms rather than on a conceptual translation of the two perspectives: "[C]ausally speaking the two theses both make reference to specific causal relations and interactions between organisms in terms of their participation in causal processes" (p. 3017). For example, in the context of individuality, an individual is a whole composed of parts, each part causally related to others by virtue of its participation in specific causal processes. Whereas, in the HPC account, members of a kind are causally linked to other members because they participate in specific homeostatic mechanisms:

> *We here argue that, if these causal processes render the species individual-like, then they also render it HPC-kind-like and* vice versa. On the one hand, the various parts of a species are kept together tightly by means of their causal relations and through their participation in specific biological processes that make the whole cohesive, integrated and individual-like. On the other hand, the cluster of co-occurring genetic and phenotypic properties partially shared by the members of the species-as-kind is kept together as a homeostatic cluster through the action of *the very same* specific biological processes. Ultimately, the same processes cause a species to be both individual-like and kind-like, hence, that entity is both an individual and a kind. (Casetta and Vecchi 2021, p. 3018; their emphasis)

But what exactly are these biological processes that make it possible to unify both categories metaphysically? According to these authors, there are developmental and genetic homeostatic mechanisms that cause phenotypic and genotypic heterogeneity in species. This uniformity produces either cohesion in the case of individuality or homeostatic stability in the case of HPC. They also noted that these processes could vary between species. For example, some species exhibit "developmental and genetic homeostasis," while others display "genetic drift, directional selection and adaptive introgression;" that is, "different species may be individual-like and kind-like in virtue of different processes" (p. 3018). Similarly, Rieppel (2009, pp. 41–42) claims that species possess specific homeostatic mechanisms that "maintain the causal integration of the processual system that is a species." Speciation occurs due to the relaxation of these homeostatic mechanisms, leading to the reintegration of a new system and consequently, the emergence of new homeostatic mechanisms. According to his argument, the key to distinguishing such homeostatic mechanisms is stasis. The mechanisms that explain stasis include "stabilizing selection, as well as genetic and developmental constraints…that may themselves be the result of stabilizing selection" among others (p. 42).[5] According to Brigandt (2009), species have diverse properties, such as common ancestry, gene flow, and ecological selection. These characteristics are relevant for the cohesion of the species: "An HPC approach views these features as the homeostatic mechanisms determining

[5] According to the H-Evo-Devo Theory, the mechanisms responsible for stasis are developmental constraints in a negative sense. See Chap. 11, Sects. 11.1.2 and 11.2.2; and Chap. 12, Sects. 12.1.2 and 12.2.2. See also below.

membership in a taxon-kind, while SAI [species as individuals] construes such features as determining the unity of the taxon-individual across space and time" (p. 86).

Notably, Nathan and Borghini (2014) argue that certain developmental processes are relevant to identifying entities as both individuals and natural kinds. In this sense, Casetta and Vecchi are correct in stating that it is by identifying the same causal processes (homeostatic mechanisms) that both ontological categories can be bound. In my view, these processes are described as developmental constraints that, as discussed in Chap. 5, are established through a procedural hierarchy. According to the conclusions of the previous section, developmental constraints are not only homeostatic mechanisms that make it possible to perceive projections for the recognition of the type (natural kind) but also the processes that collectively maintain the parts of a species as a whole (individuals). Interestingly, most proposals for this synthetic view invoke the HPC kind to avoid the essentialist approach of natural kinds. However, as argued in the previous section, there should be no hesitation to define natural kinds in the classic sense. Theoretical elements suggest that entities recognized as natural kinds possess relevant intrinsic properties, clear-cut boundaries, and even spatiotemporal recognition. This approach provides a more precise reconciliation, given that the challenges observed in natural kinds sometimes mirror those in individuality (e.g., the contrast between gradual evolution and the sudden origin of individuals). Nonetheless, it is important to establish criteria that allow for accurately identifying true evolutionary individuals (see the next section).

Moreover, identifying developmental constraints as causal mechanisms—which contribute both to projecting the type for natural kinds and to cohesion for individuality—applies not only to species but can also extend to other units. As mentioned in the previous chapter, developmental constraints operate upward and downward to construct the entire hierarchical structure for recognizing evolutionary entities. In other words, identifying this mechanism leads to considering homologues, higher taxa, species, and subspecies as units of evolution. In addition, this recognition is metaphysically supported, as these entities behave simultaneously as both natural kinds and evolutionary individuals. This dual characterization makes it possible to grasp their reality, as Brigandt (2009, p. 86) argued: "Both… agree in endorsing some sort of realism about taxa, at least species taxa. Natural kinds are assemblies of objects that are grouped according to properties that actually exist in nature, so that the boundary and unity of a natural kind are not conventional. An individual likewise has a non-arbitrary boundary and distinctive unity across time."

6.2 Criteria for Individuality

In order to metaphysically recognize the units of evolution, I introduce a set of criteria based on the hypothesis of individuality. This does not imply disregarding the consideration of natural kinds; on the contrary, as mentioned in the previous section, both categories are metaphysically compatible. In any case, describing entities as evolutionary individuals serves not only to distinguish these evolutionary entities spatiotemporally but also to identify their characteristic evolutionary patterns.

Nevertheless, before describing this set of criteria, I will clarify the theory of evolutionary individuality.

As discussed in the previous section, the thesis of individuality emerged as an alternative to treating species as natural kinds. Hence, many criteria for recognizing species as individuals opposed this ontological category (e.g., see Ghiselin 1997). However, problems started to emerge because different criteria revealed the existence of various kinds of individuals (Baum 1998; Reydon 2021). Additionally, Ereshefsky (2001, p. 112) argues that those who advocate for the view of species as evolutionary individuals actually mean that species are historical entities; in such contexts, the term "individual" becomes ambiguous.

Regardless, according to Hull (1976), the organism as an individual is a paradigmatic example of individuality: Individual organisms have proper names, are discrete entities that develop continuously in specific space and time, and so on. Once born, an individual organism persists until it dies and never appears again. Taking this into consideration, Hull posed the question: Can these same properties be applied to entities other than species? Alternatively, Mishler and Brandon (1987) assert that individuals must satisfy at least four fundamental conditions: (1) spatial boundaries, (2) temporal boundaries, (3) interactions among the parts of the entity, and (4) cohesion (where the parts of an entity uniformly respond to specific biological processes). But among these proposals, Gould's (2002) approach to individuality stands out as the most comprehensive. Interestingly, he not only suggested that these criteria could be applied to species but also to other entities in the hierarchy. Despite this proposal, there are significant issues with Gould's model that deserve consideration.

Gould (2002, pp. 602-613) outlines two categories of criteria for individuality: criteria for vernacular individuality and criteria for evolutionary individuality. The first category refers to an individual having "a discrete and definable beginning, or birth," "an equally discrete and definable ending, or death," and "sufficient stability (defined as the coherence of substance and constancy of form) during its lifetime to merit continuous recognition as the same 'thing'" (p. 602). For Gould, these criteria provide necessary conditions for identifying an entity as an individual, but they are insufficient from an evolutionary viewpoint. So, according to the second category, an evolutionary individual must also have (a) *reproduction*, individuals must be capable of producing offspring; (b) *inheritance*, which traces genealogical patterns and assesses the reproductive success of ancestors; (c) *variation*, differences among individuals; and (d) *interaction*, individuals interact with the environment and achieve reproductive success through selection. However, it is important to note that Gould applies these criteria at different levels, albeit following traditional processes (see Boucher 2017).

Although these criteria are used to identify species as units of evolution, they pose significant challenges when applied to levels other than species. For example, at the subspecies level, no recognized entities exist because "they may not persist long enough to matter in evolution, and their borders may be too 'leaky' as organisms move in and out in the absence of reproductive isolation between the parts (organisms) of different demes [populations]" (Gould 2002, p. 603). While, at the higher taxa level, neither is possible since "their parts (species) may live in such complete independence, and in such different ecologies, on distant continents," thus, they fail to meet the criterion of interaction (p. 712). Nonetheless, as noted in Chap. 4, it is not necessary to adhere strictly to a "Darwinian" perspective to support a hierarchy theory of evolution (Gontier 2010). Nor should there be a sole reliance on Darwinian individuals to identify units of evolution, as Boucher (2017) argues: "Darwinian individuality is not the same thing as metaphysical [vernacular] individuality, as plenty of things that are metaphysical individuals are not Darwinian individuals" (p. 26). Instead, any metaphysical approach must consider the typical evolutionary modes and patterns at each level: origin, stability, and extinction.

Thus, according to the H-Evo-Devo Theory, to identify units of evolution through criteria for individuality, three fundamental conditions must be recognized:

1. *Birth, discrete and definable.* Individuals must have a concrete beginning, with no fuzzy boundaries at their origin.
2. *Cohesion.* The parts of individuals behave as a coherent whole, enabling them to be distinguished spatiotemporally.
3. *Death, discrete and definable.* Individuals must disappear concretely, with no ambiguous boundaries at their extinction.

These three criteria meet the requirements proposed by various authors. For example, they allow for recognizing entities spatiotemporally and incorporate "cohesion" as a relevant component for metaphysical recognition. In addition, these criteria align with the description of the kind of entities described in Chap. 5: A developmental system maintains a specific position within the morphospace; however, it can give rise to another system occupying a different position within the morphospace (birth). Once this new system remains in a steady state, it can be recognized through time.[6] Indeed, these developmental systems exhibit a hierarchical arrangement, starting with homologues, whose variations, manifested as homogenies (character states), collectively shape the different ranks of the Linnaean hierarchy: higher taxa, species, and subspecies. In addition, these criteria describe the modes of the units of evolution that can be theoretically and empirically supported through biological processes[7] (see Table 6.3: (1) *Birth*, (2) *Cohesion*, and (3) *Death*). As mentioned earlier, given that applying criteria for individuality in a Darwinian sense is unnecessary. In this regard, I suggest the following conditions:

4. *Reproduction.* Entities can produce other entities of the same level, and, in some cases, they can also produce entities of different levels.

For example, a subspecies can give rise to another subspecies. A species can give rise to either another species, a subspecies, or a higher taxon. A higher taxon can

[6] See Chap. 5, Sect. 5.1.2.

[7] For a review of the patterns and modes of evolution according to the H-Evo-Devo Theory, see Chap. 9. For an interpretation of these modes through biological processes, see Chaps. 11 and 12.

Table 6.3 *Criteria for individuality according to the H-Evo-Devo Theory.*

Criteria	Description	Homologues	Higher taxa	Species/Genera	Subspecies
1. Birth	Individuals must have a concrete beginning, with no fuzzy boundaries at their origin.	Morphological novelty	Supraspeciation	Speciation	Subspeciation
2. Cohesion	The parts of individuals act as a coherent whole, enabling them to be distinguished spatiotemporally.	Morphological stability	Suprastasis	Specific stasis	Substasis
3. Death	Individuals must disappear concretely, with no ambiguous boundaries at their extinction.	Morphological loss	Supraextinction	Specific extinction	Subextinction
4. Reproduction	Individuals can produce other entities of the same level, and, in some cases, they can also produce entities of different levels.	M → M	HT → HT, SP	SP → HT, SP, Sub	Sub → Sub
5. Variation	Variation as an emergent property: Individuals have parts that can vary, yet they cannot lose their identity.	Forelimbs of tetrapods: of the order Primates of the family Felidae of the infraorder Cetacea of the order Chiroptera	Felidae: *Acinonyx* *Caracal* *Catopuma* *Felis* *Leopardus* *Leptailurus* *Lynx* *Prionailurus* *Puma* *Panthera*	*Panthera leo*: *P. l. azandica* *P. l. krugeri* *P. l. nubica* *P. l. persica*	*Panthera leo persica*: Intra-subspecific variation
6. Phylogeny	Individuals have a phylogenetic history that connects them both to sibling entities and to ancestral entities.	Feathers, Hair Reptilian Scales Scales of a primitive aquatic vertebrate	Canidae, Ursidae Amphicyonidae Miacidae	*Homo*, *Paranthropus* *Australopithecus* *Ardipithecus*	*G. c. peralta*, *G. c. rothschildi* *G. c. reticulata* *G. c. tippelskirchi*

(1) Birth, (2) Cohesion, and (3) Death. These first three criteria correspond to the evolutionary modes (patterns) observed in the units of evolution (for detailed information on this proposal, see Chap. 9; for insights into the biological processes and mechanisms supporting these modes, see Chaps. 11 and 12). (4) *Reproduction*. The symbols accompanying each unit indicate *M* module, *HT* higher taxa, *SP* species, *Sub* subspecies. **Homologues**. A module gives rise to another module. **Higher taxa**. A higher taxon gives rise to either another higher taxon or a species. **Species/Genera**. A species gives rise to either another species, a subspecies, or a higher taxon. **Subspecies**. A subspecies gives rise to another subspecies. (5) *Variation*. Individuals consist of parts that collectively represent variation as an emergent property. While the parts may vary, the identity of the entity remains. (6) *Phylogeny*. Examples of phylogenetic trees based on this criterion of individuality. The lines represent the connection between ancestors and offspring, not the gradual change from one taxon and another as assumed in the traditional view. In this framework, ancestral individuals do not disappear to give rise to offspring individuals; instead, ancestral individuals can coexist with offspring individuals during the same period. In subspecies, the complete names of taxa are *Giraffa camelopardalis tippelskirchi*, *Giraffa camelopardalis reticulata*, *Giraffa camelopardalis peralta*, and *Giraffa camelopardalis rothschildi*. For further information, see the text

give rise to either another higher taxon or a species. A homologue can give rise to another homologue (see Table 6.3: (4) *Reproduction*).

5. *Variation as an emergent property*. Entities possess variation as an emergent property. Although their parts may vary, entities cannot lose their identity.

The issue here is not in considering an individual as a "Darwinian variant" (an entity subject to selection),[8] but in recognizing variation as an emergent property of evolutionary individuals. For example, subspecies are composed of polymorphic forms as variants; species are composed of subspecies as variants, and a higher taxon is composed of species as variants. Even when considering homologues as units of evolution, examples of homology can be understood as a reflection of their variation as an emergent property—e.g., forelimbs of tetrapods correspond to arms/hands in humans, claws in lions, fins in dolphins, wings in bats, and so on. Therefore, variation is understood as the constitution of an individual assembled from varying parts. While these parts can vary, an individual maintains its identity through cohesion (see Table 6.3: (5) *Variation*).

6. *Phylogeny*. Entities have a phylogenetic history that connects them both to sibling entities and to ancestral entities.

For example, hypothetically, if the intricate branches of humans' phylogenetic tree could be accessed, *Homo sapiens* would be encountered alongside other individuals like *Homo neanderthalensis* as the sister taxon. Further back, *Homo heidelbergensis* is found as the ancestral node of both, and *Homo antecessor* as the ancestral individual of this latter taxon, and so forth.[9] Moving to another level, *Homo* is observed alongside other individuals such as *Paranthropus*, and *Australopithecus* as the ancestral node for both, and so on. While this example mainly pertains to higher taxa, these genera could also be interpreted as species.[10] However, this proposal applies equally to higher taxonomic levels (e.g., families, orders, classes, etc.). Similarly, analogous criteria can be applied to homologues, and numerous studies on serial homology could prove invaluable here. For example, reptilian scales originated from an epidermal structure present in an aquatic vertebrate. Feathers and mammalian hair, in turn, evolved from reptilian scales, and mammalian hair eventually gave rise to mammary glands (see Table 6.3: (6) *Phylogeny*).[11]

---

[8] See Chap. 4, Sect. 4.1.3.

[9] It is important to clarify that this scenario is speculative. It is plausible that, before *Homo heidelbergensis*, there might be another undiscovered ancestral individual in the fossil record that connects *Homo sapiens* and *Homo neanderthalensis* as the sibling groups.

[10] See Chap. 7, Sect. 7.2.

[11] See Chap. 12, Sect. 12.2.1. Specifically, consult Figure 1 in DiFrisco et al. 2023, where the authors illustrate a character tree of epidermal appendages.

Finally, what is particularly interesting about this perspective is that ancestral individuals do not necessarily go extinct when giving rise to offspring individuals. For instance, my parents did not vanish the moment I was born. Divergence, as a principle of individuality, requires a discrete origin rather than continuous anagenetic change.[12]

In summary, units of evolution are discrete entities identifiable through specific evolutionary patterns. In other words, they are distinguished metaphysically by criteria of evolutionary individuality: They must possess a discrete and definable birth, cohesion, and a definable end. Importantly, these criteria correspond to patterns and modes of evolution perceived across different levels. Additionally, other significant criteria, such as reproduction, variation, and phylogeny, serve to identify the units of evolution. Table 6.3 summarizes these requirements.

6.3 Conclusions

Following the H-Evo-Devo Theory, the metaphysical recognition of units of evolution relies on considering two ontological categories that have been widely discussed by philosophers of biology: natural kinds and individuals. With respect to natural kinds, the membership of a kind is established by relevant intrinsic properties. According to the H-Evo-Devo Theory, these properties are developmental constraints that project the perception of a stable body plan, i.e., the type. However, under this framework, the natural kind perspective does not contradict the stance of the individuality thesis; both categories are metaphysically compatible, as they are based on the same causal process. Developmental constraints serve as the mechanism that explains how an underlying body plan defines kind membership (natural kinds) and why an entity can be recognized as the same thing spatiotemporally, maintaining its parts cohesively as a whole (individuals). The fact that these entities are simultaneously both natural kinds and individuals makes their metaphysical reality accessible. Finally, providing complementary metaphysical support for recognizing the units of evolution requires establishing criteria for individuality: birth, cohesion, death, reproduction, variation, and phylogeny. These criteria facilitate the identification of specific evolutionary patterns and modes. Therefore, according to the H-Evo-Devo Theory, homologues, higher taxa, species, and subspecies are entities that truly exist in nature.

References

Amundson R (2005) The changing role of the embryo in evolutionary thought: roots of evo-devo. Cambridge University Press, Cambridge

Austin CJ (2019) Essence in the age of evolution a new theory of natural kinds. Routledge, New York

[12] Indeed, as I argue in Chap. 9, gradualism as anagenesis is not a genuine mode in nature.

Baum DA (1998) Individuality and the existence of species through time. Syst Biol 47:641–653
Bird A (2018) The metaphysics of natural kinds. Synthese 195:1397–1426
Boyd R (1991) Realism, anti-foundationalism and the enthusiasm for natural kinds. Philos Stud 61:127–148
Boyd R (1999) Homeostasis, species, and higher taxa. In: Wilson RA (ed) Species: new interdisciplinary essays. MIT Press, Cambridge MA, pp 141–185
Boyd R (2021) Rethinking natural kinds, reference and truth: towards more correspondence with reality, not less. Synthese 198:2863–2903
Boucher SC (2017) Gould on species, metaphysics and macroevolution: a critical appraisal. Stud Hist Phil Biol Biomed Sci 62:25–34
Boulter SJ (2012) Can evolutionary biology do without Aristotelian essentialism? R Inst Philos Suppl 70:83–103
Brigandt I (2009) Natural kinds in evolution and systematics: metaphysical and epistemological considerations. Acta Biotheor 57:77–97
Brigandt I (2021) Typology and natural kinds in evo-devo. In: Nuño de la Rosa L, Müller GB (eds) Evolutionary developmental biology. Springer, Cham, pp 483–493
Brigandt I (2024) Biological species. In: Koslicki K, Raven MJ (eds) The Routledge handbook of essence in philosophy. Routledge, New York, pp 276–290
Burbrink FT, Crother BI, Murray CM, Smith BT, Ruane S, Myers EA, Pyron RA (2022) Empirical and philosophical problems with the subspecies rank. Ecol Evol 12:e9069. https://doi.org/10.1002/ece3.9069
Casetta E, Vecchi D (2021) Species are, at the same time, kinds and individuals: a causal argument based on an empirical approach to species identity. Synthese 198:3007–3025
Coleman KA, Wiley EO (2001) On species individualism: a new defense of the species-as-individuals hypothesis. Philos Sci 68:498–517
Craver CF (2009) Mechanisms and natural kinds. Philos Psychol 22:575–594
Devitt M (2008) Resurrecting biological essentialism. Philos Sci 75:344–382
Devitt M (2010) Species have (partly) intrinsic essences. Philos Sci 77:648–661
Devitt M (2018) Historical biological essentialism. Stud Hist Phil Biol Biomed Sci 71:1–7
Devitt M (2021) Defending intrinsic biological essentialism. Philos Sci 88:67–82
Devitt M (2023) Biological essentialism. Oxford University Press, Oxford
DiFrisco J, Love AC, Wagner GP (2023) The hierarchical basis of serial homology and evolutionary novelty. J Morphol 284:e21531. https://doi.org/10.1002/jmor.21531
Dumsday T (2012) A new argument for intrinsic biological essentialism. Philos Q 62:486–504
Dupré J (1993) The disorder of things: metaphysical foundations of the disunity of science. Harvard University Press, Cambridge MA
Dupré J (2022) (Some) species are processes. In: Wilkins J, Zachos FE, Pavlinov IY (eds) Species problems and beyond: contemporary issues in philosophy and practice. CRC Press, pp 279–292
Ereshefsky M (2001) The poverty of the Linnaean hierarchy: a philosophical study of biological taxonomy. Cambridge University Press, Cambridge
Ereshefsky M, Matthen M (2005) Taxonomy, polymorphism, and history: an introduction to population structure theory. Philos Sci 72:1–21
Ghiselin MT (1966) On psychologism in the logic of taxonomic controversies. Syst Zool 15:207–215
Ghiselin MT (1974) A radical solution to the species problem. Syst Zool 23:536–544
Ghiselin MT (1988) The individuality thesis, essences, and laws of nature. Biol Philos 3:467–474
Ghiselin MT (1997) Metaphysics and the origin of species. State University of New York Press, Albany, New York
Godman M (2021) Scientific realism with historical essences: the case of species. Synthese 198:3041–3057
Godman M, Papineau D (2020) Species have historical not intrinsic essences. In: Bianchi A (ed) Language and reality from a naturalistic perspective, Philosophical studies series 142. Springer, Cham. https://doi.org/10.1007/978-3-030-47641-0_18

Gontier N (2010) Evolutionary epistemology as a scientific method: a new look upon the units and levels of evolution debate. Theory Biosci 129:167–182
Gould SJ (2002) The structure of evolutionary theory. The Belknap Press of Harvard University Press, Cambridge, MA
Griffiths PE (1999) Squaring the circle: natural kinds with historical essences. In: Wilson RA (ed) Species: new interdisciplinary essays. MIT Press, Cambridge MA, pp 209–228
Hall BK (1992) Evolutionary developmental biology. Chapman and Hall, London
Hull DL (1976) Are species really individuals? Syst Zool 25:174–191
Hull DL (1978) A matter of individuality. Philos Sci 45:335–360
Keller RA, Boyd RN, Wheeler QD (2003) The illogical basis of phylogenetic nomenclature. Bot Rev 69:93–110
Kitcher P (1984) Species. Philos Sci 51:308–333
Kitts DB, Kitts DJ (1979) Biological species as natural kinds. Philos Sci 46:613–622
Kripke SA (1980) Naming and necessity. Harvard University Press, Cambridge, MA
Kunz W (2012) Do species exist?: principles of taxonomic classification. Wiley-Blackwell, Weinheim
LaPorte J (2004) Natural kinds and conceptual change. Cambridge University Press, Cambridge
Magnus PD (2011) Drakes, seadevils, and similarity fetishism. Biol Philos 26:857–870
Maxwell SJ, Congdon BC, Rymer TL (2020) Essentialistic pluralism: the theory of spatio-temporal positioning of species using integrated taxonomy. Proc R Soc Qld 124:81–97
Millikan RG (1999) Historical kinds and the "special sciences". Philos Stud 95:45–65
Mishler BD, Brandon RN (1987) Individuality, pluralism, and the phylogenetic species concept. Biol Philos 2:397–414
Müller-Wille S (2011) Making sense of essentialism. Crit Q 53:61–77
Nathan MJ, Borghini A (2014) Development and natural kinds: some lessons from biology. Synthese 191:539–556
Okasha S (2002) Darwinian metaphysics: species and the question of essentialism. Synthese 131:191–213
Pavlinov IY (2023) The species problem a conceptual history. CRC Press, Boca Raton
Putnam H (1975) Mind, language and reality, Philosophical papers, vol 2. Cambridge University Press, Cambridge
Reydon TAC (2003) Species are individuals–or are they? Philos Sci 70:49–56
Reydon TAC (2005) Species as units of generalization in biological science: a philosophical analysis. Ph. D. thesis. Leiden University, Faculty of Science
Reydon TAC (2009) Species and kinds: a critique of Rieppel's "one of a kind" account of species. Cladistics 25:660–667
Reydon TAC (2021) On radical solutions in the philosophy of biology: what does "individuals thinking" actually solve? Synthese 198:3389–3411
Reydon TAC (2022) 'Species' as a technical term multiple meanings in practice, one idea in theory. In: Wilkins J, Zachos FE, Pavlinov IY (eds) Species problems and beyond: contemporary issues in philosophy and practice. CRC Press, pp 65–89
Rieppel O (2006) 'Type' in morphology and phylogeny. J Morphol 267:528–535
Rieppel O (2007) Species: kinds of individuals or individuals of a kind. Cladistics 23:373–384
Rieppel O (2008) Origins, taxa, names and meanings. Cladistics 24:598–610
Rieppel O (2009) Species as a process. Acta Biotheor 57:33–49
Rieppel O (2013) Biological individuals and natural kinds. Bio Theory 7:162–169
Richards RA (2010) The species problem: a philosophical analysis. Cambridge University Press, Cambridge
Rosenberg A (1985) The structure of biological science. Cambridge University Press, Cambridge
Sober E (1980) Evolution, population thinking, and essentialism. Philos Sci 47:350–383
Sober E (2000) Philosophy of biology. Westview Press, Colorado
Tahko TE (2024) Natural kind essentialism. In: Koslicki K, Raven MJ (eds) The Routledge handbook of essence in philosophy. Routledge, New York, pp 156–168
Walsh D (2006) Evolutionary essentialism. Br J Philos Sci 57:425–448

Wilkins JS (2009a) Defining species: a sourcebook from antiquity to today. Peter Lang

Wilkins JS (2009b) Species: a history of the idea. University of California Press, Berkeley, Los Angeles, London

Wilkins JS (2013a) Biological essentialism and the tidal change of natural kinds. Sci Educ 22:221–240

Wilkins JS (2013b) Essentialism in biology. In: Kampourakis K (ed) The philosophy of biology: a companion for educators. Springer, Dordrecht, pp 395–419

Wilson RA (1999) Realism, essence, and kind: resuscitating species essentialism? In: Wilson RA (ed) Species: new interdisciplinary essays. MIT Press, Cambridge MA, pp 187–207

Wilson RA, Barker MJ, Brigandt I (2007) When traditional essentialism fails: biological natural kinds. Philos Top 35:189–215

Winsor MP (2003) Non-essentialist methods in pre-Darwinian taxonomy. Biol Philos 18:387–400

Winsor MP (2006) The creation of the essentialism story: an exercise in metahistory. Hist Philos Life Sci 28:149–174

Witteveen J (2020) Linnaeus, the essentialism story, and the question of types. Taxon 69:1141–1149

Zachos FE (2016) Species concepts in biology: historical development, theoretical foundations and practical relevance. Springer, Switzerland

Units of Evolution (I): Subspecies and Species

7

Abstract

This chapter introduces a new theoretical framework that supports the recognition of subspecies and species as units of evolution. It begins by exploring the challenges of recognizing subspecies and species as evolving entities in their traditional sense. For example, under this view, defining and delimiting subspecies is a challenging endeavor, while species are associated with numerous concepts, making it difficult to reconcile them with a general notion of "species." Nevertheless, the H-Evo-Devo Theory offers a framework that grants subspecies and species a distinct ontological reality. According to this model, subspecies are natural groups that collectively share a common body plan but are separated from each other by small morphological gaps in simple characters as well as by the fact that they live in different environments. Subspecies do not represent the early stages of the origin of species; rather, they are groups that diversify to adapt the species to different environments, thus protecting species from extinction. Meanwhile, species are discrete entities that exhibit the first step in morphological gaps in complex characters. More importantly, they are distinguished from other taxa by being the least inclusive body plan within a hierarchy of body plans. This perspective aims to provide the long-awaited general notion of species, offering a definitive resolution to the species problem that has perplexed evolutionary biologists for years.

Keywords

H-Evo-Devo Theory · Units of evolution · Subspecies · Species · Taxonomy · Intraspecific variation · Parapatric populations · Allopatric populations · Species problem · Biological Species Concept · Phylogenetic Species Concept · Species monism · Species pluralism · Typology · Developmental constraints · Saltational evolution · Speciation · Microevolution · Macroevolution

C. Ochoa, *Hierarchical Evolutionary-Developmental Theory*, Evolutionary Biology – New Perspectives on Its Development 9,
https://doi.org/10.1007/978-3-031-88723-9_7

Most biologists take for granted that subspecies and species play a significant role in the theory of evolution; however, defining and recognizing them poses many challenges. These issues stem from the widespread acceptance of traditional models of evolution and taxonomy, namely the Modern Synthesis and phylogenetic systematics. In contrast, the H-Evo-Devo Theory offers a more consistent framework for recognizing subspecies and species as true units of evolution.

Section 7.1 advocates for including subspecies as units of evolution. It begins by describing subspecies according to the traditional view, under which they are not considered real entities because of the ambiguity of their identification. Consecutively, it delves into the H-Evo-Devo Theory, which suggests that subspecies are real entities with active roles in evolution. According to this view, subspecies are groups that share a common body plan but are separated from each other by small morphological gaps in simple characters as well as by the fact that they live in different environments. Section 7.2 advocates for species as units of evolution. It begins by outlining the problems associated with the two most popular concepts of species: the Biological Species Concept and the Phylogenetic Species Concept. Subsequently, it explores the possibility of developing a general concept of species. Finally, it presents the H-Evo-Devo Theory perspective, which contends that species are discrete entities recognized as the least inclusive body plan within a hierarchy of body plans.

7.1 Subspecies as Units of Evolution

The overarching approach discussed here concerns patterns of variation at the lower scale of evolution. This involves recognizing continuous and slightly discontinuous variations within populations below the species level, which are related in one way or another to different geographically separable varieties within the species. Such variations can manifest in various forms: continuous (clines), transitional, discontinuous (categorical), or mosaic patterns. The issue of taxonomy comes into play here in terms of how phenomena of intraspecific variation are classified. More importantly, the question arises of how these phenomena are interpreted to better understand the origin and stability of biological diversity at the lower level of the hierarchy.

The term "subspecies" is traditionally defined as those subgroups that comprise a species, more generally, as geographical variation within a species (Mayr 1982). However, contemporary discussions around the term "subspecies" highlight several properties of this category:

1. They are not reproductively isolated sufficiently to be attributed to the species category (O'Brien and Mayr 1991; James 2010; Remsen 2010; Taylor et al. 2017).
2. They are divided by genetic differences (Taylor et al. 2017).
3. They are distinguished from each other by one or more heritable phenotypic traits (Patten 2010, 2015; Remsen 2010; Wallin et al. 2017).

4. They represent geographically isolated lineages with specific local adaptations (Thorpe 1987; Braby et al. 2012; Taylor et al. 2017).
5. They are also distinguished by behavioral, ecological, and other differences (Braby et al. 2012).

Nevertheless, although some biologists accept the notion of subspecies as a fundamental concept to understand the evolutionary process in its traditional sense (Mayr 1982; Patten 2010; Braby et al. 2012; Vinarski 2015), many argue that the boundaries between subspecies are vague and their recognition as a taxonomic category is arbitrary (Wilson and Brown 1953; Gillham 1956; Cracraft 1983, p. 171; Fitzpatrick 2010; Jorgensen et al. 2012; Smith 2016, p. 4; Padial and De la Riva 2021; Reydon and Kunz 2021; Burbrink et al. 2022; Prates et al. 2023; DeSalle and Tattersall 2025). For example, the following points have been mentioned:

(a) Subspecies lack taxonomic significance in considering the speciation process (Burbrink et al. 2000; Zink 2004; Jorgensen et al. 2012; Prates et al. 2023; Burbrink et al. 2022). Since divergence occurs continuously, any cut-off within this continuous differentiation is arbitrary (Skalski et al. 2008; Winker 2010; Padial and De la Riva 2021).
(b) Many taxonomists diagnose subspecies based on minor differences, such as the color of a character (Thorpe 1987; Burbrink et al. 2000; Patten 2015).
(c) The diagnosis of subspecies often relies on heritable traits, but boundaries can be obscured by factors like hybridization and phenotypic plasticity (Huang et al. 2000).
(d) Subspecies' boundaries are often related more to practical concerns (e.g., conservation programs) than to theoretical considerations (Zink 2004; Haig et al. 2006; Fitzpatrick 2010; Taylor et al. 2017; Berrilli et al. 2024).
(e) No species concepts adequately capture subspecies definition (Winker 2010, p. 17); consequently, like species, multiple notions of subspecies exist, making it difficult to identify them through a general concept (Reydon and Kunz 2021).
(f) Many subspecies diagnosed historically and currently have been identified using imprecise methods and various criteria (Fitzpatrick 2010; Schiaffini 2020; Padial and De la Riva 2021; Prates et al. 2023; Farquhar et al. 2024).

Certainly, the traditional interpretation of evolution—which assumes that species arise gradually, generally through the geographic isolation of peripheral populations—not only complicates the application of the term "subspecies" but also hinders a proper understanding of intraspecific variation and its role in evolution. However, the H-Evo-Devo Theory offers a different perspective on variation patterns below the species level. Here, the concept of subspecies is used to designate the groups that share the least inclusive body plan (species) but are distinguished from each other by small morphological gaps in simple characters—such as color, pattern, size, proportion, etc.—as well as by the fact that they live in different environments. From this point of view, subspecies are not incipient species or models representing the origin of species through gradual and continuous differentiation;

instead, they are groups that diversify to adapt the species to different environments, thus protecting species from extinction. In the following pages, I will examine the problem of subspecies according to the traditional view in more detail. Subsequently, under the H-Evo-Devo Theory, I will argue in favor of subspecies as real units of evolution.

7.1.1 Subspecies Under the Traditional View

Taxonomists use the notion of subspecies to describe the patterns of geographic variation. Even so, for most evolutionists, subspecies represent not only intraspecific variation but also early stages of divergence that may lead to the origin of species. This is why subspecies are referred to as "incipient species," i.e., there is no clear distinction between the categories of species and subspecies at certain stages of divergence. Descimon and Mallet (2009, p. 221) also refer to these entities as "bad species" because they "seem at that stage of divergence at which species fusion is doubtful." In short, most evolutionary biologists argue that this state of ambiguity is part of the evolutionary process, as it reflects the continuous differentiation of biological units. Populations go through numerous stages of separation, initially being recognized as subspecies until they eventually reach the species level.[1]

But are subspecies real in this context? When considering patterns of geographic variation, biologists identify two types of populations related to the notion of subspecies. The first pertains to populations exhibiting a continuous geographic range (clinal variation) or "parapatric subspecies." The second refers to populations that are geographically isolated from each other or "allopatric subspecies." The key difference between these two types lies in the fact that parapatric subspecies have a continuous separation with a contact zone where individuals interbreed, while allopatric subspecies are completely geographically separated from each other despite their potential ability to interbreed (Braby et al. 2012, p. 700).

In a classic paper, Wilson and Brown (1953) addressed the issues of parapatric subspecies in their specimens. They stated that classifying groups as subspecies based on their members' geographical distribution is very challenging. When there is free interbreeding between two or more populations that are geographically close to each other, the ambiguity in recognizing some populations as separate groups becomes more evident. They noted that "the only way to resolve this situation taxonomically is to establish an arbitrary lower limit above which populations will be formally recognized as subspecies" (p. 103). In other words, confusion arises from the difficulty of delimiting two populations when both are integrated in a clinal manner (see Fig. 7.1a). Accordingly, most authors argue against considering parapatric populations as real entities due to the vague divisions in continuous intraspecific variation (Thorpe 1987; Frost and Hillis 1990; Zink 2004; James 2010; Patten 2010; Remsen 2010; Braby et al. 2012; Hillis 2020; Farquhar et al. 2024).

[1] According to many authors, although subspecies are considered incipient species, this does not mean that all subspecies will necessarily become species.

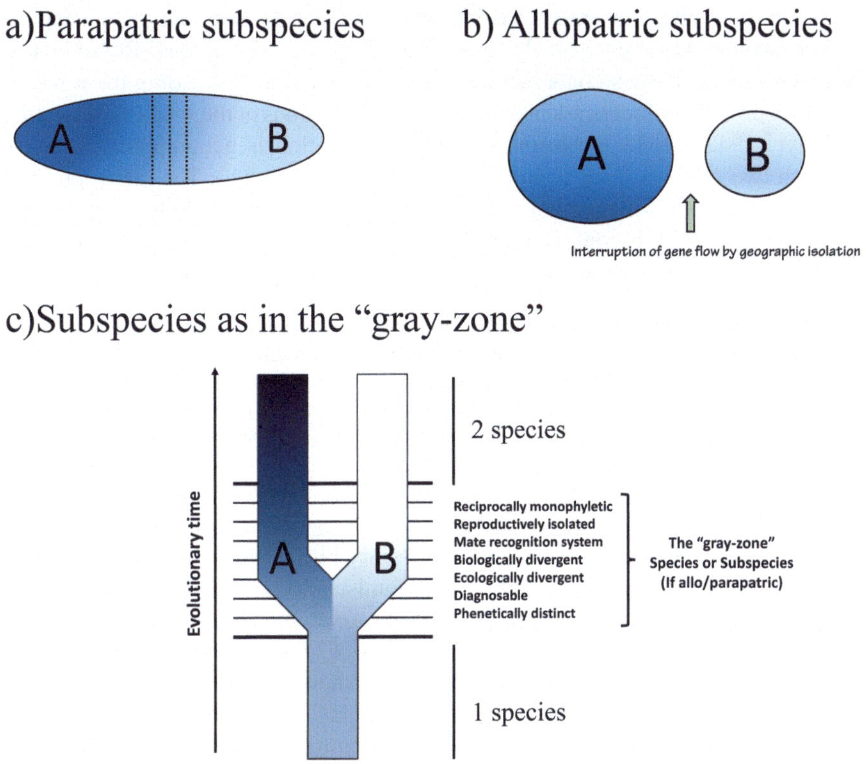

Fig. 7.1 *Different interpretations of subspecies according to the traditional view.* Populations "A" and "B" are considered potential subspecies under these perspectives. (**a**) Parapatric subspecies. "A" and "B" exhibit continuous gradations or a cline. The issue lies in establishing clear boundaries between the two underlying populations when none exist; thus, any split is arbitrary (depicted as intermittent lines). (**b**) Allopatric subspecies. "A" and "B" are geographically isolated. Because there is a lack of genetic exchange between them, they might also be classified as species. (**c**) Subspecies in the "gray zone" (according to Wallin et al. (2017); adapted from de Queiroz (2007), with permission from Oxford University Press). During speciation, "A" and "B" gradually acquire certain properties over time until they become fully differentiated species. These intermediate phases have been used to recognize them as subspecies. However, de Queiroz's original proposal considers subspecies and species as entities of the same kind. See the text for more details

On the other hand, many biologists believe that this phenomenon of continuity could provide insights into how species emerge over time, where the two extreme populations potentially represent two incipient species through parapatric speciation. For instance, Mallet (2013, p. 47) claims that many populations separated by an interbreeding contact zone exhibit significant morphological, behavioral, and even genetic differences, thereby qualifying as true subspecies. In order to standardize this differentiation, some biologists have established statistical methods to diagnose populations as subspecies (e.g., Amadon 1949; Patten and Unitt 2002; Skalski et al. 2008; Patten 2010; Patten and Remsen 2017). Still, because there is a continuum in characters, undoubtedly, the division remains arbitrary (Winker 2010).

Given these disagreements, O'Brien and Mayr (1991) state that the concept of subspecies should be exclusively applied to allopatric populations. Reproductive isolation ensures the genetic structure of these populations, providing them with a certain evolutionary independence from other subdivisions of the species. According to Mayr (1942, 1963), the primary step for speciation is geographical isolation between two or more sister populations (allopatric speciation): The interruption of gene flow between two populations by geographical barriers would foster the origin of new species.[2] In this context, reproductive isolation—by geographical barriers—is an important step to producing species. But, as Braby et al. note:

> [It] has been noted that delimiting species boundaries and discriminating between species or subspecies status for some geographically isolated populations is somewhat arbitrary because the populations are evolving through time… In other words, species may be viewed as an arbitrary cut-off along a branch in the tree of life, and allopatric subspecies are simply a point along the continuum of population differentiation comprising nested smaller interbreeding groups within species. (Braby et al. 2012, p. 708)

Keeping this in mind, it is no wonder that many geographically isolated subspecies have been classified as species by taxonomists (see Fig. 7.1b). For example, Amadon and Short (1976) highlight that: "[T]he taxonomist is faced with the task of evaluating numerous taxa that are still geographically isolated—are these species or subspecies?" (p. 161). Clarke and Okasha (2013, p. 63) illustrate this point with the giant panda, *Ailuropoda melanoleuca*, which is known for having two geographically isolated populations. One of these varieties has been recognized as a subspecies, *Ailuropoda melanoleuca qinlingensis*, distinguished from the other variety (the commonly known panda) by certain morphological features: a smaller skull, larger molars, and a coat of dark-brown and light-brown fur, unlike the more familiar black and white pattern. The subspecies rank has been assigned because of the potential for interbreeding and yielding fertile offspring between the two varieties. However, since this panda subspecies is morphologically distinct and geographically separated from the common pandas (with no observed interbreeding), Clarke and Okasha suggest that this population could hypothetically be ranked as a species rather than a subspecies.

Regarding this case, the conceptualization of subspecies also depends on the species concept being applied (Jorgensen et al. 2012; Torstrom et al. 2014). For some biologists, "subspecies" is a useful term within the Biological Species Concept (James 2010; Patten 2010, 2015). For example, as illustrated in the example above, subspecies can be distinguished from species because these groups can potentially interbreed; in contrast, groups that cannot interbreed would be classified as different species. Nonetheless, taxonomists generally do not conduct experiments to determine interbreeding capabilities among populations. Instead, many assume that

[2] While this divergence model is widely accepted by most biologists, I challenge the theoretical basis of Mayr's allopatric speciation in this work (see Chap. 10, Sect. 10.1). Additionally, the issues related to this model in the context of the species concept are discussed in detail in the next section of this chapter (Sect. 7.2).

reproductive isolation is somehow related to morphological divergence (see Burbrink et al. 2022). Additionally, when assuming that the key difference lies in the presence of gene flow between subspecies and that no gene flow exists between species, any allopatric population must be ranked as a species.[3]

But following the Phylogenetic Species Concept, the notion of subspecies is not straightforward either. Remsen (2010) maintains that since subspecies are the smallest diagnosable units within the phylogenetic tree—i.e., they are terminal taxa within the line of genealogical relationships—they can also be considered species under this view. Interestingly, most taxonomists who endorse the Phylogenetic Species Concept do not include subspecies in their hierarchy. Nevertheless, some authors argue that, although the definition of subspecies is traditionally based on the Biological Species Concept, the practices to delimit both subspecies and species boundaries are grounded in the phylogenetic criterion (Jorgensen et al. 2012; Tobias et al. 2010; Burbrink et al. 2022). In so doing, many taxonomists employ molecular methods to recognize "true species"—a diagnosis achieved when populations are genetically distinct to a sufficient extent—ensuring their true monophyletic history as well (Tautz et al. 2003; Torstrom et al. 2014). However, Mousseau and Sikes (2011) demonstrate that, in some cases, allopatric populations exhibit genetic distance even if they show no significant phenotypic differences among them.

In addressing the subspecies issue, Braby et al. (2012) and Wallin et al. (2017) endorse de Queiroz's Unified Species Concept.[4] According to this view, subspecies are situated in a gray zone, sharing some properties with species but lacking others that would be needed to be considered fully differentiated species (see Fig. 7.1c). The issue is that, for de Queiroz (2020), under the Unified Species Concept, both subspecies and species are entities of the same kind, namely, "population-level lineage." In other words, subspecies do not represent a taxonomic rank below the species level, as both subspecies and species are segments of a lineage.[5] Therefore, "subspecies are incompletely separated lineages within a more inclusive lineage" (p. 460). In response, Hillis (2021) states that by not specifying the rank, many populations within the species could be considered as species, leading to taxonomic

[3] See also Sect. 7.2.1.

[4] The Unified Species Concept, also known as the General Lineage Species Concept, was proposed by de Queiroz as a solution to the species problem. According to de Queiroz (2005, 2007), this concept lies in recognizing common elements among the different species concepts. For him, the primary property of "species" is their identification as "separately evolving lineage segments." The secondary properties are those emphasized by various species concepts (e.g., ecological differences, reproductive isolation, reciprocal monophyly, a unique combination of characters, and so on). Under this conception, de Queiroz claims that these properties are acquired at different times during the speciation process. In other words, when an ancestral lineage splits into two, the distinctive properties of the species gradually emerge until two lineages are recognized as phenotypically distinguishable.

[5] In a previous work, de Queiroz (2005) stated that the species category is not a taxonomic rank; for example, a species is an entity defined by certain properties, which can comprise other species with different properties, etc. Notably, this stance implies the absence of a true hierarchical structure.

confusion.[6] Accordingly, subspecies would need to be defined as incompletely separated lineages within the species. Nevertheless, Prates et al. (2023) state that in taxonomic practice, it is challenging to diagnose whether lineages are completely or incompletely separated lineages. Hence, the subspecies problem persists, even with this new alternative.

The concept of subspecies appears to depend on how intraspecific variation patterns are interpreted and which evolutionary divergence models are favored. Given that these interpretations stem from the traditional view, I agree with Fitzpatrick (2010) that efforts to categorize multiple processes—spanning various configurations, stages, and degrees of divergence—under a single category are unfruitful, particularly when the differentiation degree depends on the distinct patterns of geographic variation. Moreover, the notion of gradual evolution at lower levels complicates the identification of subspecies as units of evolution. This alleged gradual evolution, as proposed by the traditional view, creates ambiguity in both kinds of populations. In parapatric populations—represented by clines—it becomes challenging to establish a clear-cut distinction between two underlying populations. Whereas in allopatric populations, the boundaries between subspecies and species remain vague, as these populations gradually lose their potential for interbreeding. Despite these challenges, the H-Evo-Devo Theory proposes an alternative perspective, which not only supports the idea of subspecies as units of evolution but also offers an interpretation that explains the patterns of geographical variation at the lower levels of evolution.

7.1.2 Subspecies According to the H-Evo-Devo Theory

According to the H-Evo-Devo Theory, *subspecies are groups that share the species body plan but are separated from each other by small morphological gaps in simple characters such as color, pattern, size, and proportion, as well as by the fact that they live in different environments*. Unlike the traditional view, the H-Evo-Devo Theory does not view subspecies as incipient species nor support traditional geographical models of speciation (e.g., allopatric speciation). Instead, it suggests that these divisions merely reflect the phenomenon of diversification and adaptation of the species to different environments, thus protecting species from extinction.

It is worth noting that the term "subspecies" carries hierarchical implications, being one of the ranks in the Linnaean classification system. However, from a traditional hierarchical perspective, the phylogenetic tree is seen as a genealogical hierarchy that shows continuous evolution through the transmission of genetic information over time. If the phylogenetic tree reflects the transmission and fixation of genes within a continuum, then the challenges of conducting accurate taxonomic analyses become apparent, especially when attempting to identify discrete entities (Franz 2005). For example, while many biologists have taken for granted that subspecies boundaries can be delineated by genetic distances (e.g., Braby et al. 2012;

[6] See also de Queiroz's response (2021).

Smith et al. 2014; Taylor et al. 2017), inconsistencies have been noted when comparing molecular data to previous identifications based on morphological criteria (Torstrom et al. 2014; Zink 2004; Phillimore and Owens 2006; Prates et al. 2023; Farquhar et al. 2024). Some argue that genetic differences alone are insufficient to diagnose subspecies (Tautz et al. 2003; Patten 2010; Braby et al. 2012; Torstrom et al. 2014; Vinarski 2015). For example, critics, such as Rosel et al. (2017), challenge such molecular methods, highlighting the overreliance on mitochondrial DNA sequences and the neglect of nuclear DNA. Likewise, Wilson (2011) argues that relying solely on molecular data for classifying organisms would result in incomprehensible phylogenies, as traditional taxonomy involves a more conventional classification. Franz (2005) also points out the problematic nature of molecular-based phylogenies, particularly in assigning discrete names: "New tools to infer molecular phylogenies are reaching beyond the limits for verbal diagnoses of taxa… This option is not favored by the Linnaean system where names are defined by individual types plus a diagnosis" (p. 497).

So, many biologists assume that the separation among subspecies is primarily determined by heritable characters (Braby et al. 2012; Patten 2010, 2015; Remsen 2010; Wallin et al. 2017). Conversely, several studies have attempted to revise previous taxonomic classifications, highlighting that many traits, initially used to diagnose certain groups as subspecies, emerged not solely by direct heredity but also by phenotypic plasticity (e.g., Joyce et al. 2009; Vane-Wright and Tennent 2011; Gruber et al. 2013; Barrientos-Villalobos et al. 2018; Jeratthitikul et al. 2019). These observations have raised concerns about the reliance on molecular methods alone for diagnosing subspecies (Patten and Remsen 2017). It is important to note that this approach, which suggests that evolutionary change is solely dependent on genetic variation, should not be the exclusive basis for defining natural group boundaries across any taxonomic scale.

Nevertheless, the H-Evo-Devo Theory proposes an alternative approach to understanding the phenomenon of intraspecific variation, particularly in entities referred to as subspecies. As discussed in Chap. 5, the ranks of the Linnaean system are considered real insofar as they reflect distinct steady states in character states at different scales. In the case of subspecies, these character states manifest as simple characters, such as color, pattern, size, proportion, etc. Subspecies are identified because their character states are steadily maintained at specific positions within the morphospace. In simpler terms, the subspecies recognition results from developmental constraints operating at small scales, particularly in stabilizing simple character states over time.

This assumption underlying the identification of small morphological gaps—which differentiate these groups—is closely tied to distinct geographical environments in which they reside (Skalski et al. 2008; Patten 2010; Winker 2010). For example, according to Winker (2010), populations exhibiting clinal variation in one or more characters should be ranked as "true subspecies" only if they exhibit gaps in other characters, such as "step clines." However, in cases of strict clines or when populations are interconnected by a continuous gradient, any attempt to establish a cut-off among populations implies an ambiguous division. Taxonomy requires discrete entities; thus, such divisions are not considered appropriate for ranking, i.e., they do not represent natural entities. Instead, clines or polymorphic forms separated by continuous gradations reflect variation within subspecies as an emergent property,[7] which is likely due to phenotypic plasticity as reaction norms.

Of course, this does not mean that phenotypic plasticity precludes the recognition of subspecies, as proposed in the traditional model. According to the H-Evo-Devo Theory, genetic distance is not a criterion for diagnosing subspecies; instead, subspecies should be diagnosed by their morphological gaps, which result from discontinuous changes in ancestral characters. Therefore, a subspecies is identified by the presence of certain stable characters, which result from developmental constraints acting upon them. Now, consider that developmental constraints encompass not only cellular and molecular interactions but also external inputs such as physical forces and environmental factors (see Chap. 5, Fig. 5.4). For example, a developmental system allows for environmental inputs that result in either small continuous variations as reaction norms or discontinuous variations as polyphenisms. Indeed, in this sense, phenotypic plasticity confers advantages to subspecies by preventing extinction through reaction norms and facilitating evolutionary divergence through polyphenisms.[8] In short, I find it reasonable to diagnose subspecies based on polyphenisms because these entities only reflect how the species adapt to various environments, irrespective of whether these characters are induced by environmental stimuli.

But what support exists for the idea that subspecies are units of evolution according to the H-Evo-Devo Theory? I will use chimpanzee evolution as an example to clarify this point. Generally, two species of chimpanzees are recognized: *Pan troglodytes* (common chimpanzee) and *Pan paniscus* (bonobo or pygmy chimpanzee). *Pan troglodytes* is found throughout equatorial Africa and, according to genetic and geographic analysis (Gonder et al. 2011), comprises four subspecies: *Pan t. verus* (western chimpanzee), *Pan t. ellioti* (Nigeria-Cameroon chimpanzee), *Pan t. troglodytes* (central chimpanzee), and *Pan t. schweinfurthii* (eastern chimpanzee). According to some authors, because differences in morphological traits and behavior among the subspecies of *Pan troglodytes* are minimal and even their genetic distinctness is analogous to the differences between human populations, the division between subspecies is arbitrary (e.g., Shea and Coolidge 1988; Fischer et al. 2006; Fünfstück et al. 2015). Likewise, other authors point out that *Pan t. verus* should be classified as a different species due to its genetic distinctness and geographical isolation from other subspecies (e.g., Morin et al. 1994).

Despite differing opinions on the common chimpanzee, most biologists agree on a distinct separation between the pygmy chimpanzee (*Pan paniscus*) and the common chimpanzee (*Pan troglodytes*), considering genetic, morphological, and behavioral data (Shea and Coolidge 1988; Gonder et al. 2011; Hey 2010; Brand et al. 2022). Interestingly, there is a minority viewpoint suggesting that both chimpanzees

[7] Variation as an emergent property is one of the criteria presented in this book to identify evolutionary individuals, as discussed in Chap. 6, Sect. 6.2.

[8] See Chap. 11, Sect. 11.1.1.

should be classified as subspecies rather than species, given their morphological gaps (Schultz 1969), observed hybridization in captivity (Vervaecke and Van Elsacker 1992), and other evolutionary considerations (Horn 1979). Keeping this in mind, three possible scenarios can be considered:

1. *Subspecies within Pan troglodytes are real*; however, their diagnosis relies on the phylogenetic criterion provided by genetic data and geographical distribution. In this case, the morphological criterion plays a secondary role. Under this view, *Pan t. verus* is an example of an "incipient species" in the process of differentiation. *Pan paniscus* would have passed through this transitional stage and is now recognized as a well-differentiated species.
2. *Subspecies within Pan troglodytes are arbitrary*, as genetic and geographical data alone are insufficient to categorize them as distinct and separate groups; nevertheless, *Pan paniscus (bonobo) is still recognized as a distinct species*, making both species monotypic (without subspecies).
3. *Pan paniscus and Pan troglodytes are recognized as subspecies rather than species*. I will explain this scenario within the context of great apes' taxonomy. Figure 7.2 illustrates ten closely related hominid taxa. Each group comprises a pair that shares the same underlying type. So, the primary classification criterion is based on the observation that the different groups are separated by significant morphological gaps, allowing them to be classified into five types:
   - Type 1: *Australopithecus*
   - Type 2: *Homo*
   - Type 3: *Pan*
   - Type 4: *Gorilla*
   - Type 5: *Pongo*

It is important to note that these discontinuities are accompanied by phylogenetic relationships. Using the genus *Homo* as a reference point, to its right, *Homo* is more closely related to the genus *Pan*, an extant group. To its left, it is even more closely related to the genus *Australopithecus*, despite being extinct. An evolutionary hypothesis suggests that *Australopithecus* may be the ancestral genus of *Homo*. Now, the pair of taxa within each type is also separated by small morphological gaps, though not large enough to constitute a distinct type. This separation indicates variation within the type. Thus, according to the H-Evo-Devo Theory, the types represented as genera can also be species, as species represent the least inclusive type within a hierarchy of types, while variations within the type represent subspecies.

Taking this into account, the proposal that chimpanzees are subspecies rather than species is supported by the absence of morphological gaps below both *Pan paniscus* and *Pan troglodytes* (assuming that commonly identified varieties within each taxon are arbitrary) despite their separation dating back one million years. Notably, there are small morphological gaps between the two chimpanzee taxa. But at a higher taxonomic level, the genus *Pan* exhibits significant morphological gaps

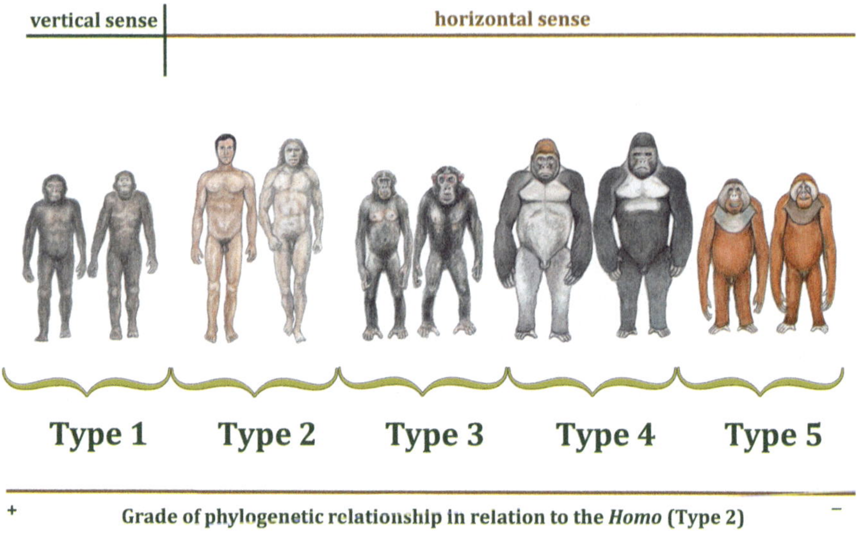

Fig. 7.2 *Species/genera boundaries delineated by significant morphological gaps in hominids.* The upper line represents the distinction between extinct (left) and extant (right) taxa, corresponding to the vertical and horizontal senses, respectively. The lower line illustrates the decreasing degree of phylogenetic relationship with *Homo* as one moves from left to right. This image illustrates the recognition of the first step in significant morphological gaps, serving as the primary diagnosis criterion to identify the least inclusive type within a hierarchy of types. Note that each underlying type comprises a pair of inclusive taxa. According to the H-Evo-Devo Theory, types are species, and the taxa within each type are subspecies. The figure shows adult male hominids (from right to left, *Australopithecus* afarensis, *Australopithecus africanus*, *Homo sapiens*, *Homo neanderthalensis*, *Pan paniscus*, *Pan troglodytes*, *Gorilla gorilla*, *Gorilla beringei*, *Pongo pygmeus*, *Pongo tapanuilensis*). Type 1: *Australopithecus*. Type 2: *Homo*. Type 3: *Pan*. Type 4: *Gorilla*. Type 5: *Pongo*. (Illustrations drawn by Aldi De Oyarzabal Salcedo)

compared to other closely related apes.[9] Thus, despite the distinctions between the common chimpanzee and the pygmy chimpanzee, a common type can be recognized between them.

This example aims to demonstrate that the issue of subspecies is most evident in scenarios one and two, which rely on traditional interpretation. The groups referred to as subspecies of *Pan troglodytes* are defined by ambiguous criteria, complicating the category's use when considering the scale of variation. However, in scenario three, if bonobos and common chimpanzees are recognized as subspecies rather than separate species, then there is an argument in favor of subspecies as entities that truly exist in nature. These entities are identified in nature because they exhibit small morphological gaps that separate them. But the reality of subspecies is further

[9]While the fossil record provides some insight into the taxonomic structure, unfortunately, the chimpanzee fossil record is quite sparse (McBrearty and Jablonski 2005). Yet, the understanding of the way the genus *Pan* evolved becomes clearer considering that closely related genera, such as *Ardipithecus*, exhibit morphological gaps in the fossil record.

supported by some ontological criteria. As discussed in Chap. 6, developmental constraints serve as the causal mechanism responsible for identifying units of evolution as both natural kinds and evolutionary individuals, thus justifying their reality. Below, I provide some brief clarifications regarding their ontological determination in this context.

Evolutionary Individuals As discussed in the previous chapter, individuals are identified by three main criteria: origin, stability, and extinction. Regarding origin, my primary goal is to advocate for an appropriate level on which Darwinian mechanisms can be applied.[10] These mechanisms can account for the origin of small morphological gaps in simple characters that do not correspond to the body plan.[11] The next step is the stability of these simple characters, which is explained through developmental constraints operating at the lowest hierarchical level, allowing for their spatiotemporal identification. Lastly, these units have the potential to disappear. Since these entities are fundamentally adaptive, their disappearance is caused by factors that alter both their local environment and their population structure.[12] Taking these patterns into consideration, if subspecies behave as evolutionary individuals, then these entities truly exist in nature.

Natural Kinds From the H-Evo-Devo Theory view, subspecies are not body plans; rather, they are variations observed within a body plan. While the previous chapter emphasized body plans as paradigmatic examples of natural kinds, subspecies also qualify as natural kinds despite not being body plans. This is because the causal mechanism (causal kind) that allows the generalization of simple characters is rooted in developmental constraints. This mechanism is precisely what makes it possible to differentiate their rank within the Linnaean hierarchy. Thus, when developmental constraints maintain complex characters or body plans, species or higher taxa can be identified. Conversely, when developmental constraints stabilize simple characters, subspecies can be identified. It is important to note that distinguishing groups as distinct entities relies on intrinsic properties (developmental constraints) rather than relational properties (e.g., interbreeding or genealogical connections). This distinction serves to identify them as natural kinds (see Chap. 6). Nonetheless, it is also valid to consider other additional criteria—such as ecological and behavioral differences—to further distinguish subspecies. In any case, if subspecies exemplify natural kinds, then they are real.

---

[10] When I refer to Darwinian mechanisms, I am not alluding to the traditional mechanisms provided by population genetics or Mayr's models of speciation. Instead, I am specifically referring to the dispersal model as discussed in Chap. 11, Sect. 11.1.1.

[11] According to the H-Evo-Devo Theory, complex characters—or those associated with the body plan—are generated by supraDarwinian mechanisms (hopeful monster model); see Chap. 11, Sect. 11.2.1 and Chap. 12, Sect. 12.1.1.

[12] For more details on these mechanisms, see Chap. 11, Sect. 11.1.

In summary, developmental constraints create the causal mechanism that allows for the recognition of a natural distinction among entities identified as subspecies, insofar as they function simultaneously as both natural kinds and evolutionary individuals. According to H-Evo-Devo Theory, subspecies represent natural groups that collectively share a common body plan (at the species level) but are distinguished from each other by small morphological gaps in simple characters—such as color, pattern, size, proportion, etc.—as well as by the fact that they live in distinct environments. Contrary to the common view, they do not represent models of the origin of species but rather groups that diversify and adapt species to different ecological circumstances, thereby avoiding the extinction of the species. These distinctive roles that subspecies play in evolution set the H-Evo-Devo Theory apart from other traditional evolutionary models. Therefore, subspecies are recognized as legitimate units of evolution.

7.2 Species as Units of Evolution

Currently, most biologists would not hesitate to assert that species play an active role in evolution. However, when asked what species are, these biologists will provide different definitions and even use various criteria to delimit them. This raises the question: Are there multiple biological entities corresponding to the term "species," or is there a single entity to which the general term "species" can be applied? Furthermore, if species obey a unique criterion as a taxonomic rank, how can the species category be distinguished from other taxonomic categories (e.g., subspecies, genera, families, etc.)? Biologists and philosophers acknowledge a problem regarding species that has both epistemological and practical implications. For most, this problem is primarily philosophical rather than empirical, involving the development of different species concepts (Pigliucci 2003; Reydon 2004; Stamos 2003; Richards 2010; Kunz 2012; Zachos 2016, 2022; Sigwart 2018; Pavlinov 2023; Wilkins 2023). As a result, researchers face the challenge of consolidating multiple concepts into a general framework where units of evolution can be recognized as species (Pavlinov 2023; Wilkins 2022, 2023). There is a consensus that this is an endless problem that remains one of the most relevant issues in the fields of evolution and taxonomy to date (Zachos 2016; Sigwart 2018; Pavlinov 2023; Wilkins 2023). As Wilkins (2023) recently stated: "The nature of species is far from settled even now."

Despite efforts to understand this issue from its historical foundations (Wilkins 2009; Pavlinov 2023), in my opinion, the conflict primarily lies in how evolution is explained. As Dupré (1993, p. 42; his italics) correctly indicates: "*[S]pecies* is a theoretical term, and the way to understand such a term is to look at the theory (or theories) in which it is embedded." Throughout this book, I argue that the theory of evolution needs a refresh of both the evolutionary processes and the identification of the units of evolution (taxonomy). Currently, evolutionary processes are generally explained based on the Modern Synthesis framework, while taxonomy has been interpreted not only from this framework but also from the perspective of phylogenetic systematics, founded by Willi Hennig. These traditional frameworks gave rise to the two most well-known concepts in this controversy: the Biological Species Concept (BSC) and the Phylogenetic Species Concept (PSC). Therefore, part of this section aims to discuss these concepts and highlight their shortcomings from different perspectives.

In light of this, since the H-Evo-Devo Theory proposes radical changes to both the structure of evolution and taxonomy, I present a radical solution to the species problem as well. My intention is not to propose adding a new concept of species to the list of previously failed concepts. On the contrary, my main goal is to establish a theoretical and practical foundation for the term "species" within this new model's framework. According to the H-Evo-Devo Theory, species are discrete entities that exhibit the first step in significant morphological gaps and are distinguished from other taxa by being the least inclusive body plan within a hierarchy of body plans. In this context, the species acquires a universal notion that can be applied generally across all multicellular groups. Thus, this work supports not only the monism of the species but also the monism of the taxonomic structure overall.

7.2.1 Species Concepts and the Species Problem

Although the term "species" is commonly used in biology, the issue is that biologists from various disciplines define it in different ways (Zachos 2016; Sigwart 2018; Pavlinov 2023; Wilkins 2023). Furthermore, analysis of the different species concepts reveals their incompatibility (Richards 2010; Pavlinov 2023). For instance, Mayden (1997) identifies at least 22 different species concepts, while Zachos (2016) found 34, each categorizing biodiversity under distinct criteria. According to Zachos (2016), all species concepts capture certain biological realities, whereas Kunz (2012) claims these concepts are merely human constructs.

I consider that the species problem arises from the radical changes in taxonomy and evolution that took place in the mid-twentieth century, influenced by the interpretations of the Modern Synthesis and phylogenetic systematics. Therefore, in this section, I will briefly discuss the two most prominent concepts from these traditional frameworks: the Biological Species Concept (BSC) and the Phylogenetic Species Concept (PSC).[13] Throughout this discussion, I will highlight their shortcomings and argue that, although these concepts initially set aside the morphological aspect, they ultimately rely on it to delineate species boundaries more precisely.

The Biological Species Concept (BSC) According to Mayr (2000), biological species are defined as "groups of interbreeding natural populations that are reproductively isolated from other such groups" (p. 17). That is, members of the same

[13] While this does not imply that other concepts are unworthy of discussion, many of them emerge due to the problems exhibited by both the BSC and the PSC. Also, this section goes beyond merely reviewing all species concepts. For a more detailed examination of species concepts and their associated issues, including further discussion of the BSC and PSC, see Zachos (2016).

species are groups of individuals that can interbreed with each other but cannot with individuals from other groups. An important criterion is that such members produce fertile offspring as well. It is crucial to note that this concept only applies to groups of individuals with the capacity for sexual reproduction who coexist in the same space (sympatric) and at the same time (synchronic). Despite its apparent simplicity, this concept is structured from the perspective of population genetics, as reproductive isolation is related to the lack of gene flow among populations, which defines their boundaries. In this context, it is important to stress that the BSC uses the degree of reproductive isolation rather than the degree of morphological discreteness to delimit species boundaries; nevertheless, BSC advocates presuppose a correlation between the two. In any case, some issues with the BSC are outlined below:

(1) *The meaning of reproductive isolation to delineate species, particularly concerning allopatric populations*

At first glance, the concept of reproductive isolation presents no issues, except when attempting to specify whether it is intrinsic (e.g., postzygotic) or extrinsic (e.g., ecological, behavioral, etc.). Theoretically, according to the BSC, two populations are classified as different species when their reproductive isolation is complete, implying that populations have ceased to exchange genes. Nevertheless, numerous studies have demonstrated that reproductive isolation, along with a decrease in gene flow, is a gradual process (Wang et al. 2019, 2020, 2022), making it difficult to measure in practice (Westram et al. 2022). Furthermore, reproductive isolation is generally inferred without directly examining gene flow (Wang et al. 2020; Burbrink and Ruane 2021), further complicating its assessment in allopatric populations (Wu et al. 2022).

According to the BSC, morphological criteria are excluded from species delimitation, which exacerbates the issue, as such populations may exhibit phenotypic divergence despite evidence of gene exchange (Dupré 1999, p. 8), or show no phenotypic divergence even with significant genetic distances between them (Mousseau and Sikes 2011). Regardless, diagnosing allopatric populations as different species relies on extrinsic rather than intrinsic criteria for reproductive isolation since demonstrating intrinsic isolation presents significant challenges (Ehrlich and Raven 1969; Mallet 1995; Hey 2001, p. 164; Zachos 2016, p. 113).

(2) *The morphological criterion behind the interbreeding criterion for recognizing species*

While Mayr consistently emphasized that the BSC does not entail a morphological criterion—particularly because he struggled to shake off the typological implication of the species concept, which was supposedly based on anti-evolutionary foundations—the reality is that many criteria for identifying biological species do rely on morphological discreteness. BSC advocates often assume that morphological gaps correlate with genetic isolation or, more specifically, with the degree of infertility among groups (Zachos 2016, p. 109). Mayr himself employed morphological criteria to identify biological species, acknowledging the practical challenges in assessing the degree of reproductive isolation (Donoghue 1985; Hey 2001, p. 165). Nevertheless, it is important to note that morphological discreteness and the degree of reproductive isolation do not always correlate (Kitcher 1984; Kluge 1990, p. 425; Donoghue 1985; Dupré 1999, p. 6). Indeed, in practice, species delimitation generally relies solely on the morphological criterion, thereby challenging the practical applicability of the BSC (Wang et al. 2019).

(3) *The recognition of the species in the fossil record*

Determining whether two groups in the fossil record are reproductively isolated is impossible; as a result, their identification relies more on morphological data (Kitcher 1984). In this context, Sepkoski (2016) claims that neontologists and paleontologists are referring to different entities when using the term "species." However, Kitcher (1984) points out that supporters of the BSC assume a correlation between morphological change and the degree of reproductive isolation, although this inference may not always be appropriate (see point 2 above). So, paleontologists are compelled to study and identify species exclusively through morphological data:

> There is a perfectly legitimate paleontological question which focuses on the rates and patterns of morphological diversification within evolving lineages, and paleontologists pursue this question by dividing lineages into species according to morphological changes. To insist that they should always formulate their inquiries by using the biological species concept is to make them take a risky trip around Robin Hood's barn. (Kitcher 1984, p. 317)

In conclusion, according to the BSC, the meaning of reproductive isolation is ambiguous because it does not specify whether such isolation is intrinsic or extrinsic. The degree of reproductive isolation does not necessarily correlate with morphological discreteness, and the practical assessments of reproductive isolation are problematic. Furthermore, the BSC has both practical and epistemological flaws, as it does not apply to groups of organisms that inhabit different spaces and times. Interestingly, the only way to infer the definition of biological species is through morphological data. Therefore, despite the long-standing establishment of the BSC as the "yardstick" for the species concept, it faces significant challenges that hinder the recognition of the true nature of species.

The Phylogenetic Species Concept (PSC) The PSC is primarily based on the method of phylogenetic systematics (cladistics), characterized by recognizing species as fundamental taxonomic units in cladistic analysis. Species are viewed as basal (the least inclusive) groups within the phylogenetic classification. According to cladistics, the phylogenetic tree is structured into various branches representing different lineages that evolve over time. The tips of these branches represent extant species, connected through genealogical relationships. Hence, members of a species are those descending from a common ancestral line. Interestingly, although there are different versions of this concept, all are rooted in the principles of phylogenetic systematics pioneered by Willi Hennig. The issues associated with this concept are as follows:

(1) *Different versions of the concept with no agreement on defining what species are*

The PSC is notably ambiguous because different authors employ distinct species concepts within the framework of cladistics without reaching a consensus (Richards 2010; Zachos 2016). Ereshefsky (2001, p. 90) identifies at least two schools within cladism: one defines the concept as diagnosable, focusing on patterns (historical record) (e.g., Cracraft 1987; Nixon and Wheeler 1990; Wheeler and Platnick 2000), while the other defines species as monophyletic groups based on the processes that cause monophyly (e.g., Mishler and Donoghue 1982; Mishler and Brandon 1987; Mishler and Theriot 2000). Although both schools are founded on Hennig's principles, the arguments for recognizing species within this framework differ significantly.

(2) *The problem of defining species by the diagnosability criterion*

According to the diagnosability version, a species is defined as "the smallest diagnosable cluster of individual organisms within which there is a parental pattern of ancestry and descent" (Cracraft 1983, p. 170). The problem with this concept lies in how to distinguish two groups as different species within the phylogenetic tree. For instance, according to one of the principles of phylogenetic systematics, two species are considered distinct not because of their differences in morphological characters but because they exhibit independent historical continuity through lines of common descent (de Queiroz and Gauthier 1992). Additionally, this perspective views the branches of the phylogenetic tree as continuous lines without clear boundaries, making it difficult to recognize discrete entities. In other words, there are no clear boundaries that taxonomically delineate groups (Kunz 2012, p. 36). But how can species be diagnosed?

Dupré (1999, p. 9) argues that if reproductive isolation were the primary factor leading to lineage divergence, we would resort to the BSC, despite the problems it entails (see above). By contrast, in practice, taxonomists diagnose species based on either morphological or genetic criteria. However, using molecular data alone to delimit species, as seen previously with subspecies, proves problematic (see also Franz 2005; Wilson 2011; Wheeler 2024). Morphological data is more reliable; yet this contradicts a fundamental principle of phylogenetic systematics, emphasizing classification based on the historical continuity of genealogical connections rather than morphological criteria alone. Besides, this definition does not consider the importance of distinguishing between interspecific and intraspecific variations since, under this criterion, any subspecies or intraspecific variation would be considered a species (see Ereshefsky 1989).

(3) *The problem of defining species through the criterion of monophyly*

According to the monophyly version of the PSC, a species is "the least inclusive taxon recognized in a formal phylogenetic classification… [O]rganisms are grouped into species because of evidence of monophyly" (Mishler and Theriot 2000, pp. 46–47). Here, it is important to emphasize that the term "monophyly" not only has multiple meanings (Rieppel 2010) but is also ambiguous (Ereshefsky 1989; Rieppel 2010). Likewise, Brent D. Mishler, a prominent advocate of this concept, embraced species pluralism by asserting that species recognition could be defined based on various aspects depending on the researcher's criteria. He stressed that the only consistently established criterion was monophyly (see Mishler and Donoghue 1982; Mishler and Brandon 1987). Despite this, given that various levels of monophyly exist within the phylogenetic tree, Mishler recently claimed (2021, 2022) that the species rank should be eliminated.

In conclusion, within the PSC framework, the definition of species remains unclear due to a lack of consensus among cladistics schools. While species are generally defined as the minimal unit of phylogenetic analysis, they are difficult to recognize within cladistic arrangements. For example, different criteria (i.e., reproductive isolation or morphological similarities) are used to diagnose species, and many subspecies can be identified as species. Additionally, the term "monophyly" is ambiguous, and the species rank, as a monophyletic group, lacks clear attributes that distinguish it from other ranks in the hierarchy. Overall, these diverse definitions lack agreement on the patterns and processes necessary to reliably identify species as fundamental units of evolution.

7.2.2 Species Category: Is It Possible to Recognize a General Notion of Species?

Another crucial aspect of the ongoing discussion about the species problem involves identifying species as a taxonomic rank within the Linnaean hierarchy. To understand this issue, philosophers suggest clarifying the distinction between species taxa and the species category. Species taxa are groups of organisms that exist in nature irrespective of their rank (e.g., *Homo sapiens* and *Pan paniscus*). In contrast, the species category is a set of properties that groups of organisms must possess to be assigned the rank of "species" and distinguished from other ranks in the Linnaean hierarchy, such as subspecies, genus, family, and so on. For example, a taxonomist may recognize dogs as a natural lineage but might hesitate to assign them the rank of either the species category (*Canis familiaris*) or the subspecies category (*Canis lupus familiaris*) (Ereshefsky 2001).

Another important consideration is the philosophical debate surrounding species realism. Two main philosophical stances address this subject: monism and pluralism. Species monism asserts the existence of a universal concept of species applicable to all groups of organisms. In contrast, species pluralism contends that all definitions of species are valid, with each representing distinct entities classified as species (Reydon 2022). Indeed, the implications of this debate also prompt the question of whether there is a natural arrangement that hierarchically represents the phylogenetic tree (Velasco 2008) or whether there are multiple natural classifications depending on which species concept is applied (LaPorte 2005).

In this context, according to Ereshefsky (1998, 1999, 2001), all species concepts independently attempt to describe the properties that members of species taxa share to categorize them into one specific rank rather than another. The challenge arises because different species concepts formulate different criteria for recognizing species taxa within a hierarchical taxonomy. Thus, Ereshefsky argues that, while species taxa are real, the species category lacks a common set of properties that distinguishes it from other categories. This line of reasoning is known as eliminative pluralism and underscores various philosophical discussions, with species pluralism at the center of the debate.

For example, as Kitcher (1984) highlighted, there are numerous ways to delimit species taxa, rendering the species category heterogeneous. Mishler and Brandon (1987) have further maintained that there is no universal criterion for the species category, as taxa can be ranked as species for different reasons. Similarly, Reydon (2005) stated that different concepts correspond to various ontological kinds, challenging their placement in a universal category.[14] More recently, Minelli (2021) has asserted that species pluralism is here to stay, suggesting that this phenomenon must be embraced as the prospect of identifying a general species concept diminishes. In addition, Wilkins (2022) posits that species is neither a real entity nor a distinct rank; rather, he claims it is an epistemic notion sustained by diverse disciplinary traditions, each attempting to describe natural entities within their respective domains. In this vein, Mishler (2021, 2022) advocates for eliminating the species category, alongside other ranks in the taxonomic hierarchy. Interestingly, while most philosophers of biology embrace species pluralism to address the species problem, some stress the importance of positing a general notion of species (e.g., Hull 1989, pp. 120–121), at least as a pragmatic framework within taxonomy—a view supported by Dupré (1999) and Brigandt (2003).

Different proposals have emerged as solutions to this issue. The simplest one suggests that, among the multiple concepts analyzed, the one that best describes the natural hierarchical arrangement should be chosen. However, as noted by Cracraft (2000), proponents of each species concept tend to believe theirs is the best, yet the notion of what constitutes "best" remains subjective. Likewise, Minelli (2021) argues that this challenge of determining a universal species concept cannot be easily resolved through authority, consensus, or an expectation of developing better criteria for species identification:

> I am not proposing here the development of multiple parallel taxonomies. My aim here is to call attention to the fact that for a large number of taxa we cannot hope, at the moment at least, for a universal consensus on a given classification because, in so far as distinct

[14] Reydon (2005) emphasizes that this perspective is not equivalent to species pluralism; instead, it is a homonymic term indicating that the entities are ontologically independent.

> research agendas are pursued and explicitly taken as foundation for the taxonomic work, different legitimate classifications will necessarily result. (Minelli 2021, p. 11)

In any case, as emphasized by Ereshefsky throughout his writings, the fundamental question is: What unifying properties does the species category possess to differentiate it from other taxonomic categories? According to the BSC, the boundaries between species and other categories are defined by interbreeding among their members. Species represent the biological entities defined by the capacity for interbreeding; subspecies are the subgroups within species that interbreed; and higher taxa are groups of species that do not interbreed, representing the historical record of species (Lee 2003). Nonetheless, there are evident difficulties in recognizing species as a category under this concept. First, as seen in the previous section, distinguishing species from subspecies, especially in allopatric populations, proves highly challenging. Second, the boundaries above the species level are unclear because reproductive isolation occurs to varying degrees. According to Lee (2003), the only "solid" argument attempting to salvage the species category, under this criterion, is that "all other proposed species criteria for delimiting the species rank (inclusiveness) fare even worse in this respect" (p. 184).

In the diagnosable version of the Phylogenetic Species Concept (PSC), species are defined as the smallest populations that are diagnosably distinct (Cracraft 1997). However, this concept lacks a definitive criterion and the evolutionary mechanisms necessary to reliably delineate these small lineages within a presumed continuum. A criterion employed in phylogenetic systematics is the time at which different taxa diverge, which can be corroborated through their genetic differences. For instance, higher taxa represent a more distant time of separation from the last common ancestor (de Queiroz and Gauthier 1992). Nevertheless, Baum (2009, p. 76) states that there is no species rank based on the time of separation from the last common ancestor because different parts of the genome exhibit different histories and, therefore, different times of separation from the last common ancestor. By contrast, in the monophyly version of PSC, species rank is not specifically defined; it is simply stated that a species should be the least inclusive monophyletic group. Indeed, as mentioned above, proponents of this concept (e.g., Mishler 1999, 2010, 2021, 2022) argue that the species rank lacks any special universal criterion. Another important point of the PSC is that none of its versions explicitly addresses the subspecies rank. In this regard, Dupré (1999) points out the existence of clearly distinguishable groups of organisms below the species level, suggesting that there is no reason to assume these groups are not units of evolution: "There is no doubt that such groups are often clearly distinguishable, and indeed for many purposes classification at this level is the most important" (p. 11).[15]

Another approach to unification is to assume that all species concepts are valid and, when considered collectively, can construct the species category. Many authors have explored this possibility from various perspectives. For instance, Wilson

[15] See the discussion of subspecies in Sect. 7.1.

(1999) through the HPC kind,[16] Pigliucci (2003) via Wittgenstein's "family resemblance,"[17] and Mayden (1997, 1999), de Queiroz (2005, 2007), Novick and Doolittle (2021), and Pavlinov (2022, 2023) through hierarchical theories of concepts, among others.

It is crucial to note that most philosophers assume that all species concepts are theoretically valid. That is, they study, analyze, and debate these concepts because they consider them scientifically legitimate, despite their discrepancies. The issue is that, as previously discussed, the two most relevant concepts in biological discussions about the species problem, the BSC and the PSC, pose many difficulties. In my opinion, the problem is that these concepts are based on inappropriate theoretical frameworks of evolution and taxonomy. For example, first, the BSC is built upon Mayr's speciation models, which assume that the interruption of gene flow leads to divergence. However, as I analyze in detail in Chap. 10, Mayr's models are outdated in light of recent data on evolutionary divergence. Second, the PSC's drawbacks stem from the inadequacies of phylogenetic systematics principles within the context of hierarchy theory. Williams (1992), for example, emphasizes inconsistencies in the foundational book of cladistics resulting from Hennig's attempt to reconcile his "hierarchical" model with the Linnaean hierarchy.[18]

Interestingly, both the BSC and the PSC, despite rejecting the morphological aspects within their theoretical frameworks, ultimately resort to this criterion for species delineation, as discussed earlier. Ironically, Oderberg (2007) claims that, despite considering the morphological criterion secondary for recognizing species boundaries, "morphology creeps into virtually every major way in which species are identified in contemporary systematics. It finds its way into ecological tests, reproductive criteria, cladistic analysis, genetic identification, mate recognition, and more" (p. 235). Similarly, Wheeler (2024) points out that modern systematics has become excessively reliant on molecular methods. He argues that describing species solely at the molecular level, while relegating morphology, has significant consequences for taxonomy: "To abandon or minimize morphology now is to fundamentally alter the core mission of systematics" (p. 92).

In this regard, it is worth emphasizing Devitt's (2008, 2023) perspective on traditional species concepts, which claims that neither the BSC nor the PSC alone can explain the identity of a particular species, e.g., why an organism is a lion and not a tiger. Instead, the morphological criterion plays a vital role and significantly contributes to the science of classification. In response to these challenges, some authors have begun to admit the significance of morphological "similarities" as a criterion for delineating species (Zachos 2016, pp. 157–161). Likewise, this criterion is also

[16] In the philosophy of biology, the concept of *Homeostatic Property Cluster Kind*, emerging in response to the anti-essentialist movement, seeks to establish the reality of taxonomic units through homeostatic mechanisms that facilitate their recognition. For further details on this line of reasoning, see Chap. 6.

[17] According to Wittgenstein's "family resemblance," any property of the species can acquire greater relevance based on the study group of interest.

[18] See Chap. 8, Sect. 8.1.1.

valuable for determining their ontological status (Hey 2001, p. 173). In any case, in practice, the morphological criterion stands as the most widely used to date (Aldhebiani 2018; Dantas-Torres 2018).

Skepticism toward the morphological criterion likely arose due to the success of the Modern Synthesis framework in classification. The architects of the Modern Synthesis believed that a radical change in taxonomic interpretations was necessary, inasmuch as old taxonomic practices were rooted in typological thinking based on essentialism and species fixity, rendering them incompatible with evolution.[19] Similarly, Willi Hennig founded phylogenetic systematics to eliminate the classical taxonomy embedded in typologism. As a result, these radical changes in evolution and taxonomy affected the very notion of species. The morphological criterion was then relegated to be incorporated into another kind of species concept: the Morphological Species Concept, which supposedly delimits species solely based on their "similarity" of traits. Additionally, the morphological criterion fell into the category of the Phenetic Species Concept, derived from Sokal and Sneath's (1963) numerical taxonomy, which initially proposed a method for quantifying morphological similarity through computational analysis.

In addition, when considering the concept of morphological "similarity," critics interpret these similarities too literally. For example, they argue that in groups exhibiting sexual dimorphism, the male and female might be classified as different species under this criterion. Likewise, for such groups of organisms with discontinuous life cycles or those exhibiting polymorphisms, delimitation becomes fruitless. Unfortunately, these arguments reveal a lack of appreciation for the morphological program since concepts emphasizing "morphological similarity" do not demonstrate true integration with current Evo-Devo aspects.

Interestingly, some authors have emphasized that, despite debates in academic settings, scientific practices support the species category as a reflection of the natural world (Currie 2016; Hazelwood 2021; Amitani 2022). For example, in his analysis supporting the reality of the species category, Currie (2016) asserts that paleontologists do not delimit species solely based on morphological "similarity." Instead, they develop morphological hypotheses informed by an understanding of the nature of variation, including models of ontogenetic development, sexual dimorphism, polymorphism, life cycles, etc. Paleontologists, therefore, delimit species only when morphological variation exceeds the normal range for the group in question: "Paleobiologists delineate species membership in terms of the processes which explain variation" (p. 802). Importantly, Currie states that paleontologists make this distinction without adhering to any particular species concept, i.e., they are indifferent to contemporary species concepts. Instead, they use the criteria available to them through biological research methodologies. Considering this, Currie concludes: "The success of paleobiological practice in species delineation gives reason to think that the species category (not simply the categories picked out by various specifications) exists… even if the category is not unified" (p. 810).

[19] However, as discussed in Chap. 2, this interpretation is inaccurate because typologism had a positive influence on the synthesis of taxonomy, morphology, and evolution.

In conclusion, the search for a unified species concept within biology faces skepticism due to the diverse proposals from different taxonomic schools on the nature of species. Species pluralism operates on the premise that "all species concepts are valid" despite lacking consensus on their interpretation. Both the BSC and the PSC highlight certain properties shared among species taxa. However, neither of these concepts clearly delineates how to identify the species category and distinguish it from other categories in the phylogenetic tree. The premise that "all concepts are valid" raises the question of whether they collectively define the species category or are all distinct ontologically. From my perspective, these traditional concepts are grounded in inadequate interpretations of the processes and patterns that explain biological evolution. Under these species concepts, the morphological criterion has been put aside despite being crucial for researchers in species delineation. Moreover, the issue lies in the misinterpreted framework of the morphological criterion, particularly in light of advances in Evo-Devo.

The reality of the species category appears like a spectrum, especially when the theories that underpin traditional concepts of species are set aside. As Currie (2016) correctly emphasizes, scientific practices grounded in a morphological program speak for themselves, claiming this reality without waiting for philosophers and biologists to reach a consensus on defining what species are. In this regard, as explored below, this study aims to propose a model where this reality can rest comfortably.

7.2.3 Species According to the H-Evo-Devo Theory

Initially, the term species emerged from the Linnaean system to designate a specific rank characterizing a fundamental unit within the taxonomic hierarchy. Similarly, following the basic principles of hierarchical theory, this word can be useful in identifying the focal level within a triadic representation (see Chap. 5, Fig. 5.3). So, according to the H-Evo-Devo Theory, *species are discrete entities that exhibit the first step in significant morphological gaps and are distinguished from other taxa by being the least inclusive body plan within a hierarchy of body plans*. Taking this description into account, the species category can be distinguished from other categories within the Linnaean system for the following reasons:

- Downwards, subspecies do not represent a body plan; instead, they are adaptive variations of the species body plan reflected in simple characters such as color, proportion, size, etc.
- Upwards, higher taxa (families, orders, classes, etc.) exhibit greater morphological gaps than species and represent more inclusive body plans than the species body plan.

It is important to note that this delineation is not based on a simple morphological rule-of-thumb methodology. Critics, for example, might argue that interpreting taxonomic ranks as morphological scales could result in multiple levels, potentially leading to ambiguous placement of species within these ranks (e.g., Lee 2003, p. 180; Mishler 1999, p. 308). Nevertheless, in most groups of multicellular organisms, distinguishing a minor body plan is straightforward. For example, Fig. 7.2 illustrates ten closely related hominid taxa. As mentioned in the previous section, the primary criterion for classifying species is based on the observation that different groups are separated by significant morphological gaps. The figure shows that each group is a pair with the same underlying type. This delineation reveals five distinct types that, although designated as genera, can also be considered species. In fact, one of this study's conclusions is that genera and species are ontologically similar. Thus, on some occasions, species may take the form of subspecies, and genera may take the form of species. This is not an ontological issue but rather a matter of recognition, as many of the taxa currently classified as species could be considered subspecies under this perspective.

So, species diagnosis can be addressed at different scales. For example, when comparing a mouse and a person, a taxonomist will classify them as belonging to different species. Progressively, various groups can be compared across distinct phylogenetic scales until the most closely related taxa are reached, as illustrated in Fig. 7.2. In this example, when comparing the genealogical relationships of these hominids with *Homo*, morphological gaps can be observed that separate the genus *Homo* (type 2) from both the genus *Pongo* (type 5), which is the most distant, and the genus *Gorilla* (type 4), which is more closely related. Additionally, morphological gaps separate *Homo* from *Pan* (type 3), even though they are the most closely related. Although *Homo* and *Pan* are not direct ancestors or descendants of each other—rather, they are sibling taxa—morphological gaps also exist between *Homo* and *Australopithecus* (type 1) despite the possibility of a direct ancestry and descent relationship between them. In this sense, species are recognizable because they mark the first step in significant morphological gaps within a hierarchy that encompasses various scales of morphological discontinuities.

Those familiar with the gradual change of the entities may find this fact difficult to appreciate, as they would expect a chain of intermediate forms connecting the ancestral genus to the offspring genus within the hypothetical continuous phylogenetic tree. However, the H-Evo-Devo Theory supports a theory of saltational evolution that explains the origin of these morphological gaps (see further below). As discussed in Chap. 6, units of evolution are discrete evolutionary individuals, each with a separate evolutionary history, some representing ancestors, others descendants, and others sibling taxa.

Overall, the H-Evo-Devo Theory maintains a boundary that distinguishes species from subspecies. This distinction is based on the type of variation and the mechanisms involved in the origin, stability, and extinction of evolutionary entities. For example, species are body plans, whereas subspecies are not; rather, they are characterized by simple characters such as color, proportion, and size. This model also proposes a distinction between species and higher taxa. As previously mentioned, while both species and higher taxa are fundamentally body plans and share similarities in their mechanisms, the difference lies in the scale. Higher taxa exhibit larger distances within the morphospace, represent more inclusive body plans, and show

Table 7.1 *Differences between the units of evolution within the Linnaean hierarchy according to the H-Evo-Devo Theory*. These differences justify the ontological value of these entities as both taxa and categories

Unit of evolution	Morphological gap	Type of variation	Type (body plan)	Mechanisms
Subspecies	Small (in simple characters)	Simple characters such as color, proportion, size, etc.	Not enough to recognize a type	*Origin:* Darwinian mechanisms *Stability:* Developmental constraints acting on simple characters *Extinction:* Factors that alter their local habitat and their population structure
Species/Genera	Significant	Complex characters	The least inclusive type in the hierarchy	*Origin:* Supradarwinian mechanisms *Stability:* Developmental constraints acting on complex characters *Extinction:* Loss of parts and mass extinctions
Higher taxa	More significant	Complex characters	More inclusive type in the hierarchy than species/genera	*Origin:* Supradarwinian mechanisms *Stability:* Developmental constraints acting on complex characters *Extinction:* Loss of parts and mass extinctions

greater discontinuities in their mechanisms than species (see Table 7.1). But how can the reality of the species be justified? According to the H-Evo-Devo Theory, developmental constraints carry an ontological role for species. They provide the basis for recognizing the projection of a certain type into a species (natural kinds) and are simultaneously the cause of species cohesion (individuals). With this in mind, I will now provide some brief clarifications on their ontological identification.

Natural Kinds It is important to note that associating species with body plans is challenging, as this notion has typically been applied to higher taxonomic levels. The current consensus stems from associating the term "species" with the taxonomy founded by the Modern Synthesis and phylogenetic systematics, leading to the development of the BSC and the PSC, respectively. Nevertheless, as discussed in Chap. 2, the concept of species was historically linked to the recognition of the type. While the type concept is now embraced through terms such as *Bauplan* or body plan, the phylogenetic tree is established as a hierarchy of types in which the least inclusive type can be perceived, as discussed above. Consequently, under this interpretation, species are types or body plans, albeit the smallest in the hierarchy. If this is the case, species effectively behave as paradigmatic examples of natural kinds in the traditional sense, with developmental constraints serving as the causal mechanism (causal kind), allowing us to perceive the least inclusive type, hence affirming their reality.[20]

Evolutionary Individuals As seen in the previous chapter, for an entity to qualify as an evolutionary individual, it must have a concrete beginning. But if species originate through slow and continuous steps, considering them as evolutionary individuals becomes problematic. The key issue is that most biologists assume the gradual origin of species, explaining the observed morphological gaps as emerging from a continuous process where intermediates are eliminated by natural selection. Yet, supposing that these gaps result from supra-Darwinian processes—namely, saltational evolution[21]—if this is the case, there are compelling reasons to believe that species have well-defined beginnings.

Following the conventional model of speciation, most biologists take for granted that the origin of species is explained by the same mechanisms that give rise to small and peripheral populations. In this regard, it is assumed that macroevolutionary patterns and processes are merely an accumulation of microevolutionary events. Two main issues arise from this perspective. First, Ernst Mayr's geographical models of speciation are based solely on genetic mechanisms and ecological-geographical situations, without any integration of morphology and embryology (Minelli and Fusco 2012).[22] Second, the persuasion that macroevolutionary mechanisms are the same as those observed at the microevolutionary level has persisted by tradition, but there is nothing in the current evolutionary theory to justify such a premise.[23]

Nevertheless, the H-Evo-Devo Theory takes an unorthodox stance: macroevolutionary patterns and processes are irreducible to microevolution. In other words, both levels are distinct and propose their own mechanisms for the origin of biological entities, such as species and subspecies, respectively. Subspecies emerge through Darwinian mechanisms, whereas species arise through supraDarwinian mechanisms. By Darwinian mechanisms, I mean those observed in a geographical dispersal model. By supraDarwinian mechanisms, I refer to those associated with the establishment of the prospecies (commonly known as hopeful monsters).[24] Therefore, drawing a dividing line between the two levels concerning the mechanisms of origin can lead to an ontological hypothesis about the rank distinction:

---

[20] For more details defending the consideration of species as natural kinds, see Chap. 6, Sect. 6.1.2.

[21] See Chap. 9, Sect. 9.2; Chap. 11, Sect. 11.1.1; Chap. 12, Sect. 12.1.1.

[22] See Chap. 10, Sect. 10.1.

[23] See Chap. 10, Sect. 10.2.

[24] See Chap. 11.

Darwinian mechanisms generate discontinuities, but these processes operate on simple characters, not involving the origin of a body plan. In contrast, supraDarwinian mechanisms produce significant morphological gaps, which explain the origin of body plans (types).

Although the question of origin is crucial, it might not be sufficient to consider species as units of evolution. Certainly, "cohesion" is an equally relevant criterion (Kendig 2022). In light of the BSC, many biologists began searching for alternative explanations to support species cohesion without reaching a consensus (see Templeton 1989; Ereshefsky 2001, pp. 115–116). However, according to the H-Evo-Devo Theory, the cause of species cohesion is attributed to developmental constraints acting on complex characters. This process is crucial, as these steady states allow species to be recognized as cohesive spatiotemporal entities, providing evidence of their reality.

In conclusion, in this framework, developmental constraints cause species cohesion and uniformity by maintaining all their parts (subspecies) within the same structural framework. A species emerges when one of its parts breaks from these developmental constraints (birth criterion), resulting in an entity with a different position within the morphospace (reproduction criterion). Species possess phylogenetic connections with both ancestral and sibling species, being independent entities themselves (phylogeny criterion). Species can diversify, giving rise to subspecies (variation criterion), each adapting to a particular environment while always maintaining the species body plan. Subspecies are confined to a common type, never exceeding its boundaries as species (natural kinds). Thus, developmental constraints allow species to be recognized over time (cohesion criterion) until they become extinct (death criterion). With this in mind, species are discrete entities with homeostatic properties (developmental constraints) that validate them as real units of evolution. In this sense, species simultaneously behave as evolutionary individuals (cohesive entities) and natural kinds (types) through the same mechanism.

7.3 Conclusions

As discussed, the current notions of "subspecies" and "species" emerged from the taxonomy founded by the Modern Synthesis and phylogenetic systematics. Yet these theoretical frameworks lack true integration; even their derivative concepts—the BSC and the PSC—contradict each other in several aspects. Moreover, these models fail to incorporate contemporary biological insights from Evo-Devo. Consequently, a series of issues hinders the development of a coherent hierarchy theory of evolution. For example, the term "subspecies" is arbitrary—despite its attributed significance—and attempts to apply this concept to the diverse phenomena of intraspecific variation have proven ineffective. Similarly, different species concepts fail to offer clear definitions or criteria for the recognition of subspecies. Regarding the term "species," neither the BSC nor the PSC adequately delineates species boundaries. Both concepts lack precise criteria for distinguishing and

Table 7.2 *Comparison of the different interpretations of subspecies and species according to the Biological Species Concept (BSC), the Phylogenetic Species Concept (PSC), and the H-Evo-Devo Theory*

Unit of evolution	Biological Species Concept (BSC)	Issues associated with the BSC	Phylogenetic Species Concept (PSC)	Issues associated with the PSC	H-Evo-Devo Theory
Subspecies	Groups within species that interbreed with each other They reflect the early stages of divergence that may lead to the origin of species	The boundary between subspecies and species becomes ambiguous due to the absence of interbreeding (or gene flow) in allopatric populations	The smallest diagnosable unit within the phylogenetic tree	Many cladistics approaches do not consider them The distinction between subspecies and species is unclear; both represent the smallest taxonomic units	Groups that share the least inclusive body plan (species) but are distinguished from each other by small morphological gaps in simple characters such as color, pattern, size, proportion, etc. as well as by the fact that they live in different environments They are groups that diversify to adapt the species to different environments, thus protecting species from extinction
Species	Groups of interbreeding natural populations that are reproductively isolated from other groups	There is ambiguity in the meaning of reproductive isolation There are problems with allopatric populations (as discussed above in subspecies issues) Morphological discreteness, rather than complete reproductive isolation, facilitates species identification It is impossible to identify species based on the fossil record	The smallest diagnosable unit within the phylogenetic tree (diagnosability version) Groups are classified as species based on evidence of monophyly (monophyly version)	There are different versions of PSC without consensus on their definition It is not clear how to identify two distinct species within the phylogenetic tree Many subspecies and polymorphisms can be classified as species under the diagnosability criterion The term "monophyly" is ambiguous Different scales of monophyly contribute to the ambiguity of the species rank	Groups that exhibit the first step in significant morphological gaps and, more importantly, are the least inclusive body plan within a hierarchy of body plans

Note that the third and fifth columns highlight the primary problems or issues associated with traditional concepts in defining these units of evolution unifying groups under the designation of "species." As a result, the existing hierarchical models remain confusing and contentious.

However, the H-Evo-Devo Theory challenges these traditional views by proposing a perspective on taxonomy and evolution that may seem new but is, in fact, quite old. In this framework, traditional concepts of "subspecies" and "species" vanish along with all their associated problems (see Table 7.2.). In addition, it introduces new discussions aimed at understanding not only the nature of subspecies and species but also other evolutionary entities—e.g., higher taxa and homologues (see next chapter). According to the H-Evo-Devo Theory's structure, subspecies are groups characterized by being separated from each other by small morphological gaps in simple characters as well as by the fact that they live in different environments. They represent parts of species that diversify and adapt to different environments, thereby protecting species from extinction. Meanwhile, species are discrete groups that represent the first step in significant morphological gaps. More importantly, they are the least inclusive body plan within a hierarchy of body plans. In both cases, these entities are dynamic and discrete, evolving quasi-autonomously. If I am correct, this model not only has the potential to resolve over a century of debate but also introduces new challenges to the understanding of evolution and taxonomy.

References

Aldhebiani AY (2018) Species concept and speciation. Saudi J Biol Sci 25:437–440

Amadon D (1949) The seventy-five percent rule for subspecies. Condor 51:250–258

Amadon D, Short LL (1976) Treatment of subspecies approaching species status. Syst Zool 25:161–167

Amitani Y (2022) Is the species problem that important? In: Wilkins J, Zachos FE, Ya PI (eds) Species problems and beyond: contemporary issues in philosophy and practice. CRC Press, pp 39–63

Barrientos-Villalobos J, Schmitter-Soto JJ, de los Monteros AJ E (2018) Several subspecies or phenotypic plasticity?: a geometric morphometric and molecular analysis of variability of the mayan cichlid *Mayaheros urophthalmus* in the Yucatan. Copeia 106:268–278

Baum DA (2009) Species as ranked taxa. Sys Biol 58:74–86

Berrilli E, Gambioli B, Bombi P, Garzia M, Muraro M, Pardo C, Reale M, Sherpa S, Ficetola GF, Vignoli L, Salvi D (2024) Subspecies inflation hampers conservation efforts: a case study on wall lizards. Biol J Linn Soc 143. https://doi.org/10.1093/biolinnean/blae001

Braby MF, Eastwood R, Murray N (2012) The subspecies concept in butterflies: has its application in taxonomy and conservation biology outlived its usefulness? Biol J Linn Soc 106:699–716

Brand CM, White FJ, Rogers AR, Webster TH (2022) Estimating bonobo (*Pan paniscus*) and chimpanzee (*Pan troglodytes*) evolutionary history from nucleotide site patterns. Proc Natl Acad Sci USA 119:e2200858119. https://doi.org/10.1073/pnas.220085811

Brigandt I (2003) Species pluralism does not imply species eliminativism. Philos Sci 70:1305–1316

Burbrink FT, Crother BI, Murray CM, Smith BT, Ruane S, Myers EA, Pyron RA (2022) Empirical and philosophical problems with the subspecies rank. Ecol Evol 12:e9069. https://doi.org/10.1002/ece3.9069

Burbrink FT, Lawson R, Slowinski JB (2000) Mitochondrial DNA phylogeography of the polytypic North American rat snake (*Elaphe obsoleta*): a critique of the subspecies concept. Evolution 54:2107–2118

Burbrink FT, Ruane S (2021) Contemporary philosophy and methods for studying speciation and delimiting species. Ichthyol Herpetol 109:874–894

Clarke E, Okasha S (2013) Species and organisms: what are the problems? In: Bouchard F, Huneman P (eds) From groups to individuals: evolution and emerging individuality. MIT Press, Cambridge MA, pp 55–76
Cracraft J (1983) Species concepts and speciation analysis. Curr Ornithol 1:159–187
Cracraft J (1987) Species concepts and the ontology of evolution. Biol Philos 2:329–346
Cracraft J (1997) Species concepts in systematics and conservation biology—an ornithological viewpoint. In: Claridge MF, Dawah HA, Wilson MR (eds) Species: the units of biodiversity. Chapman and Hall, New York, pp 325–339
Cracraft J (2000) Species concepts in theoretical and applied biology: a systematic debate with consequences. In: Wheeler QD, Meier R (eds) Species concepts and phylogenetic theory: a debate. Columbia University Press, New York, pp 3–14
Currie AM (2016) The mystery of the Triceratops's mother: how to be a realist about the species category. Erkenn 81:795–816
Dantas-Torres F (2018) Species concepts: what about ticks? Trends Parasitol 34:1017–1026
de Queiroz K (2005) A unified concept of species and its consequences for the future of taxonomy. Proc Calif Acad Sci 56:196–215
de Queiroz K (2007) Species concepts and species delimitation. Syst Biol 56:879–886
de Queiroz K (2020) An updated concept of subspecies resolves a dispute about the taxonomy of incompletely separated lineages. Herpetol Rev 51:459–461
de Queiroz K (2021) Response to criticisms of an updated subspecies concept. Herpetol Rev 52:773–776
de Queiroz K, Gauthier J (1992) Phylogenetic taxonomy. Annu Rev Ecol Syst 23:449–480
DeSalle R, Tattersall I (2025) Race realism goes both ways. Hist Philos Life Sci 47. https://doi.org/10.1007/s40656-024-00658-y
Descimon H, Mallet J (2009) Bad species. In: Settele J, Shreeve T, Konvička M, Van Dyck H (eds) Ecology of butterflies in Europe. Cambridge University Press, Cambridge, pp 219–249
Devitt M (2008) Resurrecting biological essentialism. Philos Sci 75:344–382
Devitt M (2023) Biological essentialism. Oxford University Press, Oxford
Donoghue MJ (1985) A critique of the Biological Species Concept and recommendations for a phylogenetic alternative. Bryologist 88:172–181
Dupré J (1993) The disorder of things: metaphysical foundations of the disunity of science. Harvard University Press, Cambridge MA
Dupré J (1999) On the impossibility of a monistic account of species. In: Wilson RA (ed) Species: new interdisciplinary essays. MIT Press, Cambridge MA, pp 3–22
Ehrlich PR, Raven PH (1969) Differentiation of populations. Science 165:1228–1232
Ereshefsky M (1989) Where's the species?: comments on the phylogenetic species concepts. Biol Philos 4:89–96
Ereshefsky M (1998) Species pluralism and anti-realism. Philos Sci 65:103–120
Ereshefsky M (1999) Species and the Linnaean hierarchy. In: Wilson RA (ed) Species: new interdisciplinary essays. MIT Press, Cambridge MA, pp 285–305
Ereshefsky M (2001) The poverty of the Linnaean hierarchy: a philosophical study of biological taxonomy. Cambridge University Press, Cambridge
Farquhar JE, Prates I, Doughty P, Rabosky DL, Chapple DG (2024) Morphological and genetic data challenge species and subspecies in the Lerista microtis group (Squamata: Scincidae). Zootaxa 5437:336–362
Fischer A, Pollack J, Thalmann O, Nickel B, Pääbo S (2006) Demographic history and genetic differentiation in apes. Curr Biol 16:1133–1138
Fitzpatrick JW (2010) Subspecies are for convenience. Ornithol. Monogr 67:54–61
Franz NM (2005) On the lack of good scientific reasons for the growing phylogeny/classification gap. Cladistics 21:495–500
Frost DR, Hillis DM (1990) Species in concept and practice: herpetological applications. Herpetologica 46:86–104
Fünfstück T, Arandjelovic M, Morgan DB, Sanz C, Reed P, Olson SH, Cameron K, Ondzie A, Peeters M, Vigilant L (2015) The sampling scheme matters: *Pan troglodytes troglodytes* and

P. t. schweinfurthii are characterized by clinal genetic variation rather than a strong subspecies break. Am J Phys Anthropol 156:181–191
Gillham NW (1956) Geographic variation and the subspecies concept in butterflies. Syst Zool 5:110–120
Gonder MK, Locatelli S, Ghobrial L, Mitchell MW, Kujawski JT, Lankester FJ, Stewart C-B, Tishkoff SA (2011) Evidence from Cameroon reveals differences in the genetic structure and histories of chimpanzee populations. Proc Natl Acad Sci USA 108:4766–4771
Gruber K, Schöning C, Otte M, Kinuthia W, Hasselmann M (2013) Distinct subspecies or phenotypic plasticity? Genetic and morphological differentiation of mountain honey bees in East Africa. Ecol Evo 3:3204–3218
Haig SM, Beever EA, Chambers SM, Draheim HM, Dugger BD, Dunham S, Elliott-Smith E, Fontaine JB, Kesler DC, Knaus BJ, Lopes IF, Loschl P, Mullins TD, Sheffield LM (2006) Taxonomic considerations in listing subspecies under the U.S. Endangered Species Act. Conserv Biol 20:1584–1594
Hazelwood CC (2021) The species category as a scientific kind. Synthese 198:3027–3040
Hey J (2001) Genes, categories, and species: the evolutionary and cognitive causes of the species problem. Oxford University Press, Oxford
Hey J (2010) The divergence of chimpanzee species and subspecies as revealed in multipopulation isolation-with-migration analyses. Mol Biol Evol 27:921–933
Hillis D (2020) The detection and naming of geographic variation within species. Herpetol Rev 51:52–56
Hillis D (2021) New and not-so-new conceptualizations of species and subspecies: a reply to the "it's species all the way down" view. Herpetol Rev 51:49–50
Horn AD (1979) The taxonomic status of the bonobo chimpanzee. Am J Phys Anthrop 51:273–282
Huang Y, Morrison GR, Brelsford A, Franklin J, Jolles DD, Keeley JE, Parker VT, Saavedra N, Sanders AC, Stoughton TR, Wahlert GA, Litt A (2000) Subspecies differentiation in an enigmatic chaparral shrub species. Am J Bot 107:1–18
Hull D (1989) The metaphysics of evolution. State University of New York Press, Albany
James FC (2010) Avian subspecies: introduction. Ornithol. Monogr 67:1–5
Jeratthitikul E, Phuangphong S, Sutcharit C, Prasankok P, Kongim B, Panha S (2019) Integrative taxonomy reveals phenotypic plasticity in the freshwater mussel *Contradens contradens* (Bivalvia: Unionidae) in Thailand, with a description of a new species. Syst Biodivers 17:134. https://doi.org/10.1080/14772000.2018.1554607
Jorgensen MC, Sierwald P, Mason-Gamer RJ (2012) A review of subspecies recognition in polydesmidan millipedes (Diplopoda) with a revision of the subspecies of Euryuridae (Xystodesmoidea). Zool Scri 42:317–326
Joyce DA, Dennis RLH, Bryant SR, Shreeve TG, Ready JS, Pullin AS (2009) Do taxonomic divisions reflect genetic differentiation? A comparison of morphological and genetic data in *Coenonympha tullia* (Müller), Satyrinae. Biol J Linn Soc 97:314–327
Kendig C (2022) Metaphysical presuppositions about species stability: problematic and unavoidable. In: Wilkins J, Zachos FE, Ya PI (eds) Species problems and beyond: contemporary issues in philosophy and practice. CRC Press, pp 293–313
Kitcher P (1984) Species. Philos Sci 51:308–333. Ehrlich
Kluge AG (1990) Species as historical individuals. Biol Philos 5:417–431
Kunz W (2012) Do species exist?: principles of taxonomic classification. Wiley-Blackwell, Weinheim
LaPorte J (2005) Is there a single objective, evolutionary tree of life? J Philos 102:357–374
Lee MSY (2003) Species concepts and species reality: salvaging a Linnaean rank. J Evol Biol 16:179–188
Mallet J (1995) A species definition for the Modern Synthesis. Trends Ecol Evol 10:294–299
Mallet J (2013) Subspecies, semispecies, superspecies. In: Levin SA (ed) Encyclopedia of Biodiversity, vol 7, 2nd edn. Academic Press, Waltham MA, pp 45–48
Mayden RL (1997) A hierarchy of species concepts: the denouement in the saga of the species problem. In: Claridge MF, Dawah HA, Wilson MR (eds) Species: the units of biodiversity. Chapman and Hall, New York, pp 381–424

Mayden RL (1999) Consilience and a hierarchy of species concepts: advances toward closure on the species puzzle. J Nematol 31:95–116
Mayr E (1942) Systematics and the origin of species. Columbia University Press, New York
Mayr E (1963) Animals species and evolution. Harvard University Press, Cambridge
Mayr E (1982) Of what use are subspecies? Auk 99:593–595
Mayr E (2000) The biological species concept. In: Wheeler QD, Meier R (eds) Species concepts and phylogenetic theory: a debate. Columbia University Press, New York, pp 17–29
McBrearty S, Jablonski NG (2005) First fossil chimpanzee. Nature 437:105–108
Minelli A (2021) Taxonomy needs pluralism, but a controlled and manageable one. Megataxa 1:9–18
Minelli A, Fusco G (2012) On the evolutionary developmental biology of speciation. Evol Biol 39:242–254
Mishler BD (1999) Getting rid of species? In: Wilson RA (ed) Species: new interdisciplinary essays. MIT Press, Cambridge MA, pp 306–315
Mishler BD (2010) Species are not uniquely real biological entities. In: Ayala FJ, Arp R (eds) Contemporary debates in philosophy of biology. Wiley-Blackwell, Oxford, pp 110–122
Mishler BD (2021) What, if anything, are species? CRC Press, Boca Raton
Mishler BD (2022) Ecology, evolution, and systematics in a post-species world. In: Wilkins J, Zachos FE, Ya PI (eds) Species problems and beyond: contemporary issues in philosophy and practice. CRC Press, pp 177–190
Mishler BD, Brandon RN (1987) Individuality, pluralism, and the phylogenetic species concept. Biol Philos 2:397–414
Mishler BD, Donoghue MJ (1982) Species concepts: a case for pluralism. Syst Zool 31:491–503
Mishler BD, Theriot EC (2000) The phylogenetic species concept (*sensu* Mishler and Theriot): monophyly, apomorphy, and phylogenetic species concepts. In: Wheeler QD, Meier R (eds) Species concepts and phylogenetic theory: a debate. Columbia University Press, New York, pp 44–54
Morin PA, Moore JJ, Chakraborty R, Jin L, Goodall J, Woodruff DS (1994) Kin selection, social structure, gene flow, and the evolution of chimpanzees. Science 265:1193–1201
Mousseau T, Sikes DS (2011) Almost but not quite a subspecies: a case of genetic but not morphological diagnosability in *Nicrophorus* (Coleoptera: Silphidae). Biol J Linn Soc 102:311–333
Nixon KC, Wheeler QD (1990) An amplification of the phylogenetic species concept. Cladistics 6:211–223
Novick A, Doolittle WF (2021) 'Species' without species. Stud Hist Philos Sci A 87:72–80
O'Brien SJ, Mayr E (1991) Bureaucratic mischief: recognizing endangered species and subspecies. Science 251:1187–1188
Oderberg DS (2007) Real essentialism. Routledge, New York
Padial JM, De la Riva I (2021) A paradigm shift in our view of species drives current trends in biological classification. Biol Rev 96:731–751
Patten MA (2010) Null expectations in subspecies diagnosis. Ornithol Monogr 67:35–41
Patten MA (2015) Subspecies and the philosophy of science. Auk 132:481–485
Patten MA, Remsen JV (2017) Complementary roles of phenotype and genotype in subspecies delimitation. J Hered 108:462–464
Patten MA, Unitt P (2002) Diagnosability versus mean differences of sage sparrow subspecies. Auk 119:26–35
Pavlinov IY (2022) The species problem from a conceptualist's viewpoint. In: Wilkins J, Zachos FE, Pavlinov IY (eds) Species problems and beyond: contemporary issues in philosophy and practice. CRC Press, pp 245–278
Pavlinov IY (2023) The species problem: a conceptual history. CRC Press, New York
Phillimore AB, Owens IPF (2006) Are subspecies useful in evolutionary and conservation biology? Proc R Soc B 273:1049–1053
Pigliucci M (2003) Species as family resemblance concepts: the (dis-)solution of the species problem? BioEssays 25:596–602

Prates I, Doughty P, Rabosky DL (2023) Subspecies at crossroads: the evolutionary significance of genomic and phenotypic variation in a wide-ranging australian lizard (*Ctenotus pantherinus*). Zool J Linnean Soc 197:768–786

Remsen JV (2010) Subspecies as a meaningful taxonomic rank in avian classification. Ornithol Monogr 67:62–78

Reydon TAC (2004) Why does the species problem still persist? BioEssays 26:300–305

Reydon TAC (2005) On the nature of the species problem and the four meanings of species. Stud Hist Phil Biol Biomed Sci 36:135–158

Reydon TAC (2022) 'Species' as a technical term multiple meanings in practice, one idea in theory. In: Wilkins J, Zachos FE, Pavlinov IY (eds) Species problems and beyond: contemporary issues in philosophy and practice. CRC Press, pp 65–89

Reydon TAC, Kunz W (2021) Classification below the species level: when are infraspecific groups biologically meaningful? Biol J Linn Soc 134:246–260

Richards RA (2010) The species problem: a philosophical analysis. Cambridge University Press, Cambridge

Rieppel O (2010) Species monophyly. J Zool Syst Evol Res 48:1–8

Rosel PE, Taylor BL, Hancock-Hanser BL, Leslie MS, Berta A, Cipriano F, Parson KM, Viricel A, Vollmer NL, Martien KK (2017) A review of genetic markers and analytical approaches that have been used for delimiting marine mammal subspecies and species. Mar Mamm Sci 33:56–75

Schiaffini MI (2020) Are subspecies (of *Eira barbara*) real? J Mammal 101:1410–1425

Schultz AH (1969) The skeleton of the chimpanzee. In: Bourne GH (ed) The chimpanzee, vol 1. Karger, Basel and New York, pp 50–103

Sepkoski D (2016) The "species concept" and the beginnings of paleobiology. In: Allmon WD, Yacobucci MM (eds) Species and speciation in the fossil record. The University of Chicago Press, Chicago and London, pp 9–27

Shea BT, Coolidge HJ (1988) Craniometric differentiation and systematics in the genus *Pan*. J Hum Evol 17:671–685

Sigwart JD (2018) What species mean: a user's guide to the units of biodiversity. CRC Press, Boca Raton

Skalski JR, Townsend RL, McDonald LL, Kern JW, Millspaugh JJ (2008) Type I errors linked to faulty statistical analyses of endangered subspecies classifications. J Agric Biol Environ Stat 13:199–2020

Smith IA (2016) The intrinsic value of endangered species. Routledge, New York

Smith DG, Ng J, George D, Trask JS, Houghton P, Singh B, Villano J, Kanthaswamy S (2014) A genetic comparison of two alleged subspecies of Philippine cynomolgus macaques. Am J Phys Anthropol 155:136–148

Sokal RR, Sneath PHA (1963) Principles of numerical taxonomy, vol 12. W. H, Freeman, San Francisco, p 190

Stamos DN (2003) The species problem: biological species, ontology, and the metaphysics of biology. Lexington books, Lanham

Tautz D, Arctander P, Minelli A, Thomas RH, Vogler AP (2003) A plea for DNA taxonomy. Trends Ecol Evol 18:70–74

Taylor BL, Perrin WF, Reeves RR, Rosel PE, Wang JY, Cipriano F, Baker CS, Brownell RL (2017) Why we should develop guidelines and quantitative standards for using genetic data to delimit subspecies for data-poor organisms like cetaceans. Mar Mamm Sci 33:12–26

Templeton AR (1989) The meaning of species and speciation: a genetic perspective. In: Otte D, Endler JA (eds) Speciation and its consequences. Sinauer Associates, Sunderland, Massachusetts, pp 3–27

Thorpe RS (1987) Geographic variation: a synthesis of cause, data, pattern and congruence in relation to subspecies, multivariate analysis and phylogenesis. Boll Zool 54:3–11

Tobias JA, Seddon N, Spottiswoode CN, Pilgrim JD, Fishpool LDC, Collar NJ (2010) Quantitative criteria for species delimitation. Ibis 152:724–746

Torstrom SM, Pangle KL, Swanson BJ (2014) Shedding subspecies: The influence of genetics on reptile subspecies taxonomy. Mol Phylogenet Evol 76:134–143

Vane-Wright RI, Tennent WJ (2011) Colour and size variation in *Junonia villida* (Lepidoptera, Nymphalidae): subspecies or phenotypic plasticity? Syst Biodivers 9:289–305

Velasco JD (2008) Species concepts should not conflict with evolutionary history, but often do. Stud Hist Phil Biol Biomed Sci 39:407–414

Vervaecke H, Van Elsacker L (1992) Hybrids between common chimpanzees (*Pan troglodytes*) and pygmy chimpanzees (*Pan paniscus*) in captivity. Mammalia 56:667–669

Vinarski MV (2015) The fate of subspecies category in zoological systematics. 2. The present. Biol Bull Rev 5:405–414

Wallin H, Kvamme T, Bergsten J (2017) To be or not to be a subspecies: description of *Saperda populnea lapponica* ssp. n. (Coleoptera, Cerambycidae) developing in downy willow (*Salix lapponum* L.). ZooKeys 691:103–148

Wang X, Guo Z, He Z, Xu S, Shao S, Li S, Yang M, Chen Q, Zhong C, Wu Z, Duke NC, Shi S (2019) In the absence of reproductive isolation—extensive gene flow after speciation. bioRxiv. https://doi.org/10.1101/622019

Wang X, He Z, Shi S, Wu C-I (2020) Genes and speciation: is it time to abandon the biological species concept? Natl Sci Rev 7:1387–1397

Wang X, He Z, Guo Z, Yang M, Xu S, Chen Q, Shao S, Li S, Zhong C, Duke NC, Shi S (2022) Extensive gene flow in secondary sympatry after allopatric speciation. Natl Sci Rev 9. https://doi.org/10.1093/nsr/nwac280

Westram AM, Stankowski S, Surendranadh P, Barton N (2022) What is reproductive isolation? J Evol Biol 35:1143–1164

Wheeler Q (2024) Species, science and society: the role of systematic biology. Routledge, London and New York

Wheeler QD, Platnick NI (2000) The phylogenetic species concept (*sensu* Wheeler and Platnick). In: Wheeler QD, Meier R (eds) Species concepts and phylogenetic theory: a debate. Columbia University Press, New York, pp 55–69

Wilkins JS (2009) Species: a history of the idea. University of California Press, Berkeley, Los Angeles, London

Wilkins JS (2022) The good species. In: Wilkins J, Zachos FE, Pavlinov IY (eds) Species problems and beyond: contemporary issues in philosophy and practice. CRC Press, pp 105–124

Wilkins JS (2023) Understanding species. Cambridge University Press, Cambridge

Williams PA (1992) Confusion in cladism. Synthese 91:135–152

Wilson RA (1999) Realism, essence, and kind: resuscitating species essentialism? In: Wilson RA (ed) Species: new interdisciplinary essays. MIT Press, Cambridge MA, pp 187–207

Wilson JJ (2011) Taxonomy and DNA sequence databases: a perfect match? Terr Arthropod Rev 4:221–236

Wilson EO, Brown WL (1953) The subspecies concept and its taxonomic application. Syst Zool 2:97–111

Winker K (2010) Subspecies represent geographically partitioned variation, a gold mine of evolutionary biology, and a challenge for conservation. Ornithol Monogr 67:6–23

Wu S, Wang Y, Wang Z, Shrestha N, Liu J (2022) Species divergence with gene flow and hybrid speciation on the Qinghai-Tibet Plateau. New Phytol 234:392–404

Zachos FE (2016) Species concepts in biology: historical development, theoretical foundations and practical relevance. Springer

Zachos FE (2022) Critique of taxonomic reason(ing): nature's joints in light of an 'Honest' Species Concept and Kurt Hübner's historistic philosophy of science. In: Wilkins J, Zachos FE, Pavlinov IY (eds) Species problems and beyond: contemporary issues in philosophy and practice. CRC Press, pp 315–340

Zink RM (2004) The role of subspecies in obscuring avian biological diversity and misleading conservation policy. Proc R Soc Lond B 271:561–564

Units of Evolution (II): Higher Taxa and Homologues

8

Abstract

This chapter introduces a new theoretical framework that supports the recognition of higher taxa and homologues as units of evolution. It begins by exploring the challenges of recognizing higher taxa and homologues as evolving entities in their traditional sense. For example, traditionally, higher taxa are seen as entities reflecting the historical record of species, lacking evolutionary processes, while homologues are defined as any element that allows the inference of a genealogical connection. However, the H-Evo-Devo Theory offers a framework that grants higher taxa and homologues a distinct ontological reality. According to this model, higher taxa are groups distinguished by great distances within the morphospace. Specifically, they are more inclusive body plans than the species body plan within a hierarchy of body plans. Higher taxa are regarded as active evolutionary entities, whose study helps to enhance the understanding of processes and patterns at the higher levels of the hierarchy. Meanwhile, homologues are defined as morphological modules; that is, they are quasi-autonomous parts of the organism that are highly integrated in developmental terms and maintained phylogenetically over a long time. Homologues are higher-level entities, representing the final products of development that integrate all levels of hierarchical organization. This perspective aims to resolve the debate between systematists and morphologists by unifying the concept of homology through a morphological approach.

Keywords

H-Evo-Devo Theory · Units of evolution · Higher taxa · Homologues · Phylogenetic systematics · Evo-Devo · Typology · *Bauplan* · Body plan · Natural kinds · Evolutionary individuals · Morphological gaps · Homology · Synapomorphy · Modularity · Developmental constraints · Saltational evolution · Megaevolution · Modular-evolution · Linnaean hierarchy

C. Ochoa, *Hierarchical Evolutionary-Developmental Theory*, Evolutionary Biology – New Perspectives on Its Development 9,
https://doi.org/10.1007/978-3-031-88723-9_8

Many biologists take for granted that higher taxa do not exist as independent entities, arguing that these groups lack their own evolutionary processes and depend on what occurs at the lower levels. In other words, higher taxa are viewed as historical constructs that merely reflect genealogical relationships between species. Nevertheless, I challenge this perspective. While acknowledging that some higher taxa are historical entities, I argue that most of them can be regarded as true units of evolution. Another key issue is homology, which originally referred to the correspondence of morphological structures. However, this concept expanded to include attributes of organisms at different levels of organization. Under the traditional framework, homology is defined as the same features shared among two or more species due to inheritance from a common ancestor. Nevertheless, I will demonstrate that this traditional definition is contentious. Instead, the concept of "homology" aligns more effectively with the theory of modularity, making it possible to assert that homologues are, indeed, units of evolution.

Section 8.1 advocates for higher taxa as units of evolution. It begins by describing higher taxa according to the traditional framework, a view under which higher taxa are not real and lack evolutionary processes, merely representing the historical record of species. It demonstrates that this perspective is rooted in Willi Hennig's and Ernst Mayr's interpretations of phylogenetic systematics and the Modern Synthesis, respectively. Consecutively, it delves into the H-Evo-Devo Theory, which proposes that higher taxa are real entities with active roles in evolution. According to this view, higher taxa are distinguished from each other by great distances within the morphospace, but, specifically, they are more inclusive body plans than species/genus body plan within a hierarchy of body plans. Section 8.2 advocates for homologues as units of evolution. It begins with the development of the concept of homology, particularly focusing on the traditional view of phylogenetic systematics. It then highlights the shortcomings of this concept from this perspective. Finally, it explores the conceptualization of homologues according to the H-Evo-Devo Theory, contending that homologues are equivalent to the modules of morphology, that is, homologues are quasi-autonomous parts that are highly integrated in developmental terms and maintained phylogenetically over a long time.

8.1 Higher Taxa as Units of Evolution

Traditionally, higher taxa are defined as groups comprising two or more species, reflecting the historical record among species. According to this perspective, taxonomical categories—such as genus, family, order, and class—indicate the timing and the points of divergence among related species in a phylogenetic tree. Despite their significance in emphasizing phylogenetic relationships, most biologists agree that they lack any meaningful role as agents of evolution. Some objections to higher taxa as units of evolution include the following:

1. Higher taxa represent the genealogical relationships among "species." In other words, given that species are considered true units within a hierarchical system,

higher taxa are no more than the evolutionary history of species (Eldredge and Cracraft 1980; Wiley 1981, p. 75; Eldredge 2001, p. 344; Wiley and Lieberman 2011, p. 73; Tattersall 2012).

2. Higher taxa emerge due to the processes tracing back to speciation. If it were demonstrated that no other processes could explain their occurrence except for traditional mechanisms operating at lower levels, there would be no reason to consider them real (Mayr 1963, p. 601; Eldredge and Cracraft 1980; Wiley 1981, p. 75; Ghiselin 1987, p. 141; Arthur 2003, p. 257; Wiley and Lieberman 2011, p. 73; Cartmill 2012; Tëmkin and Eldredge 2015, p. 195).
3. Higher taxa lack cohesion; that is, they do not have integrative processes whereby the parts act as a coherent whole (e.g., gene flow, genetic homeostatic, or similar selection pressure) (Eldredge and Cracraft 1980; Wiley 1981, p. 75; Lee 2003, p. 184).
4. Higher taxa do not consist of groups of interbreeding organisms; therefore, they are not reproductive units (Hennig 1966; Eldredge and Cracraft 1980; Lee 2003).
5. Higher taxa are not interactors; they lack a process analogous to the selection acting upon units at a lower level in the hierarchy. Therefore, they are not evolving entities (Gould 2002, pp. 712–713).

However, the H-Evo-Devo Theory presents a significantly different interpretation regarding the nature of higher taxa. According to this theory, although higher taxa reflect the genealogical relationships between their parts, a notable feature is that they do not form a continuum but are instead characterized by exhibiting greater morphological gaps among them (Kemp 2016). Another characteristic is that they are identified through a body plan, also known as "type" or "*Bauplan*." This recognition is possible because their parts exhibit a highly integrated common structural pattern that appears to be unchanged yet allows for certain ranges of variation as long as they do not exceed these boundaries (Raff 1996; Hall 1999; Rieppel 2006; DiFrisco and Wagner 2022).

In short, within the framework of the H-Evo-Devo Theory, higher taxa are more inclusive body plans than the species/genus body plan within a hierarchy of body plans. The significance of this perspective lies in viewing higher taxa not merely as abstract designations that indicate the genealogical nexus between species but as active evolutionary entities whose study enhances the understanding of processes and patterns at the higher levels of the hierarchy. The next pages explore the traditional view of higher taxa, followed by the H-Evo-Devo Theory's proposal that considers higher taxa true units of evolution.

8.1.1 Higher Taxa Under the Traditional View

The argument that higher taxa should be considered non-natural entities with no evolutionary significance other than historical representations of species lacks a solid foundation and emerges from the interpretations of Willi Hennig's phylogenetic systematics and Ernst Mayr's evolutionary view.[1] Unfortunately, both perspectives have exerted a profound influence on current views regarding the relationship between taxonomy and evolution.

From the perspective of phylogenetic systematics, as I have indicated throughout this book, the phylogenetic tree is constructed at a single level by depicting a continuous branching sequence. It is important to note that this approach does not consider any hierarchical structure of real inclusion but rather presents a positional structure (see Williams 1992; Valentine and May 1996; Valentine 2004; Vasilyeva and Stephenson 2008; Brower 2023). Interestingly, when depicting phylogenetic trees under a positional structure, critics of the Linnaean hierarchy argue that ranks lack ontological meaning (e.g., Ereshefsky 2001). Regarding this, some point out that categories can indicate times of evolutionary divergence (e.g., de Queiroz and Gauthier 1994). For instance, a genus emerges when an ancestral species gives rise to one or more descendant species; thus, a genus denotes that those species are more closely related to each other.

Now, considering that speciation processes occur continuously over time, categories—such as family, order, and class—indicate the absolute time by which species diversify. The longer the time that has passed since the first divergence point, the higher the rank of the category. Put differently, the more distant the genealogical connections of a species are, the more inclusive the category. For example, if two species diverged a long time ago, the group that includes them could be considered a family or an order, whereas if the divergence occurred more recently, it might only be classified as a genus (Fig. 8.1).[2] Nevertheless, according to Ereshefsky (2001, pp. 213–215), the problem is that these splits have occurred numerous times, making the 21 ranks in the Linnaean hierarchy insufficient. In addition, for Ereshefsky, this designation of ranks differs significantly from the original position of the Linnaean system, which was structured to support essentialism and creationism.[3] Therefore, since the ranks in the Linnaean system lack any biological significance, some scholars have advocated for their abandonment (e.g., Ereshefsky 2001; Zachos 2011; Mishler 2022). In fact, new nomenclature rules have been developed without consideration of ranks, despite the acknowledgment that higher taxa are hierarchically related (see Cantino and de Queiroz 2020; Laurin 2024).

Currently, most researchers consider phylogenetic systematics the most accurate approach for representing taxonomic exercises. However, Williams (1992) highlights inconsistencies in phylogenetic systematics due to Henning's—this school's founder—confused attempt to equate his "divisional hierarchy" with the Linnaean

[1] As discussed in Chap. 7, these perspectives have muddled not only the recognition of higher taxa but also that of subspecies and species.

[2] Additional criteria for establishing higher ranks from this perspective include (a) the presence of numerous species within a particular group, (b) the discovery of a sister group within an already established clade, (c) the existence of paraphyletic ancestral groups due to plesiomorphies, and d) uncertain relationships between certain members and other groups (see Smith 1994).

[3] As discussed in Chap. 2, this accusation is false. The Linnaean system was established to describe a hierarchy of types that underlies evolutionary causes as well.

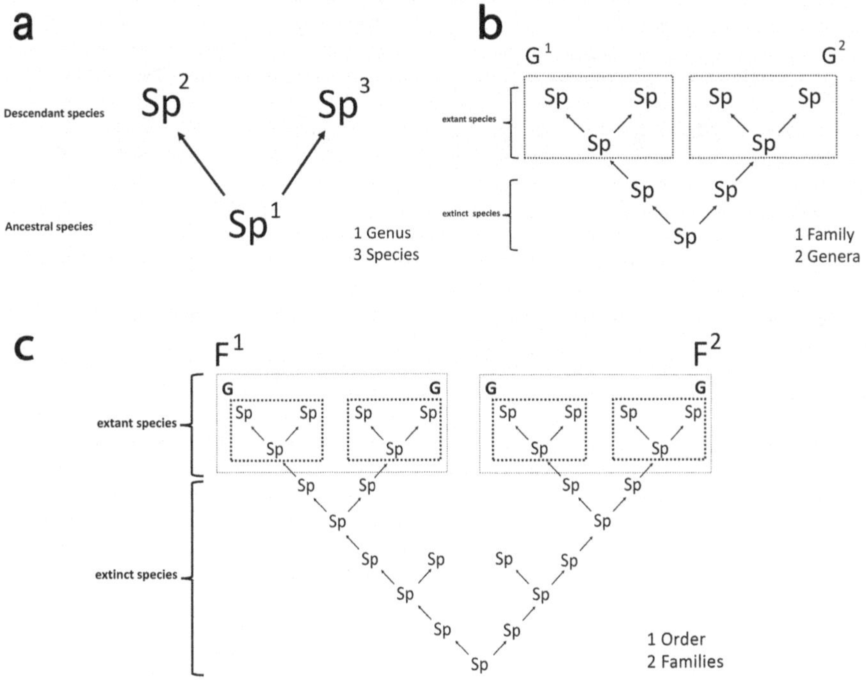

Fig. 8.1 *The origin of higher taxa according to the traditional view of phylogenetic systematics.* Higher taxa arise sequentially through the process of species divergence forming a continuum on the phylogenetic tree. (**a**) A genus arises when an ancestral species gives rise to one or more descendant species. Therefore, a genus denotes that those species are more closely related to each other. (**b**) As time progresses, new species arise while others disappear. During this process, at the two ends of the set of extant species, two genera are formed; thus, a family originates. (**c**) The origin of an order occurs when two families arise from the two ends of the set of extant species, and so on. Therefore, the boundaries of higher taxa are established by the absolute time of species divergence. The longer the time that has passed since the first divergence point, the higher the rank of the category

hierarchy. According to Williams, Hennig's divisional hierarchy encompasses a sequence of ancestors and descendants in a unidirectional, temporal, and non-inclusive relationship of connections between species. As is well known, Hennig asserted that the best classification was achieved by indicating only genealogical relationships between groups as the basis for an evolutionary explanation, disregarding the methods of "idealistic morphology" (Hamilton 2014).

In constructing his taxonomic viewpoint, Hennig proposed a supposedly hierarchical scheme of inclusion, suggesting that the boundaries of higher taxa were defined by the ancestral species including all its descendant species (Fig. 8.2). But it is important to note that this is a false hierarchy because ancestral species do not hierarchically include their descendant species, just as a mother does not hierarchically include her daughters (see Chap. 4, Fig. 4.3). Williams further notes that Hennig's divisional hierarchy depended entirely on knowledge of ancestral species.

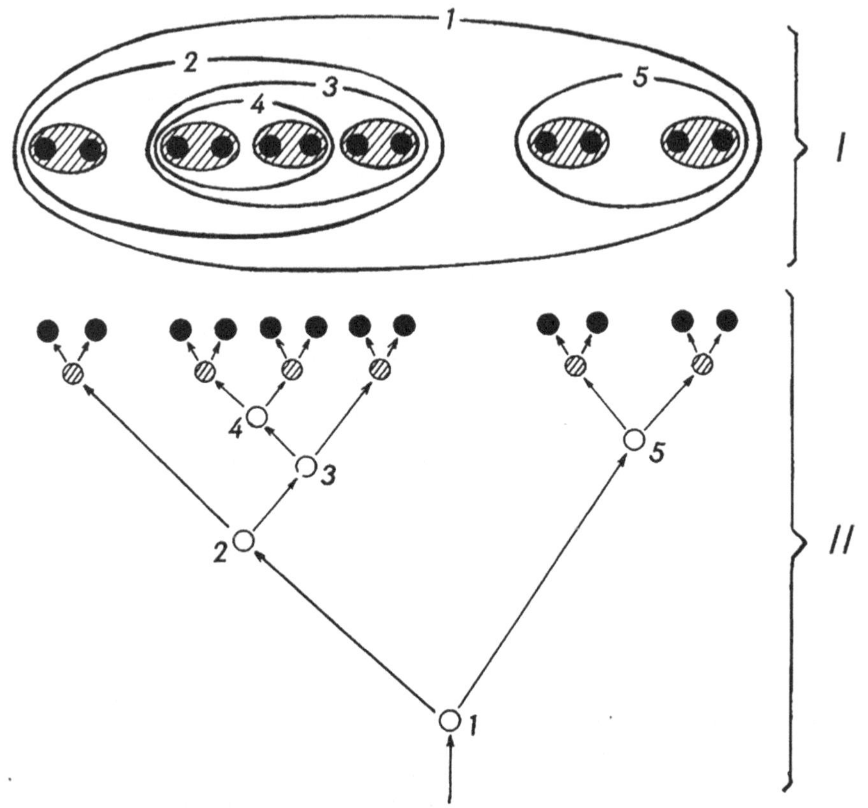

Fig. 8.2 *The boundaries of higher taxa according to Hennig*. The differently colored circles represent species. White and hatched circles represent ancestral species, while black circles represent extant species. I. The phylogenetic tree is viewed as a nested compositional hierarchy, where the boundaries of higher taxa are delineated. For example, 1 includes 2 and 5, and 2 includes 3 and 4. Hatched circles include black circles. II. The phylogenetic tree is shown in its traditional form. Circle number 1 is the earliest ancestral species, which includes all descendant species. Circle number 2 is an ancestral species that includes all its descendant species, along with ancestral species 3 and all its descendant species, and so on. Hatched circles denote the ancestral species of black circles. Even though this scheme is the most accepted representation of the phylogenetic tree in taxonomy, Williams (1992) argues that this representation is overly confusing. In short, Hennig's phylogenetic tree lacks a true hierarchy because ancestral species do not include their descendant species, just as a mother does not hierarchically include her daughters (see also Fig. 4.3 in Chap. 4). (From Hennig 1966)

For example, Hennig recognized methodological difficulties in identifying such species, so he only contemplated the hierarchy using extant species. Nonetheless, extant species alone do not reflect any hierarchical relationship under Hennig's original criterion. Williams argues that, in this context, it is hard to understand why

Hennig used the time of origin as a criterion for establishing ranks, as his scheme ultimately excluded the entire series of ancestral species identified over time:

> If Hennig relegates ancestral species to the paleontologist and deals only with recent species, then he cannot talk about time because the existence of species across time has been banned from the scheme altogether. He is no longer speaking of "recency of common ancestry"... as he claims but, rather, similarity of characters—precisely the basis for biological taxonomy that his system was developed to counteract. (Williams 1992, p. 145)

Certainly, in the absence of a morphological criterion that establishes genuine boundaries of inclusiveness, Hennig's phylogenetic systematics model becomes rather ambiguous for defining the limits of higher taxa. Surprisingly, this inconsistent interpretation of taxonomy and classification continues to be embraced today despite its numerous flaws (see Williams 1992).

With respect to the traditional evolutionary view, it is not entirely accurate to say that the architects of the Modern Synthesis relegated the evolutionary role of higher taxa. As explored in Chap. 3, while they all start from the premise that the mechanisms occurring at the microevolutionary level are sufficient to explain macroevolution, they endeavored to develop a hierarchy theory of evolution, providing different interpretations of categories above the species level. Yet the implications of Ernst Mayr's systematics during the development of the Modern Synthesis are evident. In his early writings, Mayr sought to establish the foundations of the "new systematics," focusing on the study of the lowest level of the hierarchy. According to Mayr's original argument, the evolution of populations is a continuous process. Nonetheless, the first degrees of discontinuity occur due to isolation mechanisms acting on these populations. Therefore, the taxonomic structure at all levels—subspecies, species, genus, family, etc.—forms a continuous sequence with minimal discontinuities:

> In conclusion we may say that all the available evidence indicates that the origin of the higher categories is a process which is nothing but an extrapolation of speciation. All the processes and phenomena of macroevolution and of the origin of the higher categories can be traced back to intraspecific variation, even though the first steps of such processes are usually very minute. (Mayr 1942, p. 298)

Notably, Mayr initially considered species to arise in the same way through this continuum. However, in later publications, he changed his mind, coming to view species as cohesive entities whose reality relies on their ability to interbreed. Specifically, a species maintains cohesion through gene exchange within the population, preserving its distinctiveness via isolation mechanisms that prevent the entry of genetic material from other populations. Mayr stated that this distinction delineates the boundaries between species and higher taxa, as the latter do not exchange genes and thus lack cohesion. Interestingly, Hennig echoed a similar argument, proposing that species are reproductive communities whose members exhibit a reticulating tokogenetic relationship, contrasting with higher taxa characterized by an exclusively phylogenetic "hierarchical" relationship. But, as discussed in the previous chapter (Chap. 7, Sect. 7.2), this stance encounters many difficulties in delineating boundaries between species and higher taxa, which extends to challenges explaining events of evolutionary divergence, etc.

Given the success of the Modern Synthesis and phylogenetic systematics, many contemporary scholars advocate for the position that species are units of evolution while higher taxa are not. Nevertheless, it is important to emphasize Ereshefsky's (1991) insights on this matter, as the belief that higher taxa are less real than species presents numerous challenges. For example, many biologists argue that since species are "reproductive units" (i.e., groups of interbreeding organisms), they are units of evolution. In contrast, higher taxa are not units of evolution because they are not reproductive units. However, Ereshefsky points out that "those processes which can cause a lineage to evolve do not require such lineages to be reproductive units" (pp. 88–89). In addition, it is commonly mentioned that species are units of evolution because they possess cohesion, while higher taxa lack such a process; therefore, they are not units of evolution. But Ereshefsky maintains that the nature of this species cohesion is far from clear. Even gene flow—considered the main cause of species uniformity—is controversial (see Ehrlich and Raven 1969; Barker 2007; Barker and Wilson 2010). Alternatively, some biologists claim that species are units of evolution because all their members evolve uniformly, giving rise to new species. If species have an exclusive process that allows them to be recognized as units of evolution, namely, speciation, then higher taxa are not units of evolution because their origin goes back to speciation. Yet "such speciation events do not occur in entire species but only in their founding population... Just as speciation events do not occur in entire higher taxa, speciation events do not occur in entire species either" (Ereshefsky 1991, pp. 93–94).

While Ereshefsky (1991) does not endorse the assumption that higher taxa are units of evolution, he claims that species face the same difficulties as higher taxa. According to Ereshefsky's perspective, both categories are arbitrary and represent historical events that go back to the lowest level, namely, the continuous origin of geographic populations. Consequently, their differences are only a matter of degree: "higher taxa and many species fall into the same ontological category, namely that of being a historical entity" (p. 95). Although I agree with Ereshefsky's conclusions, ultimately, some authors have considered higher taxa as real entities for several reasons (e.g., Schaefer 1976; Brothers 1985; Smith 1994; Hey 2001, p. 174; Krassilov 2004; Valentine 2004, pp. 448-449; Kemp 2007, 2016; Humphreys and Barraclough 2014; Barraclough and Humphreys 2015; Caponi 2020, 2023; Devitt 2023). Some argue that they behave as evolutionary individuals (Hull 1976, p. 183; Brothers 1985; Coleman and Wiley 2001; Caponi 2023); others propose that they are HPC natural kinds (Boyd 1999, 2010, 2021; Rieppel 2005b) or both at the same time (Brigandt 2009). Supporting this view, as seen below, the H-Evo-Devo Theory considers that most higher taxa are real and, thereby, true units of evolution.

8.1.2 Higher Taxa According to the H-Evo-Devo Theory

Recently, philosopher Michael Devitt (2023) critiqued the widespread skepticism regarding higher taxonomic categories, arguing that these categories serve as an approximate delineation of the levels of a hierarchy of explanatory taxa: "The ubiquitous talk of the higher categories in biology is not 'just pragmatic'; it serves a theoretical purpose" (p. 2). This perspective aligns particularly well with the H-Evo-Devo Theory, considering that the hierarchy of types (a hierarchy of explanatory taxa) can be translated through the ranks of the Linnaean hierarchy. As Devitt points out, the species category can be used as a reference to determine the subsequent categories. In the case of the H-Evo-Devo Theory, species are the least inclusive body plan. While the other categories are not universally designated to specific ranks, they indicate approximate levels in the hierarchy of types.

So, according to the H-Evo-Devo Theory, *higher taxa are groups distinguished by great distances within the morphospace, but, specifically, they are more inclusive body plans than the species/genus body plan within a hierarchy of body plans.* The Linnaean hierarchy is useful for representing these body plans, as it delineates ranks of hierarchical inclusion (e.g., genera within a family, families within an order, orders within a class). In this regard, the ranks—not merely the names assigned to the categories—seem more real than arbitrary. Importantly, these entities have an active role in evolution, exhibiting specific processes and patterns that manifest at higher levels of the hierarchy (megaevolution).

But how does the H-Evo-Devo Theory support the idea that higher taxa are units of evolution? Notably, developmental constraints have an ontological relevance to higher taxa. First, they serve as the causal mechanism that explains the cohesion of their parts, leading to their recognition as evolutionary individuals. Second, they allow for perceiving the projection of body plans, thus manifesting as natural kinds. Therefore, acknowledging that higher taxa represent body plans (types or *Baupläne*) makes it possible to deduce their existence. To further elucidate this argument and advocate for higher taxa as units of evolution, I will now offer some brief clarifications on this ontological identification.

Evolutionary Individuals As discussed in Chap. 6, individuals are identified by three main criteria: origin, stability, and extinction. Now, considering the origin, as observed earlier, some biologists claim that higher taxa can be traced back to speciation, forming a continuum. However, Kemp (2016) critiques phylogenetic systematics, since "at least above the species level the whole phylogenetic tree is a single continuum, of which all supraspecific taxa consist of samples taken from the branches. From this perspective, higher taxa could be claimed to have no natural existence because they have no natural boundaries" (pp. 14–15). Kemp contends that phylogenetic systematics fails to differentiate between higher and lower taxa, mainly because it overlooks large morphological distances between taxa. According to him, the primary criterion of the Linnaean system for assigning ranks is the magnitude of morphological gaps. Thereby, a higher taxon is defined as "a cluster of species that is separated by a significantly large distance from other such clusters in morphospace" (p. 11).

In short, Kemp asserts that the reality of higher taxa depends on recognizing the morphological gaps that separate taxa from each other: "These phenotypic gaps are themselves real phenomena because they result from evolutionary processes and not from arbitrary human choice. From this perspective supraspecific taxa are also therefore real entities: they consist of clusters of species separated by these natural gaps" (p. 15). How do these groups traverse great distances through the morphospace to attain different categories within the hierarchy? In other words, what kind of evolutionary processes produce such "natural gaps"? As previously discussed, a consensus assumes that higher taxa acquire definition when various closely related species arise. Nevertheless, no one would argue that species acquire definition when one or more descendant subspecies arise from an ancestral subspecies. One or more descendant species emerge from a part that belongs to an ancestral species, and their recognition as a new species only becomes meaningful when a discontinuity separating the ancestral species from the descendant species is observed. Similarly, this logic applies to higher taxa. For instance, if a genus or family contains only one species (monotypic taxa), its definition becomes meaningful when considering the magnitude of the morphological gaps that separate one taxon from another taxon, rather than merely focusing on the genealogical connections among species, as in phylogenetic systematics.

In addition, it is important to note that the size of these morphological gaps is associated with the notion of "disparity," which some authors have used to recognize specific patterns of the evolutionary process at higher levels of the hierarchy (e.g., Wills 2001; Erwin 2007; Hopkins and Gerber 2017). For example, there is little evidence in the geological record that these morphological gaps have formed through a continuous process over geological time; rather, large distances in morphospace appear early in evolution.[4]

So, if the species and higher taxa emerged through gradual evolutionary processes, then Ereshefsky's (1991) proposal that they only represent the historical record that goes back to the population level would be correct. Furthermore, if the phylogenetic tree forms a continuum, distinguishing between descendant and ancestral species would be impossible, and recognizing the origin of higher taxa would be even further out of reach. Thus, neither species nor higher taxa would be considered evolutionary individuals. At best, according to the punctuated equilibrium, species might behave as evolutionary individuals, but higher taxa would still be excluded. But if species and higher taxa arose through Supradarwinian mechanisms, that is, saltational evolution,[5] then there would be strong reasons to consider them as evolutionary individuals, as their origins would be discrete.

[4] This pattern observed in the origin of higher taxa is undoubtedly a prominent feature in the fossil record, underscoring the significance of these entities in shaping biological disparity and diversity. See Chap. 12, Sect. 12.1.

[5] See Chap. 9, Sect. 9.2; Chap. 11, Sect. 11.1.1; Chap. 12, Sect. 12.1.1.

Throughout this discussion, I have stressed that the boundaries between species and higher taxa relate to the magnitude of morphological gaps (i.e., species exhibit smaller discontinuity than higher taxa). Moreover, as mentioned in Chap. 7, species and genera are ontologically similar, which might explain why many researchers have placed too much confidence in the reality of genus (e.g., Boyd 1999, p. 182; Barraclough and Humphreys 2015; Jablonski 2017, p. 429). Other taxonomic categories could be hierarchically established by the scale of magnitude of change, but their category diagnosis—such as family, order, and class—could be arbitrary.[6]

Another contentious issue in the individuality approach concerns cohesion. From the traditional standpoint, as previously noted, higher taxa are not considered units of evolution due to the absence of known processes that uniformly integrate their parts. Many biologists take for granted that this process is the gene flow and argue that species exhibit cohesion because there is interbreeding among their parts (and thus an exchange of genetic material). In contrast, higher taxa lack cohesion and are viewed merely as genealogical connections among species (e.g., Lee 2003). Likewise, if cohesion is attributed to common selective pressure, higher taxa also fail to behave uniformly, as species within a higher taxon can establish themselves in diverse environments (see Gould 2002, pp. 712–713). However, traditional views have invoked inappropriate causes to explain cohesion. From the perspective of the H-Evo-Devo Theory, developmental constraints are the mechanism that explains the cohesion of the units of evolution. These constraints produce the morphological stability of entities at different levels, reflecting a hierarchical classification system. Most importantly, these processes allow evolutionary entities to be identified spatiotemporally, which is why they are considered evolutionary individuals in the traditional sense. Moreover, as mentioned in Chap. 6, this mechanism also makes it possible to perceive these entities as natural kinds.

Natural Kinds An aspect of paramount importance for recognizing units of evolution, which has long been overlooked, is the study of body plans or types, commonly known as "typological thinking." Regarding taxonomy, Willi Hennig's phylogenetic systematics is noted for developing a methodology that departs from "ideal morphology" (Hamilton 2014). With respect to evolutionary theory, Ernst Mayr was a prominent advocate who popularized typological thinking as an outdated approach rooted in species fixity. Within the philosophy of biology, these conceptions were linked to essentialism, a perspective that viewed species and higher taxa as natural kinds. While this view recognized these taxonomic entities as real, it was seen as incompatible with the evolutionary perspective. Therefore, to avoid the essentialist baggage of old taxonomy, it was suggested that the philosophy of natural kinds needed to be restructured within the framework of HPC kinds

[6] However, I suspect that the category of "Phylum" in animals (or "Division" in plants) is real, mainly because it is the most inclusive and distinctive category of the major groups. The characters identified to designate this category are those that appear the earliest during ontogeny (e.g., phylotypic stage) and are therefore the most conserved in terms of evolutionary stability. For a different perspective on this discussion, see DiFrisco and Wagner 2022.

(another alternative is the thesis of individuality). However, as discussed in Chaps. 2 and 6, typologism played a crucial role in the development of the Natural System, and none of the criticism against the type concept, from both biological and philosophical perspectives, is entirely justified.

The body plan—also known as "*Bauplan*" (plural *Baupläne*), coined by Joseph Henry Woodger—refers to the common pattern of highly conserved characters shared by a group of closely related organisms, which differs from the common patterns of other phylogenetically separated groups (see Fig. 8.3) (Raff 1996; Hall 1999; Willmore 2012). These morphological steady states are taxonomically expressed through the Linnaean hierarchy, which is reflected as a hierarchy of body plans. It is worth noting that some biologists apply this concept only to the phylum rank in animals, or even to other higher categories such as class, order, and family. But, a notable aspect of the body plan is that it entails a hierarchy of inclusion (Raff 1996; Hall 1999). According to the H-Evo-Devo Theory, the body plan extends to the species level, excluding only the subspecies level, considering that subspecies are distinguished by minor differences such as color, size, proportion, and behavior.[7]

Additionally, as previously mentioned, even though traditional taxonomy and evolution rejected the notion of "body plan" due to its "typological" connotation, some authors have emphasized the need to reformulate it from the Evo-Devo viewpoint (e.g., Raff 1996; Hall 1999; Rieppel 2006; Brigandt 2017; Witteveen 2018; DiFrisco and Wagner 2022). For example, according to Raff (1996), this highly integrated structural pattern—which remains stable over geological time—reflects an underlying developmental and genetic program. In this regard, Hall (1999, p. 106) states: "The maintenance of *Baupläne* (and presumably their phylogenetic

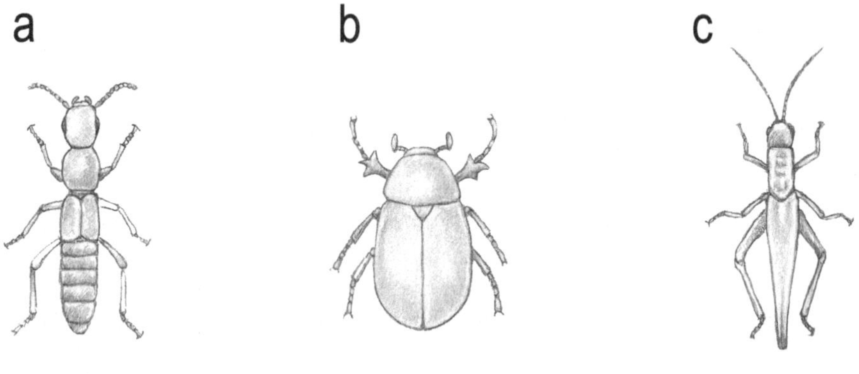

Fig. 8.3 *The body plan, or Bauplan, refers to the common pattern of highly conserved characters shared by a closely related group of organisms*. (**a**) Insect body plan. (**b**) Coleoptera body plan. (**c**) Orthopteran body plan. (Illustrations drawn by Aldi De Oyarzabal Salcedo)

[7]For an interpretation of both species and subspecies according to the H-Evo-Devo Theory, see Chap. 7.

origins) and their resistance to change are fundamentally a result of internal factors acting as developmental constraints." As concluded in Chap. 6, evolutionary entities can be recognized through the perspective of natural kinds, considering that developmental constraints are the mechanisms that explain the kind of regularities that make it possible to recognize such underlying body plans. It is important to emphasize that the concept of body plan does not imply a mystical idea of morphology, nor does it denote immutable abstractions of morphological characters in a non-historical sense. Instead, it is a concept that enables the identification of processes and regular patterns produced by developmental rules, as reflected in a taxonomic hierarchical structure applied to multicellular organisms (see also Riedl 1978; Rieppel 2006; Brigandt 2017). Therefore, from the perspective of the H-Evo-Devo Theory, developmental constraints endow higher taxa with certain morphological stability—a sort of supraspecific stasis—enabling their identification spatiotemporally. Simultaneously, this is the homeostatic mechanism by which biological essences or natural kinds can be perceived. Thus, a body plan is expressed as a highly integrated common structural pattern that appears to remain unchanged while allowing for certain ranges of variation, as long as they do not exceed its boundaries.

In summary, according to the H-Evo-Devo Theory, higher taxa are more inclusive body plans than the species/genus body plan within a hierarchy of body plans. A higher taxon stems from a part of the population that breaks its developmental constraints, resulting in a large morphological gap within the morphospace (birth criterion). This higher taxon maintains a phylogenetic relationship with other higher taxa (phylogeny criterion). It may expand and diversify, giving rise to one or more species (variation criterion) but maintaining a common body plan at its level (natural kinds). This projection of the body plan is due to its developmental constraints, which simultaneously make it possible to recognize them as the same entity throughout geological time (cohesion criterion). Eventually, the higher taxon may decline and become extinct (death criterion). While some higher taxa seem arbitrary (e.g., because their initial identification was made using traditional cladistic methods), surely many higher taxa are also real entities in nature. As Kemp (2016, p. 15) comments: "[H]igher taxa are indeed real entities and therefore their origin is worthy of investigation."

8.2 Homologues as Units of Evolution

Morphologists emphasize that organisms are composed of the same "building blocks," which function as phylogenetically distinguishable evolving entities. These entities are known as "homologues," and are identified by the term "homology." Unfortunately, traditional hierarchical evolutionary models have overlooked the significance of the organism's variational and structural properties, prioritizing instead an evolutionary perspective focused on the historical continuity of genes.[8] In

[8] See Chap. 3, Sect. 3.1 and Chap. 4, Sect. 4.2.

contrast, the H-Evo-Devo Theory restores the organism's structural properties in evolution and assesses their historical continuity through the study of homologues as morphological modules. According to this theory, homologues are quasi-autonomous parts of the organism that are highly integrated in developmental terms and maintained phylogenetically over a long time. Therefore, homology is the correspondence of homologues in different lineages or within the same organism, suggesting a common embryological pattern. In this context, homologues behave as both evolutionary individuals and natural kinds; hence, they are units of evolution.

It is important to emphasize that this model seeks to resolve the ongoing debate between systematists and morphologists by unifying the concept of "homologue" through the theory of modularity. In the following pages, I will examine the discussions surrounding the concept of homology as perceived by phylogenetic systematics and traditional morphology, highlighting the shortcomings of phylogenetic systematics. Then, I will provide arguments for supporting homologues as units of evolution according to the H-Evo-Devo Theory.

8.2.1 Homology: An Unsolved Problem? Systematics Versus Morphology

One contentious issue regarding homology stems from differing interpretations across various schools of comparative biology, for example, phylogenetic systematics and traditional morphology. What are the underlying reasons for these differences? What significance does the term "homology" hold within these fields? As many biologists know, the primary aim of phylogenetic systematics is to reconstruct the evolutionary history of taxa by defining monophyletic groups, which are established through the recognition of homologous characters (Patterson 1982; Eldredge and Cracraft 1980; Nelson 1994). As Patterson (1982, p. 21) states, "homology is the relation which characterizes monophyletic groups." A key criterion for identifying homologous is the "similarity" of features among different taxa (Eldredge and Cracraft 1980). It is important to note that this similarity of features encompasses not only organs or morphological characters, as established in traditional morphology, but also includes physiological, biochemical, genetic, and behavioral similarities, among others (Eldredge and Cracraft 1980; Lauder 1994; Nelson 1994). Moreover, within this framework, the absence of one of these features is considered evidence of monophyly. Indeed, these similarities represent hypotheses of homology, as some may result from independent evolution, known as homoplasies. Consequently, the primary objective of phylogenetic systematics is to identify the greatest number of homologous characters in order to infer the "true" relationships among taxa (Wiley and Lieberman 2011).

On the other hand, morphologists have identified a different notion of homology that delves into the biological causes or mechanisms explaining the origin and stability of homologous parts rather than focusing solely on discerning genealogical connections between groups (Roth 1984; Wagner 1989a, b; Müller 2003; DiFrisco 2023). Interestingly, proponents of this concept of homology (e.g., Roth 1991;

Wagner 1989a) stress that the notion of homology within phylogenetic systematics differs significantly from the one in traditional morphology. Both approaches operate with distinct goals, definitions, and explanations. Even the recognition and comprehension of homologues have distinct sources, among other differences. But can a unified perspective be achieved between these two approaches to homology? While some scholars have proposed various arguments for reaching a definitive solution (e.g., Cracraft 2005; Brigandt 2007, 2009; Ereshefsky 2009; Pavlinov 2012; Vogt 2017; Novick 2018; Suzuki 2021; Ballego-Campos et al. 2023; Gouvêa and Brigandt 2023; Minelli 2023; Göpel 2024), I will outline some of the challenges to such unification, considering certain shortcomings in traditional phylogenetic systematics.

(1) *When homology is explained as historical continuity due to connections of ancestry and descent, issues arise*

Typically, in phylogenetic systematics (or in the traditional evolutionary viewpoint), homology is defined in terms of "common ancestry": The characters of organisms in descendants are similar because they inherited them from a common ancestor. Nevertheless, the historical continuity of homologous characters falls short as an explanation, as it describes neither the mechanisms that generate the characters nor the mechanisms that ensure their stability over time (Wagner 1994; Brigandt 2003, 2007; DiFrisco 2021, 2023). In addition, homology via phylogenetic systematics does not consider serial homology relevant.

In order to resolve these tensions, Novick (2018) argues that while phylogenetic systematics emphasizes that those homologous characters are inherited from a common ancestor, embryological insights provide specific details on the inheritance mechanisms; therefore, these approaches are complementary. However, reference to homology, in the context of traditional systematics, may be redundant, considering that its only contribution is mentioning that homologous structures have an evolutionary source. As Wagner (2014, p. 73) comments: "The degree of similarity [according to phylogenetic systematics] can be used as a guide to reconstruct the phylogenetic relationships among species, but beyond that it is of no theoretical relevance." In conclusion, in light of the knowledge of the processes that make possible the existence and the constancy of homologous characters, reference to "common ancestry" is unnecessary as an explanation.

(2) *The meaning of "homologues" as "features" or "attributes," which are explained by the continuity of the flow of hereditary information*

Wagner (2014, p. 73) points out that when the architects of the Modern Synthesis (e.g., G. G. Simpson and Ernst Mayr) defined homologues as "features" or "attributes" due to inheritance from a common ancestor, they were not referring to organs or any particular relevant biological unit. In fact, in phylogenetic systematics, homologous similarities also encompass physiological, biochemical, behavioral, and, above all, genetic characteristics. It is worth emphasizing that phylogenetic systematics aims to recognize the greatest number of these "attributes" to establish the relationships between groups. Thus, under this traditional framework, homology is nothing more than a description of the record of evolutionary history; therefore, any feature that has hypothetically passed its structure directly through replication is considered a homologous element. Since inheritance is the transmission of information over time, genes would presumably be the best indicators of homology between groups. The problem with assuming that everything that is a replicator is homologous leads to the difficulty of what Ereshefsky (2009) calls a "hierarchical disconnect." That is, when considering the different levels of organization—genes, developmental processes, morphological structure—there is no linear relationship connecting all levels into a single homology relationship. For example, non-homologous genes can control the development of homologous structures, and vice versa (de Beer 1971; Hall 1999; Wray 1999; DiFrisco and Jaeger 2021; DiFrisco 2023; McColgan and DiFrisco 2024).

Throughout this book, I have repeatedly noted that the traditional view has marginalized the role of the organism's structural properties by assuming that the organism is merely a vehicle that carries genetic information. From this standpoint, it seems as if genes operate independently of the embryological process. Within the traditional view, the definition of homology has been extended to biological entities such as genes or chromosomes, which are characterized by faithfully transmitting their structure to the next generations.[9] Yet organs or morphological units (modules)—such as the eye, kidney, and heart—are not replicators; rather, they develop from scratch in each generation (Sattler 1984; Wagner 1989a, b; Brigandt 2002; DiFrisco 2023). Put differently, the organs of offspring are not faithful copies of the ancestors' organs; instead, they exhibit variation across generations even though they remain constant over time. This implies that not everything considered "homologous" in phylogenetic systematics (or in the traditional view of evolution) is equivalent to what is "homologous" in morphology. Furthermore, not everything inherited from ancestors has the same nature of transmission across different levels of organization. This underscores the idea that the explanation of inheritance attributed to "features" (e.g., genes) can be separated from the explanation of the stability of morphological structures over time.[10] If this is the case, then the "hierarchical disconnect" would cease to be a concern since, in morphology, the "homologue" is an entity constructed from the different levels of organization according to the theory of modularity (see below).

[9] In some cases, replication may contain errors, thus explaining the origin of variation.

[10] I am referring to the tradition that views inheritance as the mere transmission of genes without considering developmental aspects as "inheritance thinking." In contrast, "modular thinking" is the name I give to the approach that attributes the stability of structures over time to causal developmental mechanisms. While "inheritance thinking" emphasizes the transmission of simple characters from parents to offspring, "modular thinking" focuses on explaining the stability of complex structures over millions of years.

(3) *The incongruence of the original concept of homology in phylogenetic systematics as a series of gradual character transformations*

In phylogenetic systematics, two versions of the homology concept are used (according to Patterson's terminology, 1982). On the one hand, *taxic homology* refers to a character state that is present in two lineages, allowing its diagnosis as a monophyletic group, that is, synapomorphy. On the other hand, *transformational homology* refers to the sequential transformations of a character state. Historically, Hennig claimed that homology corresponded to a series of character transformations, as he had conflicts with the notion of homology according to "idealistic morphology." Presumably, Hennig had in mind the gradual change of characters as a primary process according to Darwinian evolutionary theory. This led him to assume that he was describing the "real historical processes of evolution" as opposed to a punctual change of characters as would be the case in "idealistic morphology." Nowadays, although some scholars find transformational homology important for revealing the historical process of lineages (e.g., Grant and Kluge 2004), this view contradicts the modern method of cladism, which aims to identify monophyletic groups (Scotland 2000). For example, Patterson (1982) argued that the way to establish monophyletic groups was by identifying characters in a single state of transformation (taxic homology); thus, he equated the term "homology" with synapomorphy. Regarding this, Scotland (2000, p. 152) points out that "treating data from a non-transformational perspective, rather than being antagonistic to phylogeny reconstruction, attempts to provide a more general system of cladistic analysis that does not rely on the particular model of character evolution provided by Hennig."

Moreover, if transformational homology serves as evidence of the real historical processes of evolution, systematists must be able to trace the entire series of transformation events (Rieppel 2007). The issue with ancestral character states in systematics is that they are by no means observable (Scotland 2000, p. 151; Weston 2000, p. 135; Rieppel 2007, p. 234). In other words, the pathways of character transformation are impossible to discern: "Just as it is not possible to know true patterns of phylogeny, however, so it is not possible to know true transformation series among character states, or to know with certainly where one character ends and the next begins" (Brower 2000, p. 15). This shows that if homology in "idealistic morphology" is abstract, transformational homology is even more so.

In any case, transformational homology (in Hennig's sense) is by no means a correct description of the phenomenon of homology (Rieppel 1994, p. 90), since the most significant morphological innovations arise not from gradual changes in character states but from profound and discrete changes in ontogeny (Rieppel and Kearney 2007; Havstad et al. 2015). In short, the origin of morphological characters cannot be easily inferred from the gradual evolution of a character (Müller and Newman 1999; see Chap. 9, Sect. 9.2).

(4) *Equivalence of the term "synapomorphy" in phylogenetic systematics with the concept of "homology" in traditional morphology*

As discussed earlier, transformational homology does not adequately describe the nature of homology, yet adhering to taxic homology does not improve the situation for phylogenetic systematics. Patterson (1982) described taxic homology by equating the term "synapomorphy" with homology. However, Patterson's definition of homology is framed in the context of phylogenetic reconstruction, which proves counterproductive to the concept of homology in traditional morphology (Wagner 1989a, b; Brigandt 2002; Müller 2003). For example, the term "synapomorphy" refers to any property of an organism that aids in recognizing evidence of monophyly—genes, proteins, structures, behaviors, even the absence of an element—whereas homology in morphology simply refers to an organ of an organism (Brigandt 2002; Müller 2003). Similarly, synapomorphy corresponds to a character state rather than a character. In fact, phylogenetic systematics does not distinguish between character identity and character state,[11] which is crucial in morphology, where the focus lies on studying character identity rather than character states (Müller 2003; Brigandt 2007; Wagner 2014, p. 71). Finally, the term "synapomorphy" does not consider serial homology since this concept is irrelevant for discerning phylogenetic relationships (Roth 1984; Brigandt 2003; DiFrisco 2021).

Given these issues, it can be concluded that the conceptualizations of homology in phylogenetic systematics and traditional morphology not only differ but also prove to be completely incompatible in terms of their definitions, identifications, and explanations.[12] It is important to highlight that the concept of homology developed by the Modern Synthesis and phylogenetic systematics can now be challenged when viewed through the lens of modern evolutionary biology. This involves accepting typologism and reexamining the conceptions of inheritance and gradual evolution. An alternative approach is to reject these traditional positions that define homology in confusing and erroneous terms, and instead embrace new concepts and interpretations. With this in mind, the notion of "homologues" in morphology is the most appropriate, given that they represent morphologically distinguishable units according to the theory of modularity. Moreover, since the patterns of origin and stability of these entities can be analyzed through causal mechanisms, they emerge as ideal candidates for consideration as units of evolution.

8.2.2 Homologues According to the H-Evo-Devo Theory

According to the H-Evo-Devo Theory, *homologues are quasi-autonomous parts that are highly integrated in developmental terms and maintained phylogenetically over a long time*. They are considered the most relevant and highest-level entities, resulting from the developmental process that integrates all levels of organization. This notion of the "homologue" differs from that proposed by phylogenetic systematics (as discussed in the previous section); instead, it aligns with the biological

[11] Regarding the difference between character identity and character state, see below (Sect 8.2.2).

[12] See Göpel (2024), who reaches similar conclusions but offers a solution from a different approach.

concept of homology introduced by Roth (1984) and Wagner (1989a, b). Furthermore, the notion of homologue described here primarily relies on the concept of the morphological module as its core element.

The theory of modularity in morphology developed from the idea that the organism consists of dynamic, discrete, and semi-autonomous morphological units, which are generated through a hierarchical interrelation at different levels from the expression of specific genes (Raff 1996; Wagner and Altenberg 1996; Raff and Sly 2000; West-Eberhard 2003). According to this theory, modules are defined as those parts of an organism that are highly integrated in developmental terms but loosely connected to other parts of the same organism (Callebaut 2005; Kuratani 2009; Schlosser 2002, 2023; Ochoa and Rasskin-Gutman 2015).

Now, I will delve into the fundamental aspects of this concept of modularity. First, modules are not merely static entities; instead, they are dynamic entities that emerge from a series of interactions in the developmental process at different levels of organization (Raff 1996; Gilbert and Bolker 2001; DiFrisco and Jaeger 2021; Jaeger and Monk 2021). As discussed in Chap. 5, these interactions result in developmental constraints, indicating that modules are composed of highly integrated elements in developmental terms (Kuratani 2009; Schlosser 2023). Second, modules are quasi-autonomous. Terms such as "nearly decomposable" (Simon 1962), "quasi-independent" (Lewontin 1978), or "quasi-autonomous" (Wagner 1995) imply that while the parts are not entirely disconnected from the rest of the system, they maintain a certain degree of autonomy. Broadly speaking, modules possess not only ontogenetic independence but also evolutionary individuality (Bolker 2000; Raff and Sly 2000; Wagner 2014).

Several authors have argued for a close link between the concept of the morphological module and the concepts of character and homologue in morphology (e.g., Kuratani 2009; Wagner 2014, p. 137; DiFrisco and Jaeger 2021; Jaeger and Monk 2021; Schlosser 2023). As Rasskin-Gutman (2005, p. 218) comments: "Modularity converges in meaning with the concept of homology, but adds to it a stronger level of integration of parts… [and unlike the notion of homologue in phylogenetic systematics] modularity goes one step further, providing a more robust definition of homologues with morphological integration playing a key cohesive role." From this perspective, the terms "homologue," "module," "morphological unit," and "character identity" appear to refer to the same ontological entity.

In this context, numerous authors have stressed the significance of distinguishing between "character identity" and "character state" (Roth 1984; Müller 2003; Brigandt 2007; Wagner 2007, 2014, 2015; Havstad et al. 2015; Szucsich et al. 2011; Ochoa and Rasskin-Gutman 2015; McKenna et al. 2021). Character identity pertains to homologue, module, or morphological unit (e.g., eye, heart, lung). In contrast, character states denote attributes (qualities) or variations in shape through which character identity is manifested (e.g., color, proportion, organ size). Such collections of states are also referred to as "variational modalities" (Wagner 2014). In order to clarify these distinctions, Ochoa and Rasskin-Gutman (2015) advocate for employing the term "homology" to denote correspondence in character identity and suggest differentiating it from the term "homogeny," which indicates correspondence in a specific character state. For instance, the forelimbs of birds, bats, and horses exemplify homology since the "forelimb character" is homologous across these groups; meanwhile, the shape (a specific form) of the forelimbs of all birds is an example of homogeny, as it represents a "character state" or a "variational modality" of the forelimb as a homologue.

Another pertinent aspect regarding homologues concerns their hierarchical nature (Raff and Raff 2000; DiFrisco and Jaeger 2021; Jaeger and Monk 2021; Schlosser 2023). Wagner (2014, p. 420) maintains that homology does not imply hierarchy because homologues, as character identities, form an integrated system wherein they overlap. He further suggests that distinguishing between character identity and character state does not imply hierarchy, as two lineages can share a structure as a homologous part without sharing their character states. While I agree with Wagner's observations, employing different types of hierarchies can help clarify this ambiguity. For example, given that modules result from a series of interactions at different levels of organization in the developmental process, they can be illustrated using a procedural hierarchy (see Chap. 5, Fig. 5.4 and Table 5.2). It is important to note that the outcome of this process can be studied through two types of compositive hierarchies, considering the modules as quasi-autonomous elements.

First, a *constitutive hierarchy* can be observed where one module may comprise other modules at a lower level, which are themselves part of another module at a higher level. This involves physical relationships between semi-autonomous morphological units at different scales. For example, as illustrated in Fig. 8.4a, at a higher level, a tetrapod forelimb (α) consists of three submodules: (1) stylopod, (2) zeugopod, and (3) autopod. At a lower level, the autopod (3) includes elements forming the thumb (A) and the other fingers (B). This identification of the modules is homology. Second, modules exhibit an *aggregative hierarchy* where different variational modalities or character states are organized in a nested sequence of greater or lesser inclusiveness within a module. For instance, as depicted in Fig. 8.4b, the forelimbs (a module) exhibit a character state representing the wing shape of birds (1); another inclusive state represents the wing shape of the hummingbird (A), then another inclusive state represents the wing shape of the bee hummingbird (*Mellisuga helenae*) (A″), and so forth. Therefore, identifying a particular character state is homogeny.[13]

But are homologues real? When considering the metaphysical aspect of homologues, it is important to recognize that the cause of their existence and stability stems from a specific internal organizational process. In this case, developmental constraints act as the causal mechanism whereby a common type (sameness) among morphological units is perceived; hence, they are considered natural kinds. Likewise, this mechanism also explains the evolutionary stability of these entities spatiotemporally (cohesion); thus, they are also considered evolutionary individuals. Now I

[13] For details on different types of hierarchies, see Chap. 4, Sect.4.1.1, and for their application in the Evo-Devo perspective, see Chap. 5, Sect. 5.1.1.

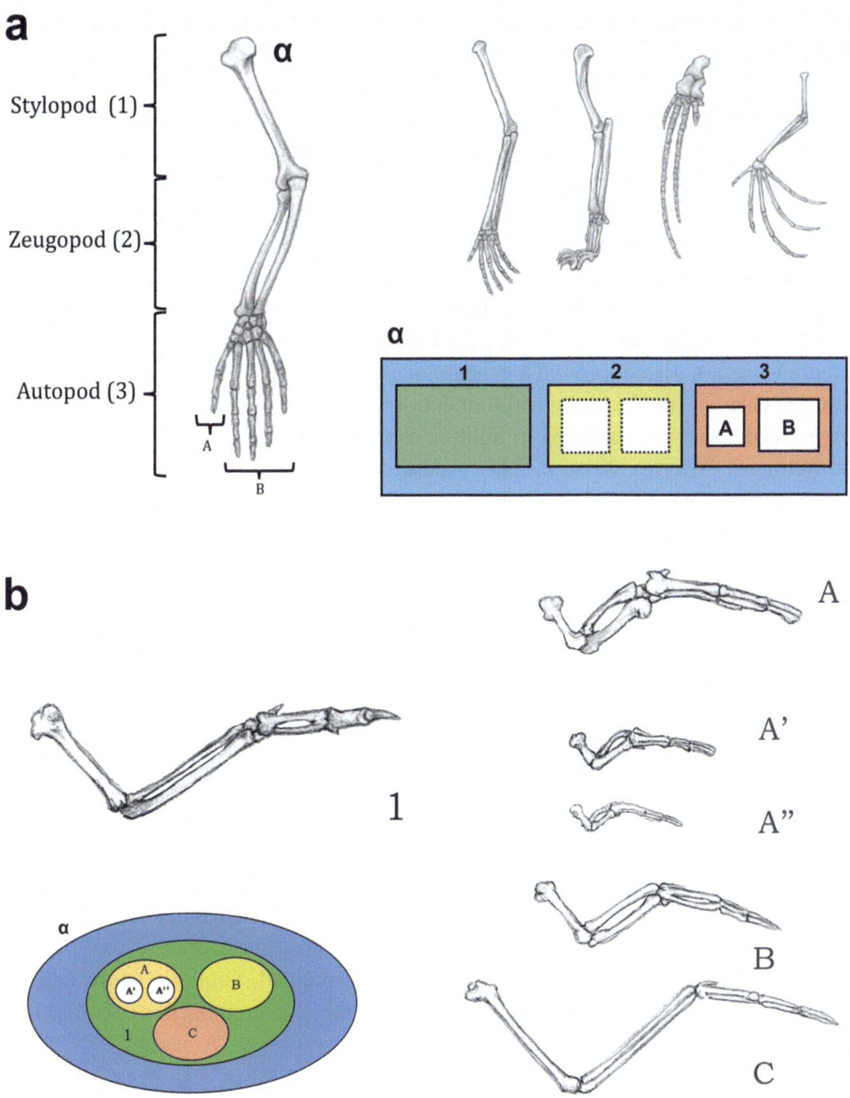

Fig. 8.4 *Two types of hierarchies are considered in the study of homology.* They illustrate the differences between character identity and character state, between homology and homogeny, and between Archetype and *Bauplan*. (**a**) *A constitutive hierarchy of modules.* The "forelimb" module, represented by the letter α, consists of modules 1, 2, and 3; and module 3 comprises submodules A and B. On the left, the Archetype of the mammalian forelimb is depicted. On the right, the variational modalities (or character states) of the module (character identity) in question are shown. This identification of the modules is homology. (**b**) *An aggregative hierarchy of variational modalities of the module.* The module, represented by the letter α, includes character state 1, which in turn encompasses character states A, B, and C. Character state A includes character states A′ and A″. The bird-forelimb *Bauplan* is represented by 1. The character states of the bird forelimb are

will provide some brief clarifications regarding their ontological determination in this context.

Natural Kinds Some philosophers of biology have suggested that homologues are parts of individuals characterized by their genealogical connections rather than natural kinds (Grant and Kluge 2004; Ghiselin 2005; Ereshefsky 2009, 2012). For example, Ghiselin (2005) argues that homologous parts are not determined by natural laws, including those outlined by the developmental approach; instead, they emerge through a series of continuous processes in the lineage, thus reflecting their historical nature: "Homology statements therefore can function as parts of an historical narrative in which mutations, recombinations, inversions, translocations and other events are traced through a genealogical nexus" (p. 99). However, it is important to consider that the biological foundations upon which Ghiselin bases his metaphysical theories are outdated in light of contemporary advances on the nature of homologues (Wagner 2014, p. 237). Additionally, as discussed in Chap. 6, traditional biology has often misinterpreted the philosophy of natural kinds, mainly due to its association with essentialism.

Yet for some time now, various authors have asserted that homologues are examples of natural kinds (Wagner 1996, 1999, 2001; Griffiths 1999; Brigandt 2002, 2003, 2009; Rieppel 2005a, 2007; Assis and Brigandt 2009; Pearson 2019; Villegas and Triviño 2023). Homologues are considered natural kinds in the sense that their members share "stable sets of dispositional properties" whose recurrence is attributed to the underlying causal homeostatic mechanisms (Wagner 2001; Pearson 2019; Villegas and Triviño 2023). Such mechanisms have been described by various authors as causal developmental processes (Hall 1999; Wagner 1999; Rieppel 2007; MacLeod 2013; Pearson 2019; DiFrisco et al. 2020; DiFrisco 2023). Under the H-Evo-Devo Theory, these common properties of homologues are explained as patterns of interactions within the developmental process, which denote developmental constraints. As clarified in Chap. 6, developmental constraints are the causal mechanisms through which the projection of a type is perceived. Similarly, in the case of homologues, it is also through developmental constraints that a common

Fig. 8.4 (continued) denoted by A (hummingbird), B (rock dove), and C (Laysan albatross). Note that A, B, and C were used to draw the bird-forelimb *Bauplan*, while A represents the hummingbird-forelimb *Bauplan*, derived from specimens A′ (calliope hummingbird) and A″ (bee hummingbird). This identification of a particular character state is homogeny
The differences between these two hierarchies serve to resolve the confusion in identifying hierarchical relationships of homology. Character identity is associated with the concept of module, while character states relate to the variational modalities of the module. Identifying the same module (e.g., stylopod), under every variety of form and function, is a relationship of homology. Identifying the same character state (e.g., the shape of the bird forelimb) is a relationship of homogeny. In short, the Archetype serves to identify modules in a system (e.g., stylopod, zeugopod, and autopod), whereas the *Bauplan* represents the stability parameters of character states (e.g., the bird-forelimb *Bauplan* and the hummingbird-forelimb *Bauplan*). (Illustrations drawn by Aldi De Oyarzabal Salcedo)

morphological unit is abstracted as "sameness." For this reason, Wake (1999, p. 45) correctly states, "homology has necessarily become an abstract concept." Therefore, it could be said that homology reverts to its old denomination through Owen's Archetype. Interestingly, Wagner (1989a) argues that this perspective is not entirely irrational, as it may seem in light of the evolutionary interpretation. Owen based his Archetype on the "laws of nature," which were responsible for generating the body parts of organisms. In the same sense, chemical elements, as examples of natural kinds, are established through "laws" that govern nature; for instance, physico-chemical mechanisms explain their atomic structure:

> Why not expect the same for the living world, as long as evolution was not an established fact? At least it could be said that the idealistic homology concept had some heuristic value since it stimulated progress in comparative anatomy and systematics; for example, it motivated the detection of the premaxillary bone in humans. (Wagner 1989a, p. 53)

This recalls Amundson's argument (2005, p. 33) that when early naturalists affirmed the reality of an entity, they did not know what that reality rested on. For many traditional historians, that reality had a metaphysical commitment to theory, for example, the Archetype as a Platonic idea or God's plan. Galileo was realistic about planetary orbits, and Mendeleev was realistic about the periodicity of elements, but neither of them knew what their causes were. Similarly, if early morphologists had in mind the type and explained it as part of natural laws—rather than as part of God's plan—this does not mean that homologues are unreal under this perspective.[14] As discussed above, the identification of natural kinds in biology is due to underlying homeostatic mechanisms working in biological processes rather than to "laws of nature" in the conventional sense (Boyd 1999; Wagner 2001). Therefore, the Archetype is updated under these terms. As Young (1993, p. 234) remarks: "[D]evelopmental processes, like assumptions of stable patterns based on adult structures, become a form of archetype."

Concerning this, it is important to consider that biologists take for granted that the set of homologues comprises a *Bauplan*. Also, this perspective implies a hierarchy, as Müller (2003, p. 59) clearly explained: "All minor clades are modifications of those major body plans, each characterized by distinct, hierarchical combinations of homologues… Evolution along a phylogenetic lineage proceeds by creating an increasing order of hierarchically nested sets of homologous building elements." Still, doubts persist regarding this hierarchical relationship between homologues and the body plan, which must also be clarified in terms of the differences between character identity and character states. Interestingly, Patterson (1982, p. 35) distinguishes two levels of the body plan, the "Archetype" and the "morphotype." The former implies the abstraction of characters without a hierarchy of groups, while the latter refers to a set of synapomorphies characterizing a particular monophyletic group. In my view, given the distinction between "homology" and "homogeny" and using the terms "Archetype" and "*Bauplan*," it could be said that there are two types

[14] For the relevance of the "type" in a historical context, see Chap. 2.

of hierarchical relationships regarding the body plan. Homology would be associated with the concept of Archetype as a set of homologues, while homogeny would relate to the notion of *Bauplan* as a set of character states. Therefore, the Archetype serves to identify modules, whereas *Bauplan* represents the stability parameters of character states (see Fig. 8.4). While I am not referring to the Archetype as natural kind, their representation is characteristic of the typological program typical in the nineteenth century but applied to the hierarchy theory of homology.[15]

Evolutionary Individuals It is noteworthy that Grant and Kluge (2004) and Ghiselin (2005) have presented homologues as evolutionary individuals according to their metaphysical interpretation. However, based on their arguments, these entities seem to be historical instead. For instance, while Grant and Kluge (2004) argue that homologues are individuals "united only by common history," they fail to address the criteria used by philosophers to identify true evolutionary individuals, for example, a defined beginning, cohesion among their parts, and a discrete end.[16] Similarly, Ghiselin (2005, p. 98) suggests that this metaphysical consideration of homologues "implies that it is logically possible to trace parts through intermediates in a series that has become transformed to such an extent that the initial wholes would seem to have nothing in common with the ones at the end of the series." But this perspective proposes no criteria for individuality either; it simply states that characters undergo a series of gradual transformations over time.[17]

Now, according to the theory of modularity, there are significant factors for considering homologues as evolutionary individuals, mainly due to their quasi-autonomous properties (Ochoa and Rasskin-Gutman 2015, p. 93). For example, Wagner (1989a, b, 2014, 2015) claims that individualized parts of development are not only autonomous from other body parts but also maintain stability and evolutionary independence over time: "developmental individuality entails evolutionary individuality." While many modular elements can be easily distinguished, identifying their boundaries can be challenging in other cases, necessitating a series of practical studies of processes to validate modularity hypotheses.[18] This is particularly evident in cases of serial homology (or repetition of parts within the same organism's body), particularly when modules emerge from the co-option of elements from a pre-existing module (e.g., gene networks, cellular tissues, even submodules). Nevertheless, a new module is identifiable because its dynamics change so radically from the ancestral module that it must be considered distinct. Another challenge lies in delineating homologous parts in terms of their modes of variation, where the origin of character identity must be clearly distinguished from the origin

[15] As discussed in Chap. 2, Sect. 2.2, naturalists such as Richard Owen and Thomas Henry Huxley applied the concept of the Archetype within an evolutionary framework and used it as a methodological guide for identifying the same structures in a higher group of organisms.

[16] See Chap. 6, Sect. 6.2.

[17] Apparently, Ghiselin advocates for Hennig's transformational homology.

[18] Regarding the difficulties related to the concept of modularity, see Zelditch and Goswami (2021).

of character state. Scholars addressing this issue have distinguished at least two types of evolutionary novelties (Müller and Wagner 1991, 2003; Müller and Newman 2005; Müller 2010; Peterson and Müller 2013; Wagner 2014, p. 126; Erwin 2021; McKenna et al. 2021; Schlosser 2023; Love 2024). This suggests a theoretical foundation for separating the origin of a module from the origin of its variational modality.[19] In short, if modules are discernible and arise discontinuously, it is possible to argue for a discrete origin of these entities.

Moreover, as discussed throughout this book, developmental constraints are mechanisms that foster cohesion. Certainly, homologues are highly cohesive entities that distinguish themselves spatiotemporally, providing sufficient evidence to consider them units of evolution. However, cohesion is also present in scales. According to DiFrisco et al. (2020), the notion of Character Identity Mechanisms (ChIMs) suggests that certain mechanisms are involved in the most conserved parts of homologues, which explains their stability over time. The main point is that modules remain stable over time regardless of the variational modalities, but these modalities also remain stable over time, albeit at a lower level. Under the H-Evo-Devo Theory, developmental constraints operate both upward and downward, reflecting this hierarchy of stability: they function at the highest level as ChIMs and at distinct lower levels as Character State Mechanisms (ChSMs), a term I use to refer to them. Although this distinction would need to be tested in practice, there is another issue to consider regarding the hierarchical nature of homologues.

DiFrisco (2023) points out that because the stability of a homologue can be decoupled at different levels of organization—morphological characters may be activated by different genes—there might not be cohesion across all levels of organization simultaneously.[20] Nevertheless, this criterion does not permit the loss of individuality. In fact, this phenomenon of "hierarchical disconnect" could be the mechanism that allows for the evolution of modules; in other words, these modular by-products would then serve as raw material for generating new modules. Therefore, both cohesion and lack thereof are significant processes in modular evolution. In any case, much exploration is still needed to understand the mechanisms of module stability. Nevertheless, it is undeniable that their internal nature makes them legitimate candidates to be considered units of evolution.

In summary, although a module emerges from the elements of a previous module, its developmental dynamics change so radically from those of the ancestral module that it must be regarded as a distinct module (birth). Modules give rise to other modules in a serial manner, each diverging and adapting to different functions (reproduction). If this is so, modules can be depicted through genealogical relationships (phylogeny). Modules exhibit high integration among their parts in such a way that their structure can be recognized spatiotemporally (cohesion) as the "same" structure under every variety of form and function (natural kind). But, despite their high level of integration, these systems still exhibit certain ranges of variation as

[19] For further details on serial homology and the distinction of different types of evolutionary novelties, see Chap. 12, Sect. 12.2.1.

[20] For details on morphological stability (stasis), see Chap. 12, Sect. 12.2.2.

character states (variation). Lastly, modules can potentially disappear (death), as the same module cannot appear twice—if all vertebrates except snakes go extinct after a mass extinction, limbs would be reported as extinct, but not the vertebral column. Considering all these criteria, there is no doubt that homologues function as units of evolution.

8.3 Conclusions

As explored in the previous chapter, traditional models of taxonomy and evolution face many difficulties when attempting to identify units of evolution within a hierarchical framework. For example, the Modern Synthesis framework promotes that higher taxa lack evolutionary significance, as their origin goes back to speciation processes. Also, according to this view, homologues are merely similar attributes due to inheritance from a common ancestor. Similarly, phylogenetic systematics propagates a fundamentally flawed notion of higher taxa, arguing that they are delimited solely by the absolute time of divergence between species relationships. In addition, under this view, homologies are merely "features" serving to identify monophyletic groups.

As noted earlier, the issue with these models lies in their neglect of the organism's variational and structural properties. They present an evolutionary viewpoint focused solely on the hereditary transmission of genes, relegating the organism to merely a vehicle that carries genes. Naturally, these models lack adequate integration of embryological phenomena and necessitate reevaluation in light of contemporary scientific advancements. The delay in such integration primarily stems from the historical misreading of typology. Hennig, for example, developed phylogenetic systematics in response to "idealistic morphology" in taxonomy, and Mayr's evolutionary biology similarly opposed typology. However, as discussed in previous chapters, typologism remains a pertinent element in the current comprehension of evolution. Perhaps this neglect partly originates from the perception that stability of form appears contrary to the notion of gradual change.

Yet this book presents an alternative view. H-Evo-Devo Theory is a model that assigns crucial significance to the organism's variational properties. Additionally, it provides an appropriate interpretation of the type concept. Specifically, it contemplates the stability of entities as a relevant evolutionary mode, acknowledging that biological entities are highly integrated systems due to specific biological processes. For example, according to the H-Evo-Devo Theory, higher taxa are interpreted as body plans that are hierarchically structured from greater to lesser inclusiveness, whereas homologues are modules or stable units identified as the "same" across different lineages. In both cases, the underlying cause for this abstract projection—which simultaneously resides in the cohesion of their parts—is attributed to developmental constraints operating hierarchically.

In addition, according to the H-Evo-Devo Theory, higher taxa not only appear real but are also recognized to play an active role in evolution, exhibiting characteristic patterns and processes that cannot be extrapolated from those observed in microevolution. Indeed, such a perspective opens up a vast research landscape for macroevolutionary studies, which has been overlooked for years (see Table 8.1 for a comparison of the nature of higher taxa according to the traditional view vs. the H-Evo-Devo Theory). Furthermore, this model unifies the concept of homology by considering the morphology module, which integrates all the different levels of organization. This concept restores the biological significance of the organism and its structural properties, providing a comprehensive and integrated discourse on the phenomena of evolution (see Table 8.2 to compare the notion of homologue between the traditional perspective and the H-Evo-Devo Theory). Considering this, it could be said that it is time to move away from outdated interpretations and begin describing evolutionary phenomena in alignment with contemporary biology. H-Evo-Devo Theory emerges as a highly plausible alternative for synthesizing taxonomy, morphology, and evolution.

Table 8.1 *Comparison of perspectives regarding higher taxa according to the traditional view and the H-Evo-Devo Theory*

Higher taxa according to	The traditional view (the Modern Synthesis and phylogenetic systematics)	H-Evo-Devo Theory
Definition	Groups that comprise two or more species, reflecting the historical record among species	The body plans that are more inclusive than the species/genus body plans within a hierarchy of body plans
Origin	Higher taxa emerge due to the processes tracing back to speciation	A higher taxon emerges when a segment of the ancestral population overcomes the developmental constraints of its previous body plan, leading to a significant disparity in the descendant population and thus giving rise to a new body plan
Cohesion	Higher taxa lack cohesion because they do not exchange genes among their constituent parts	Higher taxa cohesion is explained by developmental constraints
Boundaries	According to phylogenetic systematics, higher taxa boundaries are recognized by the absolute time of species divergence	Higher taxa exhibit natural morphological gaps among them at different scales
Role in evolution	Higher taxa play no significant role in evolution except to represent the historical record of species	Higher taxa play a very relevant role in evolution; they are active entities with specific processes and patterns that cannot be reduced to microevolution
Type of hierarchy	They are characterized by a false hierarchical structure with a positional arrangement	They are represented by an aggregative hierarchy
Metaphysical category	Historical entities	Natural kinds and individuals
Unit of evolution	No	Yes

Table 8.2 *Comparison of perspectives regarding homologues according to the traditional view and the H-Evo-Devo Theory*

Homologue according to	The traditional view (the Modern Synthesis and phylogenetic systematics)	H-Evo-Devo Theory
Definition	Attributes shared by two or more species due to inheritance from a common ancestor	Quasi-autonomous parts of the organism that are highly integrated in developmental terms and maintained phylogenetically over a long time
Explanation	The shared feature among descendants is similar due to inheritance from a common ancestor. However, this statement is incomplete, as it fails to describe the mechanisms responsible for its generation or persistence over time	Developmental constraints maintain the integration of module parts as much as possible, thereby enabling homologues to be recognizable phylogenetically
Entity to compare	Genes, proteins, biomolecules, organs, behaviors, the absence of features, or any attribute that serves to reconstruct the phylogenetic history of two or more taxa	Morphological modules
Replicator	Yes	No
Serial homology	It is not considered because this concept is not relevant for identifying the phylogenetic relationships between taxa	It is quite relevant from this perspective
Character identity or character state	Commonly, it refers to character state rather than character identity. However, the difference between both concepts is not relevant from this perspective	It refers to character identity rather than character state
Type of hierarchy	There is no mention of a nested compositional hierarchy Hierarchical disconnect: There is no linear relationship connecting all levels of organization into a single homology relationship	A procedural hierarchy shows the interactions at different levels of organization in the developmental process. This process describes a dynamic module A constitutive hierarchy represents modules at different scales, e.g., a module constitutes two or more submodules An aggregative hierarchy represents variational modalities at different scales, e.g., different character states are arranged in a nested sequence of greater or lesser inclusiveness within a module
Metaphysical category	Historical entities	Natural kinds and individuals
Unit of evolution	No	Yes

References

Amundson R (2005) The changing role of the embryo in evolutionary thought: roots of evo-devo. Cambridge University Press, Cambridge

Arthur W (2003) Micro-, macro-, and megaevolution. In: Hall BK, Olson WM (eds) Keywords and concepts in evolutionary developmental biology. Harvard University Press, Cambridge MA, pp 249–260

Assis LCS, Brigandt I (2009) Homology: homeostatic property cluster kinds in systematics and evolution. Evol Biol 36:248–255

Ballego-Campos I, Bonifácio SKV, Assis LCS (2023) A unified view of homology. Cladistics 39:398–417

Barker MJ (2007) The empirical inadequacy of species cohesion by gene flow. Philos Sci 74:654–665

Barker MJ, Wilson RA (2010) Cohesion, gene flow, and the nature of species. J Philos 107:61–79

Barraclough TG, Humphreys AM (2015) The evolutionary reality of species and higher taxa in plants: a survey of post-modern opinion and evidence. New Phytol 207:291–296

Bolker JA (2000) Modularity in development and why it matters to evo-devo. Am Zool 40:770–776

Boyd R (1999) Homeostasis, species, and higher taxa. In: Wilson RA (ed) Species: new interdisciplinary essays. MIT Press, Cambridge MA, pp 141–185

Boyd R (2010) Homeostasis, higher Taxa, and monophyly. Philos Sci 77:686–701

Boyd R (2021) Rethinking natural kinds, reference and truth: towards more correspondence with reality, not less. Synthese 198:2863–2903

Brigandt I (2002) Homology and the origin of correspondence. Biol Philos 17:389–407

Brigandt I (2003) Homology in comparative, molecular, and evolutionary developmental biology: the radiation of a concept. J Exp Zool (Mol Dev Evol) 299B:9–17

Brigandt I (2007) Typology now: homology and developmental constraints explain evolvability. Biol Philos 22:709–725

Brigandt I (2009) Natural kinds in evolution and systematics: metaphysical and epistemological considerations. Acta Biotheor 57:77–97

Brigandt I (2017) Typology and natural kinds in evo-devo. In: Nuño de la Rosa LEvolutionary developmental biology, Müller GB (eds) Springer, Cham, pp. 483–493

Brothers DJ (1985) Species concepts, speciation, and higher taxa. In: Vrba ES (ed) Species and speciation. Transvaal Museum Monograph No. 4. Transvaal Museum, Pretoria, pp 35–42

Brower AVZ (2000) Homology and the inference of systematic relationships: some historical and philosophical perspectives. In: Scotland R, Pennington T (eds) Homology and systematics: coding characters for phylogenetic analysis. Taylor and Francis, London, pp 10–21

Brower AVZ (2023) Hierarchies, classifications, cladograms and phylogeny. Cladistics 39:229–239

Callebaut W (2005) The ubiquity of modularity. In: Callebaut W, Rasskin-Gutman D (eds) Modularity: understanding the development and evolution of natural complex systems. MIT Press, Cambridge MA, pp 3–28

Cantino PD, de Queiroz K (2020) International code of phylogenetic nomenclature (PhyloCode). CRC Press, Boca Raton

Caponi G (2020) ≃Qué les pasa a los linajes? In: Combita JL, Maldonado CE (eds) Biología teórica, explicaciones y complejidad. Universidad el Bosque/BIEVEP, Bogotá, pp 403–463

Caponi G (2023) Linajes: esas cosas que evolucionan. Universidad El Bosque, Colombia

Cartmill M (2012) Primate origins, human origins, and the end of higher taxa. Evol Anthropol 21:208–2020

Coleman KA, Wiley EO (2001) On species individualism: a new defense of the species-as-individuals hypothesis. Philos Sci 68:498–517

Cracraft J (2005) Phylogeny and evo-devo: characters, homology, and the historical analysis of the evolution of development. Zoology 108:345–356

de Beer GR (1971) Homology: an unsolved problem. Oxford University Press, Oxford

de Queiroz K, Gauthier J (1994) Toward a phylogenetic system of biological nomenclature. Trends Ecol Evol 9:27–31
Devitt M (2023) The minimal role of the higher categories in biology. Biol Philos 38. https://doi.org/10.1007/s10539-022-09888-5
DiFrisco J (2021) Developmental homology. In: Nuño de la Rosa L, Müller GB (eds) . Evolutionary developmental biology. Springer, Cham, pp 85–97
DiFrisco J (2023) Toward a theory of homology: development and the de-coupling of morphological and molecular evolution. Br J Philos Sci 74:771–810
DiFrisco J, Jaeger J (2021) Homology of process: developmental dynamics in comparative biology. Interface Focus 11:20210007. https://doi.org/10.1098/rsfs.2021.0007
DiFrisco J, Wagner GP (2022) Body plan identity: a mechanistic model. Evol Biol 49:123–141
DiFrisco J, Love AC, Wagner GP (2020) Character identity mechanisms: a conceptual model for comparative-mechanistic biology. Biol Philos 35. https://doi.org/10.1007/s10539-020-09762-2
Ehrlich PR, Raven PH (1969) Differentiation of populations. Science 165:1228–1232
Eldredge N (2001) The nature and origin of supraspecific taxa revisited–with special reference to Trilobita. In: Adrain JM, Edgecombe GD, Lieberman BS (eds) Fossils, phylogeny, and form: an analytical approach. Topics in Geobiology, vol 19. Springer, New York, pp 341–375
Eldredge N, Cracraft J (1980) Phylogenetic patterns and the evolutionary process: method and theory in comparative biology. Columbia University Press, New York
Ereshefsky M (1991) Species, higher taxa, and the units of evolution. Philos Sci 58:84–101
Ereshefsky M (2001) The poverty of the Linnaean hierarchy: a philosophical study of biological taxonomy. Cambridge University Press, Cambridge
Ereshefsky M (2009) Homology: integrating phylogeny and development. Biol Theory 4:225–229
Ereshefsky M (2012) Homology thinking. Biol Philos 27:381–400
Erwin DH (2007) Disparity: morphological pattern and developmental context. Palaeontology 50:57–73
Erwin DH (2021) A conceptual framework of evolutionary novelty and innovation. Biol Rev 96:1–15
Ghiselin MT (1987) Species concepts, individuality, and objectivity. Biol Philos 2:127–143
Ghiselin MT (2005) Homology as a relation of correspondence between parts of individuals. Theory Biosci 124:91–103
Gilbert SF, Bolker JA (2001) Homologies of process and modular elements of embryonic construction. J Exp Zool (Mol Dev Evol) 291:1–12
Göpel T (2024) Synthesizing foundations in evolutionary morphology and a plea for conceptual explicitness regarding homology concepts. Front Ecol Evol 12:1343346. https://doi.org/10.3389/fevo.2024.1343346
Gould SJ (2002) The structure of evolutionary theory. The Belknap Press of Harvard University Press, Cambridge MA
Gouvêa DY, Brigandt I (2023) Against unifying homology concepts: redirecting the debate. J Morphol 284:e21599. https://doi.org/10.1002/jmor.21599
Grant T, Kluge A (2004) Transformation series as an ideographic character concept. Cladistics 20:23–31
Griffiths PE (1999) Squaring the circle: natural kinds with historical essences. In: Wilson RA (ed) Species: new interdisciplinary essays. MIT Press, Cambridge MA, pp 209–228
Hall BK (1999) Evolutionary developmental biology. Kluwer Academic Publishers, Dordrecht
Hamilton A (2014) Historical and conceptual perspectives on modern systematics: groups, ranks, and the phylogenetic turn. In: Hamilton A (ed) The evolution of phylogenetic systematics. University of California Press, Berkeley, pp 89–116
Havstad JC, Assis LCS, Rieppel O (2015) The semaphorontic view of homology. J Exp Zool B Mol Dev Evol 324:578–587
Hennig W (1966) Phylogenetic systematics. University of Illinois, Chicago
Hey J (2001) Genes, categories, and species: the evolutionary and cognitive causes of the species problem. Oxford University Press, Oxford

Hopkins MJ, Gerber S (2017) Morphological disparity. In: Nuño de la Rosa L, Müller GB (eds) Evolutionary developmental biology. Springer, Cham, pp 965–976
Hull DL (1976) Are species really individuals? Syst Zoo 25:174–191
Humphreys AM, Barraclough TG (2014) The evolutionary reality of higher taxa in mammals. Proc R Soc B 281:201132750. https://doi.org/10.1098/rspb.2013.2750
Jablonski D (2017) Approaches to macroevolution: 1. General concepts and origin of variation. Evol Biol 44:427–450
Jaeger J, Monk N (2021) Dynamical modules in metabolism, cell and developmental biology. Interface Focus 11:20210011. https://doi.org/10.1098/rsfs.2021.0011
Kemp TS (2007) The origin of higher taxa: macroevolutionary processes, and the case of the mammals. Acta Zool 88:3–22
Kemp TS (2016) The origin of higher taxa: palaeobiological, developmental, and ecological perspectives. Oxford University Press/The University of Chicago Press, Oxford
Krassilov VA (2004) Macroevolutionary events and the origin of higher taxa. In: Wasser SP (ed) Evolutionary theory and processes: modern horizons. Kluwer Academic Publishers, Dordrecht, pp 265–289
Kuratani S (2009) Modularity, comparative embryology and evo-devo: developmental dissection of evolving body plans. Dev Biol 332:61–69
Lauder GV (1994) Homology, form, and function. In: Hall BK (ed) Homology: the hierarchical basis of comparative biology. Academic Press, San Diego, pp 151–196
Laurin M (2024) The advent of PhyloCode: the continuing evolution of biological nomenclature. CRC Press, Boca Raton
Lee MSY (2003) Species concepts and species reality: salvaging a Linnaean rank. J Evol Biol 16:179–188
Lewontin RC (1978) Adaptation. Sci Am 239:212–230
Love AC (2024) Evolution and development: conceptual issues. Cambridge University Press, Cambridge
MacLeod M (2013) Limitations of natural kind talk in the life sciences: homology and other cases. Biol Theory 7:109–120
Mayr E (1942) Systematics and the origin of species. Columbia University Press, New York
Mayr E (1963) Animal species and evolution. Belknap Press of Harvard University Press, Cambridge MA
McColgan Á, DiFrisco J (2024) Understanding developmental system drift. Development 15:dev203054. https://doi.org/10.1242/dev.203054
McKenna KZ, Wagner GP, Cooper KL (2021) A developmental perspective of homology and evolutionary novelty. Curr Top Dev Biol 141:1–38
Minelli A (2023) A refreshed approach to homology-prioritizing epistemology over metaphysics. J Morphol 284:e21533. https://doi.org/10.1002/jmor.21533
Mishler BD (2022) Ecology, evolution, and systematics in a post-species world. In: Wilkins J, Zachos FE, Ya PI (eds) Species problems and beyond: contemporary issues in philosophy and practice. CRC Press, pp 177–190
Müller GB (2003) Homology: the evolution of morphological organization. In: Müller GB, Newman SA (eds) Origination of organismal form: beyond the gene in developmental and evolutionary biology. MIT Press, Cambridge MA, pp 51–69
Müller GB (2010) Epigenetic innovation. In: Pigliucci M, Müller GB (eds) Evolution: the extended synthesis. MIT Press, Cambridge MA, pp 307–332
Müller GB, Newman SA (1999) Generation, integration, autonomy: three steps in the evolution of homology. In: Bock GR, Cardew G (eds) Homology. Novartis Foundation Symposium 222. Wiley and Sons, New York, pp 65–79
Müller GB, Newman SA (2005) The innovation triad: an evodevo agenda. J Exp Zool B Mol Dev Evol 304:487–503
Müller GB, Wagner GP (1991) Novelty in evolution: restructuring the concept. Annu Rev Ecol Syst 22:229–256

Müller GB, Wagner GP (2003) Innovation. In: Hall BK, Olson WM (eds) Keywords and concepts in evolutionary developmental biology. Harvard University Press, Cambridge MA, pp 218–227
Nelson G (1994) Homology and systematics. In: Hall BK (ed) Homology: the hierarchical basis of comparative biology. Academic Press, San Diego, pp 101–149
Novick A (2018) The fine structure of 'homology'. Biol Philos 33. https://doi.org/10.1007/s10539-018-9617-3
Ochoa C, Rasskin-Gutman D (2015) Evo–devo mechanisms underlying the continuum between homology and homoplasy. J Exp Zool B Mol Dev Evol 324:91–103
Patterson C (1982) Morphological characters and homology. In: Joysey KA, Friday AE (eds) Problems of phylogenetic reconstruction. Academic Press, London, pp 21–74
Pavlinov IY (2012) The contemporary concepts of homology in biology: a theoretical review. Biol Bull Rev 2:36–54
Pearson CH (2019) Are homologies really natural kinds? Biol Philos 34:42. https://doi.org/10.1007/s10539-019-9696-9
Peterson T, Müller GB (2013) What is evolutionary novelty? process versus character based definitions. J Exp Zool B Mol Dev Evol 320:345–350
Raff RA (1996) The shape of life: genes, development, and the evolution of animal form. The University of Chicago Press, Chicago
Raff EC, Raff RA (2000) Dissociability, modularity, evolvability. Evol Dev 2:235–237
Raff RA, Sly BJ (2000) Modularity and dissociation in the evolution of gene expression territories in development. Evol Dev 2:102–113
Rasskin-Gutman D (2005) Modularity: jumping forms within morphospace. In: Callebaut W, Rasskin-Gutman D (eds) Modularity: understanding the development and evolution of natural complex systems. MIT Press, Cambridge MA, pp 207–219
Riedl R (1978) Order in living organisms: a systems analysis of evolution. Wiley and Sons, New York
Rieppel O (1994) Homology, topology, and typology: the history of modern debates. In: Hall BK (ed) Homology: the hierarchical basis of comparative biology. Academic Press, San Diego, pp 63–100
Rieppel O (2005a) Modules, kinds, and homology. J Exp Zool B Mol Dev Evol 304:18–27
Rieppel O (2005b) Monophyly, paraphyly, and natural kinds. Biol Philos 20:465–487
Rieppel O (2006) 'Type' in morphology and phylogeny. J Morphol 267:528–535
Rieppel O (2007) Homology: a philosophical and biological perspective. In: Henke W, Tattersall I (eds) Handbook of paleoanthropology. Springer, Heidelberg, pp 217–240
Rieppel O, Kearney M (2007) The poverty of taxonomic characters. Biol Philos 22:95–113
Roth VL (1984) On homology. Biol J Linn Soc 22:13–29
Roth VL (1991) Homology and hierarchies: problems solved and unresolved. J Evol Biol 4:167–194
Sattler R (1984) Homology: a continuing challenge. Syst Bot 9:382–394
Schaefer CW (1976) The reality of the higher taxonomic categories. J Zool Syst Evol Res 14:1–10
Schlosser G (2002) Modularity and the units of evolution. Theory Biosci 121:1–80
Schlosser G (2023) Rebuilding ships while at sea-Character individuality, homology, and evolutionary innovation. J Morphol 284:e21522. https://doi.org/10.1002/jmor.21522
Scotland R (2000) Homology, coding and three-taxon statement analysis. In: Scotland R, Pennington T (eds) Homology and systematics: coding characters for phylogenetic analysis. Taylor and Francis, London, pp 145–182
Simon HA (1962) The architecture of complexity. Proc Am Philos Soc 106:467–482
Smith AB (1994) Systematics and the fossil record: documenting evolutionary patterns. Blackwell Scientific Publications, Oxford
Suzuki DG (2021) Homology thinking reconciles the conceptual conflict between typological and population thinking. Biol Philos 36:26. https://doi.org/10.1007/s10539-021-09800-7
Szucsich NU, Wirkner CS, Pass G (2011) Deconstructing morphology: reply to Scholtz (2010). Acta Zool 00:1–3
Tattersall I (2012) Higher taxa: an alternate perspective. Evol Anthropol 21:221–223

Tëmkin I, Eldredge N (2015) Networks and hierarchies: approaching complexity in evolutionary theory. In: Serrelli E, Gontier N (eds) Macroevolution: explanation, interpretation and evidence. Springer, pp 183–226

Valentine JW (2004) On the origin of phyla. The University of Chicago Press, Chicago and London

Valentine JW, May CL (1996) Hierarchies in biology and paleontology. Paleobiology 22:23–33

Vasilyeva LN, Stephenson SL (2008) The Linnaean hierarchy and "extensional thinking". The Open Evolution Journal 2:55–65

Villegas C, Triviño V (2023) Typology and organismal dispositions in evo-devo: a metaphysical approach. Artefactos 12:79–102. https://doi.org/10.14201/art202312179102

Vogt L (2017) Assessing similarity: on homology, characters and the need for a semantic approach to non-evolutionary comparative homology. Cladistics 33:513–539

Wagner GP (1989a) The biological homology concept. Annu Rev Ecol Syst 20:51–69

Wagner GP (1989b) The origin of morphological characters and the biological basis of homology. Evolution 43:1157–1171

Wagner GP (1994) Homology and the mechanisms of development. In: Hall BK (ed) Homology: the hierarchical basis of comparative biology. Academic Press, San Diego, pp 273–299

Wagner GP (1995) The biological role of homologues: a building block hypothesis. N Jb Geol Paläont Abh 195:279–288

Wagner GP (1996) Homologues, natural kinds and the evolution of modularity. Amer Zool 36:36–43

Wagner GP (1999) A research programme for testing the biological homology concept. In: Bock GR, Cardew G (eds) Homology. Novartis Foundation Symposium 222. Wiley and Sons, New York, pp 125–140

Wagner GP (2001) Characters, units and natural kinds: an introduction. In: Wagner GP (ed) The character concept in evolutionary biology. Academic Press, San Diego, pp 1–10

Wagner GP (2007) The developmental genetics of homology. Nat Rev Genet 8:473–479

Wagner GP (2014) Homology, genes, and evolutionary innovation. Princeton University Press, Princeton

Wagner GP (2015) Reinventing the organism: evolvability and homology in post-Dahlem evolutionary biology. In: Love AC (ed) Conceptual change in biology: scientific and philosophical perspectives on evolution and development. Springer, Dordrecht, pp 327–342

Wagner GP, Altenberg L (1996) Perspective: complex adaptations and the evolution of evolvability. Evolution 50:967–976

Wake DB (1999) Homoplasy, homology and the problem of 'sameness' in biology. In: Bock GR, Cardew G (eds) Homology. Novartis Foundation Symposium 222. Wiley and Sons, New York, pp 24–46

West-Eberhard MJ (2003) Developmental plasticity and evolution. Oxford University Press, New York

Weston PH (2000) Process morphology from a cladistic perspective. In: Scotland R, Pennington T (eds) Homology and systematics: coding characters for phylogenetic analysis. Taylor and Francis, London, pp 124–144

Wiley EO (1981) Phylogenetics: the theory and practice of phylogenetic systematics. Wiley and Sons, New York

Wiley EO, Lieberman BS (2011) Phylogenetics: theory and practice of phylogenetic systematics, 2nd edn. Wiley-Blackwell, New Jersey

Williams PA (1992) Confusion in cladism. Synthese 91:135–152

Willmore KE (2012) The body plan concept and its centrality in evo-devo. Evo Edu Outreach 5:219–230

Wills MA (2001) Morphological disparity: a primer. In: Adrain JM, Edgecombe GD, Lieberman BS (eds) Fossils, phylogeny, and form: an analytical approach. Topics in Geobiology, vol 19. Springer, New York, pp 55–144

Witteveen J (2018) Typological thinking: then and now. J Exp Zool B Mol Dev Evol 330:123–131. https://doi.org/10.1002/jez.b.22796

Wray GA (1999) Evolutionary dissociations between homologues genes and homologues structures. In: Bock GR, Cardew G (eds) Homology. Novartis Foundation Symposium 222. Wiley and Sons, New York, pp 189–206

Young BA (1993) On the necessity of an archetypal concept in morphology: with special reference to the concepts of "structure" and "homology". Biol Philos 8:225–248

Zachos FE (2011) Linnean ranks, temporal banding, and time-clipping: why not slaughter the sacred cow? Biol J Linn Soc 103:732–734

Zelditch ML, Goswami A (2021) What does modularity mean? Evol Dev 23:377–403

Part IV

Challenging Evolutionary Statements

Patterns and Rates: Challenging Gradualism and Punctuated Equilibrium

9

Abstract

This chapter examines the traditional models that describe patterns and rates of evolutionary change. It challenges the model of gradualism as the intermediacy of transitional forms, which posits that the accumulation of small variations over geological time can explain the origin of body plans. In doing so, it addresses the issues of punctuated equilibrium, particularly challenging the terms "cladogenesis" and "anagenesis," whose conceptual distinctions remain unclear. It also demonstrates that punctuated equilibrium closely resembles phyletic gradualism in terms of evolutionary rates. Additionally, it critiques methods that attempt to quantify evolutionary rates through the assumption of gradualism. Finally, it provides reasons to support the assertion that discontinuous rates of change—rather than continuous rates of change over geological time—more effectively explain the origin of complex characters and body plans. Overall, the H-Evo-Devo Theory supports the idea that units of evolution (subspecies, species, higher taxa, and homologues) undergo precise and discrete rates and modes: an instantaneous origin, a period of stasis, and punctuated extinction. Such an assumption is supported by the fossil record and phenomena observed in embryology.

Keywords

H-Evo-Devo Theory · Tempo and mode in evolution · Evolutionary rates · Evolutionary patterns · Discontinuous variation · Continuous variation · Gradualism · Cladogenesis · Anagenesis · Phyletic gradualism · Speciation · Saltational evolution · Stasis · Extinction · Punctuated equilibrium · Evo-Devo · Evo-Devo path · Paleontology · Microevolution · Macroevolution

C. Ochoa, *Hierarchical Evolutionary-Developmental Theory*, Evolutionary Biology – New Perspectives on Its Development 9,
https://doi.org/10.1007/978-3-031-88723-9_9

Currently, most evolutionary biologists take for granted that the origin of species and higher taxa occurs through the accumulation of small individual differences over geological time. While punctuated equilibrium theory has provided insight into the rapid process of divergence and the subsequent period of stasis at the species level, doubts persist among most evolutionary biologists regarding saltational evolution. Moreover, the acknowledgment of the origin and stability of higher taxa remains unaddressed. In this chapter, I will challenge the general perspective of gradualism as the intermediacy of transitional forms. As an alternative, I will defend the stance of saltational origin in evolving entities. Overall, the H-Evo-Devo Theory supports the idea that units of evolution (subspecies, species, higher taxa, and homologues) exhibit three main patterns for their study: instantaneous origin, morphological stability, and punctuated extinction. This conception is based on evidence from the fossil record and phenomena expected from embryology.

Section 9.1 delves into the issues surrounding punctuated equilibrium, focusing on demonstrating the conceptual ambiguity of the terms "cladogenesis" and "anagenesis." Additionally, this analysis demonstrates that punctuated equilibrium is not fundamentally distinct from phyletic gradualism. Gradualism and punctuated equilibrium describe continuous rates regardless of whether they are slow or fast. Subsequently, it challenges the methods that attempt to quantify the rates of evolutionary change through the lens of gradualism. Section 9.2 presents several reasons supporting the idea that discontinuous rates of change are the most appropriate mode for explaining the origin of complex characters and body plans: (1) the measure of discontinuous rates, (2) the historical context of gradualism, and (3) embryological evidence.

9.1 Tempo and Mode in Evolution Reconsidered (Again and Again)

In 1944, the paleontologist George Gaylord Simpson introduced the terms "tempo" and "mode" to denote the patterns and rates of evolutionary change observed in the fossil record at various scales. According to Simpson, "tempo" refers to the speed of change, and he used the terms "horotely," "bradytely," and "tachytely" to indicate standard, slow, or fast rates respectively. In order to explain these patterns, he described the tempos observed at each level and formulated hypotheses about how the evolutionary processes proceed at those levels. These are the "modes":

1. *Speciation (splitting).* Subspecies and species arise from the local differentiation of two or more groups within a large population.
2. *Phyletic evolution.* Genera and families maintain a steady directional change in the entire population as a whole.
3. *Quantum evolution.* Families, orders, and classes arise rapidly from small populations.

As noted, Simpson outlined the modes according to a hierarchical point of view. For instance, the "speciation" mode (later termed "splitting") occurs at the subspecies and species level and is characterized by a horotelic rate. The "phyletic evolution" mode is distinctive to genera and families, and its rate is either horotelic or bradytelic. Meanwhile, the "quantum evolution" mode manifests above the genus level, and its rate is tachytelic. Importantly, within this framework, no part of the hierarchy excludes the tempos and modes described for each level (e.g., species can sometimes arise via quantum evolution). In this context, tempos describe the observed rates of change in the fossil record, while modes propose hypotheses for these rates, whose explanation lies in particular evolutionary mechanisms. For example, "speciation" and "phyletic evolution" occur through natural selection, while "quantum evolution" results from genetic drift.

In the 1970s, by introducing the theory of punctuated equilibrium, paleontologists Niles Eldredge and Stephen J. Gould borrowed the terms "tempo" and "mode" from Simpson, albeit with a different approach. First, they departed from Simpson's framework, which focused on recognizing higher taxa as units of evolution, and consequently restricted modes solely to the species level. Second, they introduced three modes of evolution distinct from those proposed by Simpson, which they named:

1. *Phyletic gradualism*. Species arise from the slow and steady transformation of entire populations as a whole.
2. *Speciation*. Species arise rapidly in geological time from a small local population.
3. *Stasis*. No significant changes in the most important features of species for long spans of time.

In light of this, according to Pachut and Anstey (2009), there are five modes of evolution: (1) "speciation," (2) "phylogenetic evolution," (3) "quantum evolution" (terms proposed by Simpson 1944), (4) "stasis" (according to Eldredge and Gould 1972), and (5) "random walk"[1] (according to Raup 1977). However, the terms "cladogenesis" and "anagenesis" are the two officially recognized evolution modes (Allmon 2016). Cladogenesis is considered synonymous with "speciation," while "anagenesis" is synonymous with "phyletic gradualism." In any case, these are generally defined as:

1. *Cladogenesis*: the splitting of an ancestral species into two or more descendant species.
2. *Anagenesis*: speciation that occurs in the entire population as a whole without splitting.

An intriguing observation is the emergence of various proposals and combinations concerning evolutionary modes based on this terminology, yet without a definitive

[1] The "random walk" mode indicates that the pattern shows no directional tendency; rather, the rate of change is erratic.

consensus. For instance, some researchers have proposed "budding," "bifurcation," and "anagenesis" (Wagner and Erwin 1995; Foote 1996; Silvestro et al. 2018). Others have put forth "random walk," "unbiased random walk," "anagenesis," "punctuated anagenesis," and "stasis" (Roopnarine et al. 1999; Roopnarine 2001). Or there are the terms "gradualism without stasis," "gradualism plus stasis," and "punctuation plus stasis" (Benton and Pearson 2001). Still, others have presented the combination of "punctuation," "directional evolution," "random walk," and "stasis" (Sheets and Mitchell 2001; Hunt 2008). "Gradual anagenesis," "gradual cladogenesis," "punctuated anagenesis," and "punctuated cladogenesis" have also been considered (Patzkowsky and Holland 2012, p. 132). Another proposed classification includes "punctuation," "stasis-random walk," "stasis-directional," "random walk-stasis," and "directional-stasis" (Hunt et al. 2015). Nevertheless, the question is: Why are there so many diverse modes?

The confusion surrounding the most appropriate terms to analyze patterns of evolution stems from two main sources. Firstly, it originates from the theory of punctuated equilibrium, whose advocates assume and propagate the distinction between cladogenesis and anagenesis as two distinct and opposing modes of evolution. Secondly, it arises because biologists have employed incorrect methods to measure rates of evolutionary change, particularly by deriving data from changes in extant organisms and extrapolating their findings to a macroevolutionary interpretation. However, the H-Evo-Devo Theory proposes recognizing three specific evolutionary modes for each analyzed level: sudden origin, stability, and extinction. To reach this conclusion, I will address the issue of punctuated equilibrium with a particular focus on the ambiguity of the terms "cladogenesis" and "anagenesis." Subsequently, I will briefly address the issues associated with current methods for quantifying evolutionary tempos through the lens of gradualism. Finally, I will show that both fossil record data and embryology suggest discontinuous patterns, which are the most suitable for interpreting the tempos and modes of evolution.

9.1.1 Reconsidering Punctuated Equilibrium, Particularly the Terms "Cladogenesis" and "Anagenesis"

In 1972, paleontologists Niles Eldredge and Stephen Jay Gould presented the theory of punctuated equilibrium in a chapter of the book "Models in Paleontology," edited by Tom Schopf. While this was not the first time the theory had been introduced—Eldredge had published it a year earlier in a paper for the journal *Evolution*—it was from this point onward that news of a novel "macroevolutionary" model began to circulate. The theory of punctuated equilibrium addresses the origin and development of species over geological time. It posits that most species emerge during rapid geological periods and then persist in stasis—i.e., without significant morphological changes—for most of their geological lifespan. Although the theory primarily acknowledges the evolutionary patterns displayed by species over geological time, it does not propose new mechanisms to explain these patterns other than those inferred from traditional Darwinism.

Regarding the tempos and modes in evolution, punctuated equilibrium has been the most influential model in contemporary paleontology. Nevertheless, certain contradictions can be observed in its development. What is wrong with the theory of punctuated equilibrium? In my view, Eldredge and Gould got it right in measuring tempos—rapid beginning followed by the stability of basic species traits—but they are mistaken in their description of modes. According to Eldredge and Gould, traditional paleontologists adhered to "phyletic gradualism" as a norm for interpreting the fossil record. Phyletic gradualism "holds that new species arise from the slow and steady transformation of entire populations" (p. 84). It involves the gradual separation of the descendant population from the ancestral population over time through a series of intermediate forms. These early paleontologists attributed the apparent abrupt changes in species observed in geological strata to the imperfections in the fossil record. However, Eldredge and Gould suggest an alternative interpretation of these discontinuities. They turned to the theory of allopatric speciation to explain a rapid movement in geological time in order to align with this pattern; thus, the discontinuities would be interpreted as real. They termed this mode of rapid evolution in geological time "speciation." It is important to note that the proponents of punctuated equilibrium do not deny the occurrence of phyletic gradualism; rather, they consider the model of a rapid beginning followed by a period of stasis to be more prevalent than gradual and constant change.

As previously mentioned, it is essential to note that the term "speciation" employed by Eldredge and Gould is synonymous with "cladogenesis," whereas the term "phyletic gradualism" corresponds to "anagenesis." Hence, I will use these terms interchangeably. With this in mind, I will present a series of arguments demonstrating that the distinction between cladogenesis and anagenesis employed in punctuated equilibrium is unclear. In fact, I will conclude that punctuated equilibrium and phyletic gradualism are remarkably similar models.

(1) *The term "phyletic gradualism" does not correspond to any previously established evolutionary mode that needed to be challenged*

In their 1972 study, Eldredge and Gould mention Charles Darwin as the main source of the phyletic gradualism perspective. According to them, despite identifying two evolutionary modes—"speciation" and "phyletic evolution"—Darwin "muddled this distinction and cast most of his discussion in terms of phyletic evolution. His insistence on insensibly graded sequences among fossils reflects this emphasis, for if species arise by the gradual transformation of entire populations, an even sequence of intermediates should indeed be found." Still, when Darwin discussed speciation (the splitting of lineages), "he continued to look through the glasses of transformation: he saw splitting largely as a sympatric process, proceeding slowly and gradually, and producing progressive divergence between forms." Therefore, according to Darwin, "speciation entailed the same expectation as phyletic evolution: a long and insensibly graded chain of intermediate forms" (pp. 87–89).

In a subsequent paper, Gould and Eldredge (1977) maintain their strategy of presenting "phyletic gradualism" as the less prevalent mode, based on evidence from the geological record. Nonetheless, the historical origins of "phyletic gradualism" are now attributed to the achievements of the Modern Synthesis. Specifically, these authors attribute this confusion to Simpson, from whom they likely borrowed the term "phyletic evolution" as synonymous with "phyletic gradualism."[2] Subsequently, Gould (2002, p. 562) corroborated this source, affirming that: "Simpson held strongly to this view [phyletic gradualism], and even ventured a quantitative defense, in his repeated assertions that speciation represents only a minor mode in evolution because 90 percent of important changes arise anagenetically in the phyletic mode." Meanwhile, Eldredge (2015, pp. 223–224), following an analysis of a quote from Simpson, stated: "Here we see Simpson's general assertion of the truth of gradualism as propounded by Darwin and maintained as well by Dobzhansky and Mayr when considering patterns of evolutionary change after the origin of new species through a geographic isolation event."

However, my conclusion is that Eldredge and Gould notably foisted a caricature of the history of "phyletic gradualism," crafting this concept to serve as a paradigm they sought to challenge. Løvtrop (1981), Penny (1983), Rhodes (1987), and Dennett (1995, p. 290) have noted that while Darwin was a gradualist, his gradualism did not align with the concept of "phyletic gradualism" as described by Eldredge and Gould. For example, specific excerpts from Darwin's work demonstrate that he conceived the idea of variable rates of evolutionary change, even identifying the tempos pronounced by punctuated equilibrium (origin and stability). In the fourth and fifth editions of the *Origin of Species*, respectively, Darwin wrote:

> [A]s lately insisted on by Dr. Falconer [[3]], namely, that the periods during which species have been undergoing modification, though very long as measured by years, have probably

[2] Eldredge (2015) comments that it was Gould who proposed the term "phyletic gradualism" (as well as the terms "stasis" and "punctuated equilibria" later called "punctuated equilibrium"). During the theory's development, however, Eldredge himself referred to this concept as "phyletic evolution." Indeed, he seems to be the one correlating "speciation" with "cladogenesis" and "phyletic evolution" with "anagenesis." In discussing his notes from February 1965 regarding the origin of punctuated equilibrium, he remarks: "I also maintain, in apparent homage to Mayr, that 'the fundamental unit of evolution is the species,' but (presumably following Simpson) 'speciation in its strictest sense, the formation of allopatric species from a parent species (cladogenesis) is not the model, but rather a phylogenetic evolution (or anagenesis) is probably the more common case is important, but not an essential point here'" (Eldredge 2015, p. 232). Nonetheless, in his 1970 paper, in which the punctuated equilibrium was first published, he refers to this mode as "phyletic transformation" or a "phyletic model of gradual change." Interestingly, Simpson (1953) also equates speciation with cladogenesis and phyletic evolution with anagenesis, though he does so as a suggestion by citing the terms of Bernhard Rensch. Yet he warns in his writings that Rensch's terms of cladogenesis and anagenesis carry a different meaning than the speciation and phyletic evolution described in his text (see note 1, p. 380 and note 3, p. 384).

[3] Hugh Falconer (1808–1865), a fellow paleontologist, conveyed his thoughts to Darwin in a letter dated September 24, 1869, regarding the rapid changes and stability of species over geological time: "Do not be frightened at the enclosure. I wish to set myself right by you before I go to press. I am bringing out a heavy memoir on elephants—an omnium gatherum affair, with observation on the fossil and recent species. One section is devoted to the persistence in time of the specific character of the mammoth. I trace him from before the Glacial period, through it and after it, unchangeable and unchanged as far as the organs of digestion (teeth) and locomotion are concerned. Now, the Glacial period was no joke: it would have made ducks and drakes of your dear pigeons and doves" (Falconer in Darwin and Seward 1903, p. 205).

> been short in comparison with the periods during which these same species remained without undergoing any change. (Darwin 1866, pp. 359–360)
>
> I do not suppose that the process ever goes on so regularly as is represented in the diagram [[4]], though in itself made somewhat irregular, nor that it goes on continuously; it is far more probable that each form remains for long periods unaltered, and then again undergoes modification. (Darwin 1866, p. 132)
>
> [A]lthough each species must have passed through numerous transitional stages, it is probable that the periods, during which each underwent modification, though many and long as measured by years, have been short in comparison with the periods during which each remained in an unchanged condition. (Darwin 1869, p. 418)

Similarly, Simpson's "phylogenetic evolution" does not align with Eldredge and Gould's "phyletic gradualism" (see Table 9.1). The distinction between these two perspectives primarily lies in scale. Simpson's phylogenetic evolution is not characteristic of the species level (although it does not entirely exclude it), but rather it pertains to intermediate levels such as genera and families. In Simpson's (1944, p. 202) own words: "[phyletic evolution] can give rise to new species just as well as the different mode here called speciation, but it is not the typical mode of speciation [i.e., the origin of species], and it is less clearly seen at that level than at higher levels." Despite Simpson's later publications not emphasizing this difference (e.g., Simpson 1953), it is notable that he never positioned "phylogenetic evolution" as the prevailing mode of species change. Conversely, he consistently asserts that subspecies and species originate from splitting (cladogenesis). Also, for Simpson, the "phyletic evolution" mode delineates evolutionary trends (e.g., orthogenesis) and the phenomenon of stasis (arrested evolution). Such patterns are frequently observed at these middle levels (see point 4).

Additionally, it is important to note that the term "anagenesis"—commonly used as a synonym for "phyletic gradualism"—carries a different meaning than originally intended. For instance, Bernhard Rensch (1959) used this term to indicate the progressive increase and complexity of traits in higher taxa. Meanwhile, Julian Huxley (1957) referred to anagenesis to describe the "adaptedness and biological improvement" of the lineage in question.[5] In conclusion, the phyletic gradualism (or anagenesis) depicted through punctuated equilibrium does not align with any previously established evolutionary mode that it seeks to challenge.

(2) *Considering how species evolve, there is not a clear distinction between cladogenesis and anagenesis*

Assume there is a description of an evolutionary mode that does not consider species stasis, e.g., the term "chronospecies" provides insight into this situation. So,

[4] Here, Darwin refers to the sole image found in the *Origin of Species*, illustrating branches of continuous and gradual lines extending from the ancestor to the descendants. This image has been interpreted by many researchers as a portrayal of Darwin's purported vision of "phyletic gradualism" or "anagenesis."

[5] Moreover, for Huxley, the terms "cladogenesis" and "stasigenesis" respectively represented divergence and persistence of the type.

Table 9.1 *Comparison of the tempos and modes of evolution according to* Simpson (1944, 1953) *and* Eldredge and Gould (1972). Note the similarities and differences among the modes, rates, descriptions, and levels. It is noteworthy that Simpson does not deny that different modes may occur at different levels; for instance, quantum evolution also applies at the species level. Despite Eldredge and Gould criticizing Simpson for advocating for anagenesis at the species level, in reality, both perspectives share similarities in describing lineage splitting at that level. Moreover, Eldredge and Gould's concept of phyletic gradualism diverges from Simpson's phyletic evolution when considering the species level. The only similarity between these terms lies in describing trends, but this is not what Eldredge and Gould intended to convey with phyletic gradualism as a general perspective. Indeed, Simpson's phyletic evolution in the mode of arrested evolution aligns more closely with Eldredge and Gould's stasis in terms of the bradytelic rate. In conclusion, both perspectives are not so different at the species level, but most importantly, both models support gradualism as the intermediacy of transitional forms. Further details are available in the text

Author	Mode	Tempo	Description	Level
G.G. Simpson	Speciation	Horotely	The splitting of a lineage	Subspecies and species
	Phyletic evolution	Horotely	A trend: steady directional change in the entire population as a whole	Genera and families
		Bradytely	Arrested evolution: no significant changes in the most important features of the lineage	
	Quantum evolution	Tachytely	The lineage arises rapidly from a small population	Families, orders, and classes
Eldredge and Gould	Phyletic gradualism (Anagenesis)	Horotely	Steady transformation of the entire population as a whole	Species
	Speciation (Cladogenesis)	Tachytely (at the macroevolutionary scale)	The lineage arises rapidly from a small population	Species
		Horotely (at the microevolutionary scale)	The splitting of a lineage	
	Stasis	Bradytely	No significant changes in the most important features of the species	Species

during the early development of phylogenetic systematics, Willi Hennig (1966) depicted a diagram illustrating the perpetual and continuous change of species over time (Fig. 9.1). He wrote:

> Species are relatively stable complexes that persist over long periods of time, but they are not absolutely permanent. Modern genetics tell us how new species arise… new gaps arise

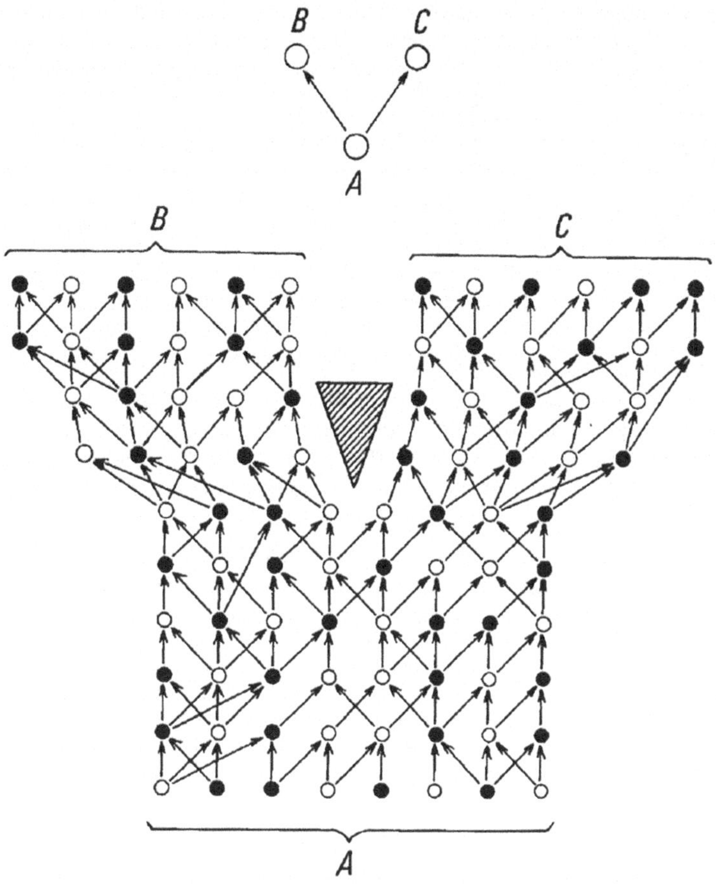

Fig. 9.1 *Speciation according to Hennig*. "A" represents the ancestral species that gives rise to the descendant species "B" and "C." This scheme illustrates the ambiguity of the terms "cladogenesis" and "anagenesis." While the image can be represented as a case of cladogenesis, it actually involves two events of anagenesis. For further details, see the text. (From Hennig 1966)

> in the genealogical relations as a result of reproductive isolation. These gaps delimit new species. As Dobzhansky and many others assume, geographical barriers play the chief role in this isolation. From these empirical facts regarding the origin of species we may assume that the gaps between existing species arose in the same way at some time in the past. This means, therefore, that we would find the gaps separating existing species bridged over if we could go back into the past. If, for example, we follow backward the stream of genealogical relations among the individuals of species B, and do the same for species C, at some time in the past we would arrive at one and the same species A, which we would call the common stem species of B and C. (Hennig 1966, p. 19–20)

In short, while geographic barriers facilitate divergence, the ancestral species (A) gradually transforms over time, giving rise to two distinct species (B and C). Turning back to punctuated equilibrium, recall that Eldredge and Gould (1972) recognized two modes of speciation: (1) "splitting" (cladogenesis) and (2) "phyletic gradualism" (anagenesis). Regarding this, theoretically, cladogenesis involves the splitting of an ancestral species into two descendant species, whereas anagenesis entails a transformation in the entire population as a whole without splitting. However, according to the contemporary understanding of speciation, anagenesis is not possible since species arise from a portion of the population, rather than the entire population as a whole (Kluge 1990, p. 422; Ereshefsky 1991). For example, consider the three geographical speciation models: allopatric speciation, parapatric speciation, and sympatric speciation. As seen in Fig. 9.2a, although gradualism—as intermediacy of transitional forms—is expected, in none of these cases does the ancestral species (A) disappear to give rise to the descendant species (B).

Using the Hennig diagram highlights this issue starkly. For example, anagenesis implies that the ancestral species "A" evolves as a whole without splitting, giving rise to the descendant species "B" while gradually disappearing (Fig. 9.2c). In contrast, cladogenesis entails that the ancestral species "A" splits, giving rise to both the descendant species "B" and "C" (Fig. 9.2d). However, the broader picture reveals that cladogenesis (bifurcation) comprises two events of anagenesis (Fig. 9.3c). On one side, "A" leads to "B," while on the other, "A" leads to "C." Without either "B" or "C," such scenarios would be interpreted as anagenesis. Aware of this issue, Gould (2002, p. 777) argued that the distinction between cladogenesis and anagenesis should be viewed in terms of tempo. In the view of punctuated equilibrium, speciation (cladogenesis) occurs rapidly, whereas in the phyletic gradualism paradigm (anagenesis), species originate slowly. Nonetheless, this perspective contradicts his later assertion that speciation over geological time could also be explained by "rapid anagenesis" (see point 4). Elsewhere, Gould (2002, pp. 793–796) suggests that the authenticity of cladogenesis rests on the persistence of the ancestor (budding). Although the ancestral species "A" gives rise to the descendant species "B," note that "A" and "B" coexist spatiotemporally (Fig. 9.2b). Consequently, Gould (2002, pp. 817–818) commends Wagner and Erwin's (1995) clarification of this distinction, introducing the terms "bifurcation" (two sibling species emerge from an ancestral species that disappears) and "budding" (two or more sibling species coexist with the ancestral species). But this discrepancy poses challenges in paleontological practice since, if the ancestor is not found in the same timeline as the descendant, then "bifurcation" must be assumed rather than "budding" (see Pennell et al. 2014). However, as previously mentioned, such "bifurcation" actually represents two cases of anagenesis. Here, it is important to remember that anagenesis is not a process that occurs in nature. Instead, a new species arises from a part of the population rather than the entire population as a whole (Fig. 9.2a). Therefore, the only valid evolutionary mode to describe the origin of species is through speciation by splitting (see Fig. 9.2b).

(3) *There is no clear distinction between cladogenesis and anagenesis when considering microevolutionary mechanisms*

Eldredge and Gould (1972) made it clear that macroevolutionary patterns were nothing more than a proper expression of well-known microevolutionary processes

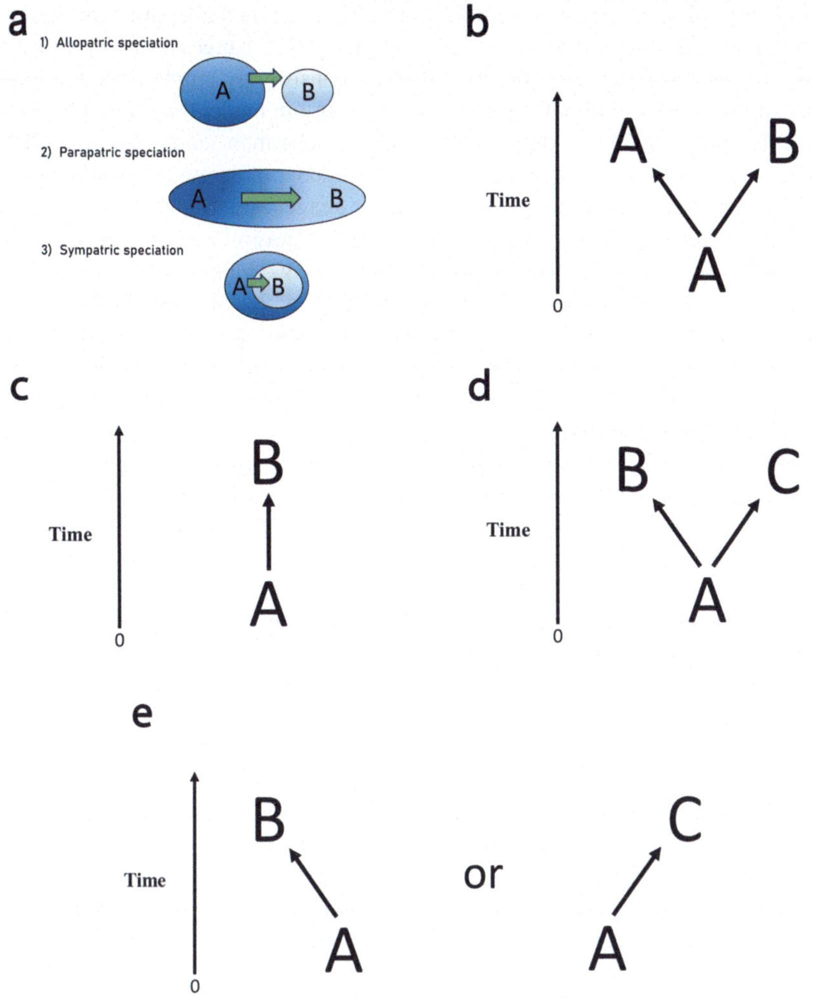

Fig. 9.2 *Graphic representations illustrating the ambiguity of the terms "cladogenesis" and "anagenesis" considering how species evolve.* (**a**) *Traditional geographical models of speciation.* Notice that in none of these cases does the ancestral species "A" disappear to give rise to the descendant species "B." (**b**) *Cladogenesis (budding).* Correct representation of cladogenesis, where the ancestral species "A" persists over time and can coexist with the descendant species "B." (**c**) *Anagenesis.* The ancestral species "A" evolves as a whole without splitting, giving rise to descendant species "B." (**d**) *Cladogenesis (bifurcation).* The ancestral species "A" disappears, giving rise to both descendant species "B" and "C" (refer also to Fig. 9.1). (**e**) *Cladogenesis as bifurcation manifests as two events of anagenesis.* The ancestral species "A" disappears, giving rise, on one side, to the descendant species "B," and on the other, to the descendant species "C."

at a geological scale.[6] In their seminal work, they promoted allopatric speciation as the microevolutionary mechanism underlying the rapid changes observed in speciation (cladogenesis). Specifically, Eldredge has consistently emphasized the importance of this process within the punctuated equilibrium model throughout his career (e.g., Eldredge 1971, 2008b, 2015, 2024; Lieberman and Eldredge 2014). Consequently, a "sympatric process" (or another geographic model) would theoretically be the microevolutionary explanation for anagenesis. Nonetheless, in discussions distinguishing cladogenesis from anagenesis, several authors apply different geographic models to both evolutionary modes (e.g., Vaupel Klein 1994, 1995; Bose and Bartholomew 2013). Even Gould (2002, p. 780; his emphasis) suggests that cladogenesis is not solely explained by allopatric speciation: "Punctuated equilibrium simply requires that any asserted mechanism of speciation, whatever its mode or style, be sufficiently rapid and localized to appear as punctuation *when scaled into geological time.*"

The issue is still far from settled, as the differences between geographic models of speciation are ambiguous, a topic I will explore further in the next chapter.[7] Likewise, Gingerich (2019, pp. 281–282) contends that while genetic isolation plays a role in subsequent morphological differentiation, it does not explain how divergence occurs after geographic isolation. He asserts that the fossil record documents morphological changes, but it alone does not say anything about the genetic mechanisms contributing to divergence: "How then is allopatric speciation to be applied to the fossil record?" Vaux et al. (2016) argue that if cladogenesis involves splitting due to the cessation of gene flow, then it is challenging to claim there was cladogenesis based solely on fossil record since it lacks information to evaluate gene flow between populations. Moreover, they suggest that if speciation ideally results from splitting due to gene flow cessation, then the distinction between cladogenesis and anagenesis becomes arbitrary because divergence can occur even without gene flow cessation. In conclusion, if the microevolutionary causes of cladogenesis or anagenesis remain poorly understood, then it follows that there are no clear distinctions between these modes of evolution.

(4) *There is no clear distinction between cladogenesis and anagenesis when considering the tempos of evolution*

In their 1977 paper, Gould and Eldredge stated that tempos "can be observed and measured," but that "modes must be inferred, usually from empirical distributions of tempos" (pp. 115–116). In doing so, they introduced punctuated equilibrium as a hypothesis about modes and emphasized the importance of paleontological data in measuring tempos. Needless to say, the relationship between tempos and modes has established evolutionary assumptions regarding how evolution proceeds, but it is important to note that Simpson's original proposal significantly differs from

[6] For example, one of Løvtrop's (1981) criticisms of punctuated equilibrium is that while it is presented as a macroevolutionary model, it is actually grounded in microevolutionary processes. According to Løvtrop, the genuine macroevolutionary processes involve saltational evolution and the origin of higher taxa.

[7] See Chap. 10, Sect. 10.1.

punctuated equilibrium regarding tempos. For Simpson, "speciation" corresponds to a standard rate (horotely), "phyletic evolution" corresponds to slow rates (bradytely and horotely), and "quantum evolution" corresponds to a rapid rate (tachytely). These tempo distributions create the picture of the mode in the adaptive zone as seen in Fig. 9.3. In contrast, for Eldredge and Gould, "speciation" exhibits a rapid rate (tachytely) on the geological time scale, but on the microevolutionary scale, the rate is standard (horotely). While "phyletic gradualism" presents standard and slow rates (horotely and bradytely), stasis exhibits a slow rate (bradytely). Therefore, concerning tempos, Simpson's "speciation" can be considered equivalent to Eldredge and Gould's "speciation" on the microevolutionary scale. Simpson's "quantum evolution" is equivalent to Eldredge and Gould's "speciation" on the macroevolutionary scale. Simpson's "phyletic evolution" as a trend is equivalent to Eldredge and Gould's "phyletic gradualism." However, Simpson's "phyletic evolution" as arrested evolution is equivalent to Eldredge and Gould's "stasis" (see Table 9.1).

Here, the differences or similarities in tempos are crucial. In his early writings on punctuated equilibrium, Eldredge (1971) emphasized that rapid speciation, through the allopatric model, should be distinguished from "gradualism": "the allopatric model, rather than gradual morphological divergence, is the more correct view of the processes underlying cases of splitting already documented by numerous workers" (p. 157). Meanwhile, in their 1972 paper, Eldredge and Gould stressed that: "by the picture of

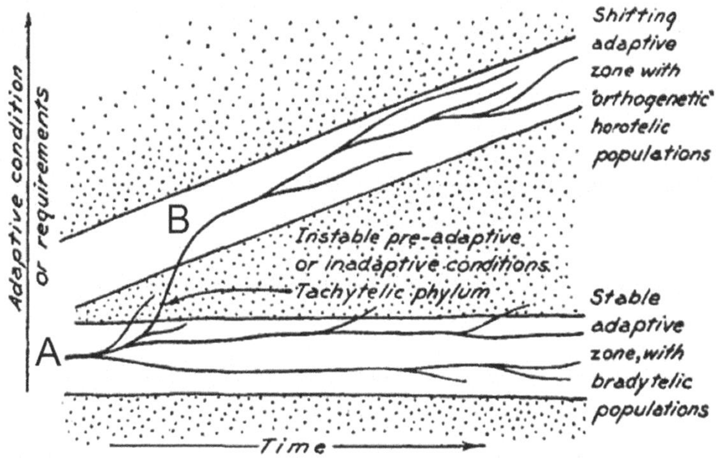

Fig. 9.3 *Simpson's diagram illustrates the distribution of different rates observed in the evolution of a lineage.* It can be observed that the ancestral lineage "A" gives rise to the descendant lineage "B" (splitting) without disappearing. Note that the descendant lineage "B" crosses the unstable adaptive zone field at a tachytelic rate, subsequently undergoing an evolutionary trend (orthogenetic) at a horotelic rate. However, the ancestral lineage "A" persists over time, experiencing a stable adaptive state (arrested evolution), at a bradytelic rate. According to Simpson, horotelic and bradytelic rates are represented through the mode of phyletic evolution (see Table 9.1). Of course, Simpson assumes gradualism as the intermediacy of transitional forms but with different evolutionary rates. Note the similarity of this scheme with the interpretation of punctuated equilibrium at the microevolutionary scale seen in Fig. 9.4. (Modified from Simpson 1944)

'phyletic gradualism'… new species arise from the slow and steady transformation," whereas "allopatric (or geographic) speciation suggests a different interpretation of paleontological data. If new species arise very rapidly in small, peripherally isolated local populations, then the great expectation of insensibly graded fossil sequences is a chimera" (p. 84). Of course, they draw a view of gradualism as being synonymous with slow and steady rates. Yet this is not entirely correct; curiously, Simpson (1984) himself mentioned this error when comparing punctuated equilibrium's mode of speciation (punctuation) with his famous "quantum evolution" mode:

> So positive a statement requires brief notice of what does claim to be a "new and general evolutionary theory" which nominally rejects the synthetic theory. This is called "punctuated equilibrium," and the synthetic theory is condemned as "gradualistic." By "punctuation" is meant essentially what is meant by "quantum evolution" in *Tempo and Mode in Evolution*, the origin of species or other taxon by exceptionally rapid evolution… More recently Gould… has redefined "punctuation" as being continuous but so rapid as to seem practically instantaneous in stratigraphic observation… Gould lumped together Darwinism, Neo-Darwinism, and Synthesis as "gradualism." As he defined "gradualism," it was a straw man for his attack. That term is not a description of any of the schools of thought that he wants to reject. "Gradualism" *sensu* Gould is one end of a continuum and "punctuated equilibrium" is the other end. (Simpson 1984, pp. xxv–xxvi)

Similarly, Eldredge (2015, p. 224; his emphasis) remarks: "Rereading Simpson's (1944) text, I was surprised to find that quantum evolution could be applied to the sudden origins of taxa at any level—*including species*." Of course, this indicates a lack of knowledge regarding Simpson's ideas during the development of punctuated equilibrium. Additionally, as previously mentioned, Simpson's concept of phyletic evolution appears to align with Eldredge and Gould's notion of phyletic gradualism regarding the horotelic rate. However, Simpson employed this mode to explain evolutionary trends observed at intermediate levels (see Table 9.1). Although Eldredge and Gould (1972) discuss the evolutionary trends in Simpson's sense, particularly when they assert that phyletic gradualism involves "a constant adjustment to a unidirectional change in one or more features of the physical environment" (p. 97), such an interpretation of trends does not necessarily reflect gradualism as a general view (see also Gingerich 2019, pp. 282–283). Moreover, Simpson's phyletic evolution is not equivalent to Eldredge and Gould's phyletic gradualism in any sense, as Simpson's phyletic evolution mode also encompasses Eldredge and Gould's stasis. For instance, in describing phyletic evolution, Simpson (1953) alluded to the concept of "arrested evolution," which occurs when a bradytelic rate is observed in evolving entities: "Bradytely is essentially a statistical effect produced by the prevalence in some groups of organisms of lines with extremely low rates of evolution, or with changes fluctuating on a small scale and not appreciably cumulative. The first point, then, is to consider the circumstances of very slow or arrested evolution in general" (p. 327). Furthermore, various scholars consider stasis as a type of gradualism, where evolutionary rates are exceedingly slow, or it represents anagenesis at a zero rate (Gingerich 2019; Vaupel Klein 1994; Allmon 2016). Allmon (2016), for instance, highlights that anagenesis comprises three sub-modes: stasis, random walk, and directional change. Indeed, this classification bears similarities to Simpson's original perspective on phyletic evolution.

Likewise, Vaupel Klein (1994) claims that punctuated equilibrium and phyletic gradualism are not contradictory but rather complementary; both involve divergence and are important to describe a complete scheme of the rates of evolution. So, following Vaupel Klein's framework (1994, 1995), Fig. 9.4 illustrates that, on a macroevolutionary scale, the exceedingly slow rate (bradytely) of (a) gradualism mirrors the stasis of (b) punctuated equilibrium (both depicted as vertical parallel lines). But when divergence occurs, in both scenarios, the ancestor survives (Sp.1 follows the parallel line with a bradytelic rate), and during splitting, the descendant lineage undergoes rapid evolution (Sp.2 separates from Sp.1 with a tachytelic rate). The distinction lies in the fact that, in the gradualism model, after this rapid movement, the descendant lineage (Sp.2) reaches equilibrium, i.e., it undergoes a bradytelic rate. Whereas in punctuated equilibrium, the descendant lineage (Sp.2) undergoes stasis. Although both models exhibit rapid movement when separating from the ancestral lineage, in punctuated equilibrium, this change is portrayed as an almost flat horizontal line, contrasting with gradualism's depiction of a slightly curved continuous line (at a horotelic rate). Nevertheless, it is important to recall that, for advocates of punctuated equilibrium, this rapid description is an expected expression on a geological scale; however, on their microevolutionary scale, the change is continuous and conventional (Fig. 9.4d): "[Punctuated equilibrium] does not deny that allopatric speciation occurs gradually in ecological time… but only asserts that this scale is a geological microsecond. Our model must be tested at the appropriate scale—by considering tempos of change in species and in the process of speciation during geological time" (Gould and Eldredge 1977, p. 121).

In this regard, it can be asserted that the authors of punctuated equilibrium admit gradualism—as intermediacy of transitional forms—despite presenting the "speciation" mode as anti-gradualist. For instance, Gould (2002, p. 776) stated: "[A] new species must pass through a short period of ambiguity during its initial differentiation from an ancestral population." This assertion contradicts his initial criticism of phyletic evolution as "a long and insensibly graded chain of intermediate forms" (Eldredge and Gould 1972, p. 89). Additionally, Gould (2002, p. 795) claimed that anagenesis can sometimes occur rapidly, thus exhibiting a punctuated tempo. Interestingly, Dawkins (1986) and Dennett (1995) had long observed that Eldredge and Gould's portrayal of gradualism was misleading. According to them, Eldredge and Gould are as gradualists as any other neo-Darwinian since they do not challenge gradualism as opposed to saltational evolution (i.e., evolution without intermediate forms) but rather oppose gradualism as "constant speedism." Furthermore, they argue that such a depiction does not align with Darwinian thought. Within this framework, rates of change are not constant but variable (see point 1). Gould (2002) once again admits this error, and in his latest definitions of phyletic gradualism, he emphasizes: "gradualism as a geological claim for slowness and smoothness (*but not constancy*) of rate" (p. 60; my emphasis). However, such a definition remains somewhat ambiguous. In summary, the differences between cladogenesis and

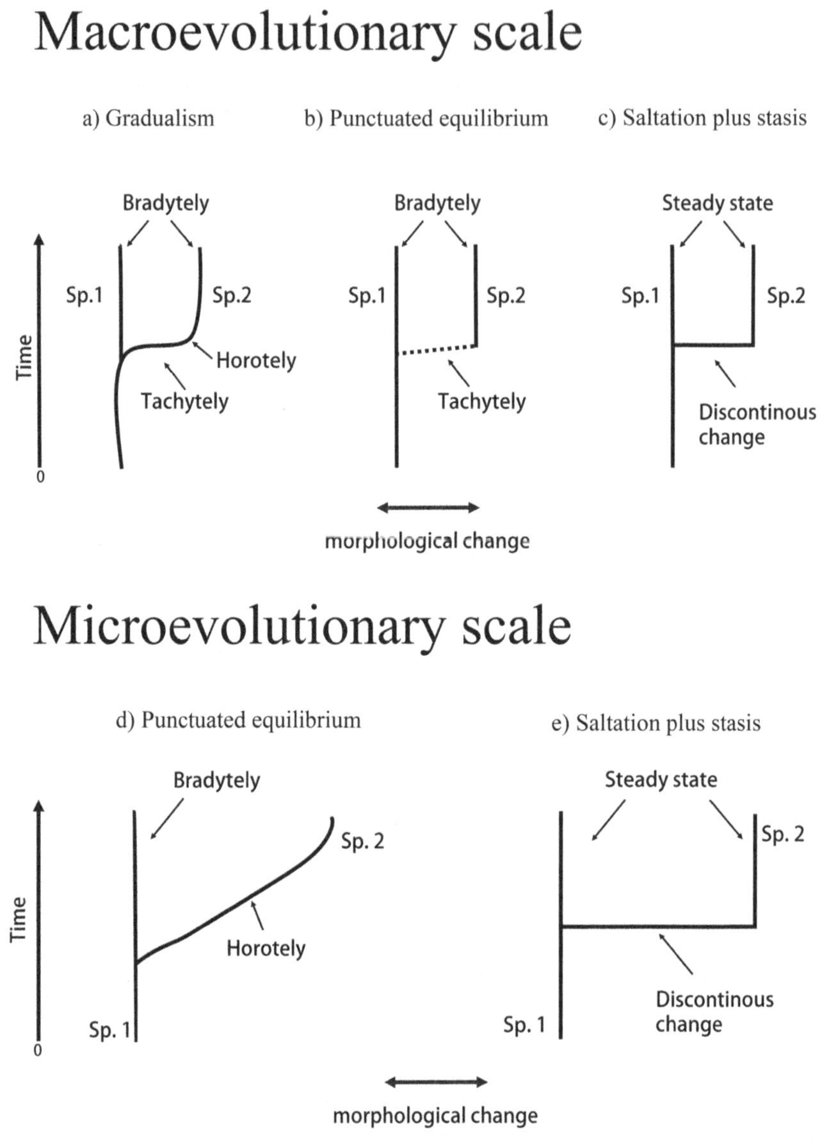

Fig. 9.4 *Distribution of evolutionary rates under three distinct evolutionary models at both macroevolutionary and microevolutionary scales*. (**a**) Gradualism. (**b**) Punctuated equilibrium. (**c**) Saltation plus stasis. The y-axis represents time, while the x-axis represents morphological change. Sp.1 is the ancestral species, and Sp.2 is the descendant species. (**a**) In the gradualism model, rates are categorized into horotely (standard), bradytely (slow), and tachytely (fast). For example, Sp.1 remains stable, maintaining a bradytelic rate. Subsequently, it splits, giving rise to Sp.2 at a tachytelic rate. Sp.2 continues evolving until it achieves equilibrium, transitioning from a horotelic to a bradytelic rate. Notably, Sp.1 does not disappear but persists in a stable state at a bradytelic rate akin to Sp.2, albeit in a different morphospace location. (**b**) In the punctuated equilibrium model at the macroevolutionary scale, Sp.1 remains in stasis at a bradytelic rate. Later, Sp.1 splits, giving rise to Sp.2 at a tachytelic rate over geological time. Once Sp.2 appears, it remains in stasis at a bradytelic rate. (**c**) The saltation plus stasis model displays discontinuous rates: a sudden

anagenesis remain unclear regarding tempos, even though tempo distribution is crucial data for formulating hypotheses about modes.

9.1.2 Reconsidering How Tempos Are Measured in Evolution

How do we quantify the tempo of evolution? The debate primarily revolves around whether evolutionary units exhibit continuous rates, thus being statistically quantifiable, or if they undergo discontinuous tempos: discrete origin, stability, and punctual extinction. Paleontological data have long been the sole source for estimating these patterns of evolution until scientists such as Philip D. Gingerich began considering other methods, particularly statistical ones, focusing on extant organisms. According to Gingerich (2019), there are at least three means to study evolutionary rates: experimental, observational, and inferential. For instance, he highlights the significance of selection experiments whose results on phenotypic change rates are transmitted from one generation to the next (e.g., Falconer 1973; Dudley and Lambert 2004; Laurie et al. 2004; Geiger and Sánchez-Villagra 2018; Lewis and Morran 2021) or field studies demonstrating change rates due to selective pressures over relatively short periods (e.g., Gilchrist et al. 2001; Bell et al. 2004; Brown and Brown 2011; Aguirre and Bell 2012; Sun et al. 2017; Rudman et al. 2022; Wagnon et al. 2022; Strickland et al. 2024), etc.

Gingerich's aim in highlighting these studies is to bolster the argument that the rates observed at the microevolutionary scale provide a more realistic depiction of evolutionary tempos compared to the patterns discerned in the fossil record, which suggest an abrupt origin followed by a period of stasis: "The rates, and the changes they represent, are all parts of one continuous distribution" (Gingerich 2019, p. 277). Moreover, Gingerich posits that rates can be measured by observing the direction of variation influenced by natural selection. The issue with this perspective arises from contemporary advancements in evolutionary developmental biology, which indicate that "units of evolution" are highly cohesive entities due to the developmental constraints operating on them. This implies a discontinuous change in the production of significant morphological innovations. It is important to note that when Gingerich delves into "variation," he only presents statistical data, lacking prior knowledge of

Fig. 9.4 (continued) change followed by a period of stability. (**d**) Punctuated equilibrium at the microevolutionary scale: Here, Sp.2 emerges from Sp.1 at a traditional horotelic rate until it achieves equilibrium again at a bradytelic rate. This observation suggests that Eldredge and Gould's punctuated equilibrium closely resembles the phyletic gradualism they criticized. Both models advocate for gradualism as the intermediacy of transitional forms at varying rates. (**e**) In the saltation plus stasis model at the microevolutionary scale, the transition from Sp.1 to Sp.2 occurs within one or a few generations. Consequently, this model exhibits a horizontal flat line, unlike the slightly curved continuous line observed in microevolutionary punctuated equilibrium. It is important to note that punctuated equilibrium necessitates species differentiation in thousands of years to attain the discontinuity required for recovering equilibrium. Further details can be found in the text

how these variations originate. For instance, contemporary knowledge suggests that many continuous variations likely stem from the phenomenon of phenotypic plasticity as a norm of reaction. Consequently, it would be erroneous to presume that all variations are "heritable" and can be filtered by natural selection. In line with this, I concur with Gould (1984) regarding the idea that the mathematical models employed to quantify tempos are mere "artifacts of the human mind" associated with the interpretation of acquired data; thus, they fail to reflect actual rates of change.

Moreover, some biologists have resorted to molecular clocks to estimate rates of change, gathering data from extant organisms without depending on the fossil record (e.g., Nee et al. 1992; Harvey et al. 1994; Thorpe et al. 2005; Lanfear et al. 2010; Ho and Duchêne 2014; Ho 2020; Duchêne et al. 2024; Layton and Wilson 2024). Other researchers have proposed the use of paleontological data alongside the molecular clock to enhance the accuracy of change rates (e.g., Liow et al. 2010; Morlon et al. 2011; Slater et al. 2012; Silvestro et al. 2018; Mitchell et al. 2019; Burin et al. 2019; Mongiardino Koch and Thompson 2021; Lindsey et al. 2024). Nevertheless, because molecular data often diverge from patterns observed in the fossil record (Bannikova and Lebedev 2022; Budd and Mann 2024), several researchers have devised methods to tackle these calibration challenges (e.g., Rabosky et al. 2014; Carruthers and Scotland 2020; Guindon 2020; Pozzi and Penna 2022; Tiley et al. 2020; Barba-Montoya et al. 2021; Wright et al. 2022; Martin et al. 2023; Zhang et al. 2025).

The main issue inherent in these considerations primarily stems from the theoretical validation underlying the concept of the molecular clock. According to this concept, the molecular clock suggests that genes change at a constant rate over time and evolutionary divergence can be estimated through phylogenetic comparisons. Nonetheless, Pulquério and Nichols (2006) argue that data obtained from the molecular clock are unreliable because different molecular techniques yield highly disparate results. Despite the proposal of new methods attempting to address this inaccuracy, there is no guarantee of their success:

> The difficulties of verifying clock-based dates mean that some discrepancies might previously have been overlooked or explained away. We await the more rigorous type of assessment with some nervousness, given that we suspect they might reveal that many past studies placed too much confidence in simple molecular clock analyses, and that their conclusions should thus be revisited. (Pulquério and Nichols 2006, p. 184)

Similarly, Schwartz and Maresca (2006) point out that estimating change rates through the molecular clock is difficult to sustain since there is no one-to-one relationship between the molecular and morphological levels. Moreover, morphological characters are certainly not in a perpetual state of transformation. Since the consideration of the molecular clock stems from a model derived from prokaryotic cells (whose reconstruction of phylogenetic trees was used to compare structural protein and DNA coding region sequences), they argue that bacteria, as unicellular organisms, do not regulate gene expression in the same way as cells in multicellular organisms. Bacteria may encode most of their genome into proteins, and although this genotype may change over time, their unicellular structure has not morphologically changed since their appearance on Earth. In contrast, multicellular organisms have a significantly different organization; they are not constructed through differences in structural genes but rather through differences in the expression of regulatory genes during embryogenesis:

> The latter implies that genetic modifications (e.g., mutations in structural proteins, or duplication/deletion events) are not sufficient to generate the morphological novelties of "higher" eukaryotes… Changes in the expression of developmentally regulated genes, either by mutations in the long promoter regions of metazoan genes and/or in transcription factors (or in their expression), are known to alter the pattern of development… and may be directly responsible for the emergence of novelties. (Schwartz and Maresca 2006, pp. 366–367)

Considering this, they conclude that: "[N]o matter how sophisticated their mathematical models, molecular systematists have not questioned the basic assumptions upon which they are based" (p. 369). However, this does not mean that the rates of evolution cannot be studied, as Stanley (1985) had already suggested:

> Although molecular clocks make a modest contribution, the fossil record is our primary source of information about rates of evolution. The notion that measurement of genotypic rates should be our foremost goal… is unsupportable. Tallies of gene differences mean very little because genes interact and vary greatly in their influence over the phenotype. The phenotype must be the primary object of concern in measuring evolutionary rate. (Stanley 1985, p. 13)

Despite the problems that the fossil record presents (for example, many groups are poorly represented; see Wagner 2000; Foote 2003), it has long served as the primary source for obtaining reliable data on phenotypic changes over time. In short, the fossil record provides direct evidence of the origin, stability, and extinction of species and higher taxa (Lieberman 2001; Jablonski and Shubin 2015; Ausich 2016). However, not all specimens from the fossil record are suitable for interpreting evolutionary tempos. For example, advancements in statistical methods in paleontology have led many researchers to bolster the belief that evolutionary sequences exhibit variable rates of change. Consequently, several authors have interpreted rates and patterns in diverse ways (e.g., Benton and Pearson 2001; Roopnarine 2001; Sheets and Mitchell 2001; Hunt 2008; Hunt et al. 2015; Hansen 2024; Duran-Nebreda et al. 2024). Yet it is worth noting that most of these studies use foraminifera as a study sample due to their abundance in the fossil record. Here, it is important to consider that foraminifera belong to the protists, i.e., they are unicellular organisms, and as previously mentioned, unicellular organisms develop very differently from multicellular organisms. On this matter, it is essential to highlight that most multicellular organisms—such as animals, plants, and fungi—exhibit discontinuous patterns of change in the fossil record. Undoubtedly, this is one of the primary concerns of Gingerich (2001, 2019), who has sought to quantify microevolutionary rates to extrapolate them to the level of macroevolution.

Particularly, numerous paleontological studies support the pattern of abrupt origin followed by the stability of change,[8] which is subsequently followed by extinction in a similarly punctuated manner.[9] In all these studies, expressions such as "sudden appearance," "sudden pulse," "sudden explosions," "very rapid evolution," "burst in evolution," "rapidly," "explosively," and "abruptly" are emphasized to indicate the origin of both species and higher taxa. Even Gingerich (2019, p. 335), skeptical of a literal reading of discontinuous patterns in the fossil record, writes: "Early bursts of evolution are common, as paleontologists and ecologists have recognized for many years." Another relevant consideration is stasis. For example, although structural stability is typically highlighted only at the species level, higher taxa similarly exhibit this pattern.[10] Therefore, expressions such as "undergone little morphological change," "arrested evolution," "steady state," "morphological persistence," etc., are commonly used to describe it.

Finally, it is noteworthy that many scientists who study evolutionary rates rarely mention extinction as an evolutionary mode. Perhaps this is because extinction was long associated with transient events related to the speciation process, in the sense of "pseudoextinction," i.e., as a population changes over time, at some point, it no longer resembles the initial population. Alternatively, the extinction of large groups was also explained by the gradual extinction of the small populations that comprised them. From this viewpoint, the massive extinctions in the fossil record were often interpreted as illusions caused by incomplete geological records. For example, Holland (2020) argues that the pulses observed during mass extinctions may have unfolded over a longer period than is commonly assumed. However, there is a high likelihood that discrete mass extinctions were real, even though some researchers do not dismiss the possibility that they may have occurred gradually (Wang et al. 2012; Nawrot et al. 2018; Wang and Zhong 2018). The solution to this paradox lies in recognizing the validity of both perspectives and acknowledging that extinctions occur hierarchically.[11] In addition, it is inferred that the sudden extinction of certain groups coincides with the instantaneous emergence of other groups to replace them (Vrba 1993; Erwin 2001; Krug and Jablonski 2012; Saupe and Myers 2021), and this occurs on different scales (Eldredge 2003, 2008a; Congreve et al. 2018). In a nutshell, from the standpoint of the H-Evo-Devo Theory, extinction is regarded as an evolutionary mode, as it indicates the death of the lineages in question.

Considering these data, it is suggested that all statistical or experimental methods that aim to quantify continuous evolutionary rates at a microevolutionary scale for multicellular organisms be abandoned. Evolutionary tempos should be researched

---

[8] Schindewolf 1950; Simpson 1953; Eldredge and Gould 1972; Ager 1976; Kennedy 1977; Paul 1977; Thomson 1977; Valentine 1977, 1980; Fagerstrom 1978; Raup 1983; Campbell and Marshall 1987; Runnegar 1987; Erwin et al. 1987; Geary 1987; McCune 1996; Carroll 1997; Schweitzer 2001; Gould 2002; Uyeda et al. 2011; Neige 2015; Kemp 2016; Yacobucci 2016; Wagner 2018; Moon and Stubbs 2020; Witts et al. 2023; Bechly 2024.

[9] Brett and Baird 1995; Bambach et al. 2004; Foote 2005; Erwin 2008; Jablonski 2005; Krug and Jablonski 2012; Neige 2015; Wang and Zhong 2018.

[10] See Chap. 12, Sect. 12.1.2.

[11] See Chap. 11, Sect. 11.2.3 and Chap. 12, Sect. 12.1.3.

using fossil record data, which reveal three distinct patterns: instantaneous origin, morphological stability, and punctuated extinction. It is not coincidental that the patterns observed in the geological record align with those predicted from an embryological perspective. As emphasized throughout this book, units of evolution are highly cohesive entities due to their developmental constraints. Under this view, evolutionary entities are hardly modifiable because of their stability in embryonic development; therefore, they necessitate discontinuous patterns of morphological change. If this is the case, then the fossil record strongly supports the patterns suggested by embryology.

9.2 Continuous or Discontinuous Rates? Why Discontinuity Is the Best Explanation for the Origin of Complex Characters and Body Plans

Despite the significant efforts made to quantify the tempo of evolution, few biologists have taken the time to consider the implications of studying the origin of phenotypic variation through the lens of Evo-Devo. Notably, many evolutionists simply assume that "complex structures" are acquired through the accumulation of small variations guided by natural selection. However, morphologists have noticed that such characters arise discontinuously during development (Salazar-Ciudad and Jernvall 2005; Theißen 2009; Frazzetta 2012; Jaeger et al. 2012; Linde-Medina and Newman 2014; Peterson and Müller 2016; Lange et al. 2018; Usai et al. 2024). Below, I will outline several reasons supporting the postulate that discontinuous rates of change offer the best explanation for the origin of complex characters and body plans.

(1) *The rate at which units of evolution emerge should be measured with respect to the magnitude of discontinuous change over the number of generations rather than continuous change over geological time*

Here, I delve into the traditional assumption of tempos. Simpson (1944) regarded the concept of "tempo" as essential for understanding the rates of change occurring at different scales. In this context, this generation of evolutionists embraced the Darwinian premise of gradualism as an intermediacy of transitional forms, albeit with varying rates of change—e.g., horotely, bradytely, and tachytely. However, according to punctuated equilibrium, Eldredge and Gould (1972) incorrectly said that "phyletic gradualism" was the notion that species evolve continuously at a perpetual rate (i.e., without stasis). In opposition to this assumption, they proposed that species evolve rapidly (tachytely) instead, followed by a period of stable change with extremely slow rates (bradytely). Nevertheless, as evidenced in the previous section, the phases of punctuated equilibrium do not differ from the gradualism they critique. Figure 9.4 shows that saltational evolution (saltation plus stasis) displays vertical and horizontal flat lines in morphological change at both the macroevolutionary (Fig. 9.4c) and microevolutionary scales (Fig. 9.4e). At the

Table 9.2 *Modes of evolution in relation to tempo, time, or number of generations in which change occurs, type of change, and sequential transitions*. Modes: phyletic gradualism and punctuation (according to Eldredge and Gould 1972); parasaltation and saltation (according to Bateman and DiMichele 2002); and stasis (general view). Tempo: slow, rapid, instantaneous, and steady state. Time in years or the number of generations. Type of change: Continuity or discontinuity. Sequential transitions: "A" represents the ancestral form, "B" and "C" are transitional forms, and "D" is the final descendant form. The bottom arrow indicates whether the change occurred during phylogeny or ontogeny. This table illustrates the differences between gradualism and saltational evolution. It is important to note that in the case of parasaltation, although it is necessary to traverse morphospace through continuity, the transition occurs during ontogeny rather than phylogeny. Thus, it can be said that it also occurs discontinuously within a few generations

Mode	Tempo	Time or no. generations	Type of change	Sequential transitions
Phyletic gradualism	Slow	More than 50,000 years	Continuity (without stasis)	A → B → C → D ⟶ Phylogeny
Punctuation	Rapid	≈ 40,000 years	Continuity	A → B → C → D ⟶ Phylogeny
Parasaltation	Instantaneous	2 or more generations	Continuity—discontinuity	A → B → C → D ⟶ Ontogeny
Saltation	Instantaneous	1 generation	Discontinuity	A → D ⟶ Ontogeny
Stasis	Steady state	≈ 4 million years	No significant changes	D → D → D ⟶ Phylogeny

macroevolutionary scale, saltational evolution resembles punctuated equilibrium by delineating a sudden origin followed by a period of stasis (see Fig. 9.4b and c). But at the microevolutionary level, saltational evolution deviates from punctuated equilibrium because it maintains a horizontal flat line since the changes occur from one generation to the next (Fig. 9.4e). It is important to note that punctuated equilibrium requires species to traverse the morphospace over thousands of years in order to reach discontinuity (Fig. 9.4d).[12] Recognizing this, it is clear why the rates at which units of evolution arise should be measured in terms of the number of generations, aligning with saltational evolution, rather than in geological time, as assumed in punctuated equilibrium. With this in mind, it raises the question of whether Bateman and DiMichele's (2002) term "parasaltation"—defined as "profound phenotypic change across two to several generations"—entails gradualism.

Now, I will examine the differences in the rate of change among the different evolutionary modes outlined in Table 9.2, including Eldredge and Gould's (1972) interpretation of the distinction between "phyletic gradualism" and "punctuation"

[12] For example, according to Gould (2002, p. 768): "At an average species lifetime of four million years, a 1-percent criterion allows 40,000 years for speciation."

as valid (even though I consider it incorrect). In the transition sequence from state "A" to state "D," phyletic gradualism shows a slow rate lasting over 50,000 years, while punctuation displays a rapid rate of about 40,000 years. Meanwhile, parasaltation and saltation exhibit an extremely instantaneous change, the former occurring over two or more generations and the latter happening in only one generation. Finally, stasis represents the tempo of stability in state "D," persisting for an average of 4 million years. Notably, phyletic gradualism and punctuation entail continuity, i.e., both support gradualism as an intermediary of transitional forms. Parasaltation delineates a subtle boundary between continuity and discontinuity, whereas saltation involves discontinuity (i.e., it does not support transitional forms). However, while parasaltation achieves discontinuity through the intermediary of transitional forms, it differs from phyletic gradualism and punctuation because these "continuous" changes occur during ontogeny rather than phylogeny. Therefore, these changes are also phylogenetically discontinuous, emerging in just a few generations (see point 3).

The problem with punctuated equilibrium lies in explaining those punctuated periods, wherein it is assumed that species rapidly differentiate through a gradual process. Why would there be gradual and perpetual changes for thousands of years, only for them to abruptly cease, resulting in morphological stability that persists for millions of years without significant alterations? If gradual environmental shifts indeed cause morphological change, why does the geological record show drastic environmental changes while the species body plan remains in stasis?[13] From the perspective of gradualism, this does not make sense. The fundamental flaw in these models is that they assume that the crucial steps in evolution occur solely through the accumulation of continuous variations. Yet a closer examination of nature reveals that variation manifests discontinuously. For instance, Salazar-Ciudad and Jernvall (2005, p. 628) assert: "Evolution from complex phenotypes can be expected to be punctuated. But here [according to embryological evidence], in contrast to the punctuated equilibrium hypothesis... the punctuation could be observable at a microevolutionary scale." Hence, it is needed to measure rates of change concerning the magnitude of discontinuous change over generations, rather than solely considering the rate of continuous change over geological time, as is commonly assumed.

(2) *Gradualism is widely accepted as the most plausible explanation for the origin of complex structures; however, this belief can be challenged when considering its historical source*

The notion that the accumulation of small variations over geological time can explain the origin of complex structures dates back to Charles Darwin's *On the Origin of Species* (1859). Darwin's gradualism model is founded on the following points:

[13] Regarding the causes of stasis at the species level, see Chap. 11, Sect. 11.2.2.

- *Continuous variations*: Darwin regarded "individual differences" or "slight variations" as the raw material of evolution: "These individual differences are highly important for us, as they afford materials for natural selection to accumulate, in the same manner as man can accumulate in any given direction individual differences in his domesticated productions" (p. 45).
- *Natural selection as a creative force*: According to Darwin, natural selection constantly acts upon these slight variations, thus playing a role as a creative force: "It may be said that natural selection is daily and hourly scrutinising, throughout the world, every variation, even the slightest; rejecting that which is bad, preserving and adding up all that is good; silently and insensibly working, whenever and wherever opportunity offers, at the improvement of each organic being in relation to its organic and inorganic conditions of life" (p. 84).
- *Uniformitarianism*: Charles Lyell's principle of uniformitarianism posits that past phenomena can be elucidated by observable processes occurring today: "the present serving as the key to the past." Darwin did not hesitate to apply Lyell's uniformitarian doctrine to the action of natural selection acting cumulatively on continuous variations over geological time: "We see nothing of these slow changes in progress, until the hand of time has marked the long lapse of ages…" (p. 84).
- *Transitional forms*: Darwin needed to demonstrate the existence of transitional forms, as they provided evidence that evolution occurred gradually. However, he stated: "If it could be demonstrated that any complex organ existed, which could not possibly have been formed by numerous, successive, slight modifications, my theory would absolutely break down" (p. 189).

Interestingly, William Bateson had already overcome this Darwinian challenge in the early twentieth century. In contrast to Darwin, Bateson demonstrated that discontinuous variations adequately explained the morphological distances that separate species.[14] The following points can describe Bateson's anti-gradualist model:

- *Discontinuous variations*: Upon observing various examples regarding the nature of variation, Bateson realized that discontinuity was the norm rather than the exception. Furthermore, he developed intriguing theories regarding the origin of discontinuous characters through the lens of the phenomenon of embryology.
- *Transitional forms*: Concerning continuous variation, Bateson (1909b) questioned whether this type of variation possessed the capacity to be selected. As he embarked on his early genetic studies, he recognized that many transitional forms could be attributed to factors other than genetics, such as environmental influences: "[T]here are intermediates due to the disturbing effects of many small causes not of genetic but presumably of environmental origin" (p. 239). Alternatively, phenotypic outcomes might also display continuity due to genetic factors. Nevertheless, this continuity would arise from specific genetic effects.

[14] For a non-traditional interpretation of Bateson's ideas, see Newman (2007) and Ochoa (2021).

For example, in order to explain cases of intermediates in coloration, Bateson noted that melanin was not completely suppressed, resulting in continuous gradations of shade: "This suppressing or inhibiting effect may be total, but when it is not total some puzzling intermediate types may be constituted" (p. 238).

- *Variation as a creative force*: If the cause of intermediate forms is attributed to factors that are not entirely "heritable" (and thus they lose the ability to be selected), and if taxonomically valuable structures appear already well formed from the start (i.e., they appear discontinuously), then creativity resides in variation rather than selection: "To begin with, we must relegate selection to its proper place. Selection permits the viable to continue and decides that the nonviable shall perish... So again, as the course of descent branches in the successive generations, selection determines along which branch evolution shall proceed, but it does not decide what novelties that branch shall bring forth" (Bateson 1909a, p. 96).

If in the "continuity versus discontinuity" debate Bateson's arguments presented many points in favor of discontinuous change, why did gradualism remain the mode most accepted by the majority of evolutionists? One key event for the widespread acceptance of gradualism is attributed to the success of the Modern Synthesis. However, it is worth noting that the architects of the Modern Synthesis did not delve into the origin of variation in a strict sense, as they considered genetic mutations to be the cause of phenotypic variation. For instance, in *Genetics and the Origin of Species*, Theodosius Dobzhansky (1937, pp. 18–19) wrote that genetic mutations can produce continuous effects on the coloration, length, diameter, size, weight, and shape of fruit fly characters. If these variations are heritable and they arise repeatedly in the course of evolution, then there is no doubt that they are the most important type of variation for understanding evolution at all scales: "The high frequency of mutations producing small changes in the phenotype raises a strong presumption in favor of supposing that such mutations play a greater role in evolutionary processes than mutations with grosser effects" (p. 26).

Similarly, in *Systematics and the Origin of Species*, Ernst Mayr (1942) studied variation, but only through field observations. Here, the morphological characters he emphasizes in his research relate to "size, proportions, structure, coloration, or color pattern" (p. 20). Mayr contends that these characters manifest in nature continuously in continental zones, forming clines: "They show... gliding intergradation between the characters of the individuals of one population as well as between neighboring populations and subspecies... Continuous variation is a typical mode of geographic variation in most animals with continuous ranges in continental areas" (p. 84). Considering this, he subsequently realizes that allopatric populations exhibit slight degrees of discontinuity. Hence, he postulates that divergence—or the initial stages of morphological separation—occurs through geographic isolation:

> "The establishment of discontinuity is a continuous process." In other words, the big gaps which we find between species are preceded by the little gaps which we find between subspecies and by the still lesser gaps which we find between populations. Of course, if these

> populations are distributed as a complete *continuum*, there are no gaps. But with the least isolation, the first minor gaps will appear. (Mayr 1942, p. 159; his emphasis and quotation marks)

Clearly, Mayr does not propose a theory of variation that explains why there are morphological gaps between two groups from an internal standpoint; rather, he merely makes a misguided inference by linking divergence with geographical observations. Needless to say, punctuated equilibrium was founded upon Mayr's allopatric theory, which assumes gradualism as the intermediacy of transitional forms to explain the gaps between species. But neither Dobzhansky nor Mayr conducted a morphological study on how variation occurs. Furthermore, the morphological characters they examine (e.g., size, proportion, color, etc.) are simple characters described here as the variational properties of complex structures. In other words, these traits are far from having the potential to produce significant evolutionary innovations. This is important to note because if the belief in gradual evolution stemmed from such assumptions, then there is an urgent need to bring back this debate.

(3) *According to embryological evidence, complex characters undoubtedly arise through discontinuous changes rather than from the accumulation of small variations over geological time*

According to Anderson (1987, p. 143): "[T]he rate of evolutionary change is dependent, among other things, on the rate at which changes can be assimilated into the developmental system." In other words, any proposal regarding evolutionary rates must align with the possibilities of change that developmental systems allow. Anderson emphasizes that the problem with gradualism lies in attempting to explain the organization of body plans.

So far, no one would hesitate to admit that small variations are relevant to evolution. Indeed, West-Eberhard (2003) presents a synthesis that underscores the significance of both continuous and discontinuous variation. However, a crucial point in this discussion needs highlighting, as noted by Salazar-Ciudad and Jernvall (2005, p. 628): "Simple phenotypes can evolve in a much more gradual way. Since they exhibit a simpler relationship between phenotype and genotype, they can adapt faster and in a finer way to selective pressures that filter for largely divergent phenotypes." In contrast, complex phenotypes arise through complex and non-linear interactions during the developmental process, entailing abrupt structural transitions (Alberch 1982; Balon 2002; Salazar-Ciudad and Jernvall 2005; Jaeger et al. 2012; Linde-Medina and Newman 2014; Peterson and Müller 2016; Hagolani et al. 2021). "There is simply no other way to get from one type of dynamical behavior to another" (Jaeger et al. 2012, p. 599). As emphasized throughout this book, units of evolution exhibit high cohesion due to the developmental constraints acting upon them. Therefore, the internal properties of these entities cannot be substituted by the mere accumulation of small changes over geological time, as Darwin thought.

According to the H-Evo-Devo Theory, there is a notable division between microevolution and macroevolution concerning the type of variation involved. Regarding microevolution, while this level corresponds to simple characters, it does not imply that all these characters arise continuously. As Chap. 11 (Sect. 11.1) explores in detail, differences between subspecies are perceived through discontinuities in simple characters such as size, proportion, color, etc. Nevertheless, it is important to note that many of the character states that characterize subspecies arise not only from genetic differences but particularly from epigenetic and environmental factors. Therefore, the intermediacy of transitional forms (e.g., between one color and another) is likely due to the phenomenon of phenotypic plasticity by the norm of reaction, which does not imply differences between true subspecies.

Regarding macroevolution, complex characters can manifest in two ways: character identities or character states. The former refers to morphological modules, while the latter represents the variations within these modules. Modules—interpreted as homologous—are discrete entities. Likewise, species and higher taxa exemplify the discontinuous grades of character states across modules. In all such cases, change is undoubtedly discontinuous. I will now highlight notable morphological innovations that underscore abrupt transitions, providing insight into body plans' origin.[15]

Nowadays, some biologists have stressed the evolutionary significance of teratology[16]— referred to as Evo-Devo Path or Evolutionary Teratology—particularly because they offer insights into the origin of body plans (Diogo et al. 2015, 2017, 2019; Guinard 2015, 2020; Diaz 2020; Diogo 2020; Ponz-Segrelles et al. 2022; Soliz and Abdala 2024). In these studies, alterations can be found in the identity, quantity, location, and proportion of the characters. In the realm of Evo-Devo, these embryological changes (developmental repatterning) are known as heterotypy (change in identity), heterometry (change in quantity), heterotopy (change in location), and heterochrony (change in time).

It is important to highlight that the first three types of changes do not imply a reference to any intermediacy of transitional forms, not even in an abstract sense. Below, I examine some examples:

I. Heterotypy:
   (a) *Insect labrum*, which arises from the fusion of two pairs of endite appendages (Haas et al. 2001).
   (b) *Bird feathers*, whose internal organization occurs through a complex series of morphological and molecular innovations (Prum 2005).
   (c) *Turtle shell*, which comprises a complex network of modifications, including the scapula located within the ribcage, noticeable alterations of ribs with excessive ossification, and the arrangement into a carapace and a plastron, among others (Rieppel 2017).

[15] The mechanisms and processes within these units will receive further explanation in Chaps. 11 and 12.

[16] The term was coined by Isidore Geoffroy Saint-Hilaire during the nineteenth century. Teratology refers to abnormalities or malformations observed in particular individuals, suggesting a radical alteration in the embryonic pathway.

II. Heterometry:
   (d) *Whale and dolphin forelimb*, a result of hyperphalangy or repetition of the number of digits (Richardson and Oelschläger 2002).
   (e) *Centipede trunk segments*, whose repetitions are associated with differences between species (Minelli et al. 2009).
   (f) *Sloth cervical vertebrae*, which exceed the standard number of seven cervical vertebrae in mammals (Böhmer et al. 2018).

III. Heterotopy:
   (g) *Lacandonia schismatica*, the only plant with an inverted floral arrangement, the stamens in the center, and the carpels around them (Piñeyro-Nelson et al. 2010).
   (h) *Zea mays*, which, compared to its ancestral state, teosinte, displays differences in overall plant architecture, including the shift from male to female reproductive structures (Fedoroff 2003).
   (i) *Chameleon limbs*, with a phenotype exhibiting syndactyly and zygodactyly (Molnar et al. 2017).

IV. Heterochrony

Given that heterochrony involves changes in proportions and size differences, such variations can potentially entail continuous changes since they require transitional forms. Indeed, this mode is particularly evident at the microevolutionary scale. However, for macroevolution, cases such as:

   (j) The presence of micromelias in the *forelimbs of Tyrannosauridae and Carnotaurinae* (Guinard 2015).
   (k) The development of *bird beaks* (Linde-Medina and Newman 2014).

   These examples clearly exhibit discontinuous changes. But if these changes in proportion or size must be continuous in the sense that they imply intermediacy of transitional forms during their development, are they not examples of gradualism? I will clarify this with an example.

   (l) *Bat forelimbs.* The shape of bats has long fascinated scientists studying mammalian evolution. They appeared in the fossil record 50 million years ago without the presence of any intermediate forms (Gunnell and Simmons 2005). In order to explain the origin of bat wings, James Dale Smith (1977) speculated about certain transitional forms, suggesting that these transitions must have occurred rapidly in geological time (Fig. 9.5a). Indeed, contemporary research indicates the existence of these transitions. But if, instead of looking at phylogeny through speculative variations, we now turn to ontogeny, which reveals real variations, we will encounter a very different scenario (Fig. 9.5b). Studies on the embryonic development of bat wings reveal that, in the early stages, they share similarities with the limb development of other mammals. However, as ontogeny progresses, the digits undergo significant lengthening due to cartilage cell proliferation (Sears et al. 2006; Cooper and Sears 2013), while the interdigital membranes persist until the final development stage (Sears 2008).

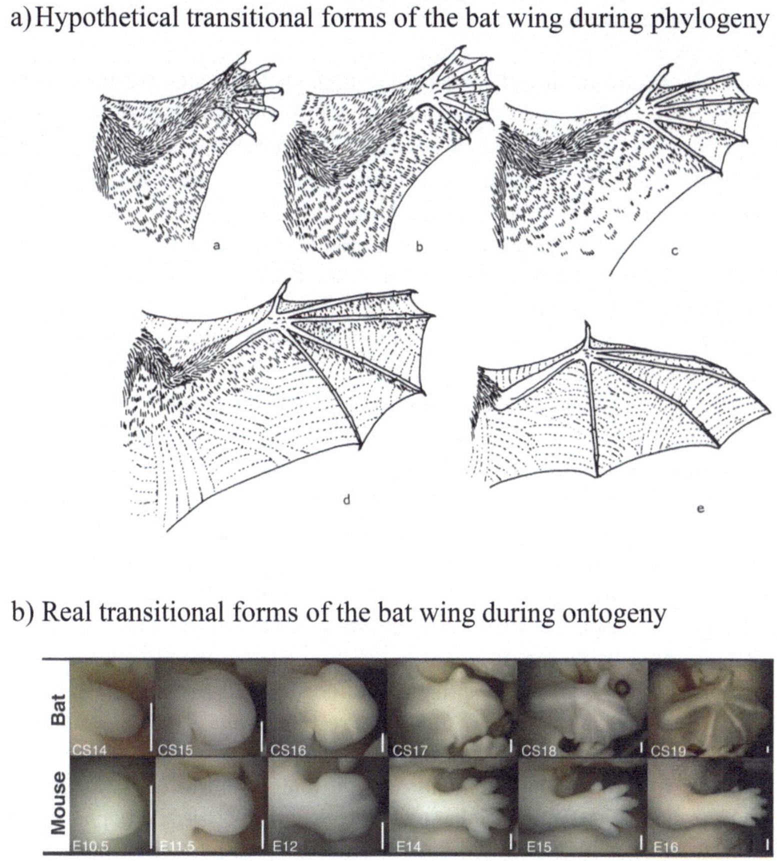

Fig. 9.5 *Transitional forms of the bat's forelimb from the perspectives of phylogeny (a) and ontogeny (b).* (**a**) Hypothetical transitional forms of the bat wing during phylogeny, according to James D. Smith: Under this interpretation, gradualism as intermediacy of transitional forms is assumed. Here, it is important to note that such phylogenetic variations remain entirely speculative. (From Smith 1977). (**b**) Real transitional forms of the bat wing during ontogeny: The images showcase a developmental comparison between the bat's forelimb (top) and the mouse's forelimb (bottom). Initially, in the early stages of development, the bat's forelimb exhibits similarities with other mammals. However, as development progresses, the digits of the bat's forelimb elongate considerably due to the proliferation of cartilaginous cells. It is crucial to emphasize that although these changes are continuous—i.e., they show transitional forms during development—these ontogenetic variations ultimately lead to discontinuous outcomes in the final development of complex characters. (From Cretekos et al. 2008, with permission from Cold Spring Harbor Laboratory Press)

This example underscores an essential principle in embryology: small changes in developmental constituent parameters result in a discontinuous phenotypic change (Brylski and Hall 1987; Linde-Medina and Newman 2014; Peterson and Müller 2016; Lange et al. 2018). In short, to account for complex structures, according to embryological expectations, rates of evolutionary change manifest through discontinuous steps rather than accumulating slight variations over geological time, as many biologists commonly assume.

9.3 Conclusions

The main conclusion of this chapter is that gradualism, as intermediacy of transitional forms, presents numerous problems; thus, there is nothing to justify its conceptual predominance. In light of this, it can be argued that punctuated equilibrium is suitable for recognizing evolutionary tempos, but it falls short in delineating differences between modes, especially given the unclear distinction between cladogenesis and anagenesis. Both punctuated equilibrium and phyletic gradualism essentially describe the same phenomenon: gradual transitions between forms irrespective of the rate of change.

In light of phenomena observed in embryology and patterns seen in the fossil record, this paradigm of gradualism as intermediacy of transitional forms—long considered valid and almost unquestionable by evolutionary biologists—is an inadequate model. Concerning embryology, evidence suggests that complex characters emerge through discontinuous steps, and once a new structural arrangement is established, it remains unchanged due to internal organizational properties (developmental constraints). With respect to the fossil record, paleontologists have identified three fundamental patterns in multicellular organisms across geological time: an instantaneous origin, followed by stability of form and punctuated extinction events. Clearly, the fossil record strongly aligns with the patterns expected in embryology. Given this correlation, it is reasonable to consider abandoning microevolutionary methods that seek to quantify evolution through the lens of gradualism. Such methods represent a waste of time and resources that could instead be redirected toward advancing this new avenue of research.

While an evolutionary hierarchical view on tempo and mode is far from new (e.g., Simpson 1944, 1953; Schindewolf 1950; Runnegar 1987; Roopnarine 2003; Gould 2002; Allmon 2016), the H-Evo-Devo Theory translates tempos into hypotheses of the modes applied to each of the units of evolution (subspecies, species, higher taxa, and homologues): instantaneous origin (saltation), structural stability (stasis), and punctuated disappearance (extinction).

References

Ager DV (1976) The nature of the fossil record. Proc Geol Ass 87:131–159

Aguirre WE, Bell MA (2012) Twenty years of body shape evolution in a threespine stickleback population adapting to a lake environment. Biol J Linn Soc 105:817–831

Alberch P (1982) Developmental constraints in evolutionary processes. In: Bonner JT (ed) Evolution and development. Springer, New York, pp 313–332

Allmon WD (2016) Coming to terms with *Tempo and Mode*: speciation, anagenesis, and assessing relative frequencies in macroevolution. In: Eldredge N, Pievani T, Serrelli E, Tëmkin I (eds) Evolutionary theory: a hierarchical perspective. The University of Chicago Press, Chicago, pp 260–281

Anderson DT (1987) Developmental pathways and evolutionary rates. In: Campbell KSW, Day MF (eds) Rates of evolution. Allen and Unwin, London, pp 143–155

Ausich WI (2016) Fossil species as data: a perspective from echinoderms. In: Allmon WD, Yacobucci MM (eds) Species and speciation in the fossil record. University of Chicago Press, Chicago and London, pp 301–311

Balon EK (2002) Epigenetic processes, when *natura non facit saltum* becomes a myth, and alternative ontogenies a mechanism of evolution. Environ Biol Fish 65:1–35

Bambach RK, Knoll AH, Wang SC (2004) Origination, extinction, and mass depletions of marine diversity. Paleobiology 30:522–542

Bannikova AA, Lebedev VS (2022) The concept of the modern molecular clock and experience in estimating divergence times of Eulipotyphla and Rodentia. Biol Bull Rev 12:3–28

Barba-Montoya J, Tao Q, Kumar S (2021) Molecular and morphological clocks for estimating evolutionary divergence times. BMC Ecol Evo 21:83. https://doi.org/10.1186/s12862-021-01798-6

Bateman RM, DiMichele WA (2002) Generating and filtering major phenotypic novelties: neoGoldschmidtian saltation revisited. In: Cronk QCB, Bateman RM, Hawkins JA (eds) Developmental genetics and plant evolution. Taylor and Francis, London, pp 109–159

Bateson W (1909a) Heredity and variation in modern lights. In: Seward AC (ed) Darwin and modern science: essays in commemoration of the centenary of the birth of Charles Darwin and of the fiftieth anniversary of the publication of the origin of species. Cambridge University Press, Cambridge, pp 85–101

Bateson W (1909b) Mendel's principle of heredity. Cambridge University Press, Cambridge

Bechly C (2024) Fossil Friday: discontinuities in the fossil record—a problem for Neo-Darwinism. Evolution News May 10 https://evolutionnews.org/2024/05/fossil-friday-discontinuities-in-the-fossil-record-a-problem-for-neo-darwinism/

Bell MA, Aguirre WE, Buck NJ (2004) Twelve years of contemporary armor evolution in a threespine stickleback population. Evolution 58:814–824

Benton MJ, Pearson PN (2001) Speciation in the fossil record. Trends Ecol Evol 16:405–411

Böhmer C, Amson E, Arnold P, van Heteren AH, Nyakatura JA (2018) Homeotic transformations reflect departure from the mammalian "rule of seven" cervical vertebrae in sloths: inferences on the Hox code and morphological modularity of the mammalian neck. BMC Evol Biol 18:84. https://doi.org/10.1186/s12862-018-1202-5

Bose R, Bartholomew AJ (2013) Macroevolution in deep time. Springer, New York

Brett CE, Baird GC (1995) Coordinated stasis and evolutionary ecology of Silurian to middle Devonian faunas in the Appalachian Basin. In: Erwin DH, Anstey RL (eds) New approaches to speciation in the fossil record. Columbia University Press, New York, pp 285–315

Brown CR, Brown MB (2011) Intense natural selection on morphology of cliff swallows (*Petrochelidon pyrrhonota*) a decade later: did the population move between adaptive peaks? Auk 128:69–77

Brylski P, Hall BK (1987) Ontogeny of a macroevolutionary phenotype: the external cheek pouches of geomyoid rodents. Evolution 42:391–395

Budd GE, Mann RP (2024) Two notorious nodes: a critical examination of relaxed molecular clock age estimates of the bilaterian animals and placental mammals. Syst Biol 73:223–234

Burin G, Alencar LRV, Chang J, Alfaro ME, Quental TB (2019) How well can we estimate diversity dynamics for clades in diversity decline? Syst Biol 68:47–62

Campbell KSW, Marshall CR (1987) Rates of evolution among Palaeozoic echinoderms. In: Campbell KSW, Day MF (eds) Rates of evolution. Allen and Unwin, London, pp 61–100

Carroll RL (1997) Patterns and processes of vertebrate evolution. Cambridge University Press, Cambridge

Carruthers T, Scotland RW (2020) Insights from empirical analyses and simulations on using multiple fossil calibrations with relaxed clocks to estimate divergence times. Mol Biol Evol 37:1508–1529

Congreve CR, Falk AR, Lamsdell JC (2018) Biological hierarchies and the nature of extinction. Biol Rev 93:811–826

Cooper LN, Sears KE (2013) How to grow a bat wing. In: Adams R, Pedersen S (eds) Bat evolution, ecology, and conservation. Springer, New York, pp 3–20

Cretekos CJ, Wang Y, Green ED, Martin JF, Rasweiler JJ 4th, Behringer RR (2008) Regulatory divergence modifies limb length between mammals. Genes Dev 22:141–151

Darwin C (1859) On the origin of species by means of natural selection, or preservation of favored races in the struggle for life. John Murray, London

Darwin C (1866) On the origin of species by means of natural selection, or preservation of favored races in the struggle for life, 4th edn. John Murray, London

Darwin C (1869) On the origin of species by means of natural selection, or preservation of favored races in the struggle for life, 5th edn. John Murray, London

Darwin F, Seward AC (eds) (1903) More letters of Charles Darwin. A record of his work in a series of hitherto unpublished letters. Volume I. John Murray, London

Dawkins R (1986) The blind watchmaker: why the evidence of evolution reveals a universe without design. WW Norton, New York

Dennett DC (1995) Darwin's dangerous idea: evolution and the meanings of life, vol 35. Simon and Schuster, New York, p 34

Diaz RE (2020) Evo-devo path as a bridge between evolution, morphological disparity, and medicine with comments on "hopeful monsters" in the age of genomics. Curr Mol Biol Rep 6:79. https://doi.org/10.1007/s40610-020-00131-2

Diogo R (2020) Introduction to evolutionary developmental pathology, or evo-devo-path: on Neodarwinism, natural mutants, hopeful monsters, syndromes, genomics, variations, humans, apes, chameleons, and dinosaurs. Curr Mol Biol Rep 6:11. https://doi.org/10.1007/s40610-020-00133-0

Diogo R, Smith CM, Ziermann JM (2015) Evolutionary developmental pathology and anthropology: a new field linking development, comparative anatomy, human evolution, morphological variations and defects, and medicine. Dev Dyn 244:1357–1374

Diogo R, Guinard G, Diaz RE Jr (2017) Dinosaurs, chameleons, humans, and evo-devo path: linking Étienne Geoffroy's teratology, Waddington's homeorhesis, Alberch's logic of "monsters," and Goldschmidt hopeful "monsters.". J Exp Zool B (Mol Dev Evo) 328:207–229

Diogo R, Razmadze D, Siomava N, Douglas N, Fuentes JSM, Duerinckx A (2019) Musculoskeletal study of cebocephalic and cyclopic lamb heads illuminates links between normal and abnormal development, evolution and human pathologies. Sci Rep 9:991. https://doi.org/10.1038/s41598-018-37735-9

Dobzhansky T (1937) Genetics and the origin of species. Columbia University Press, New York

Duchêne DA, Duchêne S, Stiller J, Heller R, Ho SYW (2024) ClockstaRX: testing molecular clock hypotheses with genomic data. Genome Biol Evol 16:evae064. https://doi.org/10.1093/gbe/evae064

Dudley JW, Lambert RJ (2004) 100 generations of selection for oil and protein in corn. Plant Breed Rev 24:79–110

Duran-Nebreda S, Vidiella B, Spiridonov A, Eldredge N, Bentley RA, Valverde S (2024) The many ways toward punctuated evolution. Palaeontology 67:e12731. https://doi.org/10.1111/pala.12731

Eldredge N (1971) The allopatric model and phylogeny in Paleozoic invertebrates. Evolution 25:156–167

Eldredge N (2003) The sloshing bucket: how the physical realm controls evolution. In: Crutchfield J, Schuster P (eds) Evolutionary dynamics: exploring the interplay of selection, accident, neutrality, and function. Oxford University Press, New York, pp 3–32

Eldredge N (2008a) Hierarchies and the sloshing bucket: toward the unification of evolutionary biology. Evo Edu Outreach 1:10–15

Eldredge N (2008b) The early "evolution" of "punctuated equilibria.". Evo Edu Outreach 1:107–113

Eldredge N (2015) Eternal ephemera: adaptation and the origin of species, from the nineteenth century, through punctuated equilibria and beyond. Columbia University Press, New York

Eldredge N (2024) Reflections on punctuated equilibria. Paleobiology 1–5:1. https://doi.org/10.1017/pab.2024.8

Eldredge N, Gould SJ (1972) Punctuated equilibria: an alternative to phyletic gradualism. In: Schopf TJM (ed) Models in paleobiology. Freeman, Cooper and Co, San Francisco California, pp 82–115
Ereshefsky M (1991) Species, higher taxa, and the units of evolution. Philos Sci 58:84–101
Erwin DH (2001) Lessons from the past: biotic recoveries from mass extinctions. Proc Natl Acad Sci USA 98:5399–5403
Erwin DH (2008) Extinction as the loss of evolutionary history. Proc Natl Acad Sci USA 105:11520–11527
Erwin DH, Valentine JW, Sepkoski JJ (1987) A comparative study of diversification events: the early Paleozoic versus the Mesozoic. Evolution 41:117–1186
Fagerstrom JA (1978) Modes of evolution and their chronostratigraphic significance: evidence from Devonian invertebrates in the Michigan Basin. Paleobiology 4:381–393
Falconer DS (1973) Replicated selection for body weight in mice. Genet Res (Camb) 22:291–321
Fedoroff NV (2003) Prehistoric GM corn. Science 302:1158–1159
Foote M (1996) On the probability of ancestors in the fossil record. Paleobiology 22:141–151
Foote M (2003) Origination and extinction through the Phanerozoic: a new approach. J Geol 111:125–148
Foote M (2005) Pulsed origination and extinction in the marine realm. Paleobiology 31:6–20
Frazzetta TH (2012) Flatfishes, turtles, and bolyerine snakes: evolution by small steps or large, or both? Evol Biol 39:30–60
Geary DH (1987) Evolutionary tempo and mode in a sequence of the upper cretaceous bivalve *Pleuriocardia*. Paleobiology 13:140–151
Geiger M, Sánchez-Villagra MR (2018) Similar rates of morphological evolution in domesticated and wild pigs and dogs. Front Zool 15:23. https://doi.org/10.1186/s12983-018-0265-x
Gilchrist GW, Huey RB, Serra L (2001) Rapid evolution of wing size clines in *Drosophila subobscura*. Genetica 112:273–286
Gingerich PD (2001) Rates of evolution on the time scale of the evolutionary process. Genetica 112–113:127–144
Gingerich PD (2019) Rates of evolution: a quantitative synthesis. Cambridge University Press, Cambridge
Gould SJ (1984) Smooth curve of evolutionary rate: a psychological and mathematical artifact. Science 226:994–995
Gould SJ (2002) The structure of evolutionary theory. The Belknap Press of Harvard University Press, Cambridge MA
Gould SJ, Eldredge N (1977) Punctuated equilibria: the tempo and mode of evolution reconsidered. Paleobiology 3:115–151
Guinard G (2015) Introduction to evolutionary teratology, with an application to the forelimbs of Tyrannosauridae and Carnotaurinae (Dinosauria: Theropoda). Evol Biol 42:20–41
Guinard G (2020) Evolutionary teratology and the path to break through the mould of the synthesis paradigm. Curr Mol Biol Rep 6:51–61
Guindon S (2020) Rates and rocks: strengths and weaknesses of molecular dating methods. Front Genet 11:526. https://doi.org/10.3389/fgene.2020.00526
Gunnell GF, Simmons NB (2005) Fossil evidence and the origin of bats. J Mamm Evol 12:209–246
Haas MS, Brown SJ, Beeman RW (2001) Homeotic evidence for the appendicular origin of the labrum in *Tribolium castaneum*. Dev Genes Evol 211:96–102
Hagolani PF, Zimm R, Vroomans R, Salazar-Ciudad I (2021) On the evolution and development of morphological complexity: a view from gene regulatory networks. PLoS Comput Biol 17:e1008570. https://doi.org/10.1371/journal
Hansen TF (2024) Three modes of evolution? Remarks on rates of evolution and time scaling. J Evol Biol 37:1523–1537
Harvey PH, May RM, Nee S (1994) Phylogenies without fossils. Evolution 48:523–529
Hennig W (1966) Phylogenetic systematics. University of Illinois, Chicago
Ho SYW (2020) The molecular clock and evolutionary rates across the tree of life. In: Ho SYW (ed) The molecular evolutionary clock. Springer, Cham, pp 3–23

Ho SYW, Duchêne S (2014) Molecular-clock methods for estimating evolutionary rates and timescales. Mol Ecol 23:5947–5965
Holland SM (2020) The stratigraphy of mass extinctions and recoveries. Annu Rev Earth Planet Sci 48:75–97
Hunt G (2008) Evolutionary patterns within fossil lineages: model-based assessment of modes, rates, punctuations and process. In: Kelley PH, Bambach RK (eds) From evolution to geobiology: research questions driving paleontology at the start of a new century, vol 14. Paleontological Society, New Haven, pp 117–131
Hunt G, Hopkins MJ, Lidgard S (2015) Simple versus complex models of trait evolution and stasis as a response to environmental change. Proc Natl Acad Sci USA 112:4885–4890
Huxley J (1957) The three types of evolutionary process. Nature 180:454–455
Jablonski D (2005) Mass extinctions and macroevolution. Paleobiology 31:192–210
Jablonski D, Shubin NH (2015) The future of the fossil record: paleontology in the 21st century. Proc Natl Acad Sci USA 112:4852–4858
Jaeger J, Irons D, Monk N (2012) The inheritance of process: a dynamical systems approach. J Exp Zool (Mol Dev Evol) 318B:591–612
Kemp TS (2016) The origin of higher taxa: palaeobiological, developmental, and ecological perspectives. Oxford University Press/The University of Chicago Press, Oxford
Kennedy WJ (1977) Ammonite evolution. In: Hallam A (ed) Patterns of evolution: as illustrated by the fossil record. Elsevier, New York, pp 251–304
Kluge AG (1990) Species as historical individuals. Biol Philos 5:417–431
Krug AZ, Jablonski D (2012) Long-term origination rates are reset only at mass extinctions. Geology 40:731–734
Lanfear R, Welch JJ, Bromham L (2010) Watching the clock: studying variation in rates of molecular evolution between species. Trends Ecol Evol 25:495–503
Lange A, Nemeschkal HL, Müller GB (2018) A threshold model for polydactyly. Prog Biophys Mol Biol 137:1–11
Laurie CC, Chasalow SD, LeDeaux JR, McCarroll R, Bush D, Hauge B, Lai C, Clark D, Rocheford TR, Dudley JW (2004) The genetic architecture of response to long-term artificial selection for oil concentration in the maize kernel. Genetics 168:2141–2155
Layton KKS, Wilson NG (2024) Validating a molecular clock for nudibranchs-no fossils to the rescue. Ecol Evol 14:e11014. https://doi.org/10.1002/ece3.11014
Lewis JA, Morran LT (2021) Advantages of laboratory natural selection in the applied sciences. J Evol Biol 35:5–22
Lieberman BS (2001) Analyzing speciation rates in macroevolutionary studies. In: Adrain JM, Edgecombe GD, Lieberman BS (eds) Fossils, phylogeny, and form: an analytical approach, Topics in Geobiology, vol 19. Springer, New York, pp 323–339
Lieberman BS, Eldredge N (2014) What is punctuated equilibrium? what is macroevolution? a response to Pennell et al. Trends Ecol Evol 29:185–186
Linde-Medina M, Newman SA (2014) Limb, tooth, beak: three modes of development and evolutionary innovation of form. J Biosci 39:211–223
Lindsey CR, Knoll AH, Herron MD, Rosenzweig F (2024) Fossil-calibrated molecular clock data enable reconstruction of steps leading to differentiated multicellularity and anisogamy in the Volvocine algae. BMC Biol 22:79. https://doi.org/10.1186/s12915-024-01878-1
Liow LH, Quental TB, Marshall CR (2010) When can decreasing diversification rates be detected with molecular phylogenies and the fossil record? Syst Biol 59:646–659
Løvtrop S (1981) Macroevolution and punctuated equilibria. Syst Zool 30:498–500
Martin BS, Bradburd GS, Harmon LJ, Weber MG (2023) Modeling the evolution of rates of continuous trait evolution. Syst Biol 72:590–605
Mayr E (1942) Systematics and the origin of species. Columbia University Press, New York
McCune AR (1996) Biogeographic and stratigraphic evidence for rapid speciation in semionotid fishes. Paleobiology 22:34–48
Minelli A, Chagas-Júnior A, Edgecombe GD (2009) Saltational evolution of trunk segment number in centipedes. Evol Dev 11:318–322

Mitchell JS, Etienne RS, Rabosky DL (2019) Inferring diversification rate variation from phylogenies with fossils. Syst Biol 68:1–18

Molnar J, Diaz RE, Skorka T, Dagliyan G, Diogo R (2017) Comparative musculoskeletal anatomy of chameleon limbs, with implications for the evolution of arboreal locomotion in lizards and for teratology. J Morphol 278:1241–1261

Mongiardino Koch N, Thomson JR (2021) A total-evidence dated phylogeny of Echinoidea combining phylogenomic and paleontological data. Syst Biol 70:421–439

Moon BC, Stubbs TL (2020) Early high rates and disparity in the evolution of ichthyosaurs. Commun Biol 3:68. https://doi.org/10.1038/s42003-020-0779-6

Morlon H, Parsons TL, Plotkin JB (2011) Reconciling molecular phylogenies with the fossil record. Proc Natl Acad Sci USA 108:16327–16332

Nawrot R, Scarponi D, Azzarone M, Dexter TA, Kusnerik KM, Wittmer JM, Amorosi A, Kowalewski M (2018) Stratigraphic signatures of mass extinctions: ecological and sedimentary determinants. Proc R Soc B 285:20181191. https://doi.org/10.1098/rspb.2018.1191

Nee S, Mooers AØ, Harvey PH (1992) Tempo and mode of evolution revealed from molecular phylogenies. Proc Natl Acad Sci USA 89:8322–8326

Neige P (2015) Events of increased biodiversity: evolutionary radiations in the fossil record. ISTE Press, London

Newman SA (2007) William Bateson's physicalist ideas. In: Laubichler M, Maienschein J (eds) From embryology to evo-devo: a history of evolutionary development. MIT Press, Cambridge MA

Ochoa C (2021) The origins of theoretical developmental genetics: reinterpreting William Bateson's role in the history of evolutionary thought. In: Delisle RG (ed) Natural selection: revisiting its explanatory role in evolutionary biology. Springer, Cham, pp 261–290

Pachut JF, Anstey RL (2009) Inferring evolutionary modes in a fossil lineage (Bryozoa: *Peronopora*) from the Middle and Late Ordovician. Paleobiology 35(2):209–230

Patzkowsky ME, Holland SM (2012) Stratigraphic paleobiology: understanding the distribution of fossil taxa in time and space. University of Chicago Press, Chicago

Paul CRC (1977) Evolution of primitive echinoderms. In: Hallam A (ed) Patterns of evolution: as illustrated by the fossil record. Elsevier, New York, pp 123–158

Pennell MW, Harmon LJ, Uyeda JC (2014) Is there room for punctuated equilibrium in macroevolution? Trends Ecol Evol 29:23–32

Penny D (1983) Charles Darwin, gradualism and punctuated equilibria. Syst Zool 32:72–74

Peterson T, Müller GB (2016) Phenotypic novelty in evodevo: the distinction between continuous and discontinuous variation and its importance in evolutionary theory. Evol Biol 43:314–335

Piñeyro-Nelson A, Flores-Sandoval E, Garay-Arroyo A, García-Ponce B, Álvarez-Buylla ER (2010) Development and evolution of the unique floral organ arrangement of *Lacandonia schismatica*. Int J Plant Dev Biol 4:86–97

Ponz-Segrelles G, Ribeiro RP, Aguado MT (2022) Monsters reveal patterns: bifurcated annelids and their implications for the study of development and evolution. Biol Rev Camb Philos Soc 97:896–922

Pozzi L, Penna A (2022) Rocks and clocks revised: new promises and challenges in dating the primate tree of life. Evol Anthropol 31:138–153

Prum RO (2005) Evolution of the morphological innovations of feathers. J Exp Zool (Mol Dev Evo) 304B:570–579

Pulquério MJF, Nichols RA (2006) Dates from the molecular clock: how wrong can we be? Trends Ecol Evol 22:180–184

Rabosky DL, Donnellan SC, Grundler M, Lovette IJ (2014) Analysis and visualization of complex macroevolutionary dynamics: an example from Australian scincid lizards. Syst Biol 63:610–627

Raup DM (1977) Stochastic models in evolutionary palaeontology. In: Hallam A (ed) Patterns of evolution: as illustrated by the fossil record. Elsevier, New York, pp 59–78

Raup DM (1983) On the early origins of major biologic groups. Paleobiology 9:107–115

Rensch B (1959) Evolution above the species level. Columbia University Press, London

Rhodes FHT (1987) Darwinian gradualism and its limits: the development of Darwin's views on the rate and pattern of evolutionary change. J Hist Biol 20:139–157
Richardson MK, Oelschläger HHA (2002) Time, pattern, and heterochrony: a study of hyperphalangy in the dolphin embryo flipper. Evol Dev 4:435–444
Rieppel O (2017) Turtles as hopeful monsters: origins and evolution. Indiana University Press, Bloomington
Roopnarine PD (2001) The description and classification of evolutionary mode: a computational approach. Paleobiology 27:446–465
Roopnarine PD (2003) Analysis of rates of morphologic evolution. Annu Rev Ecol Evol Syst 34:605–632
Roopnarine PD, Byars G, Fitzgerald P (1999) Anagenetic evolution, stratophenetic patterns, and random walk models. Paleobiology 25:41–57
Rudman SM, Greenblum SI, Rajpurohit S, Betancourt NJ, Hanna J, Tilk S, Yokoyama T, Petrov DA, Schmidt P (2022) Direct observation of adaptive tracking on ecological time scales in Drosophila. Science 375:eabj7484. https://doi.org/10.1126/science.abj7484
Runnegar B (1987) Rates and modes of evolution in the Mollusca. In: Campbell KSW, Day MF (eds) Rates of evolution. Allen and Unwin, London, pp 39–60
Salazar-Ciudad I, Jernvall J (2005) Graduality and innovation in the evolution of complex phenotypes: insights from development. J Exp Zool (Mol Dev Evo) 304B:619–631
Saupe EE, Myers CE (2021) Macroevolution. In: Nuño de la Rosa L, Müller GB (eds) Evolutionary developmental biology. Springer, Cham, pp 149–167
Schindewolf OH (1950) *Grundfragen der Paläontologie:* geologische Zeitmessung, organische Stammesentwicklung, biologische Systematik. Schweizerbart'sche Verlagsbuchhandlung, Stuttgart
Schwartz JH, Maresca B (2006) Do molecular clocks run at all? A critique of molecular systematics. Biol Theory 1:357–371
Schweitzer CE (2001) Paleobiogeography of cretaceous and tertiary decapod crustaceans of the North Pacific Ocean. J Paleontol 75:808–826
Sears KE (2008) Molecular determinants of bat wing development. Cells Tissues Organs 187:6–12
Sears KE, Behringer RR, Rasweiler JJ IV, Niswander LA (2006) Development of bat flight: morphologic and molecular evolution of bat wing digits. Proc Natl Acad Sci USA 103:6581–6586
Sheets HD, Mitchell CE (2001) Uncorrelated change produces the apparent dependence of evolutionary rate on interval. Paleobiology 27:429–445
Silvestro D, Warnock RCM, Gavryushkina A, Stadler T (2018) Closing the gap between palaeontological and neontological speciation and extinction rate estimates. Nat Commun 9:5237. https://doi.org/10.1038/s41467-018-07622-y
Simpson GG (1944) Tempo and mode in evolution. Columbia University Press, New York
Simpson GG (1953) The major features of evolution. Columbia University Press, New York
Simpson GG (1984) Introduction: forty years later. In: Tempo and mode in evolution. Reprint. Columbia University Press, New York, pp xiii–xxvi
Slater GJ, Harmon LJ, Alfaro ME (2012) Integrating fossils with molecular phylogenies improves inference of trait evolution. Evolution 66:3931–3944
Smith JD (1977) Comments on flight and the evolution of bats. In: Hecht MK, Goody PC, Hecht BM (eds) Major patterns in vertebrate evolution. Plenum Press, New York, pp 427–437
Soliz MC, Abdala V (2024) Musculoskeletal and tendinous details of selected anomalies in the locomotor system of anurans. Anat Rec 307:3282–3305
Stanley SM (1985) Rates of evolution. Paleobiology 11:13–26
Strickland K, Matthews B, Jónsson ZO, Kristjánsson BK, Phillips JS, Einarsson Á, Räsänen K (2024) Microevolutionary change in wild stickleback: using integrative time-series data to infer responses to selection. Proc Natl Acad Sci USA 121:e2410324121. https://doi.org/10.1073/pnas.241032412
Sun Y, Li M, Song G, Lei F, Li D, Wu Y (2017) The role of climate factors in geographic variation in body mass and wing length in a passerine bird. Avian Res 8. https://doi.org/10.1186/s40657-016-0059-9

Theißen G (2009) Saltational evolution: hopeful monsters are here to stay. Theory Biosci 128:43–51
Thomson KS (1977) The pattern of diversification among fishes. In: Hallam A (ed) Patterns of evolution: as illustrated by the fossil record. Elsevier, New York, pp 377–404
Thorpe RS, Leadbeater DL, Pook CE (2005) Molecular clocks and geological dates: cytochrome *b* of *Anolis extremus* substantially contradicts dating of Barbados emergence. Mol Ecol 14:2087–2096
Tiley GP, Poelstra JW, Dos Reis M, Yang Z, Yoder AD (2020) Molecular clocks without rocks: new solutions for old problems. Trends Genet 36:845–856
Usai G, Fambrini M, Pugliesi C, Simoni S (2024) Exploring the patterns of evolution: core thoughts and focus on the saltational model. Biosystems 238:105181. https://doi.org/10.1016/j.biosystems.2024.105181
Uyeda JC, Hansen TF, Arnold SJ, Pienaar J (2011) The million-year wait for macroevolutionary bursts. Proc Natl Acad Sci USA 108:15908–15913
Valentine JW (1977) General patterns of metazoan evolution. In: Hallam A (ed) Patterns of evolution: as illustrated by the fossil record. Elsevier, New York, pp 27–57
Valentine JW (1980) Determinants of diversity in higher taxonomic categories. Paleobiology 6:444–450
Vaux F, Trewick SA, Morgan-Richards M (2016) Lineages, splits and divergence challenge whether the terms anagenesis and cladogenesis are necessary. Biol J Linn Soc 117:165–176
von Vaupel Klein JC (1994) Punctuated equilibria and phyletic gradualism: even partners can be good friends. Acta Biotheor 42:15–48
von Vaupel Klein JC (1995) Phyletic gradualism versus punctuated equilibria: why case histories do not suffice. Acta Biotheor 43:259–278
Vrba ES (1993) Turnover-pulses, the red queen, and related topics. Am J Sci 293:418–452
Wagner PJ (2000) The quality of the fossil record and the accuracy of phylogenetic inferences about sampling and diversity. Syst Biol 49:65–86
Wagner PJ (2018) Early bursts of disparity and the reorganization of character integration. Proc R Soc B 285:20181604. https://doi.org/10.1098/rspb.2018.1604
Wagner PJ, Erwin DH (1995) Phylogenetic patterns as tests of speciation models. In: Erwin DH, Anstey RL (eds) New approaches to speciation in the fossil record. Columbia University Press, New York, pp 87–122
Wagnon GS, Pletcher OM, Brown CR (2022) Change in beak overhangs of cliff swallows over 40 years: partly a response to parasites? PLoS One 17:e0263422. https://doi.org/10.1371/journal.pone.0263422
Wang SC, Zhong L (2018) Estimating the number of pulses in a mass extinction. Paleobiology 44:199–218
Wang SC, Zimmerman AE, McVeigh BS, Everson PJ, Wong H (2012) Confidence intervals for the duration of a mass extinction. Paleobiology 38:265–277
West-Eberhard MJ (2003) Developmental plasticity and evolution. Oxford University Press, New York
Witts JD, Myers CE, Garb MP, Irizarry KM, Larina E, Rashkova A, Landman NH (2023) Geographic and temporal morphological stasis in the latest Cretaceous ammonoid *Discoscaphites iris* from the U.S. Gulf and Atlantic Coastal Plains. Paleobiology 49:153–175
Wright AM, Bapst DW, Barido-Sottani J, Warnock RCM (2022) Integrating fossil observations into phylogenetics using the fossilized birth–death model. Annu Rev Ecol Evol Syst 53:251–273
Yacobucci MM (2016) Towards a model for speciation in Ammonoids. In: Allmon WD, Yacobucci MM (eds) Species and speciation in the fossil record. University of Chicago Press, Chicago and London, pp 238–277
Zhang YM, Bossert S, Spasojevic T (2025) Evolving perspectives in Hymenoptera systematics: bridging fossils and genomes across time. Syst Entomol 50:1–31. https://doi.org/10.1111/syen.12645

Divergence and Levels: Challenging Allopatric Speciation and the Extrapolationist Premise

10

Abstract

This chapter challenges traditional models of speciation and the extrapolationist premise. First, it critically examines geographic models of speciation, particularly challenging the theoretical foundations of allopatric speciation. It is argued, for instance, that the original theoretical framework of allopatric speciation—Mayr's founder effect model—is obsolete in light of the progress of contemporary science. Furthermore, it demonstrates that evolutionary divergence is not caused by geographic barriers or the interruption of gene flow. Instead, divergence is explained through the combined processes of variation and adaptation to local environments (at the microevolutionary scale). Second, this chapter delves into the question of whether macroevolution is more than the sum of microevolutionary processes. To address this issue, it explores the history of the terms "microevolution," "macroevolution," and "megaevolution," followed by an analysis of their current meanings from various perspectives. Subsequently, three possible approaches to resolve this dilemma are explored: a reductionist approach, an intermediate approach, and an irreducible approach. However, it is argued that the latter alternative is the most plausible in light of contemporary scientific knowledge. Given these challenges, the H-Evo-Devo Theory proposes a new perspective on evolutionary divergence working at multiple evolutionary levels and conceptualizes a qualitative separation between different levels of evolution.

Keywords

H-Evo-Devo Theory · Allopatric speciation · Sympatric speciation · Parapatric speciation · Speciation · Divergence · Species · Founder effect · Gene flow · Extrapolationist premise · Microevolution · Macroevolution · Megaevolution · Modular-evolution

C. Ochoa, *Hierarchical Evolutionary-Developmental Theory*, Evolutionary Biology – New Perspectives on Its Development 9,
https://doi.org/10.1007/978-3-031-88723-9_10

When addressing the topic of evolutionary divergence, most biologists recognize different modes of speciation based on geographical criteria. Among these, allopatric speciation stands out as the most widely accepted and discussed mode of divergence in evolutionary scope. However, the issue is that these modes of speciation focus on explaining divergence through genetic mechanisms in an ecological context without integrating embryonic developmental processes. Furthermore, an interesting challenge in contemporary evolutionary biology revolves around whether macroevolution is more than the sum of microevolutionary processes. The issue lies in the fact that, while biologists recognize certain macroevolutionary patterns, these are generally explained through conventional microevolutionary mechanisms—a hypothesis known as the extrapolationist premise. Consequently, evolutionary theory appears to lack a truly macroevolutionary perspective.

Given these difficulties, this chapter will challenge geographic modes of speciation and the extrapolationist premise. Following this analysis, the H-Evo-Devo Theory presents a novel perspective by proposing that evolutionary divergence occurs at multiple levels and by advocating for a qualitative distinction among the different levels of evolution. While the following chapters will explore the processes and patterns of each level in greater detail, this chapter focuses on highlighting the challenges to the most relevant premises of the Modern Synthesis.

Section 10.1 presents arguments against geographic modes of speciation, focusing especially on questioning the validity of allopatric speciation. Since the original basis of allopatric speciation rests on Ernst Mayr's founder effect model, this section first shows that this model has become obsolete in light of contemporary advancements in genetics, embryology, and biogeography. Subsequently, it demonstrates that evolutionary divergence is not caused by geographic barriers or the interruption of gene flow. Instead, evolutionary divergence is explained by the combined processes of variation and adaptation to local environments (at the microevolutionary scale). Section 10.2 examines whether microevolutionary processes are sufficient to explain macroevolutionary patterns. It begins with a historical review of the concepts of "microevolution," "macroevolution," and "megaevolution." Later, it analyzes their contemporary meanings from various perspectives. Once this groundwork is laid, I will explore three alternative approaches to address this controversy: a reducible approach, which entirely reduces macroevolution to microevolutionary processes; the intermediate approach, which recognizes macroevolutionary patterns but explains them through microevolutionary mechanisms; and lastly, the irreducible approach, which supports the existence of exclusively macroevolutionary processes that cannot be reduced to the realm of microevolution. Regarding this, I conclude that the latter alternative is the most viable from the standpoint of current science.

10.1 Checkmate on the Biogeographic Modes of Speciation

Allopatric, sympatric, and parapatric speciation are widely recognized biogeographic modes of speciation in the biology community. Allopatric speciation occurs when two or more populations of the same species become isolated due to geographic barriers. Since separation interrupts gene flow, this process is presumed to promote evolutionary divergence. In contrast, sympatric speciation is the process through which two or more populations within the species become isolated in the same area—e.g., without geographic barriers—and divergence is attained through ecological specialization. Finally, parapatric speciation refers to the process in which two or more populations of the same species gradually differentiate across a continuous environment; yet there is no complete disjunction between the populations, i.e., populations share a small fraction of their distribution areas (contact zone) where interbreeding occurs (see Chap. 9, Fig. 9.2a). Since sympatric and parapatric speciation take place in the presence of gene flow, biologists assume that allopatric speciation is the most common mode of speciation.

Interestingly, such a classification has gone almost unquestioned and has been widely embraced by most biologists. However, the central point lies in the fact that the geographical modes of speciation have been employed within the framework of the Biological Species Concept, which, as discussed in Chap. 7, presents numerous issues. Moreover, while these modes of speciation have gained broad acceptance, it is important to recognize that they focus exclusively on genetic mechanisms within an ecological context, without integrating embryonic developmental processes (Minelli and Fusco 2012). Below, I will provide some arguments against these models, demonstrating their imprecision and inadequacy. In particular, I will challenge the theoretical foundation of allopatric speciation (see Table 10.1 for a summary of these arguments).

(1) *The allopatric speciation model is outdated, considering the conceptual framework in which it was initially proposed*

Ernst Mayr, one of the principal architects of the Modern Synthesis, introduced the concepts of "allopatric speciation" and "sympatric speciation" in his book *Systematics and the Origin of Species* (1942), highlighting the importance of geographic barriers in the divergence process. Later, he refined these speciation models in his 1963 book *Animal Species and Evolution*, which remained one of the most influential works on this subject for decades. By then, Mayr had formulated a theory of speciation based on what he termed the "unity of genotype."

According to Mayr (1963), species are real entities whose reality is grounded in a natural delimitation shaped by genetic differences. For instance, while species are separated from each other by morphological gaps, their distinction is somewhat due to the genetic discontinuities acquired through reproductive isolation: "The degree of morphological difference displayed by a natural population is a by-product of the genetic discontinuity resulting from reproductive isolation" (p. 31). Although a correlation between phenotypic and genetic differences was recognized, Mayr observed that many populations could be morphologically identical yet reproductively isolated. Therefore, a true species is distinguished by its genetic discontinuities rather than its morphological gaps. In this context, Mayr conceived species as real entities separated by their genetic distances, with each species possessing a distinct gene pool that remains isolated from gene exchange with other populations: "The

Table 10.1 *Summary of the shortcomings of allopatric speciation*

Points against allopatric speciation	Summary
1. The allopatric speciation model is outdated, considering the conceptual framework in which it was initially proposed	Allopatric speciation was initially rooted in Mayr's founder effect model. However, this genetic view diverges significantly from the contemporary view of genetics. Furthermore, no empirical evidence supports that such a process occurs in nature. For example, field studies reveal that evolutionary divergence is produced by ecological differences rather than founder effect conditions. Finally, this model fails to incorporate embryonic development into its framework
2. Geographic barriers themselves are not the cause of divergence	Rather than geographic barriers, the model delineates the spatial separation between populations and the ability of the organisms to disperse. When biologists advocate for allopatric speciation as the predominant mode in nature, they restrict the scope for recognizing alternative speciation modes. For example, since ecological differences are accompanied by spatial situations, empirical cases of sympatric speciation become challenging. Still, in the marine environment, where defining geographic barriers is problematic, sympatric speciation seems to be the prevailing mode. Finally, geographic spatial context is unnecessary when the modes of speciation are considered in terms of gene flow. In such scenarios, parapatric speciation seems to be the predominant mode rather than allopatric speciation
3. The interruption of gene flow does not influence evolutionary divergence	Empirical evidence demonstrates that morphological divergence can indeed occur even in the presence of gene flow through the effects of natural selection. Moreover, gene flow might offer advantages by providing greater genetic variability upon which selection can act. It follows that populations should maintain certain levels of gene flow. In any case, the extensive genetic change does not necessarily correlate with the significant morphological change. Finally, it might be inferred that gene exchange is not only absent among allopatric populations but also among individuals within the same population
4. Divergence is not explained by geographical separation or the interruption of gene flow but rather by the combined processes of variation and adaptation to local environments	Biologists advocating for allopatric speciation admit that both genetic variations and environmental differences cause divergence. However, their discourse often emphasizes geographic barriers and the interruption of gene flow. This focus on genetic changes over ecological factors has led some researchers to coin the term "ecological speciation." Nevertheless, distinguishing clear cases of "non-ecological" speciation remains challenging. Undoubtedly, divergence is explained by the combined processes of variation and adaptation to local environments

essential characteristic of the geographical isolate is that it is separated from the rest of the species by a discontinuity" (p. 366). However, the degree of discontinuity depends on the effect and efficacy of the separation between populations by external geographic barriers; isolation will never be absolute until there is a complete cessation of gene flow. Therefore, Mayr argued that if separated populations exhibit genetic differences—each maintaining a stable gene pool—this suggests species cohesion.

Interestingly, Mayr (1963) stated that species possess a sort of "genetic homeostasis" that prevents them from diverging from the common "type."[1] This stable state of "genetic cohesion" is caused by the effects of gene exchange, which keep all populations within the species interconnected. Considering this, Mayr envisioned that populations—where gene flow had persisted—were unlikely to give rise to new species. According to his reasoning, sympatric speciation was a rare occurrence. Indeed, any speciation process was uncommon unless populations could break through the "genetic homeostasis" that bound them together. Mayr termed this disruption the "genetic revolution": "During a genetic revolution, the population will pass from one well-integrated and stable condition through a highly unstable period to another period of balanced integration" (p. 538). "A genuine genetic revolution is characterized by a breakdown of genetic homeostasis through a loss or a reconstitution of previously existing balancing systems" (p. 539). Therefore, for speciation to occur, geographic barriers (e.g., oceanic islands) must interrupt gene flow. Nevertheless, Mayr contended that the mere cessation of gene flow caused by geographic barriers was insufficient: "[W]e must realize that the physical nature of these barriers (extrinsic factors) is only one aspect of this isolation" (557). Here, the founder effect played a pivotal role. For Mayr, an effective speciation could only occur by splitting a small population from a larger ancestral one. These small and peripheral populations exhibit distinct geographic isolation—e.g., those inhabiting islands—as well as significant morphological gaps:

> Taxonomists have long been aware of the importance of these peripheral isolates and have pointed out, again and again, that major deviations from the "type" of a species will most likely occur in such populations... Peripheral isolates have two further, almost universal, characteristics. They are usually comparatively small in area, with a low absolute population size. They differ from the main body of the species population and from each other in numerous, often unique, and sometimes drastic morphological, physiological, behavioral, and other characteristics. (Mayr 1963, p. 368)

According to Mayr (1963), it is important to consider that, before the splitting, populations were genetically identical, and they coexisted in similar environments. However, once isolated, they accumulated unique genetic mutations and recombination. In addition, as the environments differed, distinct selective pressures resulted. All these factors contributed to differences in the genetic composition of offspring populations, resulting in phenotypes that differed from those of other populations. This process is noteworthy because, in small populations, genetic differences are intensified by a bottleneck process, unlike in large populations that maintain genetic stability: "The passing of the population through a bottleneck permits rather drastic shifts which would be resisted by the homeostatic system of the well-integrated continental parent population" (p. 538). At this point, Mayr claimed

[1] Indeed, it may seem odd given that Mayr was a prominent critic of typology. Nevertheless, as observed in Chap. 3, Sect. 3.2.2., Mayr's ideas are not so different from those of typologists.

that small populations were left with little genetic variability. He hypothesized that this would enable them to escape their genetic constraints more easily, perhaps thinking that populations would undergo innovative genetic mutations by preventing the entry of genes from other populations. Furthermore, upon encountering a vastly different environment, they would be exposed to more intense selective pressures, thus accelerating the divergence process.

> The founder population differs from the parental population not only in the drastic reduction of the diversity of its gene pool, but also by being exposed to a totally new constellation of environmental factors, biotic as well as genetic ones. The most important of these is the sudden conversion from an open to a closed—and at that to a *small* closed—population. It is the suddenness and completeness of this shift that is decisive. The new population is at once completely emancipated from the parental population. (Mayr 1963, p. 532; his emphasis)

In summary, Mayr argued that species possess a unity of genotype that enables them to maintain their components as a coherent whole. From this perspective, speciation was a rare process due to the difficulty populations faced in breaking free from this cohesive effect. Therefore, he turned to the founder effect: Small and peripheral populations (e.g., those reaching an island), having low genetic variability and being exposed to a different environment, would more easily disrupt that genetic homeostasis, leading to the emergence of a new species.

However, despite the apparent validity of this model to date, Plutynski (2019) highlights a significant shift in the perspective of speciation genetics from the 1960s to the present. For example, instead of adhering to the concept of "unity of genotype," contemporary discussions revolve around the genome. Genes are no longer perceived as discrete entities but as genome components with regulatory or non-regulatory functions. Additionally, she points out that only a small fraction of these genes may influence reproductive isolation: "[I]nstead of viewing the genotype as 'integrated,' or controlled by 'homeostatic' forces, it became increasingly common for evolutionary biologists to view genetic differences between species as relatively minor" (p. 572). Furthermore, Plutynski comments that this narrow view of speciation presents two shortcomings. First, it overlooks asexual species, constituting a significant portion of biological diversity. Second, these concepts pose challenges in practical application; for instance, in many cases, reproductive isolation occurs gradually: "By treating the problem of speciation as the problem of reproductive isolation, the synthesis architects narrowed and defined a research program" (p. 576). Finally, Plutynski stresses that Mayr's original argument has faced extensive questioning since the 1980s, considering that resorting to the founder effect was unwarranted due to the lack of empirical or theoretical support (e.g., see also Barton and Charlesworth 1984). Regarding this, Plutynski concludes:

> In sum, advances in molecular genetics, and the development of new tools for measuring and quantifying genetic diversity within and across populations, led to a rethinking of the genetics underlying species differences. The new data allowed evolutionary biologists to reconsider the nature of species and of speciation, thus achieving some distance from the

presuppositions of synthesis architects, particularly about "homeostasis" and "cohesion" of the gene pool. (Plutynski 2019, p. 584)

Interestingly, today, some researchers recurrently invoke the founder effect (e.g., Kolbe et al. 2012; Santos et al. 2012; Matute 2013; Blakeslee et al. 2019; Kaňuch et al. 2021; Jordan et al. 2024). However, there is still no conclusive evidence that such a process occurs in nature. For example, some studies indicate that despite evidence of some populations being isolated for millions of years and exhibiting ideal conditions according to the founder effect model, they have remained morphologically unchanged since then (Schneider and Moritz 1999; Gillespie et al. 2017). Others demonstrate that ecological differences, rather than founder effect conditions, lead to evolutionary divergence (Smith et al. 1997, 2001, 2005, 2011; Schneider et al. 1999; Pergams and Ashley 2001; Freedman et al. 2010, 2023; Yeung et al. 2011). Likewise, other studies supported by computational simulations demonstrate the improbability of such a model (Walsh et al. 2005; Yeung et al. 2011).

In any case, this process could yield counterproductive results, as small populations passing through a bottleneck experience a reduction in genetic variability, thereby limiting the options available for natural selection to act upon (Pergams and Ashley 2001; Garant et al. 2007; Moran et al. 2017). Furthermore, while islands served as the perfect scenario for Mayr to test cases of allopatry via the founder effect, it is important to note that many islands possess few populations to compete with, resulting in insufficient selective pressure for small populations. Conversely, islands boasting larger expanses of land and greater habitat diversity provide colonizing populations with more opportunities to undergo evolutionary divergence (Kisel and Barraclough 2010).[2]

Moreover, considering the advancements in Evo-Devo, it is no coincidence that most speciation studies focusing solely on genetic mechanisms lack effective integration of developmental processes (Minelli and Fusco 2012). Interestingly, it could be argued that the H-Evo-Devo Theory resonates with Mayr's model in viewing species as highly cohesive entities. But, unlike Mayr's view, these entities are not mere packages of genetic information; instead, they are complex systems produced by an interaction of processes at different scales. While there is now recognition of the existence of a genetic cohesion system—e.g., "The Genetic Regulatory Network"—this integrated genetic module is only one part of a more general module constructed from different levels of organization. A sole focus on genes as the entities responsible for the stability of evolutionary units demonstrates a lack of

[2] It is interesting to note that Charles Darwin had already pointed out this fact in the *Origin of Species*: "[A]lthough small isolated areas probably have been in some respects highly favourable for the production of new species, yet that the course of modification will generally have been more rapid on large areas; and what is more important, that the new forms produced on large areas, which already have been victorious over many competitors, will be those that will spread most widely, will give rise to most new varieties and species, and will thus play an important part in the changing history of the organic world… [To the contrary, on] a small island, the race for life will have been less severe, and there will have been less modification and less extermination." (Darwin 1859, pp. 106–107)

appreciation for the complexity of organic systems. It overlooks the different levels involved in the embryonic development of organisms. In short, allopatric speciation, based on Mayr's founder effect model, is now considered outdated in light of the scientific progress made so far.

(2) *Geographic barriers themselves are not the cause of divergence*

Most biologists take for granted that geographic barriers (e.g., a mountain, a cliff, and a river) somehow influence evolutionary divergence over time. But do geographic barriers really play a role in evolutionary divergence? Initially, the modes of speciation focused on the geographical context; for example, allopatric speciation occurs when geographic barriers divide populations, whereas sympatric speciation takes place in the same geographic location, i.e., without geographic barriers (Mayr 1942). However, these distinctions were later reconsidered in terms of gene flow. Allopatric speciation occurs in the absence of gene flow, whereas sympatric speciation occurs in the presence of gene flow. This distinction favors a third category, parapatric speciation, where there is no interbreeding between the two extremes of the population; yet a contact zone is identified between the populations in which gene flow is present (Smith 1965). In any case, theoretically, the relationship between geographic isolation and gene exchange is recognized. Once the two populations are isolated, geographic barriers prevent gene flow, allowing each population to obtain evolutionary independence (reproductive isolation), thereby altering their respective gene pools (Coyne and Orr 2004). If this is the case, it is not geographic barriers that matter; instead, it is the spatial separation between populations (which would evaluate the amount of gene exchange) and the ability of the organisms to disperse (Nosil 2012, p. 142). Considering this, the challenge arises when researchers attempt to illustrate instances of "non-allopatric" speciation.

In their highly influential book on speciation, Coyne and Orr (2004) conclude that allopatric speciation represents the null hypothesis, while sympatric speciation is the rarest mode occurring in nature. According to them, the issue lies in that sympatric speciation requires populations to diverge and become reproductively isolated through ecological specialization in the presence of gene flow. It needs many factors such as "gene number, the nature of epistasis, the strength of selection, the linkage relationships of genes producing isolating barriers, the distribution of ecological resources, the genetic basis of mate discrimination, the stability of the environment, and so on" (p. 142). Given the challenges associated with sympatric speciation, Coyne and Orr (2004) suggest that allopatric speciation provides a more satisfactory explanation, even though the facts can be interpreted through both hypotheses: "[T]he main difficulty in confirming sympatric speciation in nature is rejecting the alternatives… [W]e will deem allopatric speciation as the most likely explanation unless sympatric speciation appears more plausible" (pp. 141–142). Additionally, they limit cases of sympatric speciation through the following four criteria: (1) The species must be sympatric (however, allopatric species may originate in sympatry). (2) The species must be reproductively isolated by genetic differences. (3) The sympatric taxa must form sister groups and not result from hybridization. (4) The biogeographic and evolutionary history of the group must demonstrate that allopatric speciation was very unlikely. Considering this, they ensure that:

> [T]here is strong and pervasive evidence for allopatric speciation. Although the lack of similar evidence for sympatric speciation might reflect only the difficulty of getting good data, it seems reasonable to accept a controversial scenario only when it appears more likely than a well-established one. While most proposed cases of sympatric speciation are consistent with an alternative story involving allopatry, the converse is not true. It would be hard to argue, for example, that an island endemic whose closest relative lives on a nearby continent arose by sympatric speciation. (Coyne and Orr 2004, p. 142)

Similarly, Futuyma (2008) contends that while many examples can be explained by sympatry, theory advises invoking this hypothesis only when necessary. Considering these limitations, Bolnick and Fitzpatrick (2007) assert that it becomes clear how challenging it is to present cases of this type of speciation. In other words, the hypotheses that explain the origin of species are based on a priori reasoning that favors allopatric speciation over sympatric speciation: "If allopatry is treated as a null, ad hoc hypotheses favoring allopatry will be given more credibility than equally ad hoc stories of sympatry" (p. 463). But in populations inhabiting marine environments—where organisms have greater dispersal potential leading to large migrations—sympatric speciation seems to be the most plausible hypothesis (Crow et al. 2010). Indeed, according to various authors, cases of sympatric speciation are empirically impossible to demonstrate due to the ability of many organisms to disperse and the different environments in which they specialize (Berlocher and Feder 2002; Jiggins 2006; Bolnick and Fitzpatrick 2007; Butlin et al. 2008; Elmer and Meyer 2010; Rashkovetsky et al. 2015). For example, Fitzpatrick et al. (2008) highlight some issues in cases involving parasites with specific hosts, such as phytophagous insects, where ecological differences are associated with spatial situations (see also Jousselin et al. 2013). In contrast, they point out that, in these cases, geographic ranges overlap, but populations are separated by different ecological niches where gene exchange is unlikely to occur. The term "microallopatry" has been proposed for these cases, but its use becomes more confusing when attempting to distinguish it from true sympatry. Sun et al. (2022) argue, for example, that many cases of sympatric speciation are, in reality, instances of "microallopatric" or "microparapatric" speciation.

Furthermore, some authors suggest that, by preventing reproduction, specialization within environments has the same effect as the geographical separation of populations; both scenarios result in the cessation of gene flow (Butlin et al. 2008; Fitzpatrick et al. 2009). Others propose that evolution through ecological specialization occurs in any mode of speciation (Schluter 2001; Nosil and Harmon 2009; Nosil 2012; Jousselin et al. 2013), for example, "allopatric ecological speciation." In this context, emphasis is placed on the fact that, in many cases, it is difficult to discern whether divergence was due to geographical distances or ecological isolation (Bolnick and Fitzpatrick 2007). Even Coyne and Orr (2004, p. 101) admit that in some cases of "clinal isolation," it is impossible to determine if reproductive isolation resulted from geographical distances or local ecological differences, as both processes are intertwined. In this regard, Coyne and Orr (2004, pp. 118–120) point out that demonstrating examples of parapatric speciation in nature is very challenging, and the only way to distinguish it from allopatric speciation is by providing evidence of geographic barriers. Hence, according to Bird et al. (2012), modes of speciation are merely different scenarios in which gene flow is interrupted.

With respect to gene flow, Butlin et al. (2008) argue that sympatric and allopatric speciation represent the endpoints of a spatial continuum, where the degree of gene flow between diverging populations can be delineated (Fig. 10.1). In this regard, Gavrilets (2003) illustrates the modes of speciation in terms of gene flow rather than geographic distribution. For example, within the framework of gene exchange rates (m), sympatry is represented by an initial rate ($m = 0.5$), allopatry indicates no genetic exchange ($m = 0$), and intermediate scenarios, involving migrating populations diverging, illustrate parapatry ($0 < m < 0.5$). Notably, from this perspective, parapatric speciation emerges as the predominant mode, as "[m]ost of the parameter space is occupied by parapatric conditions… even though this mode of speciation generally receives the least discussion" (Butlin et al. 2008, p. 2997–2998). Considering this, some authors have noted that defining sympatry based on gene flow is impossible to demonstrate (Bolnick and Fitzpatrick 2007; Butlin et al. 2008; Fitzpatrick et al. 2008). This disagreement in viewpoints led Coyne and Orr (2004) to advocate for allopatric speciation as the prevailing mode, while Mallet et al. (2009) proposed a synthesis between demic and geographical definitions to bolster the credibility of cases of sympatric speciation. However, it is worth noting that allopatric speciation faces similar skepticism, as "complete allopatry is likely to be rare because migrants occasionally cross even the most extreme barriers" (Butlin et al. 2008, p. 2998). Within this framework, Wang et al. (2020) contend that:

> It is assumed that allopatric distribution prohibits gene flow and sympatric distribution permits unimpeded exchanges. However, such a correspondence is inexact as allopatry can be bridged by long-distance migration and sympatry may often be associated with microal-

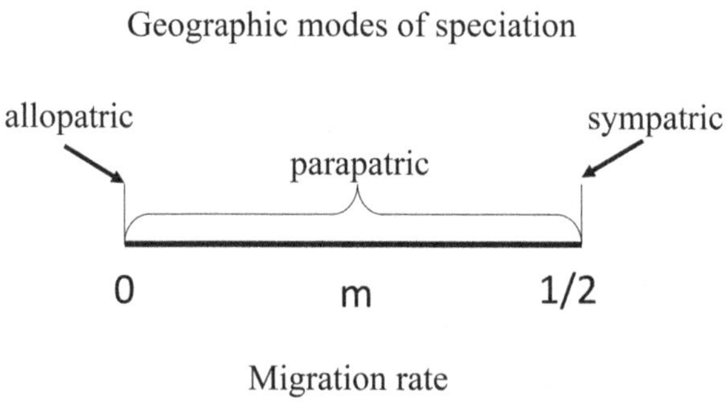

Fig. 10.1 *Modes of speciation according to Gavrilets.* The diagram illustrates a continuum of gene flow levels between diverging populations, emphasizing gene flow rather than strict geographic separation. According to this perspective, parapatric speciation appears to be the most prevalent mode. (Modified from Gavrilets 2003)

> lopatry. In either case, geography is not a good indication of gene flow, or lack thereof. No less important, if most speciation events are parapatric between adjacent populations, it would often be ambiguous whether gene flow has happened or not during speciation… With all these considerations, we do not use the geographical mode of isolation to assess gene flow. (Wang et al. 2020, p. 1391)

In conclusion, modes of speciation appear highly ambiguous when analyzed through geographical separations. Instead, they find clarity through the demic definition, which highlights parapatric speciation as the most prevalent mode. Considering this, it becomes evident that geographic barriers, seen as the spatial separation between populations, do not directly cause divergence. If they are involved, it is primarily insofar as they are accompanied by ecological differences, which do indeed partly contribute to evolutionary divergence.

(3) *The interruption of gene flow does not influence evolutionary divergence*

Many evolutionary biologists accept that the interrupting gene flow between populations fosters evolutionary divergence. The foundation of this belief rests on the theoretical framework underpinning the Biological Species Concept (BSC). Coyne and Orr (2004), for example, argue: "If one accepts some version of the biological species concept, then the central problem of speciation is understanding the origin of those isolating barriers that actually or potentially prevent gene flow in sympatry" (p. 57). Of course, one of the most efficient ways to interrupt gene flow is through the argument of allopatric speciation:

> Geographic isolation allows allopatric populations to diverge unimpeded by hybridization. Such differentiation is nearly inevitable: populations usually occupy ecologically different habitats that impose different forms of selection, *different mutations arise in different populations* (so that divergence can occur even when habitats are identical), *genetic drift fixes different genes in different isolates*, and initial evolutionary divergence can be amplified by later selection… Eventually, divergence will produce isolating barriers that allow species to remain distinct when they experience secondary contact. (Coyne and Orr 2004, p. 86; my emphasis)

In this citation, I mainly highlight the genetic factors that could theoretically distinguish it from sympatric speciation, as the latter is expected to occur solely through ecological specialization, thus counteracting the effects of gene flow presence. If this process does indeed occur in nature, then the cessation of gene flow would facilitate divergence since isolated populations develop genetic distinctions. In other words, a decrease in gene flow leads to more distinct genes, resulting in greater evolutionary divergence. Conversely, an increase in gene flow correlates with a lower number of distinct genes and, thus, less evolutionary divergence.

Nevertheless, despite the existence of an alleged mechanism explaining why the cessation of gene flow promotes divergence, sufficient evidence suggests that significant divergence can occur in the presence of gene flow through the effects of natural selection (Smith et al. 2005; Dowle et al. 2014; Liu et al. 2017; Palacios et al. 2019; Zhang et al. 2021; Wu et al. 2024). For instance, Osborne et al. (2013) investigated the divergence timeline of two species of shrubby plants within the *Senecio* genus on the volcano Mount Etna. While their findings reveal rapid divergence considering the time of volcanic formation, the presence of hybridization in both species led them to conclude that evolution through ecological specialization efficiently occurs despite gene flow. Another example can be found in the Qinghai-Tibetan Plateau sensu lato—comprising the platform, the Himalayas, and the Hengduan Mountains—which boasts a remarkable diversity of endemic species. While geographical isolation could explain this diversity, Wu et al. (2022) demonstrate that gene flow was present during speciation across all studied species. In addition, although many speciation examples focus on terrestrial systems, Potkamp and Fransen (2019) illustrate the prevalence of gene flow in most cases of speciation within the marine environment.

Some scholars go so far as to argue that gene flow offers advantages by providing greater genetic variability upon which selection can act. Consequently, in order to facilitate significant evolutionary divergence, populations should maintain certain levels of gene flow (Smith et al. 2005; Garant et al. 2007; Nolte and Tautz 2010; North et al. 2010; Lavrenchenko 2011; Smadja and Butlin 2011; Kremer et al. 2012; Sexton et al. 2014; Moran et al. 2017; Samuk et al. 2017; Zhao et al. 2024). In any case, Pennell et al. (2014) argue that the speciation process is not necessarily tied to an increase in the rate of genetic change since there is no evidence to suggest that population isolation—or any other extraordinary event—accelerates rates of genetic mutation. They also note that while significant genetic changes, such as polyploidies, could contribute to speciation, these processes do not explain all cases of divergence. Moreover, from the Evo-Devo perspective, Newman (2015, p. 435) highlights that "largescale morphological change can occur with a minimum of genetic change, while morphology can be static despite extensive genetic change."

Regardless, within the conceptual framework underpinning the Biological Species Concept, the idea that the interruption of gene flow is essential for evolutionary divergence is called into question. According to this concept, species are recognized as cohesive units due to the gene flow present within the populations comprising them; for example, the lack of gene flow between allopatric populations serves as a criterion for delineating them as distinct species. However, by justifying the species category through this concept, Lee (2003) not only acknowledges that allopatric populations within the species lack gene exchange but also states that "even in a panmictic population, every individual is not simultaneously interbreeding with every individual" (p. 185).[3] In other words, gene exchange is not solely absent among allopatric populations but also among individuals within the same population. If this is the case, the assumption that cessation of gene flow is a factor for promoting divergence becomes irrelevant, if not absurd.

[3] Lee (2003) argues that species cohesion is characterized by the potential for interbreeding rather than by the exchange of gene flow among members of a species. However, the Biological Species Concept faces numerous challenges in addressing the meaning of reproductive isolation and its relationship with gene flow (see Chap. 7, Sect. 7.2.1).

(4) *Divergence is not explained by geographical separation or the interruption of gene flow but rather by the combined processes of variation and adaptation to local environments*

When Mayr (1942) proposed allopatric speciation, he primarily considered reproductive isolation as the pivotal factor for species diagnosis, resulting from the gradual accumulation of genetic differences. Later, in 1963, Mayr expanded his concept of the species as "unity of genotype," whose cohesion depended on gene flow among their parts. He argued that geographic barriers preventing gene flow alone did not explain speciation, prompting him to adopt the founder effect model (see point 1). Mayr proposed that small populations with low genetic variability produced greater genetic differences, particularly when encountering new environmental factors, thereby accelerating evolutionary change. Such ideas, albeit not necessarily falling on the founder effect, were echoed in Coyne and Orr's (2004) account for speciation. They emphasized the relevance of genetic and ecological differences as causes of divergence in allopatry. So, if genetic and ecological differences promote divergence in isolated populations, could it not be—and I propose this as a mere hypothesis—that these genetic and ecological differences—rather than spatial isolation and interruption of gene flow—are the genuine causes of divergence?

Interestingly, most biologists who support allopatric speciation, while acknowledging geographic distances in their explanations, emphasize this dual aspect: genetic variations and environmental differences. In another vein, some biologists, dissatisfied with traditional geographic speciation models, have introduced the concept of "ecological speciation," defined as a kind of speciation that produces reproductive isolation resulting from selective divergence (Schluter 2001; Hendry 2009; Sobel et al. 2009; Nosil 2012; Beheregaray et al. 2015; Podos and Schroeder 2024). However, within such a framework, identifying examples of "non-ecological" speciation becomes challenging (Schluter 2009; Butlin et al. 2008; Sobel et al. 2009).[4] Furthermore, while these authors' emphasis on the role of the environment in speciation is crucial, such significance is contextualized within the Biological Species Concept. Yet, as discussed in Chap. 7, this perspective on species is inadequate for recognizing units of evolution. Indeed, any evolutionary model aiming to study evolutionary divergence must integrate internal and external factors not as independent processes but as a harmonious whole within the same conceptual framework. Therefore, I wholeheartedly concur with Butlin et al. (2012) and Fitzpatrick et al. (2008) when they argue that instead of solely classifying modes of speciation based on geographic spatial contexts (i.e., allopatric, sympatric, parapatric), it is preferable to center attention on the biological processes involved in evolutionary divergence (i.e., the origin of variation and the environmental influence on such variation). Indeed, the H-Evo-Devo Theory pursues these objectives, not only by endeavoring to explain divergence as a dual process between variation and environment but also

[4] Nevertheless, Czekanski-Moir and Rundell (2019) argue that this type of speciation may be more prevalent than previously assumed.

by being grounded on the premise that evolutionary divergence is inherently hierarchical.

In summary, geographic modes of speciation are remarkably ambiguous and inadequate for studying evolutionary divergence. If their support relies solely on the assumption that allopatric speciation is the most prevalent mode, their theoretical foundations are notably insufficient for the following reasons: first, the theoretical framework of allopatric speciation is outdated; second, neither geographic barriers nor the interruption of gene flow is the cause of evolutionary divergence. Instead, divergence arises from the combined processes of variation and adaptation to local environments.[5]

10.2 Micro-, Macro-, and Megaevolution: The Distinction, a Matter of Degree or Kind?

The prefixes "micro" and "macro" delineate the levels or scales at which evolutionary phenomena unfold. Typically, "microevolution" refers to changes occurring below the species level that are observable within the current timescale. In contrast, "macroevolution" encompasses changes occurring at and above the species level, necessitating a broader temporal scale. Of course, the debate around whether microevolution is distinct from macroevolution has divided evolutionary biologists for years.

The extrapolationist premise suggests that macroevolutionary patterns are comprehended through well-established microevolutionary explanations and mechanisms. However, some contend that these microevolutionary models are inadequate in fully explaining macroevolutionary processes. Interestingly, most biologists endorse the extrapolationist premise despite recognizing the importance of viewing evolution as a phenomenon operating at multiple levels. Another issue lies in the lack of a consensus definition for each level, which fosters ambiguity in the debate. This confusion is exacerbated by introducing an additional level, dubbed "megaevolution," which exclusively pertains to processes and patterns above the species level. Nevertheless, it is important to note that despite its widespread acceptance, the extrapolationist premise remains unproven. Instead, as will become clear, evidence regarding the distribution of morphological variation (e.g., morphological gaps between taxonomic groups) and patterns observed in the fossil record underscore the need to delineate distinct levels of evolution. Below, I will explore whether the distinction between micro-, macro-, and megaevolution levels is primarily quantitative or qualitative.

[5] According to the H-Evo-Devo Theory, the combined processes of variation and adaptation to local environments represent a type of divergence that occurs at the microevolutionary scale. However, at the macroevolutionary scale, an alternative explanation is needed that is different from traditional local adaptation. See Chap. 11, Sect. 11.2.1.

10.2.1 History and Definitions of Such Contemporary Conceptions

In 1927, the Russian geneticist Iurii A. Filipchenko introduced the terms "microevolution" and "macroevolution." Importantly, Filipchenko did not associate the origin of categories above the species level with the Mendelian genetic changes observed in laboratory populations. Consequently, he made a clear distinction between two levels of evolution: microevolution involves the evolution of species and "all lower subdivisions of species." Meanwhile, macroevolution encompasses the evolution of genera and "all higher systematic units" (Adams 2021, p. 206). Subsequently, Theodosius Dobzhansky, a prominent figure in the Modern Synthesis, adopted his mentor Filipchenko's terminology but assigned it a significantly different meaning. Following the extrapolationist premise, Dobzhansky (1937) interpreted these concepts in terms of timescale. That is, from his perspective, the large-scale changes assigned in taxonomy are essentially the same processes and mechanisms operating on the microevolutionary scale, only extrapolated over millions of years of evolution:

> Of course changes observed in populations may be of very different orders of magnitude, from those induced in a herd of domestic animals by the introduction of a new sire to phylogenetic changes leading to the origin of new classes of organisms. The former are obviously trifling in scale compared with the latter, and it may not be convenient to have all of them subsumed under the name "evolution." Experience seems to show, however, that there is no way toward an understanding of the mechanisms of macro-evolutionary changes, which require time on a geological scale, other than through a full comprehension of the micro-evolutionary processes observable within the span of a human lifetime and often controlled by man's will. For this reason we are compelled at the present level of knowledge reluctantly to put a sign of equality between the mechanisms of macro- and micro-evolution, and, proceeding on this assumption, to push our investigations as far ahead as this working hypothesis will permit. (Dobzhansky 1937, p. 12)

According to Erwin (2011), Dobzhansky's viewpoint echoes Lyell's uniformitarianism, which posits that major events, appearing to have occurred suddenly, actually result from the gradual accumulation of small changes over geological time. In addition, Erwin emphasizes Dobzhansky's recognition of the potential qualitative distinction between macroevolution and microevolution even though there is no empirical or experimental evidence to support it. Ultimately, Dobzhansky believed he had placed his bet on the "right horse."[6]

On the other hand, in his book *The Material Basis of Evolution* (1940), Richard Goldschmidt, a prominent critic of the Modern Synthesis, challenged the extrapolationist premise, proposing that distinct evolutionary mechanisms operate at each

[6]Adams (2021, p. 207) notes an intriguing anecdote from an interview with Dobzhansky shortly before his passing away, during which he was asked about his differing views on the concepts of microevolution and macroevolution compared to those of his mentor Filipchenko. Dobzhansky reportedly shrugged and said, almost indifferently, "He bet on the wrong horse." Another interesting observation is Dobzhansky's initial caution regarding the extrapolationist premise; however, later, he grew more confident, influenced by the subsequent works of other architects of the Modern Synthesis, notably those of the paleontologist George Gaylord Simpson (see Ochoa 2021).

level. From his viewpoint, the taxonomic scale corresponds to the morphological discontinuity obtained during divergence. In this regard, he echoed Filipchenko's approach to defining these concepts: "It will be one of the major contentions of this book to show that the facts of microevolution do not suffice for an understanding of macroevolution. The latter term will be used here for the evolution of the good species and all the higher taxonomic categories" (p. 8).

Subsequently, in line with Goldschmidt's standpoint, George Gaylord Simpson (1944) defined these concepts based on his hierarchical evolutionary view. According to Simpson, microevolution "involves mainly changes within potentially continuous populations, and there is little doubt that its materials are those revealed by genetic experimentation." In contrast, macroevolution "involves the rise and divergence of discontinuous groups, and it is still debatable whether it differs in kind or only in degree from micro-evolution" (p. 97). Once this difference was settled, Simpson posed the main challenge concerning evolutionary biology: "If the two proved to be basically different, the innumerable studies of micro-evolution would become relatively unimportant and would have minor value in the study of evolution as a whole" (p. 97). Simpson noted that geneticists focused on the lower levels of the hierarchy, while paleontologists studied the higher levels. However, his reference to higher levels does not pertain to species but rather to the higher categories of Linnaean taxonomy, such as family, order, class, and so on. Therefore, Simpson introduced an additional level, "megaevolution," which applied to the higher taxa level: "If the term 'macro-evolution' is applied to the rise of taxonomic groups that are at or near the minimum level of genetic discontinuity (species and genera), the large-scale evolution studied by the paleontologist might be called 'mega-evolution' (a hybrid word, but so is 'macro-evolution')" (p. 98).

Simpson stated that higher taxa were units of evolution whose studies could not be reducible to experimental genetics research. In fact—contrary to his later works, which favored arguments empathetic to microevolutionary studies—he maintained that paleontological data pointed to a clear distinction between two levels of evolution[7]:

> The assumption, as in Goldschmidt's work, that mega-evolution and macro-evolution are the same in all respects is no more justified than the assumption... that micro-evolution and macro-evolution differ only in degree. As will be shown, the paleontologist has more reason to believe in a qualitative distinction between macro-evolution and mega-evolution than in one between micro-evolution and macro-evolution. (Simpson 1944, p. 98)

Despite this, in later years, Mayr (1980, p. 1) emphasized that the Modern Synthesis had led to the general acceptance of two main conclusions: "[G]radual evolution can be explained in terms of small genetic changes ('mutations') and recombination, and the ordering of this genetic variation by natural selection; and the observed evolutionary phenomena, particularly macroevolutionary processes and speciation,

[7] Despite popular belief, Simpson consistently upheld this non-reductionist perspective throughout all his writings. For a more detailed discussion, see Ochoa (2021). To delve into the hierarchical viewpoints of Goldschmidt, Dobzhansky, and Simpson, see Chap. 3, Sect. 3.2.

can be explained in a manner that is consistent with the known genetic mechanisms." However, according to Adams (2021), although it is commonly believed that most architects of the Modern Synthesis supported the extrapolationist premise, figures like Julian Huxley and Bernhard Rensch maintained a cautious stance towards such conclusions. Additionally, Adams points out that while population genetics—essentially microevolutionary—was one of the cornerstones of the Modern Synthesis, it has not succeeded in resolving the issue of such a premise: "The traditional view is that population genetics… played a central role in this synthesis... But, as we have seen, population genetics did not, did not claim to, and could not prove or even address the second conclusion concerning macroevolution" (p. 216).

The primary issue was that most evolutionists took for granted that there were no special events in macroevolution. In the 1970s and 1980s, the generation of paleontologists, including prominent figures such as Niles Eldredge, Stephen Jay Gould, Steven Stanley, and Elisabeth Vrba, among others, embraced the concept of "macroevolution" to formulate theories such as punctuated equilibrium, Hierarchy Theory of Evolution, species selection, etc. Yet it is necessary to argue that this "macroevolutionary renaissance" occurs in a context vastly different from the original controversy that challenges the extrapolationist premise (Arthur 2008). Admittedly, while these theories recognize macroevolutionary patterns, they do not propose mechanisms that are different from those assumed by microevolution. Moreover, they take for granted that the differences in taxonomic scale—i.e., the morphological gaps at higher levels—are obtained through continuous change (see below).

Indeed, these differing perspectives on levels have contributed to confusion in defining contemporary concepts of "microevolution" and "macroevolution" (Hautmann 2020; Kearney et al. 2024). Some biologists make this distinction by considering Dobzhansky's definition of the timescale (e.g., Dawkins 1982, p. 289; Gingerich 1987; Bromham 2016, p. 48; Tsuboi et al. 2024). For instance, Dawkins (1982, p. 289) defines macroevolution as "the study of evolutionary changes that take place over a very large time-scale." Similarly, Gingerich (1987, p. 1053) considers that "microevolution and macroevolution are manifestations, expressed on different time scales, of common underlying processes." Other authors resort to Filipchenko and Goldschmidt's meaning based on the taxonomic scale (e.g., Arnold et al. 2001, p. 9; Callahan 2002, p. 1; Simons 2002; Levinton 2001, 2012; Arthur 2003; Jablonski 2007, 2017a; Erwin 2010; Stigall 2015, p. 301; Shpak 2019, p. 140). Meanwhile, still others define them as a combination of both views (Grantham 1995, p. 302; Grantham 2007, p. 76; Travis and Reznick 2009, p. 126; Kampourakis 2014; Huneman 2017; Toman and Flegr 2024). Nevertheless, it is important to recognize that both notions—taxonomic scale versus timescale—must be distinguished because the magnitude of variation (taxonomic scale) is not necessarily related to the timing of divergence (timescale). For example, lineages sometimes undergo minimal changes over a long period, while in other cases, they may undergo significant modifications over a short period (Arthur 2003; Erwin 2010). Finally, some authors employ the term "macroevolution" to indicate that species exhibit properties distinct from those of individuals at the population level, mainly when they discuss species selection (e.g., Grantham 2007, p. 76; Hautmann 2020; Jablonski 2017b).

Another noteworthy observation is that, while biologists currently employ the concepts of "microevolution" and "macroevolution," few have recognized a third level termed "megaevolution," which corresponds to evolution above the species level. Interestingly, the term "megaevolution" has largely fallen out of use, prompting some authors to consider the levels above the species as part of "macroevolution" (e.g., Arnold et al. 2001; Plutynski 2008; Jablonski 2017a; Vislobokova 2017; Toman and Flegr 2024). For others, the term "macroevolution" includes the species level as well as higher taxonomic levels (Erwin 2000, p. 82; Callahan 2002; Stigall 2015; Saupe and Myers 2021). Consequently, Lieberman and Eldredge (2014) advocate for reserving the term macroevolution exclusively for the species level, positing that "there is no evolutionary process operating above the species level" (p. 185).[8] However, Arthur (2021) claims that "megaevolution" constitutes a distinct level, describing it as "the evolutionary changes that produce a new body plan" (p. 75). Interestingly, although Arthur somewhat supports the extrapolationist premise, he suggests, like Simpson, that the distinction between macroevolution and megaevolution holds more significant than that between microevolution and macroevolution: "So, my view is that the debate over punctuated equilibrium was the proverbial 'storm in a teacup,' and that it distracted attention away from the more important issue of whether megaevolution is in some way different from macroevolution, which I focus on here" (p. 76).

Finally, it should be emphasized that most proponents who delineate the differences between the two levels based on the taxonomic scale do not contend that there are significant differences between microevolution and macroevolution. Nevertheless, a proper distinction between these levels hinges on the premise that, at the least, distinct mechanisms govern each. Moreover, it is not only a matter of different mechanisms but also that the outcomes of these mechanisms yield different hierarchical entities. Indeed, this distinction is underscored by some evolutionary biologists—albeit a minority—who highlight the qualitative differences between macroevolution and microevolution (e.g., Ho and Saunders 1979; Løvtrop 1981; Bateman and DiMichele 2002; Theißen 2006; Frazzetta 2012; Diogo et al. 2017; Ochoa 2024). So, the question remains open: Is macroevolution more than uninterrupted rounds of microevolution?

10.2.2 Three Possible Scenarios, but Just One Compelling Solution

Drawing from Grantham's analysis (2007), three possible scenarios emerge regarding the relationship between macroevolution and microevolution:

[8] Nonetheless, the H-Evo-Devo Theory challenges this perspective; see Chap. 8, Sect. 8.1 and Chap. 12, Sect. 12.1.

1. *The reductionist approach.* All macroevolutionary phenomena are explained simply as the sum of microevolutionary events.
2. *The intermediate approach.* Although macroevolutionary mechanisms are not entirely distinct, macroevolutionary inquiries cannot be fully reduced to microevolution. Macroevolutionary patterns are analyzed independently of the microevolutionary models that explain them. Under this framework, macroevolution and microevolution complement each other as part of the same theoretical framework.
3. *The irreducible approach.* Without question, macroevolution harbors distinct processes and patterns irreducible to microevolutionary explanations. This perspective does not imply dismissing microevolutionary investigations but rather advocates for a hierarchical evolutionary approach wherein microevolution and macroevolution (and even megaevolution) are regarded as independent fields of inquiry.

(1) *The reductionism approach*

Most biologists strongly support the extrapolationist premise. But why is the reductionist approach widely assumed to be key to a deeper understanding of macroevolutionary phenomena? Over the years, the framework of the Modern Synthesis has become deeply embedded, not only in scientific research but also in educational curricula. Many biologists educated under this perspective have adopted a lexicon of concepts and models that explain evolution. According to the framework of the Modern Synthesis, microevolutionary mechanisms are identified through population genetics (e.g., Plutynski 2008; Rolland et al. 2023; Schluter 2024). Likewise, Futuyma (2015), blinded by this orthodoxy, maintains that none of the new proposals on Evo-Devo mechanisms challenge the framework of the Modern Synthesis: "I will conclude that many of these challenges have had a positive impact on evolutionary biology, but that the fundamental principles of the ST [Modern Synthesis] remain valid, and can explain known evolutionary phenomena with only modest extension" (p. 30).

Another key issue is evolutionary uniformitarianism—or "the assumption that the [gradual] rates, mechanisms, and processes that can be observed experimentally today are sufficient to explain the sweep of evolutionary patterns through geologic time" (Erwin 2010, p. 188). This notion fostered an unwarranted confidence that prompted an embrace of the extrapolationist premise. For example, considering that large-scale patterns are related to deep time, some researchers claim that demonstrating the underlying mechanisms in macroevolution is challenging (e.g., Hendry and Kinnison 2001; Kampourakis 2014; Tsuboi et al. 2024). Consequently, there is a tendency to prioritize observations of gradual change in extant organisms, which theoretically could account for the discontinuous patterns seen in the fossil record through extrapolation (Gingerich 2019).[9] Nevertheless, Erwin (2010) points out that:

> There is no necessary reason why evolutionary uniformitarianism should be correct; it is simply an implicit assumption of most evolutionary biologists… Evolutionary biologists have tended to reject such explanations [macroevolutionary mechanisms] because they are difficult to test, not because we can demonstrate from first principles that they are untrue. (Erwin 2010, p. 188)

So, it may be relevant to ask: What kinds of macroevolutionary mechanisms would be distinct from those of microevolution? As previously mentioned, orthodox theory identifies microevolutionary mechanisms through population genetics (i.e., mutation, migration, genetic drift, and natural selection). In contrast, macroevolutionary mechanisms would be those genetic changes that yield discontinuous effects or "macromutations" as proposed by Goldschmidt (1940). However, under this perspective, the term "macromutation" often merely serves to discredit theories of saltational evolution, lacking straightforward elucidation of its meaning. Furthermore, it is important to note that these critiques largely stem from a gene-centered perspective of evolution, emphasizing genetic changes as paramount. Conversely, adopting an organism-centered perspective reveals a diverse and intricate array of mechanisms at play in the origin of units of evolution, which need not be simplistically equated solely with genetic change.[10] Notably, internal changes alone are insufficient to account for morphological divergence. Instead, this results from a nexus of interrelated processes involving the environment. Hence, mechanisms must be both intrinsic and extrinsic factors in their entirety (see below).

Regarding intrinsic changes, it is difficult to assert that population genetics mechanisms apply solely to microevolution, while Evo-Devo mechanisms are confined to macroevolution. Given that Evo-Devo mechanisms also operate at the microevolutionary scale (Nunes et al. 2013; Nicoglou 2015; Fernández-Mazueco and Glover 2017), population genetics mechanisms (e.g., genetic drift and mutation) should not be considered the sole drivers of microevolution. Moreover, such mechanisms should be reexamined in light of the understanding of the origin of variation, as they adopt a gene-centered perspective that completely ignores the organism and its structural properties. Consequently, I agree with Dietrich's (2010) critique, which asserts the absence of exclusive mechanisms for macroevolution: "[m]utation in developmentally important genes does not constitute such a macroevolutionary process that is distinct from microevolutionary processes, even if they produce macroevolutionary outcomes" (p. 172). Nonetheless, the question remains: Are the same mechanisms responsible for producing different degrees of morphological differentiation, or do these degrees themselves enable a qualitative distinction between the levels?

---

[9] However, the methodology of gradual rates of change raises several concerns; see Chap. 9, Sect. 9.1.2.

[10] Regarding this duality between a gene-centered perspective and an organism-centered perspective, see Chap. 3, Sect. 3.1.

Regardless, while the internal mechanisms that produce variation may be the same, the difference between levels could depend on the type and degree of variation generated by these processes. Arthur (2021), for example, distinguishes a qualitative distinction between two types of divergences during development: one occurring at later stages, leading to species separation, and another occurring at earlier stages, resulting in phylum-level divergence. Interestingly, for Arthur, since there are no specific genes to distinguish between these divergences and considering there are many separations between the two extremes, the hierarchical degree of morphological gaps is simply a part of a continuous process: "There is nothing magic about the phylum level of organization. It does not constitute some mystical realm in which processes fundamentally different to those of 'routine' evolution take place. Rather, it represents one end of what is probably best described as a continuum" (p. 125–126).

However, elsewhere in the text, Arthur (2021, p. 97) suggests that while some genetic changes influence pigmentation, others affect the structure of the body plan. Despite his belief that these levels of variation generally arise through gradual evolution, Arthur does not exclude the possibility of discontinuous change: "The above considerations suggest that we should not rule out a role for macromutations (in the broad sense) in the origins of animal body plans. However, we should not assume their importance either" (p. 123). Similarly, Futuyma (2015) does not deny the possibility that some genetic mechanisms involve discontinuous changes in phenotype: "Whether or not major, discontinuous single-step changes in phenotype have occurred in evolution is an empirical question… [Yet] it is also possible that it happened incrementally" (p. 40). With this in mind, there is a possibility—albeit remote from my viewpoint—that gradual changes traverse great distances in morphospace through time. Nevertheless, gradual evolution would only apply to heterochronic changes (e.g., differences in proportion and size) but never to changes that are heterotopic (change in place), heterometric (change in quantity), or heterotypic (change in type), as these modes of change lack conceivable intermediate forms.[11]

(2) *The intermediate approach*

This category encompasses punctuated equilibrium, Hierarchy Theory of Evolution, and species selection. Although some authors might place the latter within the irreducible approach (e.g., Grantham 2007), I will discuss it here as part of the intermediate approach, as it does not propose mechanisms distinct from the traditional microevolutionary framework to account for the emergent properties of species.

With respect to punctuated equilibrium, this theory stands out for revealing the patterns of the species on a geological scale (i.e., punctuation, stasis, and extinction), which are not discernible solely through the study of microevolutionary rates of change (Bokma 2015; Miller III 2015). However, this raises a fundamental question: Do these patterns arise from mechanisms different from those of

[11] See Chap. 9, Sect. 9.2.

microevolution? Erwin (2000) states that most of these discussions focus on patterns rather than processes. Hence, the main challenge facing punctuated equilibrium theory is that it assumes that no distinct mechanisms exist in macroevolution; instead, it uses the conventional microevolutionary mechanisms of speciation and extrapolates them to the geological timescale of macroevolution. This is, at least, what one of its founders says: "[T]he theory of punctuated equilibrium adopts a very conservative position. The theory asserts no novel claim about modes or mechanisms of speciation; punctuated equilibrium merely takes a standard microevolutionary model and elucidates its expected expression when properly scaled into geological time" (Gould 2002, p. 778). So, if these mechanisms are not groundbreaking, doubts arise as to whether it qualifies as a true macroevolutionary theory (Løvtrop 1981; Huneman 2017; Hautmann 2020).

An alternative perspective might suggest that, to account for the "punctuation" mode, the founder effect plays a distinct role compared to other microevolutionary processes, or that the "stasis" mode involves processes different from stabilizing selection. However, as discussed in the previous section, the founder effect presents many problems, and the mechanisms for stasis are highly varied, thus failing to present a robust challenge to avoid the extrapolationist premise.[12] Another significant point is that punctuated equilibrium is gradualist, suggesting that the discontinuities observed at the taxonomic scale may not be entirely real, as they are produced continuously and progressively by microevolutionary processes. Although initially believed to have characteristic patterns distinct from "phyletic gradualism," as discussed in the previous chapter, there are no substantial differences between both modes when comparing rates of evolutionary change.[13]

Regarding the Hierarchy Theory of Evolution, its central challenge lies in viewing genealogical entities (demes, species, and monophyletic groups) merely as carriers of hereditary genetic information. Consequently, conventional microevolutionary mechanisms align well with this framework. Moreover, as this theory adheres to the traditional perspective of the Modern Synthesis (Pievani 2016), its tendency toward reductionism is unsurprising, despite efforts to distance itself from such an approach. As Gontier (2015, p. 232) notes: "For an evolutionary scholar that wanted to work within the epistemic boundaries delineated by the Modern Synthesis, little else remained but the study of genes. The microevolutionary tradition developed by trying to incorporate lower and higher-order phenomena on the scale, but at the same time reducing these phenomena to a genetic level of analysis."[14]

Finally, concerning species selection, the debate is rooted in the dichotomy between emergent characters versus aggregate characters.[15] Fundamentally, this

[12] Indeed, some researchers face challenges in explaining stasis considering the microevolutionary studies. This problem is known as the "stasis paradox" (Haller and Hendry 2014). For a review of potential stasis mechanisms, see Chap. 11, Sect. 11.2.2.

[13] See Chap. 9, Sect. 9.1.

[14] For a more extensive critique of the Hierarchy Theory of Evolution, see Chap. 4, Sect. 4.2.

[15] See Grantham (1995), Gould (2002).

issue addresses whether species traits arise progressively through additive microevolutionary processes, or if they are emergent properties intrinsic to the species, with their origin not attributable to lower-level processes. Advocates (e.g., Grantham 1995, 2007; Hautmann 2020) claim that this phenomenon is macroevolutionary based on patterns, despite relying on microevolutionary processes to explain the origin of species properties. For example, Hautmann (2020) holds that: "The *process* of speciation in the sense of evolutionary change is microevolutionary, but the *outcome* (interspecific variation) and the *rate* of speciation are macroevolutionary" (p. 9; his emphasis and parenthesis). It is also noteworthy that when discussing true species selection for emergent characters, such characteristics are not phenotypic properties of the species (e.g., geographic range; see Grantham 2007). In this regard, by supporting the extrapolationist premise, Dietrich (2010) argues that species selection is the only candidate capable of representing a "macroevolutionary" phenomenon but that its occurrence is uncommon: "I claim that such processes are possible in the case of species selection, but are relatively rare and so are of minor evolutionary consequence when the entirety of the domain of evolutionary biology is considered" (p. 166). Or even worse, there are doubts about whether species selection is a real phenomenon, given that species parts do not interact with the environment as a whole (see Damuth 1985).[16]

In conclusion, although these theories have provided valuable insights into identifying patterns at higher levels, they do not constitute a true macroevolutionary perspective. They fail to propose distinct higher-level mechanisms beyond those established by traditional microevolutionary theory. Additionally, these perspectives overlook the differences between the levels of the taxonomic scale indicated by morphological discontinuities, which represent the original distinction. As a result, there is only one viable option worth considering.

(3) *The irreducible approach.*

As noted earlier, the distinction between macroevolution and microevolution must be contextualized within the taxonomic scale, considering that discontinuous degrees justify it. But when Levinton (2001, p. 2) defines macroevolution as "the sum of those processes that explain the character-state transitions that diagnose evolutionary differences of major taxonomic rank," he does so in the sense that it does not imply saltational evolution. Nevertheless, as explored in previous chapters, these "character-state transitions" indeed involve discontinuous changes. In discussing the significance of type in the domain of macroevolution, Levinton states:

> The problem of macroevolution has always been regarded as the problem of the origin and evolution of types and the present gulf between them. A type is a class whose members share a certain set of defining traits. Such a definition implies gaps between types, or at least discrete differences in the sets of traits that define the different types. If you don't believe in types and gaps, then you don't worry much about major evolutionary jumps, but

[16] For a more extensive critique of the Hierarchical Theory of Selection, particularly of species selection, see Chap. 4, Sect. 4.1.3.

> the belief in types, among species or among higher taxonomic constructs (e.g., baupläne) will lead you toward a deep concern about discontinuities in evolution. (Levinton 2001, p. 13)

Indeed, this leads us to revisit its definition in order to evaluate what constitutes "those processes that explain the character-state transitions" between one type and another. For Dietrich (2010) and Hautmann (2020), developmental processes are not considered relevant for qualitatively distinguishing between the levels. However, as previously mentioned, the issue pertains not only to the mechanisms involved at each level but also to their outcomes in producing discontinuities between the levels. For example, critics like Arthur (2003, 2008, 2021) and Futuyma (2015) acknowledge that developmental mechanisms may produce morphological gaps but remain skeptical about their prevalence. It is interesting to note how researchers instinctively avoid considering saltational theories. This inclination stems from the evolutionary uniformitarianism of the Modern Synthesis, which has influenced researchers' perspectives even when addressing the topic of macroevolution (Erwin 2010). Nevertheless, this view has been changing in recent years.

For example, patterns in the fossil record reveal the emergence of a hierarchical structure characterized by discontinuities at each level, mainly when these qualitative distinctions manifest in the distribution of variation. As Jablonski (2017a) points out:

> [P]rogress in modeling dynamics in the taxonomic hierarchy has derived insights from the frequency distribution of lower taxa within higher ones, including support for the claim that such distributions reflect natural evolutionary processes and not simply random agglomerations of low-ranked taxa… Sets of taxa ranked as orders, classes, and phyla exhibit different temporal and spatial patterns from those shown by species, genera, and families… a contrast seemingly consistent with suggestions that higher taxa reflect the origin of body plans while lower taxa reflect the dynamics of species formation and sorting, and that such differences may reflect different types of developmental changes. (Jablonski 2017a, p. 429)

In this same vein, Erwin (2000, 2010) argues that a hierarchy theory of evolution is grounded in the observation of morphological gaps in the fossil record, with advancements in Evo-Devo pointing to underlying mechanisms responsible for these discontinuities, particularly those related to study of the morphological innovation: "When these results are combined with new insights from the comparative study of animal development across different clades ('evo-devo') and the nature of evolutionary novelties, they do pose a challenge to microevolution as a sufficient explanation of macroevolutionary pattern" (Erwin 2010, p. 186).

Indeed, it is at this point that the contribution of the H-Evo-Devo theory to distinguish between the levels is recognized. According to this model, the nature of biological entities reveals discontinuities in the distribution of variation. Here, it is important to recognize that at least two kinds of variation can indicate a discrete distinction between the levels:

- *Simple characters*, such as color, size, and proportion.
- *Complex characters*, represented as body plans at different scales.

Considering this division, taxonomy as a tool can now be employed. As Devitt (2023) states, Linnaean categories approximately capture the levels in the taxonomic hierarchy. For instance, the conclusion of Chap. 7 is that subspecies are varieties within a species that possess simple characters. Meanwhile, species are recognized by possessing complex characters, allowing them to be conceived as the least inclusive body plan in a hierarchy of body plans. The conclusion of Chap. 8 is that higher taxa are characterized by representing more inclusive body plans than the species/genus body plan. Furthermore, in all these cases, units of evolution can be distinguished from one another through their morphological gaps. For instance, subspecies are differentiated by small discontinuous differences in simple characters, species exhibit a significant degree of discontinuity in complex characters, and higher taxa display an even greater degree of discontinuity in complex characters.

The H-Evo-Devo theory holds that the difference between microevolution and macroevolution is also justified by the distinct mechanisms involved in the origin, stability, and extinction of the units of evolution. Regarding the origin, as I have mentioned throughout this book, subspecies arise through Darwinian mechanisms or the dispersal model. In contrast, the origin of species and higher taxa occurs through supraDarwinian mechanisms or the "hopeful monster" model. Although the mechanisms underlying the cohesion of units of evolution are relatively similar, the distinction lies in the kind of variation. For instance, developmental constraints acting on simple characters maintain the cohesion of the subspecies, whereas those acting on complex characters maintain the cohesion of both species and higher taxa. As for extinction, the H-Evo-Devo Theory delineates two levels through hierarchical effects. For instance, subspecies extinction results from habitat alteration and changes in their population structure, whereas the extinction of species and higher taxa is attributed to the "loss of parts" (either of subspecies or species) and mass extinction events.

Additionally, while the H-Evo-Devo Theory clearly distinguishes between microevolution and macroevolution, the latter can also be subdivided by considering the magnitude of processes occurring at higher levels of the hierarchy. In other words, even though species and higher taxa share the same mechanisms of origin, stability, and extinction, their differing degrees of discontinuity justify their separation. Thus, the term "megaevolution" is revived within this new perspective to exclusively address patterns and processes at the level of higher taxa. Furthermore, the H-Evo-Devo Theory introduces an additional level, "modular-evolution," which—although detached from the taxonomic scale—serves by indicating the introduction of morphological characters and the changes occurring to them.[17]

In summary, the H-Evo-Devo Theory provides a theoretical framework for separating evolution into different levels. Nonetheless, the following must be considered: First, while the model of microevolution is not entirely novel, it attempts to integrate contemporary Evo-Devo knowledge to explain the origin of simple characters. Second, the most contentious aspect is that the H-Evo-Devo Theory embraces

[17] For a more detailed explanation of the processes occurring at each of these levels, see Chaps. 11 and 12.

a saltational evolution theory to account for the origin of body plans. The problem lies in that explaining body plans through the lens of gradualism poses more problems than solutions (see Chap. 9, Sect. 9.2). Also, traditional Darwinism—as the prevailing paradigm—has impeded the consideration of "unconventional" evolutionary theories. However, as forthcoming chapters will reveal, the hopeful monster model is very compelling, as it integrates different mechanisms and scenarios (both intrinsic and extrinsic) based on multidisciplinary biological knowledge (such as developmental biology, paleontology, biogeography, systematics, among others). After all, Filipchenko and Goldschmidt were right; there is no other way; this is the best horse to bet on (see Table 10.2).

10.3 Conclusions

This chapter critically examined some fundamental principles of the Modern Synthesis. First, I argued that models of geographic speciation are too ambiguous and inadequate, particularly highlighting shortcomings in the theoretical underpinning of allopatric speciation. Thus, I propose that the best way to explain microevolutionary divergence is through the combined process of variation and adaptation to local environments. Also, it is essential to recognize that divergence occurs hierarchically.

Second, addressing the inquiry into whether macroevolution is decoupled from microevolution, I asserted the inadequacy of the extrapolationist premise, which hinders a comprehensive understanding of the hierarchical nature of evolution. To establish a qualitative distinction between the levels, it is necessary to propose different mechanisms for each level and recognize that they operate on distinct kinds of variation (e.g., simple or complex characters). The issue is that current theories using the term "macroevolution" only present patterns at higher levels, yet they rely on the extrapolationist premise in order to explain those patterns. So, the only viable option is to embrace "unconventional" evolutionary theories that can account for the emergence of observed discontinuities between levels. Certainly, the H-Evo-Devo Theory fulfills this goal by establishing a qualitative distinction between levels of evolution. It proposes a framework that addresses divergence, stability, and extinction at different hierarchical levels, while also offering precise definitions for each level, identifying specific mechanisms and patterns, thereby completely challenging the extrapolationist premise.

This analysis makes it evident that traditional concepts of evolution have posed significant challenges to the development of a hierarchy theory of evolution. However, the H-Evo-Devo Theory presents a promising alternative, not only providing the possibility to investigate the true causes of evolutionary divergence but also to understand the mechanisms involved at each level, thus facilitating a better understanding of the hierarchical framework of the phylogenetic tree.

Table 10.2 *Analysis aimed at addressing the question: "Is macroevolution more than uninterrupted rounds of microevolution?"* Three possible alternatives are considered. Note the series of challenges that the traditional options (1 and 2) face. However, based on the findings of this study, option 3 appears to be the most favorable choice

Is macroevolution more than uninterrupted rounds of microevolution?			
	Option 1. The reductionist approach	Option 2. The intermediate approach	Option 3. The irreducible approach
Description	All macroevolutionary phenomena are explained simply as the sum of microevolutionary events	There are no self-governing mechanisms for macroevolution, but all the macroevolutionary studies cannot be reduced to microevolutionary studies. Both microevolution and macroevolution complement each other within the same theoretical framework	Without question, macroevolution has distinct patterns and processes irreducible to microevolutionary explanations. This perspective advocates for a hierarchical view wherein "micro-," "macro-," and even "megaevolution" are established as independent fields of inquiry
Microevolution	Short timescale	Below species level Short timescale	Subspecies level
Macroevolution	Large timescale	At and above species level Large timescale	Species level
Megaevolution	Rarely used	Rarely used	Higher taxa level
Evolutionary framework	Modern Synthesis	Punctuated equilibrium, Hierarchy Theory of Evolution, species selection	H-Evo-Devo Theory
Points in favor	The observations of gradual rates of change in extant organisms might be reliable. Gradual and cumulative changes can produce significant morphological differentiation over time. There are no mechanisms for macroevolution other than those observed in microevolutionary studies	Punctuated equilibrium and species selection have made visible the patterns of the species at the geological scale	Given that it proposes a qualitative distinction between the levels, it supports a hierarchy theory of evolution. It standardizes the terms micro-, macro-, and megaevolution to indicate the level of study of the units of evolution. It encourages a wider range of research programs to study evolution at different levels

(continued)

Table 10.2 (continued)

Is macroevolution more than uninterrupted rounds of microevolution?			
	Option 1. The reductionist approach	Option 2. The intermediate approach	Option 3. The irreducible approach
Points against	It is uncertain whether complex characters are explained in the same manner as the origin of simple characters. Biologists place confidence in evolutionary uniformitarianism because they trust the validity of the Modern Synthesis assumptions, yet these can be easily challenged. So far, there is no convincing evidence that the extrapolationist premise is true. Although the genetic and embryological mechanisms underlying microevolution are the same as those of macroevolution, other factors can make a difference, e.g., small genetic changes can result in significant morphological gaps	These theories do not consider mechanisms for macroevolution other than those of microevolution. Macroevolutionary patterns are explained through the accumulation of small variations over geological time. In this sense, it does not fully address truly macroevolution. It does not consider evolution above the species level, or it rarely does	In order to avoid the extrapolationist premise, a saltational evolution theory must be accepted. However, from a contemporary scientific standpoint, there is no compelling reason to dismiss it
Probability of success	Unlikely	Unlikely	Likely

References

Adams MB (2021) Little evolution, BIG evolution: rethinking the history of Darwinism, population genetics, and the "synthesis". In: Delisle RG (ed) Natural selection: revisiting its explanatory role in evolutionary biology. Springer, Cham, pp 195–229

Arnold SJ, Pfrender ME, Jones AG (2001) The adaptive landscape as a conceptual bridge between micro- and macroevolution. Genetica 112–113:9–32

Arthur W (2003) Micro-, macro-, and megaevolution. In: Hall BK, Olson WM (eds) Keywords and concepts in evolutionary developmental biology. Harvard University Press, Cambridge MA, pp 249–260

Arthur W (2008) Conflicting hypotheses on the nature of mega-evolution. In: Minelli A, Fusco G (eds) Evolving pathways: key themes in evolutionary developmental biology. Cambridge University Press, New York, pp 50–61

Arthur W (2021) Understanding evo-devo. Cambridge University Press, Cambridge
Barton NH, Charlesworth B (1984) Genetic revolutions, founder effects, and speciation. Ann Rev Ecol Syst 15:133–164
Bateman RM, DiMichele WA (2002) Generating and filtering major phenotypic novelties: neoGoldschmidtian saltation revisited. In: Cronk QCB, Bateman RM, Hawkins JA (eds) Developmental genetics and plant evolution. Taylor and Francis, London, pp 109–159
Beheregaray LB, Cooke GM, Chao NL, Landguth EL (2015) Ecological speciation in the tropics: insights from comparative genetic studies in Amazonia. Front Genet 5:477. https://doi.org/10.3389/fgene.2014.00477
Berlocher SH, Feder JL (2002) Sympatric speciation in phytophagous insects: moving beyond controversy? Annu Rev Entomol 47:773–815
Bird CE, Fernandez-Silva I, Skillings DJ, Toonen RJ (2012) Sympatric speciation in the post "Modern Synthesis" era of evolutionary biology. Evol Biol 39:158–180
Blakeslee AMH, Haram LE, Altman I, Kennedy K, Ruiz GM, Miller AW (2019) Founder effects and species introductions: a host versus parasite perspective. Evol Appl 13:559–574
Bokma F (2015) Evolution as a largely autonomous process. In: Serrelli E, Gontier N (eds) Macroevolution: explanation, interpretation and evidence. Springer, Cham, pp 87–112
Bolnick DI, Fitzpatrick BM (2007) Sympatric speciation: models and empirical evidence. Annu Rev Ecol Evol Syst 38:459–487
Bromham L (2016) Testing hypotheses in macroevolution. Stud Hist Philos Sci A 55:47–59
Butlin RK, Galindo J, Grahame JW (2008) Sympatric, parapatric or allopatric: the most important way to classify speciation? Philos Trans R Soc B 363:2997–3007
Butlin RK et al (2012) What do we need to know about speciation? Trends Ecol Evol 27:27–39
Callahan HS (2002) Microevolution and macroevolution: introduction. In: Encyclopedia of life science. John Wiley and Sons, Chichester, pp 1–6. https://doi.org/10.1038/npg.els.0001717
Coyne JA, Orr HA (2004) Speciation. Sinauer Associates, Sunderland MA
Crow KD, Munehara H, Bernardi G (2010) Sympatric speciation in a genus of marine reef fishes. Mol Ecol 19:2089–2105
Czekanski-Moir JE, Rundell RJ (2019) The ecology of nonecological speciation and nonadaptive radiations. Trends Ecol Evol 34:400–415
Damuth J (1985) Selection among "species": a formulation in terms of natural functional units. Evolution 39:1132–1146
Darwin C (1859) On the origin of species by means of natural selection, or preservation of favored races in the struggle for life. John Murray, London
Dawkins R (1982) The extended phenotype: the long reach of the gene. Oxford University Press, New York
Devitt M (2023) The minimal role of the higher categories in biology. Biol Philos 38. https://doi.org/10.1007/s10539-022-09888-5
Dietrich MR (2010) Microevolution and macroevolution are governed by the same processes. In: Ayala FJ, Arp R (eds) Contemporary debates in philosophy of biology. Wiley-Blackwell, Oxford, pp 169–179
Diogo R, Guinard G, Diaz RE Jr (2017) Dinosaurs, chameleons, humans, and evo-devo path: linking Étienne Geoffroy's teratology, Waddington's homeorhesis, Alberch's logic of "monsters," and Goldschmidt hopeful "monsters.". J Exp Zool B (Mol Dev Evo) 328:207–229
Dobzhansky T (1937) Genetics and the origin of species. Columbia University Press, New York
Dowle EJ, Morgan-Richards M, Trewick SA (2014) Morphological differentiation despite gene flow in an endangered grasshopper. BMC Evol Biol 14:216
Elmer KR, Meyer A (2010) Sympatric speciation without borders? Mol Ecol 19:1991–1993
Erwin DH (2000) Macroevolution is more than repeated rounds of microevolution. Evol Dev 2:78–84
Erwin DH (2010) Microevolution and macroevolution are not governed by the same processes. In: Ayala FJ, Arp R (eds) Contemporary debates in philosophy of biology. Wiley-Blackwell, Oxford, pp 180–193
Erwin DH (2011) Evolutionary uniformitarianism. Dev Biol 357:27–34

Fernández-Mazueco M, Glover BJ (2017) The evo-devo of plant speciation. Nat Ecol Evol 1. https://doi.org/10.1038/s41559-017-0110
Fitzpatrick BM, Fordyce JA, Gavrilets S (2008) What, if anything, is sympatric speciation? J Evol Biol 21:1452–1459. https://doi.org/10.1111/j.1420-9101.2008.01611.x
Fitzpatrick BM, Fordyce JA, Gavrilets S (2009) Pattern, process and geographic modes of speciation. J Evol Biol 22:2342–2347
Frazzetta TH (2012) Flatfishes, turtles, and bolyerine snakes: evolution by small steps or large, or both? Evol Biol 39:30–60
Freedman AH, Thomassen HA, Buermann W, Smith TB (2010) Genomic signals of diversification along ecological gradients in a tropical lizard. Mol Ecol 19:3773–3788
Freedman AH, Harrigan RJ, Zhen Y, Hamilton AM, Smith TB (2023) Evidence for ecotone speciation across an African rainforest-savanna gradient. Mol Ecol 32:2287–2300
Futuyma DJ (2008) Sympatric speciation: norm or exception? In: Tilmon KJ (ed) Specialization, speciation, and radiation: the evolutionary biology of herbivorous insects. University of California Press, Berkeley, pp 136–148
Futuyma D (2015) Can modern evolutionary theory explain macroevolution? In: Serrelli E, Gontier N (eds) Macroevolution: explanation, interpretation and evidence. Springer, Cham, pp 29–85
Garant D, Forde SE, Hendry AP (2007) The multifarious effects of dispersal and gene flow on contemporary adaptation. Funct Ecol 21:434–443
Gavrilets S (2003) Perspective: models of speciation: what have we learned in 40 years? Evolution 57:2197–2215
Gillespie RG, Brewer MS, Roderick GK (2017) Ancient biogeography of generalist predators on remote oceanic islands. J Biogeogr 44:1098–1109
Gingerich PD (1987) Evolution and the fossil record: patterns, rates, and processes. Can J Zool 65:1053–1060
Gingerich PD (2019) Rates of evolution: a quantitative synthesis. Cambridge University Press, Cambridge
Goldschmidt RB (1940) The material basis of evolution. Yale University Press, New Haven
Gontier N (2015) Uniting micro- with macroevolution into an extended synthesis: reintegrating life's natural history into evolution studies. In: Serrelli E, Gontier N (eds) Macroevolution: explanation, interpretation and evidence. Springer, Cham, pp 227–275
Gould SJ (2002) The structure of evolutionary theory. The Belknap Press of Harvard University Press, Cambridge MA
Grantham TA (1995) Hierarchical approaches to macroevolution: recent work on species selection and the "effect hypothesis". Annu Rev Ecol Syst 26:301–321
Grantham TA (2007) Is macroevolution more than successive rounds of microevolution? Palaeontology 50:75–85
Haller BC, Hendry AP (2014) Solving the paradox of stasis: squashed stabilizing selection and the limits of detection. Evolution 68:483–500
Hautmann M (2020) What is macroevolution. Palaeontology 63:1–11
Hendry AP (2009) Ecological speciation! Or the lack thereof? Can J Fish Aquat Sci 66:1383–1398
Hendry AP, Kinnison MT (2001) An introduction to microevolution: rate, pattern, process. Genetica 112–113:1–8
Ho MW, Saunders PT (1979) Beyond neo-Darwinism–an epigenetic approach to evolution. J Theor Biol 78:573–591
Huneman P (2017) Macroevolution and microevolution: issues of time scale in evolutionary biology. In: Bouton C, Huneman P (eds) Time of nature and the nature of time: philosophical perspectives of time in natural sciences. Springer, Cham, pp 315–357
Jablonski D (2007) Scale and hierarchy in macroevolution. Palaeontology 50:87–109
Jablonski D (2017a) Approaches to macroevolution: 1. General concepts and origin of variation. Evol Biol 44:427–450
Jablonski D (2017b) Approaches to macroevolution: 2. Sorting of variation, some overarching issues, and general conclusions. Evol Biol 44:451–475
Jiggins CD (2006) Sympatric speciation: why the controversy? Curr Biol 16:333–334

Jordan DO, Mendes CB, Andrade SCS (2024) Brazilian Northeast coast *Prosorhochmus belizeanus* (Hoplonemertea: Nemertea) populations analyses reveal colonization by founder effect. Mar Biodivers 54. https://doi-org.pbidi.unam.mx:2443/10.1007/s12526-024-01449-1
Jousselin E, Cruaud A, Genson G, Chevenet F, Foottit RG, Coeur d'acier A (2013) Is ecological speciation a major trend in aphids? Insights from a molecular phylogeny of the conifer-feeding genus *Cinara*. Front Zool 10:56
Kampourakis K (2014) Understanding evolution. Cambridge University Press, New York
Kaňuch P, Berggren Å, Cassel-Lundhagen A (2021) A clue to invasion success: genetic diversity quickly rebounds after introduction bottlenecks. Biol Invasions 23:1141–1156
Kearney M, Lieberman BS, Strotz LC (2024) Tangled banks, braided rivers, and complex hierarchies: beyond microevolution and macroevolution. J Evol Biol 37:1402–1412
Kisel Y, Barraclough TG (2010) Speciation has a spatial scale that depends on levels of gene flow. Am Nat 175:316–334
Kolbe JJ, Leal M, Schoener TW, Spiller DA, Losos JB (2012) Founder effects persist despite adaptive differentiation: a field experiment with lizards. Science 335:1086–1089
Kremer A, Ronce O, Robledo-Arnuncio JJ, Guillaume F, Bohrer G, Nathan R, Bridle JR, Gomulkiewicz R, Klein EK, Ritland K, Kuparinen A, Gerber S, Schueler S (2012) Long-distance gene flow and adaptation of forest trees to rapid climate change. Ecol Lett 15:378–392
Lavrenchenko LA (2011) Testing of alternative hypotheses for speciation: a case study on ground vertebrates in tropical mountains. Biol Bull 38:551–557
Lee, MSY (2003) Species concepts and species reality: salvaging a Linnaean rank. J Evol Biol 16:179–188
Levinton JS (2001) Genetics, paleontology, and macroevolution, 2nd edn. Cambridge University Press, New York
Levinton JS (2012) Macroevolution: overview. In: Encyclopedia of life science. John Wiley and Sons, Chichester, pp 1–9. https://doi.org/10.1038/npg.els.0001771
Lieberman BS, Eldredge N (2014) What is punctuated equilibrium? What is macroevolution? a response to Pennell et al. Trends Ecol Evol 29:185–186
Liu H, Ye Z, Liang J, Wang S, Bu W (2017) Niche divergence of two closely related *Carbula* species (Insecta: Hemiptera: Pentatomidae) despite the presence of a hybrid zone. Ecol Entomol 43:204–214
Løvtrop S (1981) Macroevolution and punctuated equilibria. Syst Zool 30:498–500
Mallet J, Meyer A, Nosil P, Feder JL (2009) Space, sympatry and speciation. J Evol Biol 22:2332–2341
Matute DR (2013) The role of founder effects on the evolution of reproductive isolation. J Evol Biol 26:2299–2311
Mayr E (1942) Systematics and the origin of species. Columbia University Press, New York
Mayr E (1963) Animals species and evolution. Harvard University Press, Cambridge
Mayr E (1980) Prologue: some thoughts on the history of the evolutionary synthesis. In: Mayr E, Provine W (eds) The evolutionary synthesis: perspectives on the unification of biology. Harvard University Press, Cambridge, pp 1–48
Miller W III (2015) Unification of macroevolutionary theory: biologic hierarchies, consonance, and the possibility of connecting the dots. In: Eldredge N, Pievani T, Serrelli E, Tëmkin I (eds) Evolutionary theory: a hierarchical perspective. University of Chicago Press, Chicago, pp 243–259
Minelli A, Fusco G (2012) On the evolutionary developmental biology of speciation. Evol Biol 39:242–254
Moran EV, Reid A, Levine JM (2017) Population genetics and adaptation to climate along elevation gradients in invasive *Solidago canadensis*. PLoS One 12:e0185539
Newman SA (2015) Development and evolution: the physics connection. In: Love AC (ed) Conceptual change in biology: scientific and philosophical perspectives on evolution and development. Springer, Dordrecht, pp 421–440

Nicoglou A (2015) Phenotypic plasticity: from microevolution to macroevolution. In: Heams T, Huneman P, Lecointre G, Silberstein M (eds) Handbook of evolutionary thinking in the sciences. Springer, Dordrecht, pp 285–318
Nolte AW, Tautz D (2010) Understanding the onset of hybrid speciation. Trends Genet 26:54–58
North A, Pennanen J, Ovaskainen O, Laine A (2010) Local adaptation in a changing world: the roles of gene-flow, mutation, and sexual reproduction. Evolution 65:79–89
Nosil P (2012) Ecological speciation. Oxford University Press, Oxford
Nosil P, Harmon L (2009) Niche dimensionality and ecological speciation. In: Butlin RK, Bridle JR, Schluter D (eds) Speciation and patterns of diversity. Cambridge University Press, New York, pp 127–154
Nunes MDS, Arif S, Schlötterer C, McGregor AP (2013) A perspective on micro-evo-devo: progress and potential. Genetics 195:625–634
Ochoa C (2021) Inertia, trend, and momentum reconsidered: G. G. Simpson—an orthogeneticist? In: Delisle RG (ed) Natural selection: revisiting its explanatory role in evolutionary biology. Springer, Cham, pp 261–290
Ochoa C (2024) Deconstructing the extended evolutionary synthesis: do we need a new theory of evolution? In: Delisle RG, Esposito M, Ceccarelli D (eds) Unity and disunity in evolutionary biology: deconstructing Darwinism. Spinger, Cham, pp 403–439
Osborne OG, Batstone TE, Hiscock SJ, Filatov DA (2013) Rapid speciation with gene flow following the formation of Mt. Etna. Genome Biol Evol 5:1704–1715
Palacios C, García-R S, Parra JL, Cuervo AM, Stiles FG, McCormack JE, Cadena CD (2019) Shallow genetic divergence and distinct phenotypic differences between two Andean hummingbirds: speciation with gene flow? Auk 136:1–21
Pennell MW, Harmon LJ, Uyeda JC (2014) Speciation is unlikely to drive divergence rates. Trends Ecol Evol 29:72–73
Pergams ORW, Ashley MV (2001) Microevolution in Island rodents. Genetica 112–113:245–256
Pievani T (2016) Hierarchy theory and the extended synthesis debate. In: Eldredge N, Pievani T, Serrelli E, Tëmkin I (eds) Evolutionary theory: a hierarchical perspective. University of Chicago Press, Chicago, pp 351–364
Plutynski A (2008) Speciation and macroevolution. In: Sarkar S, Plutynski A (eds) A companion to the philosophy of biology. Blackwell Publishing, Malden, MA, p 169
Plutynski A (2019) Speciation post synthesis: 1960–2000. J Hist Biol 52:569–596
Podos J, Schroeder KM (2024) Ecological speciation in Darwin's finches: ghosts of finches future. Science 386:211–217
Potkamp G, Fransen CHJM (2019) Speciation with gene flow in marine systems. Contrib Zool 88:133–172
Rashkovetsky E, Frenkel Z, Michalak P, Korol A (2015) Sympatric differentiation and speciation: insights from Drosophila studies. In: Pontarotti P (ed) Evolutionary biology: biodiversification from genotype to phenotype. Springer, Switzerland, pp 107–140
Rolland J, Henao-Diaz LF, Doebeli M, Germain R, Harmon LJ, Knowles LL, Liow LH, Mank JE, Machac A, Otto SP, Pennell M, Salamin N, Silvestro D, Sugawara M, Uyeda J, Wagner CE, Schluter D (2023) Conceptual and empirical bridges between micro- and macroevolution. Nat Ecol Evol 7:1181–1193
Samuk K, Owens GL, Delmore KE, Miller SE, Rennison DJ, Schluter D (2017) Gene flow and selection interact to promote adaptive divergence in regions of low recombination. Mol Ecol 26:4378–4390
Santos J, Pascual M, Simões P, Fragata I, Lima M, Kellen B, Santos M, Marques A, Rose MR, Matos M (2012) From nature to the laboratory: the impact of founder effects on adaptation. J Evol Biol 25:2607–2622
Saupe EE, Myers CE (2021) Macroevolution. In: Nuño de la Rosa L, Müller GB (eds) Evolutionary developmental biology. Springer, Cham, pp 149–167
Schluter D (2001) Ecology and the origin of species. Trends Ecol Evol 16:372–380
Schluter D (2009) Evidence for ecological speciation and its alternative. Science 323:737–741

Schluter D (2024) Variable success in linking micro- and macroevolution. Evol J Linn Soc 3. https://doi.org/10.1093/evolinnean/kzae016
Schneider C, Moritz C (1999) Rainforest refugia and evolution in Australia's wet tropics. Proc R Soc Lond B 266:191–196
Schneider C, Smith TB, Larison B, Moritz C (1999) A test of alternative models of diversification in tropical rainforests: ecological gradients vs. rainforest refugia. Proc Natl Acad Sci USA 96:13869–13873
Sexton JP, Hangartner S, Hoffmann AA (2014) Genetic isolation by environment or distance: which pattern of gene flow is most common? Evolution 68:1–15
Shpak M (2019) Macroevolution. In: Fath B (ed) Encyclopedia of ecology. Elsevier, Second edition, pp 140–145
Simons A (2002) The continuity of microevolution and macroevolution. J Evol Biol 15:688–701
Simpson GG (1944) Tempo and mode in evolution. Columbia University Press, New York
Smadja CM, Butlin RK (2011) A framework for comparing processes of speciation in the presence of gene flow. Mol Ecol 20:5123–5140
Smith HH (1965) More evolutionary terms. Syst Zool 14:57–58
Smith TB, Wayne RK, Girman DJ, Bruford MW (1997) A role for ecotones in generating rainforest biodiversity. Science 276:1855–1857
Smith TB, Schneider CJ, Holder K (2001) Refugial isolation versus ecological gradients. Genetica 112–113:383–398
Smith TB, Calsbeek R, Wayne RK, Holder KH, Pires D, Bardeleben C (2005) Testing alternative mechanisms of evolutionary divergence in an African rain forest passerine bird. J Evol Biol 18:257–268
Smith TB, Thomassen HA, Freedman AH, Sehgal RNM, Buermann W, Saatchi S, Pollinger J, Milá B, Pires D, Valkiūnas G, Wayne RK (2011) Patterns of divergence in the olive sunbird *Cyanomitra olivacea* (Aves: Nectariniidae) across the African rainforest–savanna ecotone. Biol J Linn Soc 103:821–835
Sobel JM, Chen GF, Watt LR, Schemske DW (2009) The biology of speciation. Evolution 64:295–315
Stigall AL (2015) Speciation: expanding the role of biogeography and niche breadth in macroevolutionary theory. In: Serrelli E, Gontier N (eds) Macroevolution: explanation, interpretation and evidence. Springer, Switzerland, pp 301–327
Sun N, Yang L, Tian F, Zeng H, He Z, Zhao K, Wang C, Meng M, Feng C, Fang C, Lv W, Bo J, Tang Y, Gan X, Peng Z, Chen Y, He S (2022) Sympatric or micro-allopatric speciation in a glacial lake? Genomic islands support neither. Natl Sci Rev. https://doi.org/10.1093/nsr/nwac291
Theißen G (2006) The proper place of hopeful monsters in evolutionary biology. Theory Biosci 124:349–369
Toman J, Flegr J (2024) Macro and micro evolution. In: Shackelford TK (ed) Encyclopedia of sexual psychology and behavior. Springer, Cham. https://doi.org/10.1007/978-3-031-08956-5_1365-1
Travis J, Reznick DN (2009) Adaptation. In: Ruse M, Travis J (eds) Evolution: the first four billion years. Harvard University Press, Cambridge (MA), pp 105–131
Tsuboi M, Sztepanacz J, De Lisle S, Voje KL, Grabowski M, Hopkins MJ, Porto A, Balk M, Pontarp M, Rossoni D, Hildesheim LS, Horta-Lacueva QJ, Hohmann N, Holstad A, Lürig M, Milocco L, Nilén S, Passarotto A, Svensson EI, Villegas C, Winslott E, Liow LH, Hunt G, Love AC, Houle D (2024) The paradox of predictability provides a bridge between micro- and macroevolution. J Evol Biol 37:1413–1432
Vislobokova IA (2017) The concept of macroevolution in view of modern data. Paleontol J 51:799–898
Walsh HE, Jones IL, Friesen VL (2005) A test of founder effect speciation using multiple loci in the auklets (Aethia spp.). Genetics 171:1885–1894
Wang X, He Z, Shi S, Wu C-I (2020) Genes and speciation: is it time to abandon the biological species concept? Natl Sci Rev 7:1387–1397

Wu S, Wang Y, Wang Z, Shrestha N, Liu J (2022) Species divergence with gene flow and hybrid speciation on the Qinghai-Tibet Plateau. New Phytol 234:392–404

Wu Y, Linan AG, Hoban S, Hipp AL, Ricklefs RE (2024) Divergent ecological selection maintains species boundaries despite gene flow in a rare endemic tree, *Quercus acerifolia* (maple-leaf oak). J Hered 115:575–587

Yeung CKL, Tsai P-W, Chesser RT, Lin R-C, Yao C-T, Tian X-H, Li S-H (2011) Testing founder effect speciation: divergence population genetics of the spoonbills *Platalea regia* and *Pl. minor* (Threskiornithidae, Aves). Mol Biol Evol 28:473–482

Zhang X, Rayner JG, Blaxter M, Bailey NW (2021) Rapid parallel adaptation despite gene flow in silent crickets. Nat Commun 12:50. https://doi.org/10.1038/s41467-020-20263-4

Zhao W, Gao J, Hall D, Andersson BA, Bruxaux J, Tomlinson KW, Drouzas AD, Suyama Y, Wang XR (2024) Evolutionary radiation of the Eurasian *Pinus* species under pervasive gene flow. New Phytol 242:2353–2368

Part V

The Levels of Evolution: Biological Processes

Levels of Evolution (I): Microevolution and Macroevolution

11

Abstract

According to the H-Evo-Devo Theory, macroevolution is decoupled from microevolution. Microevolution belongs to the subspecies level, whose character states correspond to simple characters, e.g., color, pattern, size, and proportion. In contrast, macroevolution belongs to the species level, whose character states correspond to complex characters or body plans. This chapter explores the biological processes surrounding the origin, stability, and extinction of both subspecies and species. Concerning their origin, subspecies arise through Darwinian mechanisms (dispersal models), whereas species emerge through supraDarwinian mechanisms (prospecies model). Regarding stasis, the stability of both subspecies and species is attributed to developmental constraints acting on simple and complex characters, respectively. Finally, differences in extinction are explained by a hierarchical structure: Subspecies extinction results from factors altering their local habitat and their population structure, while species extinction arises from the loss of parts and mass extinction events. Essentially, this chapter marks the first modern effort to distinctly delineate evolution into at least two levels: microevolution and macroevolution.

Keywords

H-Evo-Devo Theory · Microevolution · Macroevolution · Species · Subspecies · Subspeciation · Speciation · Stasis · Substasis · Species extinction · Subextinction · Darwinian mechanisms · SupraDarwinian mechanisms · Prospecies · Hopeful monster · Saltational evolution · Phenotypic plasticity · Behavioral choice · Niche construction · Natural selection · Developmental constraints · Developmental bias · Background extinction · Mass extinction

C. Ochoa, *Hierarchical Evolutionary-Developmental Theory*, Evolutionary Biology – New Perspectives on Its Development 9,
https://doi.org/10.1007/978-3-031-88723-9_11

Following the framework of the H-Evo-Devo Theory, this chapter shows the distinct processes that distinguish microevolution and macroevolution. Specifically, it delves into the mechanisms underlying the origin, stability, and extinction of both subspecies and species. Section 11.1 examines the processes related to microevolution patterns, i.e., the subspecies level. It begins by outlining *subspeciation*, which occurs through Darwinian mechanisms (speciation mode). Then, it analyzes *substasis*, whose explanation is rooted in developmental constraints. Finally, it addresses *subextinction*, caused by factors that alter their local habitat and their population structure. Section 11.2 explores the processes related to the pattern of macroevolution, i.e., the species level. Initially, it describes *speciation*, which occurs through supraDarwinian mechanisms (prospecies model). Then, it discusses *specific stasis*, which is explained by developmental constraints. Finally, it analyzes *specific extinction*, which is attributed to the loss of parts and mass extinction events.

11.1 Microevolution (Subspecies Level)

Needless to say, numerous microevolutionary models are associated with processes elucidated by population genetics. Likewise, to understand the divergence process, most studies employ solely genetic mechanisms within ecological contexts, without integrating embryonic developmental processes (Minelli and Fusco 2012).[1] However, some authors have emphasized the importance of Evo-Devo studies applied at the intraspecific level, referred to as Micro-Evo-Devo (Nunes et al. 2013; Fernández-Mazuecos and Glover 2017; Houle and Jones 2021). Given the current advancements in science, it is evident that the mechanisms operating at this level need to be reformulated, especially because classical models focus on a gene-centered perspective rather than an updated model that would adopt an organism-centered perspective. Therefore, I believe that the H-Evo-Devo Theory represents the kind of model aimed at achieving this objective.[2]

According to the H-Evo-Devo Theory, microevolution represents the lowest level of the hierarchy, which studies the processes related to the patterns of origin, stability, and extinction of subspecies. By subspecies, I refer to those groups that share a common body plan at the species level but are separated from each other by small morphological gaps in simple characters as well as by the fact that they live in different environments. Under this view, subspecies are not incipient species or models representing the origin of species through allopatric speciation. Instead, they are groups that diversify to adapt the species to different environments, thereby avoiding species extinction.

[1] For example, regarding the critique of traditional geographic speciation modes, particularly allopatric speciation, see Chap. 10, Sect. 10.1.

[2] Another model that supposedly integrates an organism-centered perspective is the Extended Evolutionary Synthesis. However, as I have demonstrated in another publication (Ochoa 2024), the structure of this framework exhibits several shortcomings.

Another important point is that, although the discontinuities among subspecies are not large enough to represent a body plan, they still allow for their identification as discrete entities, so they are real.[3] Furthermore, given that the character states assessed for subspecies diagnosis are simple characters (e.g., color, pattern, size, and proportion)—unlike complex characters, which are associated with differences at the body plan level—the H-Evo-Devo Theory posits that there is a distinct separation between microevolution and macroevolution concerning origin, stability, and extinction.[4]

Therefore, this section briefly addresses the mechanisms underlying the three evolutionary modes of subspecies. The first evolutionary mode at this level is *subspeciation*, whose involved mechanisms are those provided by the dispersal model or *Darwinian mechanisms*. Although this model is far from new, it can be complemented by advances in Evo-Devo. The second mode is *substasis*, whose mechanisms involve the well-known developmental constraints working on the fixation of simple characters. Finally, the third evolutionary mode is *subextinction*, whose causes are attributed to factors altering their local habitat and their population structure.

11.1.1 Subspeciation

According to the H-Evo-Devo Theory, the *subspeciation* mode focuses on the origin of subspecies. Following Fig. 11.1, imagine a body plan (species) consisting of a small number of individuals (a). Subsequently, this population gives rise to additional groups within the same body plan, occupying different positions within the morphospace (b). Note that these groups cannot vary beyond the boundaries of the species: subspecies only reflect variations within the species body plan; they are groups that adapt and protect species from extinction.

This microevolutionary model of divergence is explained by Darwinian mechanisms. But what do I mean by Darwinian mechanisms? As widely known among biologists, Charles Darwin (1859) explained evolutionary divergence through the dispersal model. According to his view, species arise within a specific geographic area through adaptation by natural selection (center of origin). As the population expands, offspring disperse and colonize new environments, adapting and thus producing evolutionary divergence. While this model does not exclude the possibility that divergence can occur within the same place—e.g., during an environmental shift—it emphasizes the significant role dispersal plays in evolutionary change. It is important to highlight that, although this model is far from new, it is enriched by advances from Evo-Devo, particularly those that address factors related to organism-environment interaction. As discussed elsewhere in this book, when considering the organism and its interaction with the environment, reference is made not only to the mechanism of natural selection but also to processes such as phenotypic

[3] For a detailed examination of subspecies as units of evolution, see Chap. 7, Sect. 7.1.

[4] See also Chap. 10, Sect. 10.2.2.

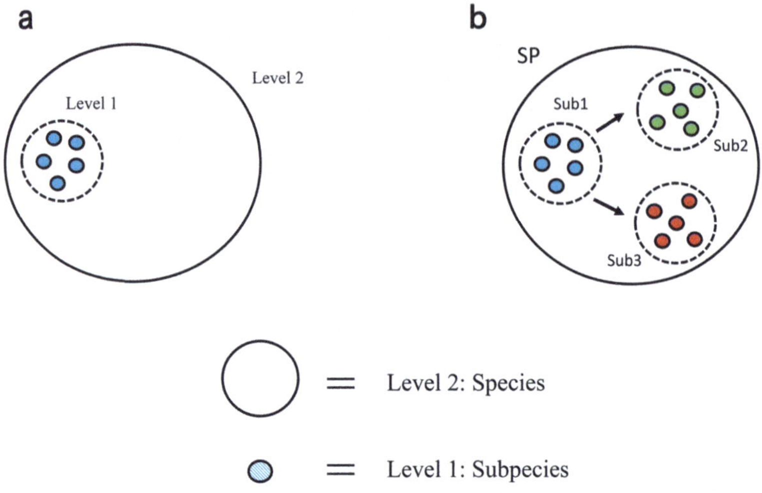

Fig. 11.1 *Subspeciation depicted in a nested compositional hierarchy, according to the H-Evo-Devo Theory.* The area inside each circle represents the morphospace (though it might also represent geographic spatial distances), and each circle indicates a form located within the morphospace. The size of the circle denotes its level: the small circle corresponds to Level 1, while the large circle corresponds to Level 2. The different colors of the small circles illustrate the gaps between subspecies. The dotted circles inside the large circle represent the subspecies boundaries delineated by developmental constraints, while the solid outline that draws the large circle denotes the species boundaries defined by developmental constraints. (**a**) One species without subspecies. It represents the founding population. (**b**) One species with three subspecies. Note that subspecies can give rise to other subspecies, but they cannot transcend their level due to the species' developmental constraints (solid outline). SP species, Sub subspecies. For more details, see the text

plasticity, niche construction, and behavioral choice. Below, I will proceed step by step.

Darwinian mechanisms are characterized by comprising two processes that simultaneously generate evolutionary divergence:

1. *Variation*. The variations studied at this level are simple characters—e.g., color, pattern, size, and proportion—which arise from various sources, including (a) genetic (or epigenetic) changes, (b) changes induced by environmental influence, (c) changes produced by physical rules, etc.
2. *Adaptation to a local environment*. Individuals disperse and occupy new environments (or modify the environment upon colonization), undergoing adaptive changes due to their interaction with the environment.

Variation In the traditional population genetics framework, genetic mutations are seen as the origin of new phenotypic variations. However, according to the H-Evo-Devo Theory, many phenotypic variations are attributed not only to changes in gene structure but also to changes in the gene regulatory functions (Wittkopp and Kalay 2012). Also, Evo-Devo studies consider not only genetic changes but also alterations occurring in the entire developmental system (Smith 2003; Arthur 2004). In this regard, it is important to highlight the role of the environment as a source of variation, given that the normal development of organisms is mediated by their direct interaction with the environment (West-Eberhard 2003; Sultan 2015; Gilbert 2016).[5] Likewise, the response of the organism to physical factors is also important (Newman and Bhat 2009; Newman 2021).

As I have emphasized throughout this book, the primary characteristics that make it possible to distinguish between lower-ranking groups morphologically are the differences in simple characters—i.e., variations within a body plan—such as changes in color, pattern, size, proportion, etc. The goal of this section is to identify the origin of these differences. Consider, for example, the coloration and patterns of bird plumage. These differences enable ornithologists to distinguish between subspecies and species and serve various functions, such as camouflage, temperature regulation, sexual attraction, protection, communication, and interaction with other individuals, etc. Additionally, these variations can occur between sexes, life cycles, and seasons. The color pigments of bird plumage mainly include melanins (eumelanin and pheomelanin), carotenoids, pterins, and psittacofulvins. These pigments are produced through various mechanisms. Melanins are synthesized in specialized cells known as melanocytes; carotenoids and pterins are acquired through the diet; and psittacofulvins are synthesized endogenously, exclusively in parrots. Eumelanin is responsible for the color black, while pheomelanin produces hues ranging from yellow to orange. Carotenoids yield yellow, orange, and red shades, while pterins and psittacofulvins produce yellow and red hues.[6] But how are these color differences attained?

When it comes to melanin, its distribution and production (developmental control) are governed by the interplay of various genes, both structural and regulatory (Yang et al. 2019; Akiyama and Kinoshita 2021; Kulikova 2021; Li et al. 2023). For instance, variations in these genes result in polymorphisms defined by feather color (Araguas et al. 2018; Yang et al. 2019; Qi et al. 2023; Rodseth et al. 2024). Moreover, as highlighted, the contribution of genes alone is insufficient to account for these differences; environmental conditions also play a significant role in melanin production. Factors such as food availability (Fríncke-Craig et al. 2015; Galván et al. 2019; Gudowska et al. 2022), oxidative stress (Galván and Alonso-Alvarez 2009; Leclaire et al. 2019), hormonal regulation (Roulin et al. 2008; Saino et al. 2013; Corbel et al. 2016), temperature (Sirkiä et al. 2010; Järvistö et al. 2015; López-Rull et al. 2023), social interactions (Jawor and Breitwisch 2003; Price 2006), and parasites (Jawor and Breitwisch 2003; Gunderson et al. 2009) are all relevant to obtaining differences in melanin regulation.

Regarding carotenoids, since they are obtained through diet, color differences are primarily determined by the availability of dietary sources (McGraw 2006; Price

[5] Of course, this difference is crucial. The Modern Synthesis focuses solely on genetic changes, thereby denying any role the environment may play in generating variation.

[6] For a comprehensive overview of the mechanisms underlying bird plumage coloration, see Akiyama and Kinoshita (2021), Price-Waldman and Stoddard (2021).

2006; Matrková and Remeš 2012; Delhey et al. 2023). However, other factors contribute to the variability in carotenoid-based coloration, including genetic control (Toews et al. 2017; Toomey et al. 2017), metabolic pathways (LaFountain et al. 2015; Toomey et al. 2022), hormonal regulation (Serra et al. 2025), feather structure and development (Landeen and Badyaev 2012), the ability of carotenoids to bind to feather proteins (Mendes-Pinto et al. 2012), the presence of parasites (Hill 2006), among others.

Furthermore, many birds' bright colors (structural coloration) result from microscopic structural and chemical changes provoked by physical-optical effects associated with light reflection (Yoshioka and Akiyama 2021; Eliason et al. 2022). As for coloration patterns such as mottling, stippling, single lacing, double lacing, autosomal barring, etc. (Fig. 11.2), the precise mechanisms underlying their formation remain unknown. Nevertheless, two theories currently explain this phenomenon: One suggests they arise from positional information provided by genetic determinants, while the other proposes self-organizing dynamics during development. The fact that both theories are supported by considerable experimental evidence suggests that they are not mutually exclusive but rather complementary (Haupaix and Manceau 2020). Concerning the self-organizing model, Lin et al. (2013) contend that the complexity of these pigment patterns results from processes occurring during melanogenesis. Factors such as differentiation timing (where the absence of progenitor melanocytes may arise from premature differentiation, or these melanocytes may remain undifferentiated in specific locations), interactions with other

Fig. 11.2 *Color patterns in feathers*. (**a**) Single lacing, (**b**) double lacing, (**c**) penciling, (**d**) autosomal barring, (**e**) stippling, (**f**) butter cup-autosomal barring modification, (**g**) mottling, and (**h**) spangling with an asymmetric pattern. (From Yoshioka and Akiyama 2021)

cells, and molecules involved in differentiation (genes, hormones) all play crucial roles. Additionally, some researchers have associated these patterns with those observed in the mathematical simulation of the Turing reaction-diffusion model. Rather than suggesting a chemical basis for the morphogen, they include processes of cell interaction (Prum and Williamson 2002; Inaba and Chuong 2020) as well as the physical and mechanical processes within cell tissues (Caballero et al. 2012; Haupaix and Manceau 2020). An alternative theory proposes that changes in the genetic regulation of melanin production—i.e., the activation and suppression of involved genes—lead to reaction-diffusion behavior (Schwochow et al. 2021; Yoshioka and Akiyama 2021).

Since heterochrony is a common evolutionary mode at this level, I will briefly use another example to illustrate the slight changes in size, proportion, and behavior between two closely related taxa. Although the common chimpanzee (*Pan troglodytes*) and the pygmy chimpanzee or bonobo (*Pan paniscus*) are recognized as distinct species, from the perspective of the H-Evo-Devo Theory, they are considered as subspecies instead since their differences are not large enough to distinguish between two body plans (for instance, humans and chimpanzees represent two distinct body plans at the same level).[7] Many authors have noted that the two groups of chimpanzees exhibit allometric differences in different parts of the body, suggesting that heterochronic changes led to divergence (e.g., Lieberman et al. 2007; Singleton et al. 2011; Robinson 2012; Behringer et al. 2014; Morimoto et al. 2014). Some researchers indicate that most of the pygmy chimpanzee's structures are paedomorphic versions of those in the common chimpanzee (e.g., Lieberman et al. 2007; Robinson 2012), while others show that the bonobo exhibits some peramorphic characters as well (Morimoto et al. 2014; Diogo et al. 2017). Overall, it has been highlighted that the differences between the two types of chimpanzees are not uniformly heterochronic, contrary to what has often been assumed (Williams et al. 2003; Boughner and Dean 2008).

Notably, these differences arise before birth (Lieberman et al. 2007; Boughner and Dean 2008), leading to the assumption that the separation between bonobos and common chimpanzees resulted from changes in genes regulating growth rates during embryonic development (Morimoto et al. 2014). For instance, Kovalaskas et al. (2020) suggest that genes related to hormonal changes may have influenced the divergence between the two groups. In this context, Behringer et al. (2014) indicate that thyroid hormone levels—which are essential for growth—decrease much later in bonobos than in common chimpanzees. This difference in timing affects not only the rate at which their phenotypic traits develop but also their behavior. In support of this, Stimpson et al. (2016) report differences in the amygdala of bonobos and common chimpanzees in terms of neuronal organization, which partly explains the differences in social behavior between the two groups. According to Kovalaskas et al. (2020), differences in oxytocin and serotonin signaling pathways also play a

[7] See Chap. 7, Sect. 7.1.2; specifically, see Fig. 7.2.

significant role in behavior.[8] Interestingly, environmental factors may have played a crucial role in the divergence of these groups, as differences in growth rates have been found between captive individuals and wild individuals (Bolter and Zihlman 2012; Hanegraef and Spoor 2024). Additionally, competition for food (Kovalaskas et al. 2020; Wrangham 2023) as well as increased resistance to environmental stress affecting the secretion rhythms of growth hormones (Behringer et al. 2014) may have been important factors when the founding species occupied new habitats.[9]

Adaptation to a Local Environment At the microevolutionary level, adaptation is a pivotal process that contributes to evolutionary divergence. A body plan must ensure its components adapt to local environments to avoid extinction. Below, I will briefly outline this process using the dispersal model. Before proceeding, I want to highlight the processes involved in the organism and its interaction with the environment, which will enhance the understanding of this model.

Certainly, natural selection stands out as one of the most important mechanisms at this level.[10] However, selection not only operates on characters generated by the expression of specific genes but can also act on phenotypes with a wide range of variations induced by the environment (genetic accommodation). In other cases, the environment can induce variations in certain favorable directions during individual development (phenotypic accommodation). Yet, over time, these initially environmental-induced characters may become fixed through internal processes along with natural selection (genetic assimilation) (West-Eberhard 2003; Nijhout et al. 2021; Scheiner and Levis 2021). As previously mentioned, it is crucial not to limit the understanding of organism-environment interaction solely to natural selection. For example, these aforementioned processes—genetic accommodation, phenotypic accommodation, and genetic assimilation—are considered within the domain of phenotypic plasticity, which several biologists have highlighted as relevant for explaining evolutionary divergence (e.g., West-Eberhard 2003; Minelli and Fusco 2012; Nonaka et al. 2015; Levis and Pfennig 2016; Dayan et al. 2019; Pfennig

[8] But this does not mean that behavioral differences are governed by a rigid set of rules dictated by variations in hormonal production rates, as Diogo (2017, p. 161) rightly points out: "[T]his behavioral choice was not passive at all: it was one among many others that were possible within the behavioral/physiological plasticity of the organisms and the environmental context in which they lived."

[9] Recently, Furuichi (2023) and Wrangham (2023) endorsed a model of bonobo evolutionary dispersal associated with the premise that behavioral differences preceded morphological changes. While it is undeniable that behavior directly influences subsequent evolution, as Diogo (2017) has illustrated with numerous examples, in this scenario, morphological changes might have been accompanied by interaction with the environment, subsequently prompting hormonal changes and, consequently, resulting in behavioral differences.

[10] According to the H-Evo-Devo Theory, natural selection operates in two distinct manners: firstly, at the microevolutionary level, involving the preservation or elimination of simple characters, as illustrated in this context, and secondly, at the macroevolutionary level, participating in the preservation or elimination of complex characters, as seen in the Sect. 11.2.1.

2021), particularly because it facilitates individuals' colonization to new environments (Hollander et al. 2014; Arendt 2015; Parsons et al. 2019).

Similarly, according to niche construction theory, organisms are active entities that modify their environment—e.g., by secreting chemicals or constructing physical elements such as soils, burrows, webs, etc.—thereby creating an environment that simultaneously influences their evolution. For example, when descendants inherit these constructed environments, they experience distinct selection pressures (Odling-Smee et al. 2003; Sultan 2015; Laland et al. 2019). Consequently, as organisms encounter new environments, niche construction fosters phenotypic plasticity, establishing an interdependence between both processes (Schwab and Moczek 2021; Brady 2023). Since niche construction is closely intertwined with these processes, it also contributes to evolutionary divergence (Sultan 2015).

Another environment-dependent process is behavioral choice, defined as the contingent behavior—whether conscious or unconscious—that organisms exhibit in response to external stimuli (Diogo 2017). It is important to note that behavioral choice is a cornerstone in organism-environment interaction since this process determines the active life of individuals—which reproduce, obtain resources like food and water, either avoid or seek predation, respond to interactions with other individuals, and more. In turn, this also alters selection pressures (Duckworth 2009; Diogo 2017). In short, (1) natural selection, (2) phenotypic plasticity, (3) niche construction, and (4) behavioral choice collectively contribute to organism-environment interaction, ultimately producing subspeciation.

To begin, consider a local population with a particular body plan, e.g., a new species without subspecies (Fig. 11.1a). While evolutionary change can occur within the same geographic area (e.g., as a response to local environmental changes), new populations generally emerge through dispersal, i.e., individuals from a particular environment disperse to occupy different spatial ranges (Fig. 11.1b). As they settle in a new environment, their interaction with it produces adaptive differences compared to the original population. Although migration entails risks—i.e., the environment may pose challenges for many individuals—some might encounter favorable conditions that promote evolutionary divergence. It is worth highlighting the various methods through which populations acquire adaptive traits. For example, when populations disperse in gradual environments, they steadily differentiate, eventually forming a clinal distribution. Although some biologists have discussed the importance of clines or continuous intermediate gradients as facilitators of evolutionary divergence (e.g., Doebeli and Dieckmann 2003; Armsworth and Roughgarden 2008), it seems to me that clines only reflect the intrinsic variation of subspecies. Moreover, this sort of variation likely stems largely from phenotypic plasticity as a norm of reaction. In this context, clines might play a beneficial role in reducing the risk of subspecies extinction. Nevertheless, I believe that discrete spaces in diverse environments, which are associated with discontinuous character states, are the best candidates for the origin of subspecies.

For example, a founder population comprised of few individuals may harbor certain characters with exaptive value, i.e., these functional traits could have originated elsewhere, likely as spandrels emerging as a byproduct of ontogeny processes

(Gould 2002). Conversely, in other scenarios, upon encountering these new environments, some individuals may acquire the same adaptive direction through phenotypic plasticity (Arendt 2015; Parsons et al. 2019), often resulting in polyphenism, i.e., discontinuous changes induced by the environment (Nijhout 2003; Pfennig and McGee 2010; Pfennig 2021, p. 67). The idea is that when these individuals face new environments—acquiring new resources, encountering other competitors, avoiding new predators, etc.—they alter their phenotype in accordance with the potential plasticity of their ancestral developmental system (Parsons et al. 2019). For instance, using the aforementioned examples of variation, birds might change the color of their plumage in response to differences in diet, temperature, interaction with other individuals, parasites, and so on. If several individuals arrive in a new environment, they will all change their phenotype simultaneously. For example, since certain pigments are acquired from diet, changes in food sources will influence the type or intensity of coloration within the colonizing population (see Price 2006). However, it should be considered that variation is not random but rather channeled according to the potential of the population's developmental system. This process, known as developmental bias,[11] operates in concert with natural selection, phenotypic plasticity, niche construction, and behavioral choice.

Another relevant point is that when individuals spread into different environments, they do not always do so randomly but rather choose based on habitat, i.e., populations can select the most favorable environments (Sultan 2015, pp. 71–72; Shirani and Miller 2024), even adjusting according to their phenotype (Edelaar et al. 2008, 2017; Camacho et al. 2020; Turko and Rossi 2022). Nonetheless, according to Diogo (2017), the concept of behavioral choice differs from the simple concept of habitat choice in the sense that individuals not only select the appropriate environment for survival but are also capable of selecting more than one option. For example, in some cases, when individuals change their behavior—by learning from or imitating others—and this behavior persists in a particular environment, it consequently leads to a favorable phenotypic change. From this, it can be inferred that if there had not been a behavioral change and persistence in a particular environment, such phenotypic variation would never have occurred.[12]

In conclusion, during the process of subspeciation, diverse explanations arise concerning the origin of phenotypic variation. These discontinuous characters are simple and do not imply the origin of a body plan. The dispersal model indicates that when individuals arrive in different environments—or when a local environment changes—they adapt to the local environment through processes related to organism-environment interaction (natural selection, phenotypic plasticity, niche construction, behavioral choice). Therefore, it could be argued that these mechanisms, along with developmental bias, serve as the primary drivers of subspeciation.

[11] Regarding the action of developmental bias, see the discussion in the following section.

[12] Indeed, this idea is part of the phenomenon known as the "Baldwin Effect." Diogo (2017) includes it in a broader model called "Organic Nonoptimal Constrained Evolution" (ONCE).

11.1.2 Substasis

According to the H-Evo-Devo Theory, *substasis* represents the mode focused on the stability of character states at the subspecies level. As depicted in Fig. 11.1, when an ancestral subspecies (Sub1) gives rise to one or more descendant subspecies (Sub2, Sub3), they occupy different positions within the morphospace, subsequently remaining fixed spatiotemporally. But what mechanisms account for this period of morphological stability at the microevolutionary level?

Given that many of the character states at this level are adaptive, it is reasonable to consider natural selection as the mechanism that stabilizes these characters when groups settle in a stable environment. So, some biologists have suggested that "stabilizing selection" alone is the cause of stasis (e.g., Estes and Arnold 2007; Haller and Hendry 2014).[13] In fact, the notion that natural selection acts as a constraint on structure underlies the idea that all directions of variation are possible. Under this scenario, natural selection channels variation toward the most adaptive state. Once adaptive stability is attained, selection eliminates any variation that deviates from that steady state. However, the notion of "random variation" can be challenged in light of the evidence surrounding the origin of variation, even at this level.

For example, consider the organism-environment interaction. Regardless of natural selection, the stability of the environment tends to maintain characters within a population, as the environment is an instructive factor in shaping the normal development of the individuals inhabiting it (West-Eberhard 2003; Sultan 2015; Gilbert 2016). Therefore, when the environment shifts or individuals migrate to new habitats, their phenotypic characters are likely to change, primarily due to phenotypic plasticity. Nevertheless, it is important to note that organismal homeostasis, shaped by environmental interaction, also underlies the argument of developmental constraints. For instance, since phenotypic plasticity is a feature of the developmental system, rather than merely being a simple direct outcome of the genotype (Sultan 2019; Walsh and Sultan 2024), the developmental system's response to the environment is channeled into a limited number of transformation pathways. That is, variation is guided by developmental bias (Parsons et al. 2019; Stansfield and Parsons 2024).

In order to elucidate this process of canalization, I will examine the possible colors of bird plumage as an example. Stoddard and Prum (2011) argue that while certain plumage colors may be disadvantageous—natural and sexual selection can either favor or eliminate some colorations—there are physical, physiological, and metabolic constraints that restrict the broad color spectrum in birds. It is well known that birds can perceive a wide range of colors; however, these options are unavailable in bird plumage due to various constraining factors. For example, carotenoid-based pigmentation only yields colors ranging from yellow to red, while melanin-based colors produce black, brown, yellow, and orange. This explains why mammals exhibit a narrow spectrum of colors, given that their pigmentation relies exclusively on melanin (Sumner and Mollon 2003). Additionally, the expression of

[13] They propose it for the species level, see Sect. 11.2.2.

carotenoids acquired through diet restricts the color range in granivorous birds, unlike frugivorous birds, which exhibit a broader spectrum of color options (Mahler et al. 2003). The physiological ability of certain individuals to absorb carotenoids from their diet is also a factor to consider (Tschirren et al. 2003). For example, Delhey et al. (2023) argue that the red color arises from metabolic processes, starting from the yellow color, which is obtained directly from the diet. Although the prevalence of red over yellow can be attributed to sexual selection, they state that internal and metabolic factors also explain its stability. Similarly, Carlson and Stoddard (2024) point out that physiological and developmental constraints influence the plumage patterns of melanerpine woodpeckers, limiting the appearance of certain shapes such as wide bars or circular spots. Furthermore, evidence suggests that the color patterns are constrained by physical mechanisms as well (Prum and Williamson 2002; Caballero et al. 2012). Regarding structural colors, Eliason et al. (2022) mention that variations at the nanoscale of feather structures affect and restrict the color spectrum within the groups in question: "For example, hydrodynamic constraints have been implicated in barb microstructure changes in penguins (e.g., flattened barbs, loss of the central vacuole) that may have excluded other mechanisms of generating blue color and led to its production via novel keratin nanofibers" (p. 2). In a nutshell, across all these instances, the distribution of plumage coloration is not random but developmentally channeled (Nordén and Price 2018).

Therefore, while natural selection acts on available variation, developmental constraints impose limitations on it (Pearce 2011). But how exactly do these processes of constraint and selection operate? The dynamic proceeds as follows: (1) Developmental bias directs variation toward certain points in the morphospace. (2) Once the variation is established at a particular point within the morphospace, developmental constraints[14] maintain fixed character states for prolonged periods.[15] (3) Finally, natural selection either preserves or eliminates that character depending on its function.[16] In this regard, I agree with Salazar-Ciudad (2006) that instead of seeing developmental constraints and natural selection as opposing forces, these mechanisms operate on entirely different aspects. In short, the stability of these simple characters resides in their structural aspect rather than their functional context. Therefore, the causes of substasis are the developmental constraints (in the negative sense) acting at the lowest levels of the hierarchy.

---

[14] In its negative sense, or in the context of the H-Evo-Devo Theory, see Chap. 5, Sect. 5.1.2.

[15] These depend on the scale of variation. For example, unlike color, changes in size or proportion may exhibit a greater degree of developmental robustness (strong developmental constraints). For instance, revisiting the case of chimpanzee evolution, Diogo et al. (2017) note that the pygmy chimpanzee has not exhibited significant changes for approximately two million years, which marks the estimated point at which the pygmy chimpanzee likely diverged from the common chimpanzee.

[16] See Fig. 11.7, which depicts this dynamic of processes at the macroevolutionary level. Also see this discussion in Sect. 11.2.2.

11.1.3 Subextinction

According to the H-Evo-Devo Theory, *subextinction* is the evolutionary mode that refers to the death of subspecies. Since subspecies are adapted to their corresponding local environments (habitats), factors that alter their local habitat and population structure are the primary causes of their extinction. Accordingly, certain ecologists and evolutionists have shown a keen interest in discerning the factors that cause the natural extinction of populations at the microevolutionary level. Below are some of these reasons:

1. *Environmental stochasticity*. Random environmental fluctuations, whether gradual or sudden, disturb the distribution and density of extant populations (Lande 1993; Fujiwara and Takada 2017), e.g., alterations in temperature, precipitation, fire frequency, are all factors that lead to extinction (Cahill et al. 2012).
2. *Habitat fragmentation and habitat loss*. Habitat fragmentation refers to the destruction of extensive habitat into small, isolated patches, whereas habitat loss signifies the complete disappearance of habitat due to destruction (Fahrig 2003). The more fragmented the habitats become as populations are distributed across these patches, the greater the risk of extinction due to environmental changes and other factors (Fagan et al. 2005; McLaughlin et al. 2002; Gaggiotti and Hanski 2004). For instance, island populations experience reduced growth due to the passage of cyclones and hurricanes, yet this does not suggest that they are not adapted to their environment. By analogy, continental populations could face similar challenges if their habitats were reduced (Simberloff 1995).
3. *Habitat quality*. This refers to the properties of a habitat that are necessary to maintain the necessary conditions for a population to increase its fitness in a particular space (Heinrichs et al. 2016). Populations inhabiting areas with lower resource quality are more prone to extinction (Robles and Ciudad 2012).
4. *Competition*. The coexistence of different populations results in competition for resources, and hence this factor potentially affects population density (Bengtsson and Milbrink 1995; Hunt and Bonsall 2009). Generally, this competition is triggered by the introduction of invasive populations, which can negatively affect native populations in a relatively short time (Gurevitch and Padilla 2004; Gilbert and Levine 2013).
5. *Population size fluctuations*. Populations randomly decrease in size. As the number of individuals declines, small populations face the challenge of individuals struggling to find mates within the same population (Allee effect). This issue is exacerbated by additional factors that further decrease their numbers (Matthies et al. 2004; Labrum and Gomulkiewicz 2020).
6. *Dispersal ability*. While dispersal is a positive factor, as it facilitates the introduction of new individuals to different environments (McCauley et al. 2014), it can be counterproductive in certain situations (Higgins 2009; Ben-Zion et al. 2012), especially for small populations (Gaggiotti and Hanski 2004, p. 361). Additionally, when populations have greatly reduced geographic ranges and face limitations in their dispersal capacity, they become highly prone to extinction (McCauley et al. 2014).

Another important point of discussion is the relationship between extinction and the origin of subspecies. According to Vrba's turnover-pulse hypothesis (1993), environmental changes lead to extinctions and migration events within lineages, thus promoting the emergence of new lineages as well. Under the H-Evo-Devo Theory, if one or several subspecies disappear due to the abovementioned factors, they will leave some vacant niches that other subspecies could occupy. In other words, the extinction of some subspecies will facilitate either the origin or introduction of others. Interestingly, ecologists use the term "metapopulation" to study this balance between extinction and colonization in extant populations (Harrison 1991; Thrall et al. 2010).

Following Eldredge's "sloshing bucket model" (2003), while the extinction of subspecies occurs in specific localities, there are different circumstances in which this process can affect higher levels within the hierarchy. For example, suppose a species with four subspecies loses one due to local environmental disturbance. In that case, hypothetically, the overall structure of the species should remain intact unless a more widespread environmental disruption also threatens the other subspecies. While the loss of the subspecies is irreparable, this does not apply at a lower level. Polymorphisms within a subspecies—typically those forming clines—can recover after a minimal environmental disturbance, as these populations can be reestablished by individuals from neighboring areas (McLaughlin et al. 2002).

11.2 Macroevolution (Species Level)

Within the framework of the H-Evo-Devo Theory, macroevolution represents a level that is distinctive and non-extrapolable when compared to microevolutionary events. Under this model, the level of macroevolution addresses the modes of origin, stability, and extinction of entities recognized as species (or genera). By species/genera, I refer to those discrete entities separated from each other by a significant degree of morphological gaps and distinguished from other taxa because they represent the least inclusive body plan within a hierarchy of body plans. Given that, under this model, species (or genera) are body plans—as opposed to the simple characters of microevolution—this level necessitates a special explanation.

Following the H-Evo-Devo Theory, this section briefly examines the mechanisms underlying the three evolutionary modes (patterns) of species/genera. The first evolutionary mode at this level is *speciation*, whose mechanisms involve the prospecies model or supraDarwinian mechanisms. The second mode is *specific stasis*, whose mechanisms are developmental constraints working on the fixation of body plan characters. Finally, the third mode is *specific extinction*, whose causes are attributed to the loss of parts and mass extinction events.

11.2.1 Speciation

According to the H-Evo-Devo Theory, *speciation* is the mode focused on the origin of species/genera. Following Fig. 11.3, imagine an ancestral species composed of three subspecies (a). In this scenario, the subspecies cannot generate a new species because they are restricted by the species' developmental constraints, i.e., these populations can only produce additional subspecies. Therefore, for a species to give rise to a new species, a radical change capable of breaking the ancestral species' developmental constraints is necessary. Once such a threshold is overcome, the new entity occupies a different position within the morphospace; this is the *prospecies* (b).

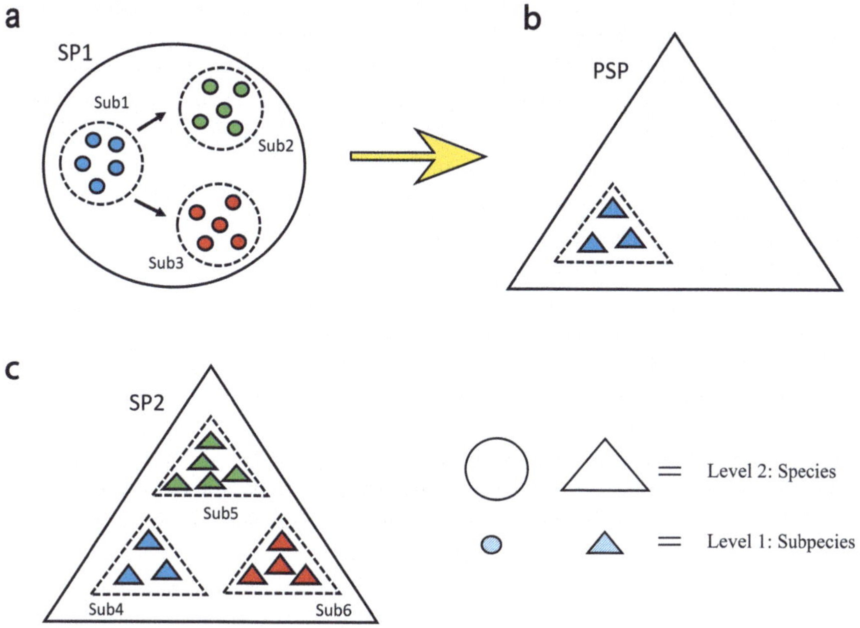

Fig. 11.3 *Speciation depicted in a nested compositional hierarchy, according to the H-Evo-Devo Theory.* The area inside each figure represents the morphospace, and each figure indicates a form located within the morphospace. The size of the figure denotes its level: the small figures correspond to Level 1, while the large figures correspond to Level 2. The color differences of the small figures illustrate the gaps between subspecies. Differences in the geometric figures—e.g., circle and triangle—illustrate the gaps between species. The dotted figures within the large figures represent the subspecies boundaries delineated by developmental constraints. The solid outline of the large figures (circle or triangle) denotes the species boundaries defined by developmental constraints as well. (**a**) One species with three subspecies. (**b**) Prospecies. (**c**) A new species with three subspecies. Note that to give rise to a new species, a radical change is necessary to break the ancestral species' developmental constraints. This process produces a prospecies, which is characterized by exhibiting a significant morphological gap compared to its ancestral population. Although a prospecies shares the properties of any species, it initially lacks subspecies. Then, to become a taxonomic species, it must subsequently produce subspecies. SP species, PSP prospecies, Sub subspecies

However, a prospecies needs to generate more subspecies in order to establish itself taxonomically as a species and ensure its survival (c).

This macroevolutionary model of divergence is explained by supraDarwinian mechanisms. But what do I mean by supraDarwinian mechanisms? As widely known among biologists, Richard Goldschmidt (1940) proposed the "hopeful monster argument" to explain speciation. According to this idea, species originate through specific genetic changes that trigger sudden transformations in the developmental system (thus rejecting the notion of genetic mutations with cumulative effects over time), leading to significant evolutionary change in a single generation. If the resulting individuals are functional upon emergence, then they represent the origin of a new lineage.

Although the hopeful monster argument introduced a novel idea in addressing the issue of the origin of species, it is worth noting that it has made minimal progress. Still, Bateman and DiMichele (1994, 2002) revisited this model, paying close attention to the establishment of hopeful monsters in the environment. A noteworthy distinction is that rather than labeling these entities as "hopeful monsters," they call them "prospecies." In particular, the H-Evo-Devo Theory adopts Bateman and DiMichele's prospecies model—albeit with some adjustments—as it aligns most closely with the explanation needed for the origin of body plans at any level.[17]

SupraDarwinian mechanisms are characterized by comprising two processes that simultaneously generate evolutionary divergence:

1. *Variation*. The variations studied at this level are complex characters or those characterizing a body plan. Initially, either genetic changes or environmentally induced changes break the developmental constraints of a previous body plan in order to give rise to a new one.
2. *Fitness to different environments*. Body plans can establish themselves in different environments, that is, they do not have a perfect fit with a particular environment. Nevertheless, they sometimes acquire key innovations allowing them to increase their fitness. Prospecies typically flourish—though not exclusively—in places with little competition, but once these populations increase in number and acquire the designation of species, they can fit into different environments by adapting their parts as subspecies.[18]

[17] Notably, I am not attempting to portray this model as an exceptional theory of speciation—particularly due to the use of the "hopeful monster" concept. On the contrary, I aim to highlight this model as a standardized mode for understanding the origin of body plans, clearly differing from the microevolutionary explanation provided by the dispersal model. Additionally, this model is not an alternative to gradual evolution, because, as discussed in Chap. 9, gradualism as the intermediacy of transitional forms does not reflect a common process in nature considering the origin of variation. It does not even explain simple characters. Both microevolution and macroevolution entail discontinuous steps, differing from each other primarily in the type of variation studied (whether simple or complex characters).

[18] Throughout this book, I have reiterated the applicability of the prospecies model to both species and higher taxa. Building on this premise, it is clear that, on occasion, certain species behave like subspecies, particularly considering that genera are very similar to species, as concluded in Chap. 7, Sect. 7.2.

Variation (on the origin of prospecies) Before proposing the type of variation considered at this level, it is necessary to clarify some important concepts based on Bateman and DiMichele's (2002) prospecies model. According to their terminology, a taxonomic species refers to "one or more (typically many more) populations separated from all other comparable populations by phenotypic, and putative genotypic, discontinuities that are believed to reflect one or more isolating mechanisms operating over a considerable period of time." In contrast, they define a prospecies as "a putatively recently evolved lineage possessing the essential intrinsic properties of a taxonomic species but yet to achieve levels of abundance and especially of longevity acceptable to practising taxonomists" (p. 112). However, according to the H-Evo-Devo Theory, species and prospecies are populations distinguished from other populations by significant morphological gaps. That is, the descendant population exhibits a large discontinuity compared to the ancestral population (see Fig. 11.3a and b). The difference between species and prospecies is that a species is distinguished by possessing subspecies, whereas a prospecies represents a potential lineage that must give rise to entities below its level to establish itself as a taxonomic species. In other words, a prospecies requires sufficient time to produce subspecies, thereby acquiring elements necessary to avoid extinction (see Fig. 11.3c).[19] Under this interpretation, both species and prospecies represent body plans. The species is an already established body plan that includes a diversity of elements or subspecies, which reduces its risk of extinction. In contrast, the prospecies is a new body plan that depends on specific conditions to establish itself and requires sufficient time to gain lower-level diversity. Therefore, the type of variation that distinguishes the level of macroevolution (and megaevolution) pertains to the type or body plan.

As discussed in other chapters, the concept of type or body plan refers to the common pattern of characters that are highly conserved due to developmental constraints. So, in order to give rise to a new body plan, a major change capable of breaking the developmental constraints of an ancestral body plan is needed. Thus, overcoming these constraints requires either a significant genetic change or a drastic environmental shift. In this regard, it has been observed that major genome rearrangements are unnecessary; instead, a change in a single gene or a small set of genes is sufficient to produce large phenotypic effects (Theißen 2006, 2009; Rubinoff and Le Roux 2008; Usai et al. 2024). For example, Shubin and Marshall (2000) argue that a few genetic changes could be responsible for the morphological differences observed between species. Similarly, the action of homeotic genes has been identified as the cause of body plan divergence. For instance, Hox genes regulate the development of body part identities in animals such as insects and vertebrates (Carroll 1995). In plants, MADS-box genes (Theißen et al. 2000) and SEPALLATA genes (Malcomber and Kellogg 2005) determine floral organ identity,

[19] In the original interpretation by Bateman and DiMichele (1994, 2002), a prospecies is a transient entity that disappears within a few generations; nevertheless, if it persists and its population increases, it then achieves the status of a taxonomic species.

as illustrated by the well-known ABC model (Bowman et al. 1991), etc. Additionally, some researchers have emphasized the role of the hybridization of entire genomes between species, which results in the "instantaneous" origin of new species (e.g., Mallet 2007; Dittrich-Reed and Fitzpatrick 2013; Vallejo-Marín et al. 2015; Usai et al. 2024).

Since complex structures arise from different genetic mechanisms, each case should be analyzed independently to understand how that variation occurred. However, it is necessary to emphasize that most of these transformations do not result solely from structural changes in the genes but from change in the regulatory region of the DNA (Wittkopp and Kalay 2012). For example, Smith (2003, p. 616) states that "shifts in timing of gene expression can produce gross morphological changes such as limb elongation, differentiation of additional serial elements, or shifts in identity of elements."

Alternatively, some biologists have proposed that environmental stress or extreme environmental changes can disrupt the normal development of organisms, resulting in morphological gaps (West-Eberhard 2003, pp. 505–507; Badyaev et al. 2005; Jaeger et al. 2012). For example, much attention has been devoted to Matsuda's (1987) work, which suggested that environmental changes alter the hormonal activity of organisms, leading to sudden transformations. These changes could subsequently be maintained through processes such as natural selection and genetic assimilation (West-Eberhard 2004). In other cases, the environment could directly influence genetic regulations (Maresca and Schwartz 2006; Jaeger et al. 2012). Regarding this, it follows that both genetic and environmental factors can simultaneously operate to produce macroevolutionary novelties (West-Eberhard 2004).

Once the developmental constraints are broken, the next step is to consider that such a change must be possible within the parameters of the developmental system, i.e., the product must be ontogenetically viable. The body plan is governed by deeply integrated developmental rules (Raff 1996), which allow it to maintain its characters fixed at a specific position within the morphospace (see Chap. 5). However, transitioning from one steady state to another within the morphospace necessitates a discontinuous traversal of the system.[20] As discussed in Chap. 9 (Sect. 9.2), since evolutionary change is discontinuous, the evolutionary rate of macroevolution should be measured with respect to the magnitude of discontinuous change over the number of generations rather than continuous change over geological time, as is commonly assumed. Hence, transformations can occur in a single generation (saltation) or across two or more generations (parasaltation) (Bateman and DiMichele 2002).

In order to understand the type of discontinuous modification within the developmental parameters, the focus could be placed on teratology, i.e., the study of abnormalities or malformations in individual organisms, indicating a radical deviation from normal developmental pathways. Currently, biologists have paid considerable attention to this topic, particularly because of its relevance to the study of macroevolution and its relationship with medicine. This emerging field is dubbed

[20] (see Chap. 5, Sect. 5.1.2 and Chap. 9, Sect. 9.2).

"Evo-Devo Path" (Diaz 2020; Diogo 2020) or "Evolutionary Teratology" (Guinard 2015, 2020). Advocates of Evo-Devo Path have particularly stressed the parallels between certain human pathologies and some key innovations seen in animal body plans. Therefore, the study of abnormalities sheds light on the formation of complex structures, encompassing those that may or may not manifest within the spectrum of possibilities offered by embryonic development (Alberch 1989; Olson 2012; Diogo et al. 2019; Cao et al. 2024). As Blumberg (2009) aptly puts it:

> The "options" that anomalies offer are embedded within the mechanisms of development. Anomalies help us to glimpse these mechanisms with a clarity that can be lost by focusing on well-formed organisms alone… That such strikingly similar forms can arise in these distantly related species testifies to the action of ancient developmental mechanisms. Moreover… even extreme anomalies such as these can be produced with surprising ease through the subtlest kinds of developmental adjustments… The boundary between novelty and anomaly is a fuzzy one. After all, bodily forms considered anomalous in one species can signify novelty in another. (Blumberg 2009, p. 11)

> Anomalies are difficult to ignore; they matter for development and they matter for evolution. (Blumberg 2009, p. 13)

What types of variations are possible within the realm of embryonic development? Developmental repatterning can be categorized into the following types: heterotypy (change in identity), heterometry (change in number), heterotopy (change in spatial arrangement), and heterochrony (change in timing). Although not all biologists employ this specific terminology, they generally recognize these types of changes under the broader term "homeosis" (transformation of one body part into another). For example, in vertebrates, these changes can manifest as alterations in identity (Guinard 2012) or in the number of vertebrae (Varela-Lasheras et al. 2011; Böhmer et al. 2018); in arthropods, there are shifts in identity (Averof and Patel 1997) or in the number of segments (Minelli et al. 2009); in plants, there are modifications in identity or the spatial arrangement of floral organs (Ronse De Craene 2003), and so forth. Particularly, it is noteworthy that heterochrony constitutes one of the main modes of macroevolutionary change (McNamara 2012), given its propensity to generate considerable morphological disparity in terms of size and proportion (McNamara and McKinney 2005; Bonett 2021; Dobreva et al. 2022; Gayford et al. 2025).

Finally, one of the most contentious issues regarding the validity of saltationism—or the credibility of the hopeful monster—lies primarily in the belief that it is unlikely for the same discontinuous variation to occur twice in the same population. In other words, for an individual (a hopeful monster) to mate and produce offspring, it must encounter at least one other individual (another hopeful monster) that exhibits the same type of variation at the same time. This is the first challenge Frazzetta (1975) presents, which he identifies as the paradigm of the "improbable." This dilemma presents a serious challenge from the gene-centered perspective, particularly given the presumption that variation occurs randomly. Nevertheless, Frazzetta (2012) introduced the "large-step assessment model," or LSA. Through computational simulations and the consideration of specific variables across three groups—marine animals, insects, and reptiles—he demonstrates that genetic variations with discontinuous effects on the phenotype, while less prevalent, can occur multiple times within the same population during certain periods throughout geological time. Frazzetta (2012) writes:

> The overall results, though derived through much simplified assessment methods, are congruent with expectations that acquisition and retention of large-effect mutations is an uncommon event. All three cases indicate that any such occurrences would take place over one hundred to one-million generations. This span is great in human observational time scales, and the phenomena would be spread over the entire planet. These two factors, occurrences that surpass human life spans, and which are dissolved into global space, reduce the opportunities for their discovery. At the same time, even the rarest computed occurrence is not "rare" in geological time. In this view, nothing in the LSA results are inconsistent with the occasional but regular production of advantageous mutations that have large, phenotypic effects, that often go unnoticed by human observation. (Frazzetta 2012, p. 54)

However, the estimate could be even greater when considering that, as mentioned earlier, there are trends in variation based on the potential forms permitted by development. For instance, Alberch (1989) noted that teratological forms exhibit developmentally well-coordinated arrangements, i.e., they are not random. Hence, given the constraints imposed by development, certain forms are more likely to appear than others. As discussed in the previous section, this phenomenon has been recognized through the concept of "developmental bias" (Arthur 2004, 2021; Uller et al. 2018; Stoltzfus 2019; Moczek 2020).[21] Essentially, this concept suggests that alterations occurring during morphological change are only channeled along certain pathways, leading closely related groups to repeatedly produce similar variations. Remarkably, this phenomenon has also been observed in distantly related species that share the same embryological framework (Diogo et al. 2019; Shkil et al. 2019). In short, descendants will likely acquire the same type of variation over time, as the pathways of change are shaped by developmental bias. Moreover, some researchers have highlighted that environmental factors may also facilitate these variations. Since a shift in the environment affects the embryonic development of individuals, channeling variation towards specific positions, this effect can simultaneously act on many individuals at the same time, especially if they persist in the same environment (West-Eberhard 2003, p. 504; West-Eberhard 2004; Badyaev et al. 2005).

In summary, a body plan exhibits high cohesion due to its developmental constraints. Thus, the emergence of a new body plan necessitates either a genetic alteration or a significant environmental shift capable of breaking the ancestral body plan's developmental constraints. This change must be ontogenetically possible and, most importantly, viable. Given the developmental system's potential to simultaneously channel the same type of variation (developmental bias) in different individuals, which can be facilitated by environmental influences, a radical change can

[21] See Fig. 11.7 for an overview of the role of developmental bias in evolution.

occur in a part of the ancestral population in a single generation (or over a few generations), giving rise to a prospecies.

Fitness to different environments (establishing prospecies) One of the most contentious issues within the prospecies model is the environmental establishment of hopeful monsters. While their origin can be explained through Evo-Devo mechanisms, the understanding of their interaction with the external environment remains limited (Bateman and DiMichele 2002; Theißen 2006). How might a hopeful monster thrive in a highly competitive environment? This marks Frazzetta's (1975) second challenge to the "improbable paradigm." In his original theory, Goldschmidt (1933, 1940) devoted considerable attention to describing the creation of hopeful monsters. However, regarding their environmental role, he wrote very little, merely suggesting that hopeful monsters might establish themselves in unoccupied niches. Bateman and DiMichele (1994) interpreted Goldschmidt's hopeful monster as possessing a high degree of fitness and capable of thriving even in highly competitive environments. In contrast, Bateman and DiMichele (1994, 2002) propose that prospecies exhibit a notably low level of fitness, implying their reliance on highly specific conditions for establishment. According to their hypothesis, prospecies typically originate in unoccupied habitats, often resulting from environmental disturbances, providing a temporary respite from selection pressures. Over time, prospecies would incrementally enhance their fitness through adaptation to environmental challenges. Considering this, although I agree with these authors that unoccupied habitats are ideal opportunities for the survival of the prospecies, there are instances where they can also establish themselves in environments characterized by conventional environmental equilibrium. Below, I will explain this in detail.

According to Rui Diogo (personal communication), since the architects of the Modern Synthesis were gradualists and adaptationists, they identified a problem surrounding how the hopeful monster could possess optimally adaptive structures. However, Diogo states that this problem is unfounded, as the structures of these individuals are not as functional as initially presumed. That is, the monster's morphology harbors numerous imperfections, necessitating only the bare minimum for survival. Indeed, under the term "etho-eco-morphological mismatches," Diogo (2017) highlights the lack of a direct correlation between ecological function and organism structure. This suggests that the phenotype is more related to phylogeny than to adaptive function. For example, Vidal-García et al. (2014) undertook a study to quantify morphological differences in the shape and size of 129 species of Australian anurans. Their goal was to investigate the link between habitat type and the body's morphological adaptation. Yet, their findings revealed that: "There was no clear relationship between body size and environmental niche… Our results suggest that phylogenetic legacy is important in the evolution of body size and shape in Australian anurans, but also that the conservative body plan of many frogs works well in a wide variety of habitats" (p. 181). Considering the above, it can be concluded that the body plan characters are not optimally adaptive; thus, such traits cannot fall into the category of the "improbable." They persist because they possess the minimal functionality required to live in different environments.

At this level, natural selection is a decisive process; however, it operates differently from what is observed at the microevolutionary level. Before analyzing this, consider the types of functions present in organic systems. According to the H-Evo-Devo Theory, two types of functions are identified: (a) specific functions, explaining adaptive characters tailored for specialized habitats, and (b) general functions, involving characters that fit in different environments. In the case of specific functions, individuals evolve as specialists, reaching a higher degree of fitness by solving functional challenges in specific environments, e.g., being cryptic, acquiring beaks capable of breaking tough seeds, changing coloration to avoid predation, and so forth. Regarding general functions, individuals evolve as generalists, fitting their characters to different environments, e.g., whales adjusted the position of their blowholes and the shape of their forelimbs to live in a diverse marine environment, birds developed forelimbs suitable for flight in any environment, and felines transformed specialized claws for hunting various prey, etc.

Considering the above, the idea is that selection operates across different levels. On the one hand, it preserves structures with specific functions; on the other hand, it maintains those with general functions. In essence, while natural selection at the microevolutionary level increases fitness for a specific environment (e.g., to acquire a particular resource), at the macroevolutionary level, it operates by increasing the fitness of organisms capable of thriving in diverse environments (e.g., by accessing various resources).

For example, in textbooks, Darwin's finches are frequently cited as prime examples of adaptive radiation, particularly regarding the correlation between beak morphology and dietary preferences, e.g., a smaller beak typically indicates a diet of small insects or soft seeds, while a larger beak suggests a preference for hard seeds. Nonetheless, De León et al. (2014) analyzed the diets of four sympatric species within the genus *Geospiza*, revealing that they were surprisingly varied, indicative of a generalist feeding strategy rather than a specialist one. However, during periods of drought, when resources became scarce, these finches resorted to consuming the type of food most compatible with their beak morphology. Following Fig. 11.4, suppose the beak of the ancestral *Geospiza* was of medium size—similar to that of *Geospiza fortis* or at least falling within the typical range of size and shapes observed in the genus (see al-Mosleh et al. 2021)—allowing it to consume a wide variety of foods (see Fig. 11.4c). At this level, natural selection primarily favors this character by potentially eliminating individuals with extremely aberrant beak shapes (see Fig. 11.4d).[22] That is, such individuals have either low fitness or die because of their inability to effectively feed with these deformities (see Van Hemert et al. 2012a). This will be the case unless such phenotypes happen to match a previously unex-

[22] Note that there are two types of filters in variant production: an internal filter influenced by developmental bias, determining what is ontologically possible, and another filter shaped by natural selection, restricting what is functionally appropriate. It is this latter filter that is exemplified in this case. See this dynamic of processes in more detail below in Sect. 11.2.2.

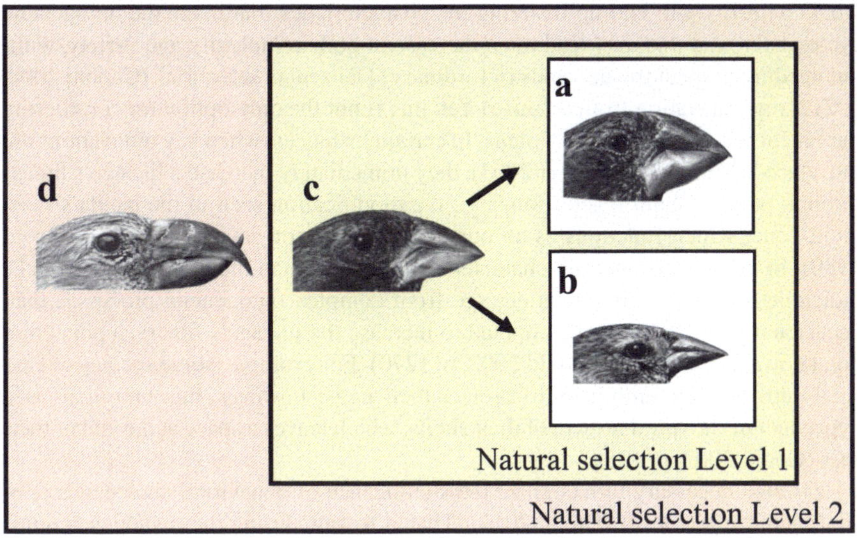

Fig. 11.4 *Depiction of the two levels at which natural selection operates.* Each square's area illustrates the forms permitted or excluded by natural selection corresponding to their respective level. For example, at the macroevolutionary level (level 2), natural selection disfavors forms within the orange area and favors those within the yellow area. In contrast, at the microevolutionary level (level 1), natural selection disfavours forms within the yellow area and favors those within the white area. (**a**) *Geospiza magnirostris*. (**b**) *Geospiza fuliginosa*. (**c**) *Geospiza fortis*. (**d**) An individual with an aberrant beak. Assume that *Geospiza fortis* (**c**) was the founding species whose beak shape suggests a generalist strategy. It should be noted that beak forms outside this range are so aberrant that these individuals cannot feed, thus decreasing their fitness or causing their demise (**d**). However, during times of drought, increased competition for food occurs, prompting natural selection to operate at the microevolutionary level, resulting in forms **a** and **b**, which suggest a specialist strategy. (*The images of Darwin's finches were sourced from Grant and Grant (2003), photographs by B. Rosemary Grant and Peter R. Grant, with permission from Oxford University Press. Image **d** has been edited to portray a potential aberrant specimen found in nature)

ploited food source (see Van Hemert et al. 2012b; Purificação 2019). Nevertheless, at a different level, when food competition intensifies, natural selection operates at the microevolutionary scale. On the one hand, it results in the emergence of *Geospiza magnirostris* (see Fig. 11.4a), with a thicker beak that enables it to crack hard seeds; on the other hand, it leads to *Geospiza fuliginosa* (see Fig. 11.4b), with a smaller beak better suited for consuming smaller seeds.[23]

Nonetheless, if prospecies can survive across different environments, it follows that under specific conditions—e.g., when abundant resources are available—selection at the microevolutionary level becomes subdued. Thus, developmental processes can generate a wider array of forms within the morphospace. This suggests

[23] This example might support a theory of "multilevel selection"; however, it is proposed within a context markedly distinct from the traditional Hierarchical Theory of Selection. See Chap. 4, Sect. 4.1.3.

"that control… will end up fostering the characteristics that make the ontogenetic cycles safer, and this may lead to an increase in both complexity and variety, without needing to meet the demands of [ordinary] Darwinian selection" (Caponi 2008, p. 212; my translation from Spanish).Yet, this is not the only option for considering the functional role within body plans. In certain instances, when key innovations are introduced (Müller and Wagner 2003), they immediately increase a lineage's fitness through positive natural selection, e.g., the modification seen in the panda's sesamoid bone, which functions as an opposable thumb for grasping bamboo (Gould 1980). In other cases, certain characters of the body plan may serve as spandrels, indicating that these structures emerge from complex ontogenetic processes that, occasionally, are functionally coopted to increase the lineage's fitness, a phenomenon known as exaptation (Gould 2002, p. 1270). For example, some species of land snails utilize their umbilicus to protect their eggs; however, this umbilicus is a byproduct of the spiral growth of their shells, which leaves a space at the end of their axis (Gould 2002, p. 1259).

It is also necessary to recognize the significance of behavioral choice as a relevant factor at this level (Diogo 2017). That is to say, during these sudden origins, there is a close codependence between morphology and behavior. For instance, Purificação (2019) observed an individual shiny cowbird (*Molothrus bonariensis*) with a deformity in its beak—a grossly elongated and downward-curved maxilla—exhibiting feeding behavior different from other individuals of its species. This bird tilted its head to one side, using the base of its beak on the grass to capture and eat ants (see Fig. 11.5).[24] Additionally, according to Blumberg (2009, pp. 105–140), all bodies are plastic and capable of self-adjustment within the potential of development. The insistence on obtaining food, evading predators, growing under certain

Fig. 11.5 *Example of the codependence between morphology and behavior.* (**a**) An individual shiny cowbird (*Molothrus bonariensis*) with an abnormally elongated beak, which is also exhibiting distinct behavior compared to its relatives. (**b**) The abnormal individual compared with a normal one. (From Purificação (2019). Photographs courtesy of Keila Nunes Purificação, used under a Creative Commons Attribution 4.0 International License)

[24] See Van Hemert (2012b) on a similar case.

temperatures, etc., causes structural adjustments during growth (phenotypic accommodation), resulting in structures that appear optimally adapted to the prevailing living conditions. However, "malformed" organisms strive to survive in a highly competitive environment, coordinating their bodies according to the flexibility allowed by their developmental system, e.g., the malformations observed in quadruped animals born with two limbs.

Furthermore, other factors need to be considered. For example, in sexual populations, those significant discontinuous changes can distinctly separate populations. After the divergence event, Schwartz (1999) suggests that descendant populations acquire morphological, physiological, and behavioral innovations so discontinuous that they prevent them from reproducing with the ancestral population. In animals, individuals from different populations no longer mate with each other because they are emitting different pheromones, adopting distinct reproductive behaviors, accessing diverse food sources, and consequently visiting different habitats, etc.[25] Moreover, some field studies suggest that prospecies settlement may be more common than previously assumed. For instance, there have been reports of homeotic plant species establishing themselves in the wild over at least 160 years (Wang et al. 2011). Also, it has been identified that some of these individuals have coexisted with the wild state for centuries (Hameister et al. 2013). Similarly, Mallet (2007) documented the successful introduction of certain hybrid plants in specific locations in recent times.

Nevertheless, there is a case that sheds light on the early stages of prospecies establishment in the environment. Some years ago, in a locality in Chiapas, Mexico, the discovery of a plant with a very peculiar morphology was reported: it was small, lacked chlorophyll, had star-shaped flowers and no leaves, among many other unique features. But *Lacandonia schismatica* surprised scientists worldwide with its inverted floral whorls, unique in the world, with stamens positioned at the center and carpels on the outside; this arrangement deviated significantly from the common pattern observed in all flowers (Martínez and Ramos 1989) (Fig. 11.6). Biologists associated this process with a homeotic/heterotopic change and, consequently, used it as indisputable evidence of saltational evolution (Piñeyro-Nelson et al. 2010).

Interestingly, Vergara-Silva et al. (2003) examined several populations of *Triuris brevistylis*, a plant closely related to *Lacandonia*, found in nearby areas. Despite being dioecious, they discovered some bisexual individuals with homeotic variations, including the inverted arrangement observed in *Lacandonia* (suggesting a predisposition in the Triuridaceae group to produce similar variations). In addition, the discovery of other specimens of *Lacandonia* in Brazil was reported. Interestingly, both "species" exhibit minimal differences in morphology (Rudall et al. 2016), suggesting they could be considered the same species. These studies shed light on the

[25] This does not mean advocating for the perspective of species as reproductive entities (biological species); rather, avoiding hybridization favors each of the species to maintain their discrete characters. Furthermore, in some cases, hybridization can be a source of variation and provide opportunities for establishing new lineages.

Fig. 11.6 *Lacandonia schismatica*. (**a**) Complete view. (**b**) Flower. Here, I interpret *Lacandonia* as a natural example of a prospecies. Note that *Lacandonia* has inverted its floral whorl compared to the common pattern, with stamens in the center and carpels on the outside (photographs courtesy of Esteban Martínez).

possible origin of *Lacandonia*, likely beginning with a small number of individuals—akin to those observed in *Triuris brevistylis*—before becoming more abundant. I hypothesize that *Lacandonia* is a natural example of a prospecies, characterized by lacking intraspecific variation. But why do the two populations become significantly geographically distant? According to Esteban Martínez (personal communication), a researcher involved in the discovery of *Lacandonia*, this plant was likely distributed throughout the Latin American corridor extending from Mexico to Brazil but remained undocumented due to environmental degradation in some countries. In any case, the identification of distinct populations suggests that the next step toward its establishment could involve the expansion and dispersion of its type to progressively generate units below its level,—and thus avoid the risk of extinction.[26]

Given the above, it is clear that prospecies establishment should not be restricted to the conditions outlined by Bateman and DiMichele (2002) solely because their structures seem to exhibit low levels of fitness according to the paradigm of the "improbable." However, their proposal makes sense, especially because prospecies have many opportunities to settle in environments with little competition. Many of these unoccupied habitats must be practically devoid of competitors because in areas where resources are scarce and competitors survive, natural selection becomes more stringent, providing greater survival opportunities to specialists.

Another interesting aspect of their proposal is the significant role of environmental catastrophes. This hypothesis is not entirely implausible, considering that, as discussed earlier, many significant morphological changes are triggered by exposure to extreme environments. Alternatively, if populations arrive in those disturbed environments (or new environments to exploit, such as newly formed islands), which are practically empty, they have many chances to maintain their structures.

[26] Indeed, in the view of Esteban Martínez (personal communication), the taxonomic level of distinction of *Lacandonia* could be placed as high as the order, or even at a higher rank.

Likely, these individuals would initiate the creation of a new ecological network alongside the other species that are migrating in. Indeed, these notions resonate with Vrba's turnover-pulse hypothesis (1993), suggesting that environmental shifts resulting in species extinction and emigration, in turn, promote the origin and immigration of other species. Similarly, this would explain why there is an increase in disparity across distinct islands[27] and why this increase in disparity is also observed after a mass extinction event.

In conclusion, prospecies emerge from small populations, typically—albeit not exclusively—establishing themselves in habitats characterized by little competition, e.g., those that are unoccupied—often following an environmental catastrophe—or in environments where resources are very abundant. Given that prospecies possess characters at the level of the body plan, it is important to emphasize that these structures do not correspond to any adaptive optimum; nevertheless, they fulfill the minimum requirements for survival in different environments, and their existence is permitted solely by macroevolutionary natural selection. Additionally, some body plan structures occasionally obtain key innovations, enhancing their immediate reproductive success through positive natural selection or exaptation. Alternatively, their structures find their morphological adjustment due to behavioral persistence. In any case, body plan structures are more linked to phylogenetic constraints than to their fit with the function. Once established in an environment, these populations need to increase in number and disperse across different environments. This proliferation often results in the emergence of lower-level entities—such as species or subspecies—which serve as mechanisms to mitigate the risk of extinction.

11.2.2 Specific Stasis

According to the H-Evo-Devo Theory, *specific stasis* refers to the mode that recognizes the stability of characters at the species/genus level. That is, when an ancestral species gives rise to one or more descendant species, each occupies a different position within the morphospace and subsequently maintains a stable form across space and time. But what exactly drives this process of morphological stability? According to this model, the cause of stasis is developmental constraints acting on body plan characters. Below, I delve deeper into the discussion of this evolutionary mode.

Despite the widespread acknowledgment of the stasis phenomenon, biologists have not yet reached a consensus on its underlying causes. Some proposed explanations include stabilizing selection (Estes and Arnold 2007; Haller and Hendry 2014; González Marín et al. 2018; Rolland et al. 2023), habitat tracking ecology (Raia et al. 2011), phenotypic plasticity (Wake et al. 1983; West-Eberhard 2003; Piras et al. 2009), adaptive optima for specialists and generalists (Zelditch et al. 2020), factors related to geographic structure within the species (Lieberman and Dudgeon 1996; Eldredge et al. 2005; Nehm 2008), genetic constraints (Blows and Hoffmann

[27] This observation led Ernst Mayr to interpret speciation through his founder effect model. However, as discussed in Chap. 10, Sect. 10.1, this model has several drawbacks.

2005; Hansen and Houle 2004), among others. Here, it is crucial to note that these hypotheses have developed within the confines of the traditional evolutionary perspective, which lacks a comprehensive hierarchical evolutionary framework.

For example, some authors have pointed out that stabilizing selection cannot account for the cause of stasis, as the species body plan persists despite environmental fluctuations over geological time (Lieberman and Dudgeon 1996; Gould 2002, p. 878; Cleaveland et al. 2022). The main issue is that microevolutionary studies—such as those in quantitative genetics and stabilizing selection in the laboratory—are opposed to the observation of stasis at the macroevolutionary level. This apparent contradiction is commonly referred to as the "stasis paradox" (Haller and Hendry 2014). While attempts have been made to address this issue (e.g., Haller and Hendry 2014; Voje et al. 2018; O'Sullivan et al. 2019; Stroud et al. 2023; Garcia-Erill et al. 2024), none of these studies offer a definitive solution. Aware of these difficulties, paleontologists argue that different populations within a species evolve independently through local adaptations. Since the cumulative impact of these independent changes does not extend to altering the fundamental properties of the species, morphological stability is maintained (Lieberman and Dudgeon 1996; Eldredge et al. 2005). Yet how and under what circumstances does a directional trend arise to break the species' homeostasis? The issue lies in the fact that, despite recognizing macroevolutionary patterns, these paleontologists remain exclusively committed to microevolutionary mechanisms, as discussed in Chap. 9.

Other options also present difficulties. For instance, when discussing phenotypic plasticity, it is suggested that stasis occurs because populations within species adapt to diverse environments through the plasticity generated by their developmental system (Wake et al. 1983). However, Monroe and Bokma (2010) criticize this view, contending that phenotypic plasticity actually promotes divergence rather than constraining it. Alternatively, Zelditch et al. (2020) propose that stasis is only explained in species with specialist morphology that exhibit ecological behaviors akin to generalists. Specialists may be vulnerable to environmental changes, whereas generalists are more adapted to a wider range of adaptive zones. Yet the absence of functional constraints may lead to a decline in the phenotypes of generalists on the adaptive plateau, rendering them unable to revert to an ancestral adaptive optimum.

In any case, the resolution to these contradictions lies in the hierarchical framework provided by the H-Evo-Devo Theory. As discussed in the previous section, species are composed of subspecies, each exhibiting independent evolutionary patterns. Of course, this view aligns with the paleontologists' argument mentioned earlier. But unlike them, the H-Evo-Devo Theory proposes that subspecies cannot escape from the developmental constraints imposed by the species. This perspective explains why both natural selection and phenotypic plasticity can be observed at the microevolutionary level without affecting the higher level. Interestingly, this aspect of phenotypic plasticity had already been noted by some authors (West-Eberhard 2003; Piras et al. 2009; Braendle and Félix 2009). Furthermore, the issue concerning generalists and specialists is addressed by recognizing that body plan characters are mostly generalists, while lower-level simple characters tend to be specialists.

Nonetheless, neither phenotypic plasticity nor the adaptive generalists' argument explains the phenomenon of stasis at the species level.

Interestingly, while explanations of stasis based on developmental constraints exist, they are relatively rare and often reduce to a form of genetic constraint (e.g., the original explanation for stasis proposed by Eldredge and Gould in 1972). Otherwise, when considered, they often operate in concert with some of the previously mentioned mechanisms (e.g., Williamson 1987; Lieberman and Dudgeon 1996). However, under the H-Evo-Devo Theory, developmental constraints are defined according to Pere Alberch's (1982) argument, which conceptualizes developmental systems as highly integrated hierarchical entities fixed at specific positions within the morphospace.[28]

There is a widespread tendency to attribute stasis to an environmental context, likely due to the framework of traditional evolutionary theory, which emphasizes the significant role of natural selection. Nevertheless, as already noted in the section on substasis, developmental constraints and natural selection play different roles. As illustrated in Fig. 11.7, developmental constraints fix an ancestral structure at a specific position within the morphospace (A). Once these developmental constraints are broken, variation is guided by developmental bias toward other positions within the morphospace (B and C). Subsequently, developmental constraints once again stabilize this new variation. Finally, natural selection determines which variations (B or C) are retained or eliminated based on their functional role. Considering this, it can be concluded that while natural selection can eliminate forms at any time after their appearance, once established, developmental constraints can maintain almost unchanged structures for millions of years until their extinction. Thus, developmental constraints, rather than natural selection, are the cause of specific stasis. If natural selection were implicated in this process, it would play a role in both directing and fixating variation, but this is not the case.[29]

For example, Wang et al. (2022) recently documented what is possibly the first appearance of the elongated sesamoid bone that served as a thumb in ancestors of modern pandas. Their observations revealed that this structure had practically not changed for over seven million years of evolution except for proportion adjustments relative to body size and the emergence of a small hook at its distal end (see Fig. 11.8). Yet, they expected that, from its origin, this structure would gradually elongate toward an optimal adaptation. They attributed the lack of continuous elongation to a balance constrained by weight-bearing requirements given that the animal had adopted a plantigrade posture while walking. Notably, their interpretation was framed within the traditional theory of evolution. They anticipated slight accumulative variations in sesamoid elongation as natural selection guided these variations toward their adaptive optimum. However, when they did not observe this pattern, they conjectured that such variations would be eliminated by natural

[28] See Chap. 5, Sect. 5.1.2.

[29] For other interpretations of these dynamics, see Arthur (2021, pp. 91–95); González-Forero (2023).

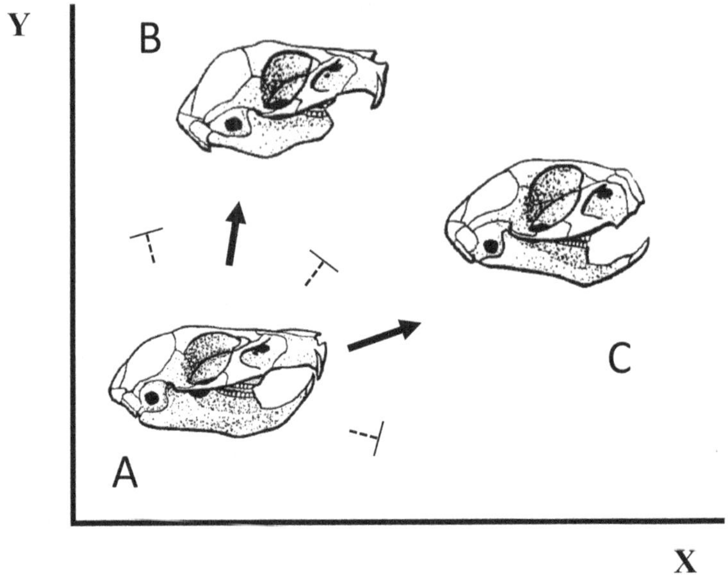

Fig. 11.7 *The dynamic of the macroevolutionary process through three main mechanisms: developmental constraints, developmental bias, and natural selection.* The diagram illustrates the distribution of guinea pig skull variations within the morphospace. The arrows indicate the possible pathways of evolutionary change. (**A**) Normal skull. (**B**) Skull with a reduction in the lower jaw. (**C**) Skull with a reduction in the upper jaw. Skull **A** represents the ancestral form, fixed in a specific position within the morphospace due to developmental constraints. Once these developmental constraints are broken, variation is channeled toward certain forms, and not others, through a mechanism known as developmental bias. Forms **B** and **C** are the options permitted by the developmental system, guided by developmental bias. These forms become fixed in new positions within the morphospace due to developmental constraints. Finally, natural selection will determine which of these forms (**B** or **C**) persists according to functional requirements. Note that these three processes have distinct roles. Developmental constraints maintain form stability, developmental bias drives variation to other positions, and natural selection acts to either retain or eliminate these forms based on their functional role. (The scheme is inspired by Alberch's (1982) representation of morphospace, as shown in Fig. 5.2a. The guinea pig skull drawings are from Wright and Wagner (1925))

selection, thereby preventing their continuous growth because of a secondary function related to locomotion.

Conversely, according to the H-Evo-Devo Theory, the elongation of the sesamoid bone appeared suddenly in the founding panda population without transitional forms. While the increase in fitness is attributed to macroevolutionary natural selection, the model predicts that the sesamoid bone did not need to undergo continuous change toward a functional trend, as developmental constraints stabilized this structure from its origin. Nevertheless, in subsequent generations, a slight adjustment took place, resulting in the emergence of the observed hook at the distal end. In this scenario, this hook appeared discontinuously, and subsequently, developmental constraints stabilized it, while natural selection enhanced the fitness of the

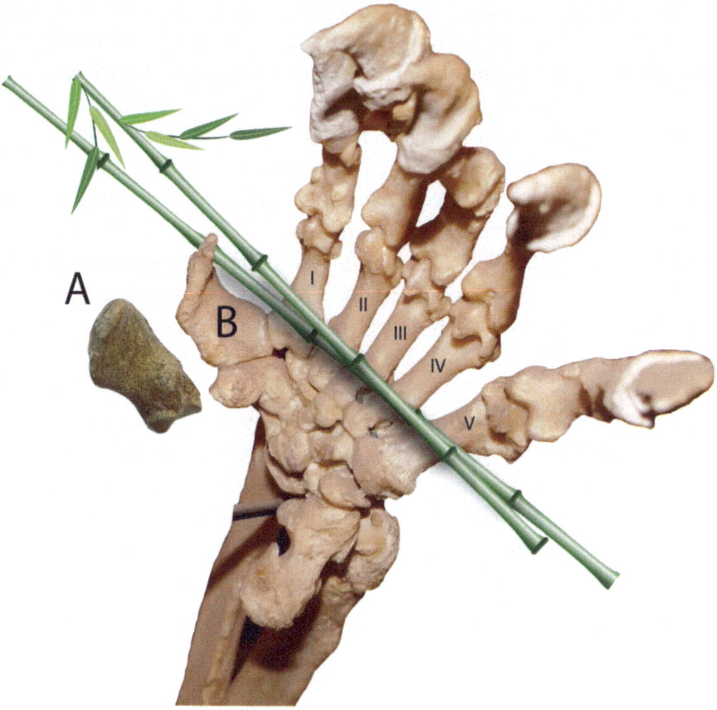

Fig. 11.8 *Comparison between the first elongated sesamoid bone of ancestral pandas* (**A**) *and the elongated sesamoid bone of modern pandas* (**B**). Note the hook at the distal end of the modern panda's sesamoid bone. (Modified from Wang et al. 2022)

individuals possessing it. Note that, in the case of the origin of the hook, both mechanisms were operating at a microevolutionary level. In any case, the fixation of the elongated sesamoid bone and the presence of the hook remained morphologically stable since their inception due to the hierarchical action of developmental constraints. Subsequently, natural selection—also acting hierarchically—increased the fitness of the individuals possessing them.

11.2.3 Specific Extinction

According to the H-Evo-Devo Theory, *specific extinction* represents the evolutionary mode indicating the death of species/genera. That is, once a species/genus arises and persists over a long period of geological time, consequently, it vanishes at any moment in its life history. The focal points of discussion within this evolutionary mode revolve around its hierarchical nature: whether processes at lower, higher, or both levels govern species extinction.

Interestingly, several authors express their concern and the urgency of incorporating a hierarchical perspective into the topic of extinction (e.g., Eldredge 2003;

Jablonski 2017; Oney et al. 2013; Congreve et al. 2018; Levin 2019). Given this, it becomes clear that species extinction is contingent upon processes operating at both lower and higher levels. Regarding the causes of subspecies extinction (as outlined in Sect. 11.1.3), they impact the species level in two ways. First, if a species lacks subspecies, then it behaves like a subspecies and is subject to the same causes affecting it. Second, if a species comprises two or more subspecies, then its extinction becomes dependent on the decline of each subspecies individually. I will term the latter phenomenon of extinction *the loss of parts*. As for the causes of higher taxa extinction (see Chap. 12, Sect. 12.1.3), they similarly affect species, particularly because such causes produce global effects. This type of extinction corresponds to the concept of *mass extinction*. Below, I delve into these two criteria with a bit more detail.

Loss of Parts (Initiating Conditions) Levin (2019) aptly proposes that species extinction is determined by the extinction of subspecies, which occupy different geographical ranges in heterogeneous environments. Similarly, other authors have highlighted that species with narrow geographical ranges are more prone to extinction, unlike species with broader geographical ranges that are more likely to survive. Accordingly, the ability to disperse and colonize new environments becomes crucial (Birand et al. 2012; Saupe et al. 2015; Chichorro et al. 2019; Yan et al. 2022). In a different vein, to emphasize the taxonomic structure through the phenomenon of extinction, Flessa and Jablonski (1985) have postulated that fossil record evidence suggests that the increased number of elements at a lower level (e.g., species) boosts the survival probability of higher levels (e.g., family). This demonstrates that a higher-ranking taxon is not built through an increase in the number of species, as is commonly assumed.[30]

As previously noted, according to the H-Evo-Devo Theory, subspecies represent the phenomenon of diversification and adaptation of species to different environments, protecting species from extinction. Therefore, the dynamics of a species' background extinction are reflected in the extinction of each of its parts as subspecies. So, the risk of a species' extinction is proportional to the number of subspecies it produces. That is, the fewer the subspecies, the greater the risk of extinction, by contrast, the more subspecies, the greater the probability of the species' survival. In any case, if a species is recent (prospecies) or loses all its subspecies—e.g., during an environmental catastrophe—then this entity will exhibit the characteristics of the subspecies; hence, it can succumb to the causes that directly affect the lower level (see Sect. 11.1.3).

[30] In the traditional view of evolution, higher taxa originate through speciation events. That is, the phylogenetic tree is structured from the bottom up: species emerge first, followed by other taxonomic categories (see Chap. 8, Sect. 8.1.1). However, the H-Evo-Devo Theory proposes an inverted structure for the phylogenetic tree, where higher taxa arose first, followed by the origin of lower-level units. For a detailed discussion of this argument, see Chap. 12, Sect. 12.1.1.

Mass Extinction (Boundary Conditions) Some researchers have emphasized the distinction between background extinction and mass extinction, particularly regarding the type of selectivity involved. For example, Jablonski (1989) notes that during periods of background extinction, genera with a greater number of species and widespread distribution may exhibit higher resistance to extinction. However, these genera sometimes perish during mass extinction events, while those with fewer species survive. This prompts the question: What properties do these less diverse genera possess that enhance their survival?

In interpreting the H-Evo-Devo Theory, it is evident that not all species extinction events can be attributed solely to the extinction of their parts as subspecies. While possessing a substantial number of subspecies ensures survival during global catastrophes, there are occasions when certain species' properties function as spandrels, providing them with an advantage during adversity (Gould 2002). Moreover, it is also essential to emphasize the role of historical contingencies in determining the winners (Gould 1989). A significant ecological catastrophe on a large scale disrupts any advantageous property that species may possess, and this happens randomly. Therefore, the causes of mass extinctions also contribute to species extinction, which impacts the variational structure of higher taxa as well.[31]

11.3 Conclusions

While there is considerable evidence to divide evolution into at least two levels, the widespread acceptance of the Modern Synthesis framework has hindered any progress in this area, particularly because it supports the extrapolationist argument.[32] Despite the significant contributions that many evolutionists have made to the subjects of microevolution and macroevolution, a clear and successful distinction between these levels has not yet been achieved. In short, traditional models of evolution have created more issues than resolutions, resulting in various contradictory explanations without offering an effective means to reconcile both levels. However, as observed, the H-Evo-Devo Theory stands as the first modern endeavor to address this challenge. Microevolution and macroevolution represent distinct levels, each with its own mechanisms and explanations of the evolutionary process. Therefore, this model fosters new research initiatives aimed at comprehending the hierarchical nature of evolution. A concise overview of the differences between microevolution and macroevolution according to this model is provided in Table 11.1 and Fig. 11.9.

[31] Delve further into this topic in Chap. 12, Sect. 12.1.3.

[32] See Chap. 10, Sect. 10.2.

Table 11.1 *Comparison between microevolution and macroevolution according to the H-Evo-Devo Theory.* In particular, this table shows the mechanisms involved in the (1) origin, (2) stability, and (3) extinction of both subspecies and species

Level	Microevolution	Macroevolution
Unity	Subspecies	Species/Genera
Type of variation	Simple characters—e.g., color, pattern, size, proportion, etc.	Complex characters or body plan characters
Type of interaction with the environment	Adaptation to a local environment	Fitness to different environments
(1) Origin	Darwinian mechanisms: Dispersal model	SupraDarwinian mechanisms: Prospecies model
Variation	Genetic changes, changes induced by environmental influence, changes produced by physical rules Developmental bias	Genetic changes, changes induced by environmental influence Developmental bias
Establishment in environment	Natural selection, phenotypic plasticity, niche construction, and behavioral choice	Natural selection, phenotypic accommodation, and behavioral choice
(2) Stasis	Developmental constraints operating on simple characters	Developmental constraints operating on complex characters
(3) Extinction	Factors that alter their local habitat and their population structure such as environmental stochasticity, habitat fragmentation, habitat loss, habitat quality, competition, population size fluctuations, dispersal ability, etc.	Loss of parts and mass extinctions

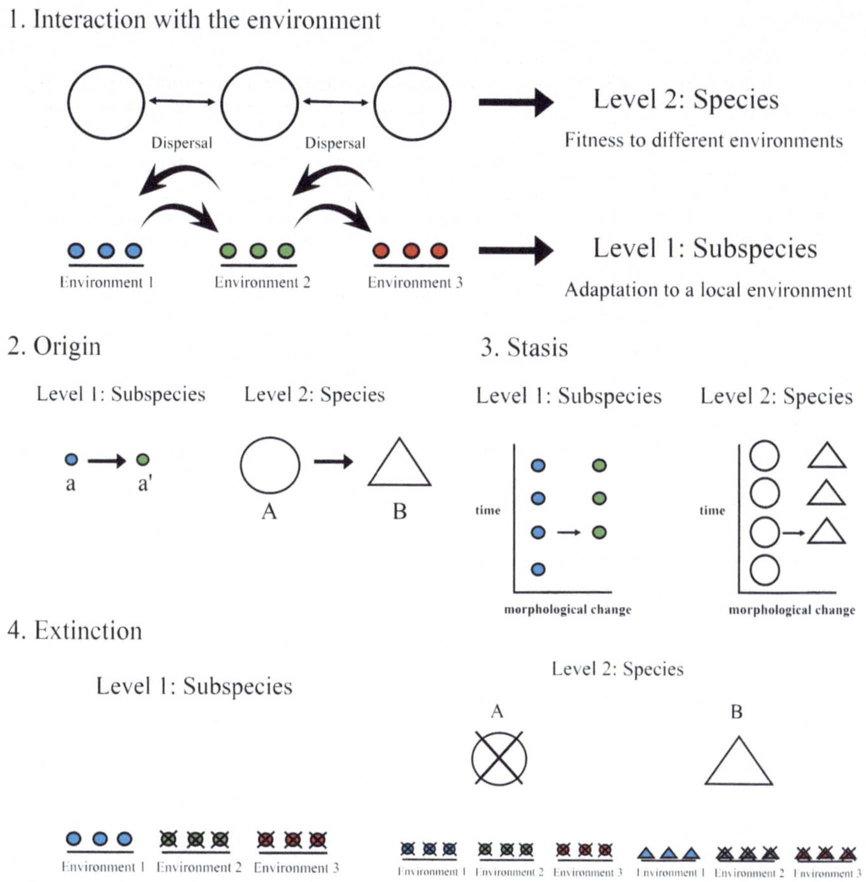

Fig. 11.9 *Comparison between microevolution and macroevolution, according to the H-Evo-Devo Theory.* Level 1: Microevolution pertains to the subspecies level. Level 2: Macroevolution corresponds to the species/genera level. **(1)** *Differences concerning interaction with the environment.* At level 1, subspecies are adapted to local environments. At level 2, species/genera can fit to different environments. **(2)** *Origin.* At level 1, changes in simple characters, such as a color shift from (a) to (a'). At level 2, changes in complex characters such as the shift from one body plan (A) to another (B). **(3)** *Stasis.* At level 1, once a color is established, it remains stable. At level 2, once a body plan change is attained, it remains stable. **(4)** *Extinction.* At level 1, subspecies disappear due to factors that alter their local habitat and their population structure. At level 2, species become extinct either because they gradually lose their parts as subspecies or due to mass extinction events (A). Alternatively, if a species has lost all its subspecies, or has recently emerged (prospecies), it behaves like a subspecies. Consequently, it is directly affected by the same causes as at the lower level (B)

References

Akiyama T, Kinoshita K (2021) Body color expression in birds. In: Hashimoto H, Goda M, Futahashi R, Kelsh R, Akiyama T (eds) Pigments, pigment cells and pigment patterns. Springer, Singapore, pp 91–126

Alberch P (1982) Developmental constraints in evolutionary processes. In: Bonner JT (ed) Evolution and development. Springer, New York, pp 313–332

Alberch P (1989) The logic of monsters: evidence for internal constraint in development and evolution. Geobios 22:21–57

al-Mosleh S, Choi GPT, Abzhanov A, Mahadevan L (2021) Geometry and dynamics link form, function, and evolution of finch beaks. Proc Natl Acad Sci USA 118:e2105957118. https://doi.org/10.1073/pnas.2105957118

Araguas RM, Sanz N, Viñas J, Vidal O (2018) MC1R polymorphism associated with plumage color variations in *Coturnix chinensis*. Anim Genet 49:475–477

Arendt JD (2015) Effects of dispersal plasticity on population divergence and speciation. Heredity 115:306–311

Armsworth PR, Roughgarden JE (2008) The structure of clines with fitness-dependent dispersal. Am Nat 172:648–657

Arthur W (2004) The effect of development on the direction of evolution: toward a twenty-first century consensus. Evol Dev 6:282–288

Arthur W (2021) Understanding evo-devo. Cambridge University Press, Cambridge

Averof M, Patel NH (1997) Crustacean appendage evolution associated with changes in Hox gene expression. Nature 388:682–686

Badyaev AV, Foresman KR, Young RL (2005) Evolution of morphological integration: developmental accommodation of stress-induced variation. Am Nat 166:382–395

Bateman RM, DiMichele WA (1994) Saltational evolution of form in vascular plants: a neoGoldschmidtian synthesis. In: Ingram DS, Hudson A (eds) Shape and form in plants and fungi. Academic Press, London, pp 61–100

Bateman RM, DiMichele WA (2002) Generating and filtering major phenotypic novelties: neoGoldschmidtian saltation revisited. In: Cronk QCB, Bateman RM, Hawkins JA (eds) Developmental genetics and plant evolution. Taylor and Francis, London, pp 109–159

Behringer V, Deschner T, Murtagh R, Stevens JMG, Hohmann G (2014) Age-related changes in Thyroid hormone levels of bonobos and chimpanzees indicate heterochrony in development. J Hum Evol 66:83–88

Bengtsson J, Milbrink G (1995) Predicting extinctions: interspecific competition, predation and population variability in experimental *Daphnia* populations. Oecologia 101:397–406

Ben-Zion Y, Fried Y, Shnerb NM (2012) Migration, coherence and persistence in a fragmented landscape. Theor Ecol 5:481–493

Birand A, Vose A, Gavrilets S (2012) Patterns of species ranges, speciation, and extinction. Am Nat 179:1–21

Blows MW, Hoffmann AA (2005) A reassessment of genetic limits to evolutionary change. Ecology 86:1371–1384

Blumberg MS (2009) Freaks of nature: what anomalies tell us about development and evolution. Oxford University Press, New York

Böhmer C, Amson E, Arnold P, van Heteren AH, Nyakatura JA (2018) Homeotic transformations reflect departure from the mammalian "rule of seven" cervical vertebrae in sloths: inferences on the Hox code and morphological modularity of the mammalian neck. BMC Evol Biol 18:84. https://doi.org/10.1186/s12862-018-1202-5

Bolter DR, Zihlman AL (2012) Skeletal development in *Pan paniscus* with comparisons to *Pan troglodytes*. Am J Phys Anthropol 147:629–636

Bonett RM (2021) Heterochrony. In: Müller GB (ed) Nuño de la Rosa L. Evolutionary developmental biology. Springer, Cham, pp 15–28

Boughner JC, Dean MC (2008) Mandibular shape, ontogeny and dental development in bonobos (*Pan paniscus*) and chimpanzees (*Pan troglodytes*). Evol Biol 35:296–308

Bowman JL, Smyth DR, Meyerowitz EM (1991) Genetic interactions among floral homeotic genes of Arabidopsis. Development 112:1–20

Brady TPA (2023) Selected look at niche construction theory including its incorporation of the notion of phenotype-mediated developmental plasticity. Biol Theory 18:20–29

Braendle C, Felix MA (2009) The other side of phenotypic plasticity: a developmental system that generates an invariant phenotype despite environmental variation. J Biosci 34:543–551

Caballero L, Benítez M, Alvarez-Buylla ER, Hernández S, Arzola AV, Cocho G (2012) An epigenetic model for pigment patterning based on mechanical and cellular interactions. J Exp Zool (Mol Dev Evol) 318:209–223

Cahill AE, Aiello-Lammens ME, Fisher-Reid MC, Hua X, Karanewsky CJ, Ryu HY, Sbeglia GC, Spagnolo F, Waldron JB, Warsi O, Wiens JJ (2012) How does climate change cause extinction? Proc R Soc B 280:20121890. https://doi.org/10.1098/rspb.2012.1890

Camacho C, Sanabria-Fernández A, Baños-Villalba A, Edelaar P (2020) Experimental evidence that matching habitat choice drives local adaptation in a wild population. Proc R Soc B 287:20200721. https://doi.org/10.1098/rspb.2020.0721

Cao D, Garai S, DiFrisco J, Veenvliet JV (2024) The logic of monsters: development and morphological diversity in stem-cell-based embryo models. Interface Focus 14:20240023. https://doi.org/10.1098/rsfs.2024.0023

Caponi G (2008) Selección interna: el control de la filogenia por la ontogenia en una perspectiva variacional. Theoria 62:196–218

Carlson ML, Stoddard MC (2024) Evolution of plumage patterns in a pattern morphospace: a phylogenetic analysis of melanerpine woodpeckers. Am Nat 203:55–72

Carroll SB (1995) Homeotic genes and the evolution of arthropods and chordates. Nature 376:479–485

Chichorro F, Juslén A, Cardoso P (2019) A review of the relation between species traits and extinction risk. Biol Conserv 237:220–229

Cleaveland C, Prothero DR, Syverson VJP (2022) How did La Brea Swainson's hawks (*Buteo swainsoni*) respond to climate change? In: Lucas SG, Blodgett RB, Lichtig AJ, Hunt AP (eds) New Mexico Museum of Natural History and Science Bulletin 90. Fossil Record 8, Albuquerque, pp 111–117

Congreve CR, Falk AR, Lamsdell JC (2018) Biological hierarchies and the nature of extinction. Biol Rev 93:811–826

Corbel H, Legros A, Haussy C, Jacquin L, Gasparini J, Karimi B, Frantz A (2016) Stress response varies with plumage colour and local habitat in feral pigeons. J Ornithol 157:825–837

Darwin C (1859) On the origin of species by means of natural selection, or preservation of favored races in the struggle for life. John Murray, London

Dayan DI, Graham MA, Baker JA, Foster SA (2019) Incorporating the environmentally sensitive phenotype into evolutionary thinking. In: Uller T, Laland KN (eds) Evolutionary causation biological and philosophical reflections. MIT Press, Cambridge MA and London, pp 81–107

De León LF, Podos J, Gardezi T, Herrel A, Hendry AP (2014) Darwin's finches and their diet niches: the sympatric coexistence of imperfect generalists. J Evol Biol 27:1093–1104

Delhey K, Valcu M, Dale J, Kempenaers B (2023) The evolution of carotenoid-based plumage colours in passerine birds. J Anim Ecol 92:66–77

Diaz RE (2020) Evo-Devo Path as a bridge between evolution, morphological disparity, and medicine with comments on "hopeful monsters" in the age of genomics. Curr Mol Biol Rep 6:79. https://doi.org/10.1007/s40610-020-00131-2

Diogo R (2017) Evolution driven by organismal behavior: a unifying view of life, function, form, mismatches, and trends. Springer, Cham

Diogo R (2020) Introduction to evolutionary developmental pathology, or evo-devo-path: on Neodarwinism, natural mutants, hopeful monsters, syndromes, genomics, variations, humans, apes, chameleons, and dinosaurs. Curr Mol Biol Rep 6:11. https://doi.org/10.1007/s40610-020-00133-0

Diogo R, Molnar JL, Wood B (2017) Bonobo anatomy reveals stasis and mosaicism in chimpanzee evolution, and supports bonobos as the most appropriate extant model for the common ancestor of chimpanzees and humans. Sci Rep 7:608. https://doi.org/10.1038/s41598-017-00548-3
Diogo R, Razmadze D, Siomava N, Douglas N, Fuentes JSM, Duerinckx A (2019) Musculoskeletal study of cebocephalic and cyclopic lamb heads illuminates links between normal and abnormal development, evolution and human pathologies. Sci Rep 9. https://doi.org/10.1038/s41598-018-37735-9
Dittrich-Reed DR, Fitzpatrick BM (2013) Transgressive hybrids as hopeful monsters. Evol Biol 40:310–315
Dobreva MP, Camacho J, Abzhanov A (2022) Time to synchronize our clocks: connecting developmental mechanisms and evolutionary consequences of heterochrony. J Exp Zool B Mol Dev Evol 338:87–106
Doebeli M, Dieckmann U (2003) Speciation along environmental gradients. Nature 421:259–264
Duckworth RA (2009) The role of behavior in evolution: a search for mechanism. Evol Ecol 23:513–531
Edelaar P, Siepielski AM, Clobert J (2008) Matching habitat choice causes directed gene flow: a neglected dimension in evolution and ecology. Evolution 62:2462–2472
Edelaar P, Jovani R, Gomez-Mestre I (2017) Should I change or should I go? phenotypic plasticity and matching habitat choice in the adaptation to environmental heterogeneity. Am Nat 190:506–520
Eldredge N (2003) The sloshing bucket: how the physical realm controls evolution. In: Crutchfield J, Schuster P (eds) Evolutionary dynamics: exploring the interplay of selection, accident, neutrality, and function. Oxford University Press, New York, pp 3–32
Eldredge N, Gould SJ (1972) Punctuated equilibria: an alternative to phyletic gradualism. In: Schopf TJM (ed) Models in paleobiology. Freeman, Cooper and Co, San Francisco California, pp 82–115
Eldredge N, Thomson JN, Brakefield PM, Gavrilets S, Jablonski D, Jackson JBC, Lenski RE, Lieberman BS, McPeek MA, Miller W III (2005) The dynamics of evolutionary stasis. Paleobiology 31:133–145
Eliason CM, Clarke JA, Kane SA (2022) Wrinkle nanostructures generate a novel form of blue structural color in great argus flight feathers. iScience 26:105912. https://doi.org/10.1016/j.isci.2022.105912
Estes S, Arnold SJ (2007) Resolving the paradox of stasis: models with stabilizing selection explain evolutionary divergence on all timescales. Am Nat 169:227–244
Fagan WF, Aumann C, Kennedy CM, Unmack PJ (2005) Rarity, fragmentation, and the scale dependence of extinction risk in desert fishes. Ecology 86:34–41
Fahrig L (2003) Effects of habitat fragmentation on biodiversity. Annu Rev Ecol Evol Syst 34:487–515
Fernández-Mazuecos M, Glover BJ (2017) The evo-devo of plant speciation. Nat Ecol Evol 1:110. https://doi.org/10.1038/s41559-017-0110
Flessa KW, Jablonski D (1985) Declining Phanerozoic background extinction rates: effect of taxonomic structure? Nature 313:216–218
Frazzetta TH (1975) Complex adaptations in evolving populations. Sinauer Associates, Sunderland
Frazzetta TH (2012) Flatfishes, turtles, and bolyerine snakes: evolution by small steps or large, or both? Evol Biol 39:30–60
Frincke-Craig M, Brown JL, Briggs CW, Poulson SR, Collopy MW, Feldman CR (2015) Relationships between plumage coloration, diet diversity, and winter body condition in the Lesser Goldfinch. J Ornithol 156:143–151
Fujiwara M, Takada T (2017) Environmental stochasticity. In: eLS. John Wiley and Sons, Chichester. https://doi.org/10.1002/9780470015902.a0021220.pub2
Furuichi T (2023) Prolonged sexual receptivity in females and its impact on the evolution of bonobos. In: Furuichi T, Idani G, Kimura D, Ihobe H, Hashimoto C (eds) Bonobos and people at Wamba: 50 years of research. Springer, Singapore, pp 545–564

Gaggiotti OE, Hanski I (2004) Mechanisms of population extinction. In: Hanski I, Gaggiotti OE (eds) Ecology, genetics and evolution of metapopulations. Academic Press, Burlington, pp 337–366

Galván I, Alonso-Alvarez C (2009) The expression of melanin-based plumage is separately modulated by exogenous oxidative stress and a melanocortin. Proc Biol Sci 276:3089–3097

Galván I, Jorge A, Nielsen JT, Møller AP (2019) Pheomelanin synthesis varies with protein food abundance in developing goshawks. J Comp Physiol B 189:441–450

Garcia-Erill G, Wang X, Rasmussen MS, Quinn L, Khan A, Bertola LD, Santander CG, Balboa RF, Ogutu JO, Pečnerová P, Hanghøj K, Kuja J, Nursyifa C, Masembe C, Muwanika V, Bibi F, Moltke I, Siegismund HR, Albrechtsen A, Heller R (2024) Extensive population structure highlights an apparent paradox of stasis in the impala (*Aepyceros melampus*). Mol Ecol 33:e17539. https://doi.org/10.1111/mec.17539

Gayford JH, Irschick DJ, Chin A, Rummer JL (2025) Heterochrony and oophagy underlie the evolution of giant filter-feeding lamniform sharks. Evol Dev 27:e12496. https://doi.org/10.1111/ede.12496

Gilbert S (2016) Developmental plasticity and developmental symbiosis: the return of Eco-Devo. Curr Top Dev Biol 116:415–433

Gilbert B, Levine JM (2013) Plant invasions and extinction debts. Proc Natl Acad Sci USA 110:1744–1749

Goldschmidt RB (1933) Some aspects of evolution. Science 78:539–547

Goldschmidt RB (1940) The material basis of evolution. Yale University Press, New Haven

González Marín A, Olave M, Avila LJ, Sites JW Jr, Morando M (2018) Evidence of body size and shape stasis driven by selection in Patagonian lizards of the *Phymaturus patagonicus* clade (Squamata: Liolaemini). Mol Phylogenet Evol 129:226–241

González-Forero M (2023) How development affects evolution. Evolution 77:562–579

Gould SJ (1980) The panda's thumb: more reflections in natural history. WW Norton and Company, New York

Gould SJ (1989) Wonderful life: the Burgess Shale and the nature of history. WW Norton and Company, New York

Gould SJ (2002) The structure of evolutionary theory. The Belknap Press of Harvard University Press, Cambridge MA

Grant BR, Grant PR (2003) What Darwin's finches can teach us about the evolutionary origin and regulation of biodiversity. Bioscience 53:965–975

Gudowska A, Janas K, Wieczorek J, Woznicka O, Płonka PM, Drobniak SM (2022) Canalised and plastic components of melanin-based colouration: a diet-manipulation experiment in house sparrows. Sci Rep 12:18484. https://doi.org/10.1038/s41598-022-21811-2

Guinard G (2012) Evolutionary concepts meet the neck of penguins (Aves: Sphenisciformes), towards a "survival strategy" for Evo-Devo. Theory Biosci 131:231–242

Guinard G (2015) Introduction to evolutionary teratology, with an application to the forelimbs of Tyrannosauridae and Carnotaurinae (Dinosauria: Theropoda). Evol Biol 42:20–41

Guinard G (2020) Evolutionary teratology and the path to break through the mould of the synthesis paradigm. Curr Mol Biol Rep 6:51–61

Gunderson AR, Forsyth MH, Swaddle JP (2009) Evidence that plumage bacteria influence feather coloration and body condition of eastern bluebirds *Sialia sialis*. J Avian Biol 40:440–447

Gurevitch J, Padilla DK (2004) Are invasive species a major cause of extinctions? Trends Ecol Evol 19:470–474

Haller BC, Hendry AP (2014) Solving the paradox of stasis: squashed stabilizing selection and the limits of detection. Evolution 68:483–500

Hameister S, Nutt P, Theißen G, Neuffer B (2013) Mapping a floral trait in Shepherds purse – 'Stamenoid petals' innatural populations of *Capsella bursa-pastoris* (L.). Flora 208:641–647

Hanegraef H, Spoor F (2024) Maxillary morphology of chimpanzees: captive versus wild environments. J Anat 244:977–994

Hansen TF, Houle D (2004) Evolvability, stabilizing selection, and the problem of stasis. In: Pigliucci M, Preston K (eds) Phenotypic integration: studying the ecology and evolution of complex phenotypes. Oxford University Press, New York, pp 130–150

Harrison S (1991) Local extinction in a metapopulation context: an empirical evaluation. Biol J Linn Soc 42:73–88
Haupaix N, Manceau M (2020) The embryonic origin of periodic color patterns. Dev Biol 460:70–76
Heinrichs JA, Bender DJ, Schumaker NH (2016) Habitat degradation and loss as key drivers of regional population extinction. Ecol Model 335:64–73
Higgins K (2009) Metapopulation extinction risk: dispersal's duplicity. Theor Popul Biol 76:146–155
Hill GE (2006) Environmental regulation of ornamental coloration. In: Hill GE, McGraw KJ (eds) Bird coloration: mechanisms and measurements. Volume I. Harvard University Press, Cambridge MA, pp 507–560
Hollander J, Verzijden M, Svensson E, Brönmark C (2014) Dispersal and phenotypic plasticity. In: Hansson L-A, Åkesson S (eds) Animal movement across scales. Oxford University Press, Oxford, pp 110–125
Houle D, Jones LT (2021) Micro-evo-devo. In: Müller GB (ed) Nuño de la Rosa L. Evolutionary developmental biology. Springer, Cham, pp 1047–1060
Hunt JJFG, Bonsall MB (2009) The effects of colonization, extinction and competition on co-existence in metacommunities. J Anim Ecol 78:866–879
Inaba M, Chuong C-M (2020) Avian pigment pattern formation: developmental control of macro- (across the body) and micro- (within a feather) level of pigment patterns. Front Cell Dev Biol 8. https://doi.org/10.3389/fcell.2020.00620
Jablonski D (1989) The biology of mass extinction: a palaeontological view. Phil Trans R Soc Lond B 325:357–368
Jablonski D (2017) Approaches to macroevolution: 2. Sorting of variation, some overarching issues, and general conclusions. Evol Biol 44:451–475. https://link.springer.com/article/10.1007/s11692-017-9434-7
Jaeger J, Irons D, Monk N (2012) The inheritance of process: a dynamical systems approach. J Exp Zool (Mol Dev Evol) 318B:591–612
Järvistö PE, Calhim S, Schuett W, Velmala W, Laaksonen T (2015) Foster, but not genetic, father plumage coloration has a temperature-dependent effect on offspring quality. Behav Ecol Sociobiol 69:335–346
Jawor JM, Breitwisch R (2003) Melanin ornaments, honesty, and sexual selection. Auk 120:249–265
Kovalaskas S, Rilling JK, Lindo J (2020) Comparative analyses of the *Pan* lineage reveal selection on gene pathways associated with diet and sociality in bonobos. Genes Brain Behav 20. https://doi.org/10.1111/gbb.12715
Kulikova IV (2021) Molecular mechanisms and gene regulation of melanic plumage coloration in birds. Russ J Genet 57:893–891
Labrum MJ, Gomulkiewicz R (2020) Metapopulation Allee effects, habitat destruction, and extinction in metacommunities. J Math Biol 80:93–110
LaFountain AM, Prum RO, Frank HA (2015) Diversity, physiology, and evolution of avian plumage carotenoids and the role of carotenoid-protein interactions in plumage color appearance. Arch Biochem Biophys 572:201–212
Laland KN, Odling-Smee FJ, Feldman MW (2019) Understanding niche construction as an evolutionary process. In: Uller T, Laland KN (eds) Evolutionary causation biological and philosophical reflections. MIT Press, Cambridge MA and London, pp 127–152
Lande R (1993) Risks of population extinction from demographic and environmental stochasticity and random catastrophes. Am Nat 142:911–927
Landeen EA, Badyaev AV (2012) Developmental integration of feather growth and pigmentation and its implications for the evolution of diet-derived coloration. J Exp Zool (Mol Dev Evol) 318:59–70
Leclaire S, Perret S, Galván I, Bonadonna F (2019) Pheomelanin-based coloration is related to individual quality and oxidative stress in blue petrels. Evol Ecol 33:873–887

Levin DA (2019) Intraspecific lineages as focal points in the extinction and persistence of species. Plant Syst Evol 305:719–726

Levis NA, Pfennig DW (2016) Evaluating "plasticity-first" evolution in nature: key criteria and empirical approaches. Trends Ecol Evol 31:563. https://doi.org/10.1016/j.tree.2016.03.012

Li R, Wang Y, Liu Y, Li D, Tian Y, Liu X, Kang X, Li Z (2023) Effects of SLC45A2 and GPNMB on melanin deposition based on transcriptome sequencing in chicken feather follicles. Animals (Basel) 13:2608. https://doi.org/10.3390/ani13162608

Lieberman BS, Dudgeon S (1996) An evaluation of stabilizing selection as a mechanism for stasis. Palaeogeogr Palaeoclimatol Palaeoecol 127:229–238

Lieberman DE, Carlo J, Ponce de León M, Zollikofer CPE (2007) A geometric morphometric analysis of heterochrony in the cranium of chimpanzees and bonobos. J Hum Evol 52:647–662

Lin SJ, Foley J, Jiang TX, Yeh CY, Wu P, Foley A, Yen CM, Huang YC, Cheng HC, Chen CF, Reeder B, Jee SH, Widelitz RB, Chuong CM (2013) Topology of feather melanocyte progenitor niche allows complex pigment patterns to emerge. Science 340:1442–1445

López-Rull I, Salaberría C, Fargallo JA (2023) Plastic plumage colouration in response to experimental humidity supports Gloger's rule. Sci Rep 13:858. https://doi.org/10.1038/s41598-023-28090-5

Mahler B, Araujo LS, Tubaro PL (2003) Dietary and sexual correlates of carotenoid pigment expression in dove plumage. Condor 105:258–267

Malcomber ST, Kellogg EA (2005) SEPALLATA gene diversification: brave new whorls. Trends Plant Sci 10:427–435

Mallet J (2007) Hybrid speciation. Nature 446:279–283

Maresca B, Schwartz JH (2006) Sudden origins: a general mechanism of evolution based on stress protein concentration and rapid environmental change. Anat Rec B New Anat 289B:38–46

Matrková J, Remeš V (2012) Environmental and genetic effects on pigment-based vs. structural component of yellow feather colouration. PLoS One 7:e36640. https://doi.org/10.1371/journal.pone.0036640

Martínez E, Ramos CH (1989) Lacandoniaceae (Triuridales): Una Nueva Familia de México. Ann Missouri Bot Gard 76:128–135

Matsuda R (1987) Animal evolution in changing environments: with special reference to abnormal metamorphosis. John Wiley and Sons, New York

Matthies D, Bräuer I, Maibom W, Tscharntke T (2004) Population size and the risk of local extinction: empirical evidence from rare plants. Oikos 105:481–488

McCauley SJ, Davis CJ, Werner EE, Robeson MS II (2014) Dispersal, niche breadth and population extinction: colonization ratios predict range size in North American dragonflies. J Anim Ecol 83:858–865

McGraw KJ (2006) Mechanics of carotenoid-based coloration. In: Hill GE, McGraw KJ (eds) Bird coloration: mechanisms and measurements. Volume I. Harvard University Press, Cambridge MA, pp 177–242

McLaughlin JF, Hellmann JJ, Boggs CL, Ehrlich PR (2002) Climate change hastens population extinctions. Proc Natl Acad Sci USA 99:6070–6074

McNamara KJ (2012) Heterochrony: the evolution of development. Evo Edu Outreach 5:203–218

McNamara KJ, McKinney ML (2005) Heterochrony, disparity, and macroevolution. Paleobiology 31:17–26

Mendes-Pinto MM, LaFountain AM, Stoddard MC, Prum RO, Frank HA, Robert B (2012) Variation in carotenoid–protein interaction in bird feathers produces novel plumage coloration. J R Soc Interface 9:3338–3350

Minelli A, Fusco G (2012) On the evolutionary developmental biology of speciation. Evol Biol 39:242–254

Minelli A, Chagas-Júnior A, Edgecombe GD (2009) Saltational evolution of trunk segment number in centipedes. Evol Dev 11:318–322

Moczek AP (2020) Biases in the study of developmental bias. Evol Dev 22:3–6

Monroe MJ, Bokma (2010) Punctuated equilibrium in a neontological context. Theory Biosci 129:103–111

Morimoto K, Ponce de León MS, Zollikofer CPE (2014) Phenotypic variation in infants, not adults, reflects genotypic variation among chimpanzees and bonobos. PLoS One 9:1–10
Müller GB, Wagner GP (2003) Innovation. In: Hall BK, Olson WM (eds) Keywords and concepts in evolutionary developmental biology. Harvard University Press, Cambridge MA, pp 218–227
Nehm RH (2008) The dynamics of evolutionary stasis and change in the "*Prunum maoense* group". In: Nehm RH, Budd AF (eds) Evolutionary stasis and change in The Dominican Republic Neogen. Springer, pp 171–191
Newman SA (2021) Inherency. In: Müller GB (ed) Nuño de la Rosa L. Evolutionary developmental biology. Springer, Cham, pp 121–131
Newman SA, Bhat R (2009) Dynamical patterning modules: a "pattern language" for development and evolution of multicellular form. Int J Dev Biol 53:693–705
Nijhout HF (2003) Development and evolution of adaptive polyphenisms. Evol Dev 5:9–18
Nijhout HF, Kudla AM, Hazelwood CC (2021) Genetic assimilation and accommodation: models and mechanisms. Curr Top Dev Biol 141:337–369
Nonaka E, Svanbäck R, Thibert-Plante X, Englund G, Brännström Å (2015) Mechanisms by which phenotypic plasticity affects adaptive divergence and ecological speciation. Am Nat 186:E000–E000
Nordén KK, Price TD (2018) Historical contingency and developmental constraints in avian coloration. Trends Ecol Evol 33:574–576
Nunes MDS, Arif S, Schlötterer C, McGregor AP (2013) A perspective on micro-evo-devo: progress and potential. Genetics 195:625–634
O'Sullivan RJ, Aykanat T, Johnston SE, Kane A, Poole R, Rogan G, Prodöhl PA, Primmer CR, McGinnity P, Reed TE (2019) Evolutionary stasis of a heritable morphological trait in a wild fish population despite apparent directional selection. Ecol Evol 9:7096–7111
Ochoa C (2024) Deconstructing the extended evolutionary synthesis: do we need a new theory of evolution? In: Delisle RG, Esposito M, Ceccarelli D (eds) Unity and disunity in evolutionary biology: deconstructing Darwinism. Spinger, Cham, pp 403–439
Odling-Smee FJ, Laland KN, Feldman MW (2003) Niche construction: the neglected process in evolution. Princeton University Press, Princeton and Oxford
Olson ME (2012) The developmental renaissance in adaptationism. Trends Ecol Evol 27:278–287
Oney B, Reineking B, O'Neill G, Kreyling J (2013) Intraspecific variation buffers projected climate change impacts on *Pinus contorta*. Ecol Evol 3:437–449
Parsons K, McWhinnie K, Pilakouta WL (2019) Does phenotypic plasticity initiate developmental bias? Evol Dev 22:56. https://doi.org/10.1111/ede.12304
Pearce T (2011) Evolution and constraints on variation: variant specification and range of assessment. Philos Sci 78:739–751
Pfennig DW (2021) Key questions about phenotypic plasticity. In: Pfennig DW, West-Eberhard MJ (eds) Phenotypic plasticity and evolution: causes, consequences, controversies. CRC Press, Boca Raton, pp 55–88
Pfennig DW, McGee M (2010) Resource polyphenism increases species richness: a test of the hypothesis. Philos Trans R Soc B 365:577–591
Piñeyro-Nelson A, Flores-Sandoval E, Garay-Arroyo A, García-Ponce B, Álvarez-Buylla ER (2010) Development and evolution of the unique floral organ arrangement of *Lacandonia schismatica*. Int J Plant Dev Biol 4:86–97
Piras P, Marcolini F, Raia P, Curcio MT, Kotsakis T (2009) Testing evolutionary stasis and trends in first lower molar shape of extinct Italian populations of *Terricola savii* (Arvicolidae, Rodentia) by means of geometric morphometrics. J Evol Biol 22:179–191
Price TD (2006) Phenotypic plasticity, sexual selection and the evolution of colour patterns. J Exp Biol 209:2368–2376
Price-Waldman R, Stoddard MC (2021) Avian coloration genetics: recent advances and emerging questions. J Hered 112:395–416
Prum RO, Williamson S (2002) Reaction-diffusion models of within-feather pigmentation patterning. Proc R Soc Lond B 269:781–792

Purificação KN (2019) A case of beak deformity in the Shiny Cowbird *Molothrus bonariensis* and a review on beak deformities in wild birds in Brazil. Rev Bras Ornitol 27:212–217

Qi Y, Zhang X, Pang Y, Yuan B, Cheng J (2023) Identification of polymorphism in the MC1R gene and its association with the melanin content in feathers of Chinese yellow quails. Braz J Poult Sci 25. https://doi.org/10.1590/1806-9061-2022-1648

Raff RA (1996) The shape of life: genes, development, and the evolution of animal form. The University of Chicago Press, Chicago

Raia P, Passaro F, Fulgione D, Carotenuto F (2011) Habitat tracking, stasis and survival in Neogene large mammals. Biol Lett 8:64. https://doi.org/10.1098/rsbl.2011.0613

Robinson C (2012) Geometric morphometric analysis of mandibular shape diversity in *Pan*. J Hum Evol 63:191–204

Robles H, Ciudad C (2012) Influence of habitat quality, population size, patch size, and connectivity on patch-occupancy dynamics of the middle spotted woodpecker. Conserv Biol 26:284–293

Rodseth E, Amar A, Sumasgutner P, Ingle RA (2024) Plumage polymorphism in the black sparrowhawk (*Accipiter melanoleucus*) is strongly associated with expression level of *agouti signalling protein*. J Hered. https://doi.org/10.1093/jhered/esae068

Rolland J, Henao-Diaz LF, Doebeli M, Germain R, Harmon LJ, Knowles LL, Liow LH, Mank JE, Machac A, Otto SP, Pennell M, Salamin N, Silvestro D, Sugawara M, Uyeda J, Wagner CE, Schluter D (2023) Conceptual and empirical bridges between micro- and macroevolution. Nat Ecol Evol 7:1181–1193

Ronse De Craene LP (2003) The evolutionary significance of homeosis in flowers: a morphological perspective. Int J Plant Sci 164:S225–S235

Roulin A, Almasi B, Rossi-Pedruzzi A, Ducrest A-L, Wakamatsu K, Miksik I, Blount JD, Jenni-Eiermann S, Jenni L (2008) Corticosterone mediates the condition-dependent component of melanin-based coloration. Anim Behav 75:1351–1358

Rubinoff D, Le Roux JJ (2008) Evidence of repeated and independent saltational evolution in a peculiar genus of sphinx moths (*Proserpinus*: Sphingidae). PLoS One 3:e4035. https://doi.org/10.1371/journal.pone.0004035

Rudall PJ, Alves M, Sajo MG (2016) Inside-out flowers of *Lacandonia brasiliana* (Triuridaceae) provide new insights into fundamental aspects of floral patterning. PeerJ 4:e1653. https://doi.org/10.7717/peerj.1653

Saino N, Canova L, Costanzo A, Rubolini D, Roulin A, Møller AP (2013) Immune and stress responses covary with melanin-based coloration in the barn swallow. Evol Biol 40:521–531

Salazar-Ciudad I (2006) Developmental constraints vs. variational properties: how pattern formation can help to understand evolution and development. J Exp Zool B Mol Dev Evol 306:107–125

Saupe EE, Qiao H, Hendricks JR, Portell RW, Hunter SJ, Soberón J, Lieberman BS (2015) Niche breadth and geographic range size as determinants of species survival on geological time scales. Glob Ecol Biogeogr 24:1159–1169

Scheiner SM, Levis NA (2021) The Loss of phenotypic plasticity via natural selection: genetic assimilation. In: Pfennig DW, West-Eberhard MJ (eds) Phenotypic plasticity and evolution: causes, consequences, controversies. CRC Press, Boca Raton, pp 161–181

Schwab DB, Moczek AP (2021) Evo-Devo and niche construction. In: Müller GB (ed) Nuño de la Rosa L. Evolutionary developmental biology. Springer, Cham, pp 1179–1192

Schwartz JH (1999) Homeobox genes, fossils, and the origin of species. Anat Rec New Anat 257:15–31

Schwochow D, Bornelöv S, Jiang T, Li J, Gourichon D, Bed'Hom B, Dorshorst BJ, Chuong CM, Tixier-Boichard M, Andersson L (2021) The feather pattern autosomal barring in chicken is strongly associated with segregation at the MC1R locus. Pigment Cell Melanoma Res 34:1015–1028

Serra L, Griggio M, Casasole G, Pirrello S, Fusani L, Pilastro A (2025) Elevated testosterone levels during moult have contrasting effects on structural and carotenoid-based plumage colours in Eurasian Blue Tits. Ibis 167:263–276

Shirani F, Miller JR (2024) Matching habitat choice and the evolution of a species' range. bioRxiv. https://doi.org/10.1101/2024.06.25.600700

Shkil F, Siomava N, Voronezhskaya E, Diogo R (2019) Effects of hyperthyroidism in the development of the appendicular skeleton and muscles of zebrafish, with notes on evolutionary developmental pathology (Evo-Devo-Path). Sci Rep 9:5413. https://doi.org/10.1038/s41598-019-41912-9

Shubin NH, Marshall CR (2000) Fossils, genes, and the origin of novelty. Paleobiology 26:324–340

Simberloff D (1995) Habitat fragmentation and population extinction of birds. Ibis 137:S105–S111

Singleton M, Rosenberger AL, Robinson C, O'Neill R (2011) Allometric and metameric shape variation in Pan mandibular molars: a digital morphometric analysis. Anat Rec 294:322–334

Sirkiä PM, Virolainen M, Laaksonen T (2010) Melanin coloration has temperature-dependent effects on breeding performance that may maintain phenotypic variation in a passerine bird. J Evol Biol 23:2385–2396

Smith KK (2003) Time's arrow: heterochrony and the evolution of development. Int J Dev Biol 47:613–621

Stansfield C, Parsons KJ (2024) Developmental bias as a cause and consequence of adaptive radiation and divergence. Front Cell Dev Biol 12:1453566. https://doi.org/10.3389/fcell.2024.1453566

Stimpson CD, Barger N, Taglialatela JP, Gendron-Fitzpatrick A, Hof PR, Hopkins WD, Sherwood CC (2016) Differential serotonergic innervation of the amygdala in bonobos and chimpanzees. Soc Cogn Affect Neurosci 11:413–422

Stoddard MC, Prum RO (2011) How colorful are birds? evolution of the avian plumage color gamut. Behav Ecol 22:1042–1052

Stoltzfus A (2019) Understanding bias in the introduction of variation as an evolutionary cause. In: Uller T, Laland KN (eds) Evolutionary causation biological and philosophical reflections. MIT Press, Cambridge MA and London, pp 29–61

Stroud JT, Moore MP, Langerhans RB, Losos JB (2023) Fluctuating selection maintains distinct species phenotypes in an ecological community in the wild. Proc Natl Acad Sci USA 120:e2222071120. https://doi.org/10.1073/pnas.2222071120

Sultan SE (2015) Organism and environment: ecological development, niche construction, and adaptation. Oxford University Press, New York

Sultan SE (2019) Genotype-environment interaction and the unscripted reaction norm. In: Uller T, Laland KN (eds) Evolutionary causation biological and philosophical reflections. MIT Press, Cambridge MA and London, pp 109–126

Sumner P, Mollon JD (2003) Colors of primate pelage and skin: objective assessment of conspicuousness. Am J Primatol 59:67–91

Theißen G (2006) The proper place of hopeful monsters in evolutionary biology. Theory Biosci 124:349–369

Theißen G (2009) Saltational evolution: hopeful monsters are here to stay. Theory Biosci 128:43–51

Theißen G, Becker A, Di Rosa A, Kanno A, Kim JT, Münster T, Winter K-U, Saedler H (2000) A short history of MADS-box genes in plants. Plant Mol Biol 42:115–149

Thrall PH, Burdon JJ, Murray BR (2010) The metapopulation paradigm: a fragmented view of conservation biology. In: Young AG, Clarke GM (eds) Genetics, demography and viability of fragmented populations. Conservation Biology 4. Cambridge University Press, pp 75–96

Toews DPL, Hofmeister NR, Taylor SA (2017) The evolution and genetics of carotenoid processing in animals. Trends Genet 33:171–182

Toomey MB, Lopes RJ, Araújo PM, Johnson JD, Gazda MA, Afonso S, Mota PG, Koch RE, Hill GE, Corbo JC, Carneiro M (2017) High-density lipoprotein receptor SCARB1 is required for carotenoid coloration in birds. Proc Natl Acad Sci USA 114:5219–5224

Toomey MB, Marques CI, Araújo PM, Huang D, Zhong S, Liu Y, Schreiner GD, Myers CA, Pereira P, Afonso S, Andrade P, Gazda MA, Lopes RJ, Viegas I, Koch RE, Haynes ME, Smith DJ, Ogawa Y, Murphy D, Kopec RE, Parichy DM, Carneiro M, Corbo JC (2022) A mechanism for red coloration in vertebrates. Curr Biol 32:4201–4214

Tschirren B, Fitze PS, Richner H (2003) Proximate mechanisms of variation in the carotenoid-based plumage coloration of nestling great tits (*Parus major* L.). J Evol Biol 16:91–100

Turko AJ, Rossi GS (2022) Habitat choice promotes and constrains phenotypic plasticity. Biol Lett 18:20210468. https://doi.org/10.1098/rsbl.2021.0468

Uller T, Moczek AP, Watson RA, Brakefield PM, Laland KN (2018) Developmental bias and evolution: a regulatory network perspective. Genetics 209:949–966

Usai G, Fambrini M, Pugliesi C, Simoni S (2024) Exploring the patterns of evolution: core thoughts and focus on the saltational model. Biosystems 238:105181. https://doi.org/10.1016/j.biosystems.2024.105181

Vallejo-Marín M, Buggs RJA, Cooley AM, Puzey JR (2015) Speciation by genome duplication: repeated origins and genomic composition of the recently formed allopolyploid species *Mimulus peregrinus*. Evolution 69:1487–1500

Van Hemert C, Handel CM, O'Brien DM (2012a) Stable isotopes identify dietary changes associated with beak deformities in Black-capped Chickadees (*Poecile atricapillus*). Auk 129:460–466

Van Hemert C, Handel CM, O'Hara TM (2012b) Evidence of accelerated beak growth associated with avian keratin disorder in Black-capped Chickadees (*Poecile atricapillus*). J Wildl Dis 48:686–694

Varela-Lasheras I, Bakker AJ, van der Mije SD, Metz JAJ, van Alphen J, Galis F (2011) Breaking evolutionary and pleiotropic constraints in mammals: on sloths, manatees and homeotic mutations. EvoDevo 2:11

Vergara-Silva F, Espinosa-Matías S, Ambrose BA, Vázquez-Santana S, Martínez-Mena A, Márquez-Guzmán J, Martínez E, Meyerowitz EM, Alvarez-Buylla (2003) Inside-out flowers characteristic of *Lacandonia schismatica* evolved at least before its divergence from a closely related taxon, *Triuris brevistylis*. Int J Plant Sci 164:354–357

Vidal-García M, Byrne PG, Roberts D, Keogh JS (2014) The role of phylogeny and ecology in shaping morphology in 21 genera and 127 species of Australo-Papuan myobatrachid frogs. J Evol Biol 27:181–192

Voje KL, Starrfelt J, Liow LH (2018) Model adequacy and microevolutionary explanations for stasis in the fossil record. Am Nat 191:509–523

Vrba ES (1993) Turnover-pulses, the red queen, and related topics. Am J Sci 293:418–452

Wake DB, Roth G, Wake MH (1983) On the problem of stasis in organismal evolution. J Theor Biol 101:211–224

Walsh DM, Sultan SE (2024) The higher-order norm of reaction biological agency and adaptive phenotypic response. In: Fábregas-Tejada A, Baedke J, Prieto GI, Radick G (eds) The riddle of organismal agency new historical and philosophical reflections. Routledge, London, pp 114–130

Wang Y-Q, Melzer R, Theißen G (2011) A double-flowered variety of lesser periwinkle (*Vinca minor fl. pl.*) that has persisted in the wild for more than 160 years. Ann Bot 107:1445–1452

Wang X, Su DF, Jablonski NG, Ji X, Kelley J, Flynn LJ, Deng T (2022) Earliest giant panda false thumb suggests conflicting demands for locomotion and feeding. Sci Rep 12:10538. https://doi.org/10.1038/s41598-022-13402-y

West-Eberhard MJ (2003) Developmental plasticity and evolution. Oxford University Press, New York

West-Eberhard MJ (2004) Ryuichi Matsuda: a tribute and a perspective on pan-environmentalism and genetic assimilation. In: Hall BK, Pearson RD, Müller GB (eds) Environment, development, and evolution: toward a synthesis. The MIT Press, Cambridge MA, pp 109–116

Williams FL, Godfrey LR, Sutherland MR (2003) Diagnosing heterochronic perturbations in the craniofacial evolution of *Homo* (Neandertals and modern humans) and *Pan* (*P. troglodytes* and *P. paniscus*). In: Thompson JL, Krovitz GE, Nelson AJ (eds) Patterns of growth and development in the genus *Homo*. Cambridge University Press, Cambridge, pp 295–319

Williamson PG (1987) Selection or constraint?: a proposal on the mechanism for stasis. In: Campbell KSW, Day MF (eds) Rates of evolution. Allen and Unwin, London, pp 129–142

Wittkopp PJ, Kalay G (2012) *Cis*-regulatory elements: molecular mechanisms and evolutionary processes underlying divergence. Nat Rev Genet 13:59–69

Wrangham R (2023) Hypotheses for the evolution of bonobos: self-domestication and ecological adaptation. In: Furuichi T, Idani G, Kimura D, Ihobe H, Hashimoto C (eds) Bonobos and people at Wamba: 50 years of research. Springer, Singapore, pp 521–544

Wright S, Wagner K (1925) Types of subnormal development of the head from inbred strains of Guinea pigs and their bearing on the classification and interpretation of vertebrate monsters. Am J Anat 54:383–447

Yan C, He F, He J, Zhang Z (2022) The relationship between local and regional extinction rates depends on species distribution patterns. Ecography 2. https://doi.org/10.1111/ecog.05828

Yang C-W, Ran J-S, Yu C-L, Qiu M-H, Zhang Z-R, Du H-R, Li Q-Y, Xiong X, Song X-Y, Xia B, Hu C-M, Liu Y-P, Jiang X-S (2019) Polymorphism in *MC1R*, *TYR* and *ASIP* genes in different colored feather chickens. 3 Biotech 9:203. https://doi.org/10.1007/s13205-019-1710-z

Yoshioka S, Akiyama T (2021) Mechanisms of feather structural coloration and pattern formation in birds. In: Hashimoto H, Goda M, Futahashi R, Kelsh R, Akiyama T (eds) Pigments, pigment cells and pigment patterns. Springer, Singapore, pp 343–366

Zelditch ML, Li J, Swiderski (2020) Stasis of functionally versatile specialists. Evolution 74:1356–1377

Levels of Evolution (II): Megaevolution and Modular-Evolution 12

Abstract

According to the H-Evo-Devo Theory, megaevolution belongs to the higher taxa level, and modular-evolution pertains to the homologue level. This chapter explores the biological processes surrounding the origin, stability, and extinction of higher taxa as well as the origin and stability of homologues as modules of comparative anatomy. Concerning the origin of higher taxa, the taxonomic structure in which they arise is built from the top-down, i.e., higher-level body plans originate first, followed sequentially by lower-level body plans. Additionally, the sudden origin of these entities at various taxonomic ranks (as hopeful monsters) is highlighted. Regarding stability, higher taxa exhibit supraspecific stasis, a phenomenon described in other studies under the term "living fossils," which is explained by developmental constraints. With respect to higher taxa extinction, this process results from mass extinctions and the loss of parts. Meanwhile, modules are produced through the phenomenon of co-option, and the cause of their stability is attributed to developmental constraints acting at the deepest level of the hierarchy. Overall, this chapter specifically emphasizes the biological causes at higher levels of the hierarchy.

Keywords

H-Evo-Devo Theory · Megaevolution · Modular-evolution · Mosaic evolution · Higher taxa · Homologue · Modularity · Supraspeciation · Morphological novelty · Morphological innovation · Suprastasis · Morphological stability · Supraextinction · Hopeful monster · Saltational evolution · Mass extinction · Homology · Serial homology · Character identity · Character state · Developmental constraints

C. Ochoa, *Hierarchical Evolutionary-Developmental Theory*, Evolutionary Biology – New Perspectives on Its Development 9,
https://doi.org/10.1007/978-3-031-88723-9_12

Given the framework of the H-Evo-Devo Theory, this chapter explores the processes involved in megaevolution and modular-evolution. Specifically, it delves into the origin, stability, and extinction of higher taxa, as well as the origin and stability of homologues. Section 12.1 examines the processes related to megaevolution, i.e., at the higher taxa level. It begins with a description of *supraspeciation*, particularly analyzing the taxonomic structure related to the origin of higher taxa, which occurs from the top-down rather than the bottom-up. In other words, according to this model, more inclusive body plans arise first, followed by less inclusive ones. Additionally, this section highlights the sudden origins of these entities at various taxonomic ranks (as hopeful monsters). Subsequently, it analyzes *suprastasis*, whose mechanisms are developmental constraints. Finally, it addresses *supraextinction*, which is attributed to both mass extinctions and the loss of parts. Section 12.2 explores the processes related to modular-evolution, i.e., at the homologue level. It begins by describing the origin of modules or *morphological novelty*, which generally arise through the process of co-option. The debate addressed here focuses on the difference between "novelty" and "innovation." Lastly, it examines the mode of stability of homologues or *morphological stability*, which is explained by developmental constraints acting at the deepest level of the hierarchy. However, the discussion will focus on the distinction between the stability of the module and the stability of one of its variational modalities.

12.1 Megaevolution (Higher Taxa Level)

According to the H-Evo-Devo Theory, megaevolution refers to the evolution of higher taxa. As concluded in Chap. 8, higher taxa are groups distinguished by great distances within the morphospace, but, specifically, they are more inclusive body plans than the species/genus body plan within a hierarchy of body plans. Importantly, the H-Evo-Devo Theory contrasts with the traditional view, which treats higher taxa as merely historical entities emerging from speciation processes. Instead, it proposes that higher taxa are real entities with a distinct evolutionary role, representing the higher evolutionary units in the hierarchy. Moreover, higher taxa exhibit specific patterns such as instantaneous origin, stability, and discrete extinction. These characteristics make it possible to study them spatiotemporally, akin to species and subspecies.

This section briefly examines the processes involved in the three evolutionary modes (patterns) of higher taxa. The first mode at this level is *supraspeciation*, whose mechanisms involve the prospecies model or supraDarwinian mechanisms. However, since these mechanisms were covered in the previous chapter, this discussion will focus on the taxonomic structure related to the origin of higher taxa. According to this model, the phylogenetic tree is built from the top-down, i.e., higher-level body plans appear first, followed sequentially by those of lower levels. Additionally, this section highlights the sudden origin of higher taxa through this taxonomic structure. The second evolutionary mode is *suprastasis*, which allows for the recognition of higher taxa spatiotemporally. Notably, other authors have described this mode using the concept of "living fossils." Here, I conclude that the causes of this evolutionary mode are developmental constraints acting hierarchically on body plans. Finally, the third evolutionary mode is *supraextinction*, whose causes are attributed to mass extiction events and the loss of parts (e.g., species).

12.1.1 Supraspeciation

According to the H-Evo-Devo Theory, *supraspeciation* is the mode that focuses on the origin of higher taxa. Particularly, this mode describes how most higher taxa originate instantaneously, followed by a prolonged persistence of stasis. It is important to emphasize that the mechanisms involved in supraspeciation are the same as those in speciation, i.e., the supraDarwinian mechanisms (prospecies model) account for divergence in both macroevolution and megaevolution. The distinction between these levels lies in the magnitude of morphological discontinuity obtained in the organism: smaller for species/genera and larger for higher taxa. Furthermore, while both levels pertain to the origin of body plans, higher taxa are characterized by being more inclusive body plans than the species/genus body plan. Since these processes were detailed in the previous chapter,[1] this section will focus on the structure of the phylogenetic tree, which is constructed from the top-down rather than the bottom-up according to this model. Finally, I present specific examples to illustrate the sudden origin observed across different taxonomic ranks.

Remarkably, higher taxa do not originate from the bottom-up, as generally assumed. Instead, higher-level entities emerge first in the taxonomic structure, followed by subsequent lower-level entities. As seen in Chap. 8, the prevailing belief about the taxonomic structure holds that higher taxa arise through the process of speciation. According to this view, the phylogenetic tree is constructed at a single level in a continuous and branching sequence between ancestors and descendants. As species diverge over time, differences between them emerge, such that the more distant the genealogical connections, the more inclusive the category (e.g., family, order, class). Therefore, the morphological gaps between species arise from divergence time and the elimination of many transitional species. In short, this traditional model assumes that lower taxa (subspecies or species) emerge first, and their accumulation over time sequentially produces higher taxa (genera, families, orders, classes, and so forth). However, according to the fossil record, "there are clues suggesting that just the reverse was true—that most higher taxa were built from the top down rather than bottom up" (Erwin et al. 1987, p. 1183).

Interestingly, paleontologists have long recognized that higher-ranking taxonomic units emerged earlier than lower-ranking taxonomic ones (Erwin et al. 1987; Foote 1993; Holman 1996; Valentine 2004; Erwin 2007; Hughes et al. 2013; Oyston et al. 2015). In other words, given that the magnitude of the morphological gap is associated with taxonomic rank (Erwin 2007), the fossil record indicates that the origin of a phylum (in animals) or a division (in plants)—representing a greater

[1] See Chap. 11, Sect. 11.2.1.

morphological disparity—occurred first, followed by the emergence of classes, orders, families, and so on (Erwin et al. 1987). Valentine (2004) describes this phenomenon more clearly:

> Thus the record makes it appear that phyla evolved first; then classes evolved, as major architectural themes based on the phylum bodyplans; then orders evolved, as modifications of class bodyplans; and so on. This record runs counter to what might be expected during the origin of phyla, which would be the divergences of two lineages from common ancestors, at first at the species level only. Then as time passed their differences would become more pronounced, the two lineages becoming as distinctive as are average genera, and then as are average families, then as orders, and so forth. In fact such stages of divergence must have occurred during the history of Metazoa, but we don't see them. (Valentine 2004, p. 444)

Interestingly, Rabosky et al. (2012) argued that, according to the traditional model, older higher taxa should contain more species than more recent ones. However, after analyzing a large number of higher taxa in multicellular organisms, they found no evidence to support this prediction. They conclude that the age of a higher taxon is unrelated to species richness. Furthermore, some authors argue that an increase in the number of species does not necessarily coincide with an increase in morphological disparity (MacDougall et al. 2019; Beech and Lamsdell 2021). For example, Clark et al. (2023) noted that the body plans of plants occupied their current positions within the morphospace since their origin, and it was not observed that these groups reached these morphological distances cumulatively through ancestral species: "The phylomorphospace constructed from both fossil and extant taxa shows that fossil taxa do not alter the fundamental pattern of morphospace occupation seen in extant taxa and none lie beyond the regions of morphospace occupation circumscribed by the extant phylomorphospace" (p. 1622). Similarly, López-Martínez et al. (2024, p. 1353) highlight that "angiosperm flowers also followed this pattern, reaching their maximal disparity in the Early Cretaceous, meaning that major structural variations … were explored early in the history of the group. Interestingly, this occurred when angiosperm species richness was still low compared with other land plants."

In light of this pattern, several researchers emphasize that disparity (distance between higher taxa) and diversity (distance between species) are decoupled (e.g., Gould 1989; Foote 1993; Valentine 2004, pp. 443–444; Melzer and Theißen 2016; Minelli 2016; Oyston et al. 2015; Alhajeri and Steppan 2018; Romano 2021; Allaire et al. 2023).[2] Considering this, Melzer and Theißen (2016) state that a body plan with a high degree of developmental robustness (strong developmental constraints), which provokes invariability at that level, has the potential to promote greater variability at another level by generating a larger number of species (see also Minelli 2016; Oyston et al. 2015; Rutishauser 2016). For example, when describing the unusual plant groups Lentibulariaceae and Podostemaceae,[3] Rutishauser (2016)

[2] Interestingly, this difference in levels is also related to extinction patterns, see Sect. 12.1.3.

[3] See further on in this chapter for the origins of this distinctive plant family.

notes that, despite these groups diverging from the basic angiosperm body plan, their current body plans are morphologically very rigid:

> The switches to new and unique vegetative bauplans in Lentibulariaceae and Podostemaceae probably facilitated rather than prevented the evolution of species diversity. *Utricularia* (including *Polypompholyx*) contains 230 species, being by far the largest genus of all carnivorous angiosperms ... The river-weeds (Podostemaceae) with a total of approx. 54 genera and approx. 310 species are by far the largest family of truly aquatic angiosperms. (Rutishauser 2016, p. 828)

Of course, the H-Evo-Devo Theory aligns perfectly with these interpretations. This can be seen, for example, when analyzing the origin of a body plan within a phylogenetic context. Following Fig. 12.1, imagine an ancestral body plan consisting of three taxa (Fig. 12.1a). A new body plan emerges when a population undergoes a significant discontinuous change that breaks the developmental constraints of the previous body plan (Fig. 12.1b). Note that this new body plan lacks taxonomic

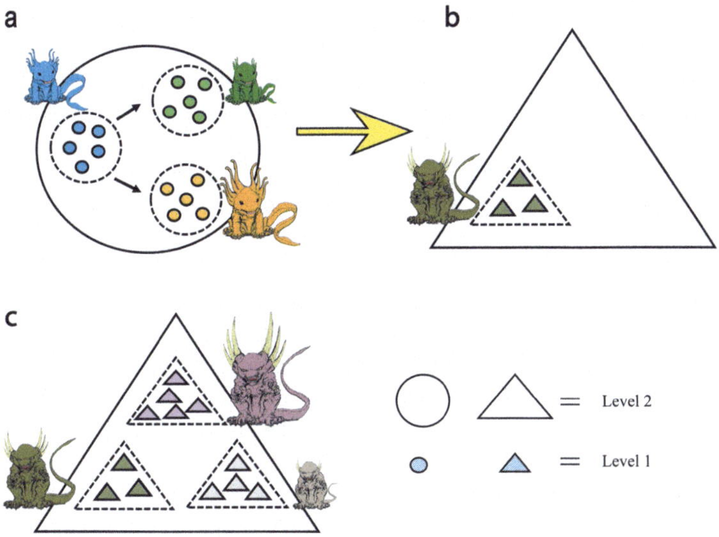

Fig. 12.1 *The origin and diversification of a body plan depicted in a nested compositional hierarchy, according to the H-Evo-Devo Theory*. The area inside each figure represents the morphospace, and each figure indicates a form located within the morphospace. (Note that these figures come with a fictitious character.) The size of the figure denotes its level: the small figures correspond to Level 1, while the large figures correspond to Level 2. The different colors of the small figures illustrate the gaps between body plans at Level 1. Differences in the geometric shapes—i.e., circles and triangles—illustrate the gaps between body plans at Level 2. The dotted figures inside the large figure represent the body plan boundaries at Level 1, delineated by developmental constraints. In contrast, the solid outline that draws the large figures (circles or triangles) denotes the body plan boundaries at Level 2, defined by developmental constraints. (**a**) An ancestral body plan with three taxa. (**b**) The origin of a new body plan occurred when a population underwent discontinuous changes that broke the developmental constraints of the previous body plan. (**c**) Once established, the new body plan generated additional lower-level taxa, thereby gaining taxonomic recognition. (Illustrations drawn by Oscar Morales Arteaga)

diversity. Nevertheless, this population can be regarded as a potential supraspecific entity, as it only needs to produce taxa at a lower level to gain such taxonomic recognition. Once this is accomplished, it can be recognized as a diversified body plan (Fig. 12.1c).[4] Thus far, I have avoided using taxonomic ranks; the purpose here is to identify these ranks through the ontological exercise described below.

In line with this argument and as illustrated in Fig. 12.2, three taxa representing the ancestral body plan can be identified. According to Linnaean nomenclature, they can be named (1) *Amicus caeruleus*, (2) *Amicus viridis*, and (3) *Amicus flavus*. Now, imagine that *Amicus caeruleus* is the ancestral taxon that gave rise to a new taxon, represented here as (4). If (4) is discontinuous with respect to the common ancestor, how should it be named using the Linnaean nomenclature? Is it necessary to use the same genus, or can a new genus be proposed for this taxon? (see Fig. 12.2 and create a name for the taxon 4). If you placed it in the same genus (i.e., *Amicus*), you have applied the traditional principle of phylogenetic systematics, which focuses exclusively on genealogical relationships. However, this approach overlooks a crucial point being developed here on the morphological aspect. Conversely, if you placed it in a new genus, then you have correctly applied this metaphysical exercise by considering morphological discontinuity as an important variable.

To clarify this further, I named taxon (4) *Monstrum sperans*. It is important to note that the proposal for a new genus was not based on its close phylogenetic relationship with other species but on the magnitude of the morphological gap, which showed a radical structural change relative to its ancestor. Now, following Fig. 12.3, I named the descendant taxa (5) *Monstrum cornuatus* and (6) *Monstrum parvum*. As it seems, the genus appeared before diversifying into different species. Furthermore, if the initial discontinuity is even greater, these entities could represent not just a different genus but also a different family, order, class, or phylum. As emphasized in the previous chapter, a prospecies lacks lower-level elements, and to be recognized as a taxonomic species, it must produce subspecies. The same rule applies to higher taxa; in this sense, a prospecies (hopeful monster) would not only be a potential species but also a potential genus, family, order, class, etc.

Curiously, many researchers have unconsciously encountered similar situations. For example, Alejandro Oceguera-Figueroa,[5] from UNAM,[6] received a specimen of a leech from Peru while pursuing his PhD at the American Museum of Natural History. The leech was reported to parasitize the nasal cavities of children who played in the rivers. Upon dissecting the specimen, Alejandro discovered unusual characteristics that had never been seen before in a leech. Most leeches have three jaws with rows of teeth, whereas this new leech, later named *Tyrannobdella rex*, had a single jaw with eight very large teeth. Additionally, its male and female reproductive organs were extremely reduced, among other unique features (Fig. 12.4). Alejandro recounts that the novel characteristics of *Tyrannobdella rex* made it

[4] This description corresponds to the origin of taxonomic entities through supraDarwinian mechanisms or the prospecies model, as detailed in Chap. 11, Sect. 11.2.1.

[5] Personal communication.

[6] Universidad Nacional Autónoma de México.

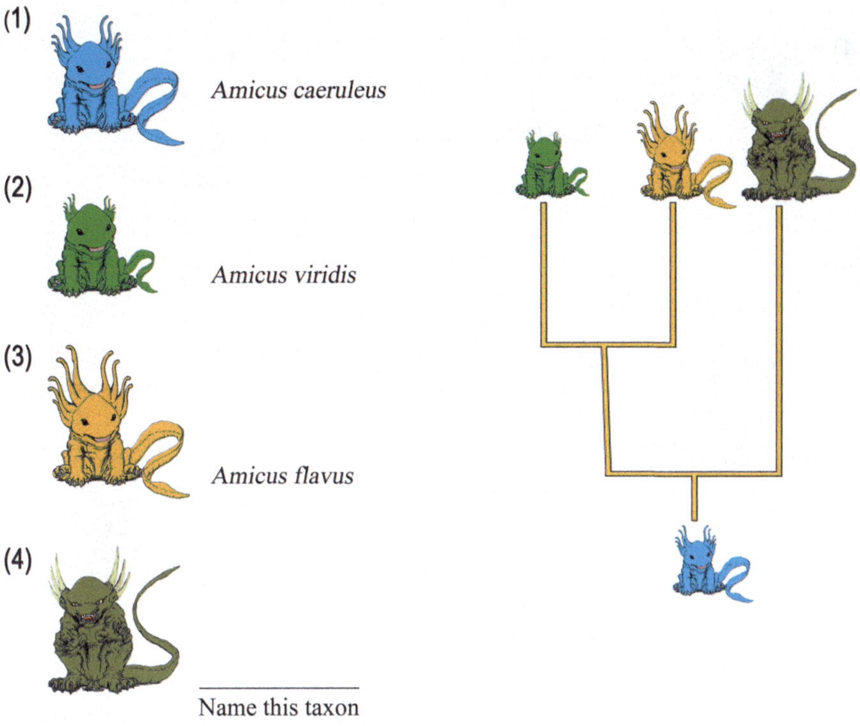

Fig. 12.2 *Ontological exercise on applying the Linnaean system to an inverted taxonomic structure according to the H-Evo-Devo Theory.* This exercise involves a morphological and phylogenetic comparison of taxa identified during the origin of a body plan. On the left, the proposed names for the studied taxa are presented according to the Linnaean nomenclature. On the right, a phylogenetic tree illustrates the genealogical relationships among the taxa. According to this exercise, the following names are proposed within the ancestral body plan: (1) *Amicus caeruleus*, (2) *Amicus viridis*, (3) *Amicus flavus*. Note that *Amicus caeruleus* is the ancestral taxon that gave rise to (4), which underwent a discontinuous change. Based on this information, how should taxon (4) be named using the Linnaean nomenclature? Create a name for taxon (4) and continue reading the text (see Fig. 12.1 for further details). (Illustrations drawn by Oscar Morales Arteaga)

challenging for his team to place it within the phylogenetic tree of leeches. However, through genetic analysis and other characteristics, it was classified within the Praobdellidae family and identified as a sibling species to *Pintobdella chiapasensis*. Interestingly, Alejandro and his team assigned the leech to a different genus due to its novel characteristics, as there were no other morphologically similar species to relate it to the supposed continuity tree (see Phillips et al. 2010).[7]

Similarly, Rubinoff and Le Roux (2008) performed a genetic analysis on sphinx moths in both the genus *Proserpinus*, which includes seven species, and the closely

[7] Despite this, Alejandro does not view the phenomenon as evidence of saltational evolution. Instead, he proposes that there may be (or may have been) undiscovered species that could represent the transitional forms explaining the discontinuities between the genera.

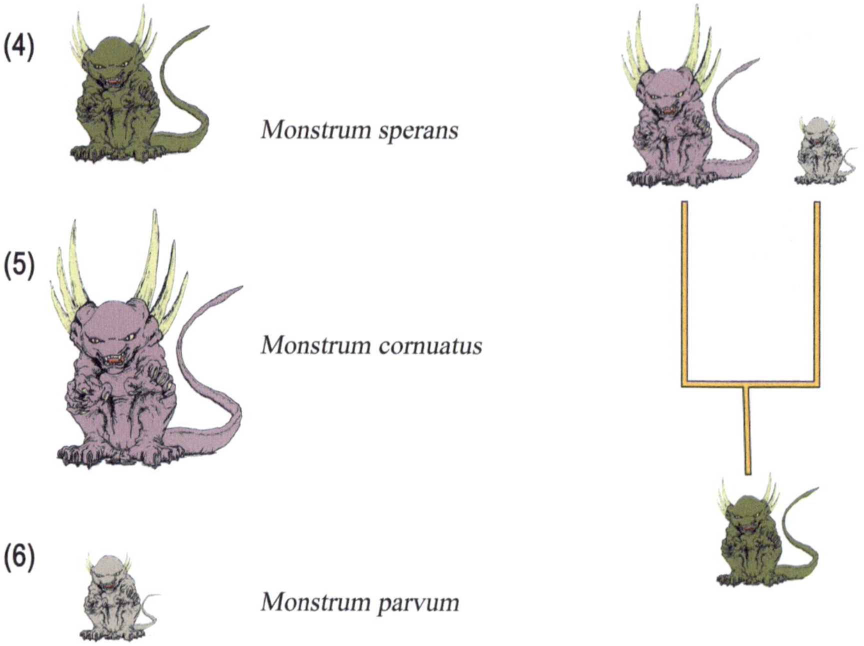

Fig. 12.3 *Morphological and phylogenetic comparison of three taxa during the diversification of a body plan*. This figure continues the ontological exercise introduced in Fig. 12.2, which supports the perspective of an inverted taxonomic structure according to the H-Evo-Devo Theory. On the left, the taxa observed within the new body plan are named as follows: (4) *Monstrum sperans*, (5) *Monstrum cornuatus*, and (6) *Monstrum parvum*. On the right, a phylogenetic tree illustrates the genealogical relationships among the taxa. Note that *Monstrum sperans* is the ancestral taxon. According to this model, the new body plan represented by the genus *Monstrum* originated from the ancestral body plan *Amicus*, as shown in Fig. 12.2. Subsequently, this genus produced other species, thereby diversifying the body plan (see Figure 12.1c). Under this top-down dynamic, these taxa could represent not only different genera but also different higher taxonomic categories, such as families, orders, classes, and phyla (for additional details, see the main text). (Illustrations drawn by Oscar Morales Arteaga)

related genus *Arctonotus*, which comprises only one species. They point out that *Arctonotus* is easily distinguished from *Proserpinus* by certain discontinuous physiological and morphological characters. Nevertheless, their results revealed virtually no genomic divergence between the two genera. In light of this, they interpret these findings as evidence of saltational evolution, suggesting that *Proserpinus* might be the ancestral genus. Otherwise, a gradual, albeit rapid, genetic change would have led to consistent genetic separation in *Arctonotus*, which was not observed.

In short, the higher-level body plans emerge first, followed sequentially by the lower-level body plans. This principle predicts the arrangement of the phylogenetic tree; both the fossil record and the pattern seen in certain taxa that show discontinuous morphological innovations support this view. Considering this, I will now

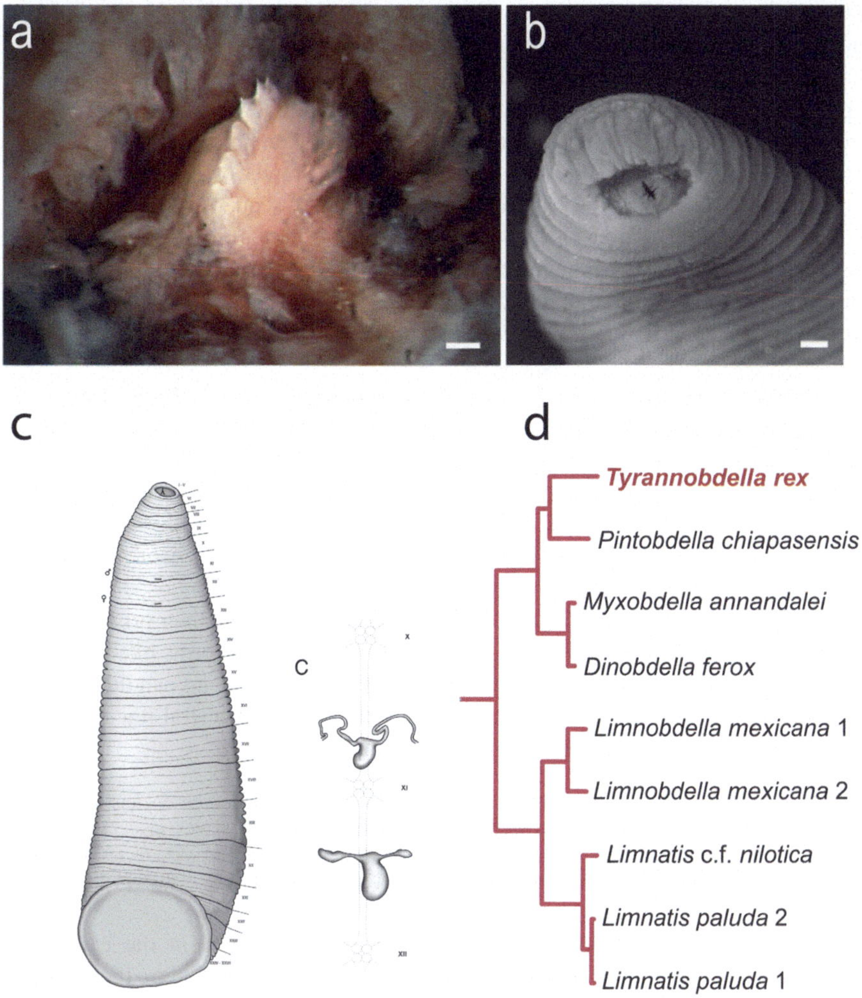

Fig. 12.4 *Anatomy and phylogenetic relationships of Tyrannobdella rex.* (**a**) View of the single dorsal jaw of *Tyrannobdella*, showing its long teeth. (**b**) View of the anterior sucker of *Tyrannobdella*, where the dorsal jaw is visible. (**c**) The whole body of the *Tyrannobdella*, with its reproductive organs illustrated on the right side. (**d**) The phylogenetic tree depicts the relationship that *Tyrannobdella* has with other taxa. Note that *Tyrannobdella* is more closely related to the *Pintobdella chiapasensis*. The authors chose to classify it under a different genus due to its distinctive characteristics. (From Phillips et al. 2010)

examine examples across the taxonomic scale, illustrating the sudden appearance of higher taxa.

On the Origin of Phyla (or Divisions) The category of phylum (in animals) or division (in plants) is the most inclusive rank within the hierarchy of body plans.[8] As noted earlier, the characters that allow their recognition at this taxonomic rank appear first. Major animal groups that define the phylum level emerged so abruptly in the fossil record, around 500 million years ago, that this event is called the "Cambrian Explosion." This does not mean that all phyla originated exclusively during the Cambrian; some continued to appear in subsequent periods (Valentine 2004). Importantly, animals achieved remarkable morphological disparity immediately after their early origin. Interestingly, Gould (1989) suggests that there was initially more variety of animal body plans than species. In contrast, today a greater number of species are constrained within a limited number of surviving body plans.

In plants, at the division level, the earliest groups appeared in the Early Ordovician and resembled extant bryophytes (Graham et al. 2000). Subsequently, during the Silurian period, the first vascular plants emerged. In the Early Devonian, these plants underwent a sudden increase in disparity, including the appearance of many contemporary groups (Kenrick and Crane 1997). For this reason, this event is considered analogous to the Cambrian Explosion in animals (Harrison and Morris 2017). Today, one of the most diverse plant groups is the angiosperms, which achieved significant morphological disparity immediately upon their origin (López-Martínez et al. 2024). Researchers generally agree that angiosperms likely emerged from a group of gymnosperms (Soltis et al. 2008). However, no candidate taxon, whether contemporary or fossil, clearly connects to their ancestral lineage (Wang 2018).[9] In this context, Liu and Wang (2015) state that:

> It is logical and widely believed that angiosperms are derived from their ancient gymnospermous ancestors and there should be a series of transitional stages in between … Searching for such intermediate entities is so challenging that Tom Harris characterised the history of this research as an 'unbroken record of failure.' (Liu and Wang 2015 p. 707)

Similarly, a significant diversity of angiosperms is observed in the early Late Cretaceous. Nevertheless, evidence from the Jurassic period, including the record of flowers and other characteristic structures, suggests an earlier appearance (Liu and Wang 2015; Wang 2018; Cui et al. 2022; Frolov and Enushchenko 2022; Han et al. 2023; Fu et al. 2024). According to Wang (2019, p. 578), this implies that, immediately following their appearance, "some angiosperms may have flourished in certain niches although still rare and ecologically minor in the vegetation."

---

[8] In the Linnaean classification system, the level above the phylum is the Kingdom (and there is also an even higher level called Domain). Nonetheless, in the traditional view, these categories simply reflect genealogical connections among organisms and may not necessarily correlate with identifying a body plan. For instance, while all animal phyla share a common ancestor and are thus classified within the same Kingdom (Animalia); each of these groups exhibits a distinct body plan.

[9] Yet, some conjectures suggest a connection between certain structures of a group of fossil gymnosperms and those of angiosperms (see Herman 2024).

These cases of immediate disparity in the early history of major animal and plant groups, followed by an increase in diversity, present a significant challenge to the conventional model regarding *initial diversification followed by an increase in disparity over geological time*. But this does not mean that the origin of a higher taxon at the phylum (or division) level occurred exclusively during the origin of the first animals or plants. While less common, the origin of a phylum or division could theoretically happen at any point in history. For example, in 1995, Funch and Kristensen reported the discovery of an animal living in the mouths of shrimp that was so distinct from other animals that they classified it into a new phylum, *Cycliophora*: "We believe that all the above features of this previously undescribed species warrant the recognition of a new phylum" (p. 711). Notably, the description of this new phylum also demonstrates innovation in the inclusive hierarchical sequence of Linnaean ranks: "new class," "new order," "new family," "gen. Nov." (new genus), "sp. nov." (new species):

Phylum	Cycliophora
Class	Eucycliophora
Order	Symbiida
Family	Symbiidae
Genus	*Symbion*
Species	*Symbion pandora*

Shortly thereafter, these researchers made a similar discovery, this time of a different animal inhabiting the hot springs of Greenland. According to Kristensen and Funch (2000): "*Limnognathia maerski* cannot be included in any existing group without difficulty. Therefore, we have established a new class, Micrognathozoa, for the new freshwater animal" (p. 46). Although this new animal was classified as a new class—due to its ancestral affinities with other groups—its jaw exhibits such an exceptional level of complexity that it might even suggest a difference at the phylum level: "The complexity of the jaw of *Limnognathia* is greater than that of jaws of any invertebrate group and far beyond the complexity of Gnathostomulida" (p. 32).

Given the above, the traditional view might suggest that these are ancient groups, remnants of a time when they were likely much more diverse but have since declined. While the H-Evo-Devo Theory does not dismiss this possibility, it also proposes an alternative interpretation: These could be newer groups that simply need to diversify downward to fill the lower Linnaean ranks by producing less inclusive, innovative characters. Although there is limited information available to test this hypothesis, this pattern is at least observable in the fossil record, as evidenced by the origin of the earliest animal and plant groups.

On the Origin of Classes While the level of phylum (or division) may be considered real—since it represents the highest level of the hierarchy—the categories below this rank are undoubtedly artificial (except for the species/genus level, see Chap. 7). This is because there is no ontological argument that establishes a standardized scale of discontinuity for certain levels. Nevertheless, it is important to emphasize that the level recognized as "class" exhibits significant disparity, closely approaching that of the phylum. For example, all arthropods (insects, crustaceans, chelicerates, myriapods, etc.) belong to the phylum Arthropoda because they share a basic common body plan, e.g., an exoskeleton, segmentation, and jointed appendages. However, this body plan also exhibits a wide variety of inclusive body plans that are highly conspicuous. Although these varieties are generally recognized at the subphylum level, from a certain perspective, they could also be categorized as "class." Nevertheless, that is not the focus of this discussion. Given this scale of disparity, the important question to ask is how did the different classes of arthropods arise? And even more importantly, how long did it take?

Most arthropods emerged during the Cambrian Explosion in the fossil record, including arthropod body plans with no extant representatives (Gould 1989). This suggests that their origin was instantaneous and likely due to changes in the expression of Hox genes, which are associated with the differentiation of various body regions (Hughes and Kaufman 2002). For example, the origin of insects dates back to the Devonian. Nonetheless, no organism in the fossil record connects insects to previously existing arthropods, even though molecular data indicate that insects have a close relationship with crustaceans, particularly the branchiopods (Glenner et al. 2006).[10] This gap is not the only intriguing aspect of this issue. Ronshaugen et al. (2002) demonstrate that the simple alteration in the Hox gene, Ultrabithorax (Ubx), is associated with the transition from a crustacean-like ancestor with multiple limbs to the hexapod body plan of insects. They conclude: "The successive removal of Ser/Thr residues might quantitatively influence repression function and morphology, allowing viable microevolutionary steps toward 'hopeful monsters' with macroevolutionary alterations in body shape" (p. 917). Similarly, Cirripedes—crustaceans with a radically different morphology from other crustaceans, whose affinity is only revealed in their larvae—appear in the fossil record until the mid-Carboniferous, after which they began to diversify (Chan et al. 2021). Interestingly, Mouchel-Vielh et al. (1998) found that the three orders of Cirripedes lack the abdominal-A (abdA) gene, associating this absence with the lack of abdominal segments, a notable feature of their body plan. In light of this, Géant et al. (2006) question whether this pattern of homeotic gene loss in Cirripedes represents an example of saltational evolution. If this is the case, Cirripedes could be considered "hopeful monsters."

On the Origin of Orders The order rank could represent the relevant body plans situated in the middle of the hierarchy. This does not imply that the category of "order" is real. As I have emphasized multiple times, Linnaean ranks in the higher taxa denote levels of inclusivity of body plans. For example, in the Late Devonian, the appearance of tetrapods marked a significant innovation within the group of craniates (vertebrates). From this basic body plan, a wide variety of less inclusive body plans arose (e.g., mammals, birds, turtles, lizards), which subsequently gave

[10] For an alternative interpretation, see Schwentner et al. 2017.

rise to other even less inclusive body plans, each representing a structural variety deserving recognition.

Snakes possess a highly distinctive body plan. Their axial skeleton consists of numerous vertebrae forming an elongated trunk, and they lack limbs, although some species exhibit vestiges of them. Despite these characteristics indicate a radical change from the tetrapod body plan, they are not truly unique to the group. Trunk elongation and limb reduction have occurred multiple times in the evolution of tetrapods, particularly in the lizard lineage (Leal and Cohn 2018). Therefore, some researchers have emphasized that the true distinguishing feature of snakes is the shape of their skulls (Cundall and Irish 2008; Caldwell 2020). Although their affinities with other groups, both extinct and extant, remain uncertain (Head et al. 2022), full-bodied snakes appear in the fossil record in the Late Cretaceous. Interestingly, these specimens exhibit fully developed but reduced hindlimbs. However, fragments of jaws and skulls from the Middle Jurassic suggest an earlier origin (Caldwell 2020). It is important to note that these anatomical elements are still present in modern snakes. Furthermore, no intermediate steps are observed in the fossil record that indicates the construction of the snake body plan (Cundall and Irish 2008; Caldwell 2020).

Moreover, in a laboratory study, Cohn and Tickle (1999) demonstrated that the expression of Hox genes in snakes expands along the body axis, transforming the entire body into a thoracic identity region similar to that observed in other vertebrates.[11] This process results in the loss of the forelimbs. They further indicate that the hindlimbs were lost through a different genetic mechanism. These changes, along with the conservation of hindlimb vestiges, align with the specimens found in the fossil record. Based on these findings, Cohn and Tickle conclude:

> Progressive expansion of Hox gene expression domains along the body axis can account for the major morphological transitions in snake evolution… Such higher-order genetic changes could have resulted in sudden anatomical transformations, rather than gradual changes, during snake evolution, a hypothesis which can be tested by the fossil record. (Cohn and Tickle 1999, pp. 477–478).

Another example to consider is the evolution of bats, which display a highly distinctive body plan. Their forelimbs feature greatly elongated digits connected by a long interdigital skin membrane (patagium) extending along the body in addition to other unique skeletal novelties. These characteristics are clearly specialized for flight. Still, the phylogenetic relationships of bats with other orders remain uncertain. Neither morphological observations nor molecular studies have been able to place bats precisely on the mammalian phylogenetic tree (Gunnell and Simmons 2005; Alexander 2015, pp. 131–132). Furthermore, bats appeared suddenly in the fossil record around 50 million years ago, in the Eocene, with no evidence of transitional forms linking them to their non-flying ancestors (Gunnell and Simmons 2005).

[11] For further details on the genetic mechanisms behind trunk elongation in snakes, see Leal and Cohn (2018).

Regarding the embryological basis of these transformations, Sears et al. (2006) found that during the early stages of embryonic development, the size of bat digits is very similar to that of mice digits. However, as development progresses, bat digits begin to elongate disproportionately (see Chap. 9, Fig. 9.5b). This change is attributed to the increased proliferation and differentiation of cartilage cells, driven by alterations in the regulation of *Bmp2*. Additionally, Sears (2007) points out that bat membranes result from *BMP* suppression due to *Gremlin* expression and an increase in fibroblast growth factors (FGFs). These factors work together to retain the interdigital membranes, which are typically eliminated in other vertebrates during development.[12] In light of these findings, Sears (2007, p. 6) concludes that "small changes in the expression of genes critical to many aspects of development have driven large changes in bat wing morphology."

On the Origin of Families. The family is the taxonomic rank that represents the least inclusive level among the higher taxa and is close to the species/genus level in the hierarchy of body plans. While this category may not be entirely real, its proximity to the lower levels offers valuable insight into the evolutionary dynamics within the hierarchy. For example, Valentine (1990) points out that because the fossil record for families is more complete than that for species, families can serve as a guide for understanding the evolutionary patterns of species. This observation implies that families might be studied as units of evolution in their own right. Furthermore, despite their closeness to the lower body plan levels, families also exhibit significant morphological gaps, as will be discussed further below.

The families Lentibulariaceae (bladderworts) and Podostemaceae (river-weeds) exhibit unique body plans that deviate significantly from the typical angiosperm structure. For example, it is difficult to recognize common plant organs such as leaves, stems, and roots in these plants. These transformations are linked to adaptations to aquatic environments from terrestrial origin (Rutishauser 2016). In addition, Lentibulariaceae has no representatives in the fossil record, and molecular data is problematic for determining its relationship with other plants due to the high rate of genetic heterogeneity among its members (Cieslak et al. 2005). Regarding genetic analysis, Carretero-Paulet et al. (2015) indicate that the simple gain and loss of function of regulatory genes contributed to the specialized body plan of *Utricularia gibba*. Similarly, the relationship of the Podostemaceae family with other plant groups remains uncertain due to their lack of a fossil record (Mohan Ram and Sehgal 1992). However, their unusual body plan is known to result from changes in the regulation of specific genes during development (Katayama et al. 2011, 2013). In light of these findings, Rutishauser (2016, p. 828) comments that "[b]ladderworts and river-weeds provide nice examples of hopeful monsters."

An example in animals is the chameleon, belonging to the family Chamaeleonidae. The members of this group exhibit characteristics that are significantly different from those of other lizard groups. They are notable for their ability to move their eyes independently, their capacity to change color according to their environment,

[12] For further details on the genetic mechanisms that led to the development of bat characters, see Cooper and Sears (2013).

and their modified tongue, which rapidly extends with a sticky tip to catch prey. But, a particularly distinctive feature is their highly modified autopod[13] which, in most species, allows them to grasp tree branches (Diaz Jr and Trainor 2015; Molnar et al. 2017). Chameleons originated in the early Miocene, and their relationship with other groups of Squamata is uncertain, although it is generally believed that agamas are their closest relatives (Tolley et al. 2013). Early chameleons are identical to modern ones, and there is a significant gap in the fossil record connecting them to agamas or their relatives (Moody and Roček 1980; Tolley and Herrel 2013).[14] Regarding the specialized modifications of the autopod, it is emphasized that chameleons retain the basal structures of the tetrapod autopod (Molnar et al. 2017), including the same pattern of genes active during limb development, and their discontinuous differences are attributed solely to changes in regulation (Diaz Jr and Trainor 2015). Consequently, Molnar et al. (2017) argue that chameleons could represent "hopeful monsters."

In conclusion, a clear pattern emerges across different taxonomic levels: higher taxa originate discontinuously—as hopeful monsters—followed by the immediate stability of the main framework. In other words, when a new body plan arises, despite its significant departure from the ancestral body plan, it achieves embryological stability that is strong enough to secure a new position within morphospace. Initially represented by a small number of individuals, its population gradually expands until it becomes quite abundant. Moreover, this new body plan can facilitate the evolution of less inclusive body plans (e.g., species). Thus, higher-level body plans originate first, followed subsequently by lower-level body plans, rather than the reverse.

12.1.2 Suprastasis

According to the H-Evo-Devo Theory, *suprastasis* refers to the mode of the stability of body plan characters at higher taxonomic levels such as family, order, class, and phylum. Once these high-level taxonomic structures emerge and become stabilized, they tend to remain unchanged for extended geological periods. Although these structures exhibit some variability, they do so within well-defined limits. It is crucial to emphasize that this enduring constancy of the body plan across space and time is attributable to developmental constraints operating at the higher levels of the Linnaean hierarchy, as presented in this section. Below, I will break this down step-by-step.

[13] The autopod is characterized by an opening that separates the phalanges into opposing directions (zygodactyly), forming two bundles of fused digits (syndactyly). Additionally, some mesopodial elements are reduced, while others are elongated, shaping a ball-and-socket joint at the base where the autopod connects with the zeugopod. This configuration provides enhanced rotational abilities (Diaz Jr and Trainor 2015).

[14] However, a neonatal specimen from the Mid-Cretaceous provides evidence of an unknown group that appears to have a close relationship with chameleons, based on both morphological and phylogenetic analysis (see Daza et al. 2016).

Interestingly, most biologists still fail to recognize stasis at the supraspecific level. However, this pattern of stability is quite evident, particularly when examining the constancy of body plans over geological time. The issue is that many biologists are so entrenched in traditional evolutionary ideas that these clear patterns often go unnoticed. As Gould (2002, p. 939) notes: "The predominant pattern of life's history cannot fail to be instructive—and such non-trending [supraspecific stasis] may well mark a norm of this magnitude, even if heretofore hidden in plain sight because we also see with our minds, and conventional concepts can be more blinding than mere ocular failure." Therefore, the task at hand is to make the visible more visible.

Despite the apparent lack of discussion on supraspecific stasis—likely because evolutionary biologists have predominantly focused on species, often deeming higher taxa as having limited evolutionary significance—this pattern has been part of evolutionary debates, primarily when considering the controversial concept of "living fossils." Generally, living fossils are defined as lineages that have persisted through long geological intervals with minimal morphological change—that is, their extant representatives are virtually identical to those found in the fossil record—and that have exhibited low species diversity (Turner 2019). Examples of such groups include horseshoe crabs, ginkgo trees, tuataras, tadpole shrimps, and coelacanths (Fig. 12.5). Still, the classification of organisms as "living fossils" has been based on various criteria, without including low diversity as a requirement (Werth and Shear 2014; Lidgard and Love 2018; Lidgard and Kitchen 2023; Sterner 2023). Instead, the most significant criterion for this concept is the recognition of stasis or low rates of morphological change (Cavin and Guinot 2014; Kin and Błażejowski 2014; Lidgard and Love 2018; Watkins 2021; Bicknell et al. 2022; Cavin and Alvarez 2022). For this reason, some authors argue that this phenomenon warrants special attention within the framework of evolutionary biology research (Lidgard and Love 2018, 2021, 2023; Lidgard and Kitchen 2023).

It is important to emphasize that this type of stasis differs from the pattern observed at the species level. Several biologists—including those who have critiqued the concept of "living fossils"—have noticed this distinction, highlighting that stability (stasis) at the supraspecific level is separate from that observed at lower levels (e.g., Suno-Uchi et al. 1997; Casane and Laurenti 2013; Bicknell et al. 2022; Goto et al. 2022). For instance, Suno-Uchi et al. (1997) discuss the stasis seen in tadpole shrimps (Triops):

> The … results indicate that most of the morphological differences are not associated with speciation (lineage separation). These long-term morphological stases are not species-level properties. Some morphological characters are well preserved beyond the taxonomic and phylogenetic species level, and some show significant variation within phylogenetic species. Gould and Eldredge … claimed that stasis can be recognized as a meaningful and predominant pattern within the history of a species. The present study, however, indicates that stases are important evolutionary patterns, but they are not species-level properties. (Suno-Uchi et al. 1997, p. 456)

Fig. 12.5 *Groups categorized as "living fossils." Extant specimens are shown above, compared with their fossil record counterparts below.* (**a**, **b**) *Triops*. (**c**, **d**) *Ginkgo*. (**e**, **f**) Xiphosurans. These groups are characterized by their morphological stability over millions of years. However, as discussed in the text, this pattern is not unique to these groups but is a general characteristic of all higher taxa. This pattern can be referred to as suprastasis. (From Lidgard and Love 2018, photographs by James Lamsdell, with permission from Oxford University Press)

Another important point is to question whether groups commonly cited as living fossils (such as coelacanths, tadpole shrimps, horseshoe crabs, and ginkgo trees) represent rare cases of morphological stability compared to other groups. Many researchers tend to view these examples as anomalies, seeking explanations that portray supraspecific stasis as an exception rather than a general pattern applicable to all groups (e.g., Cavin and Alvarez 2022; Bicknell et al. 2022). However, a closer examination of the fossil record reveals that these instances of supraspecific stasis are not exceptions but rather part of a common pattern. For example, in 1984, Eldredge and Stanley edited a book titled *Living Fossils*, presenting 34 cases that illustrate the phenomenon of "arrested evolution." The book aimed to gather various researchers, primarily paleontologists, to discuss this phenomenon and assess whether this pattern necessitated a special explanation: "The only criterion for a group's inclusion here is that a living species must be anatomically very similar (bordering on identity) to a fossil species which occurs very early in the history of the lineage" (Eldredge and Stanley 1984, p. 3).

Given the stringent evidence, it is not surprising that all authors categorized their respective taxa as living fossils using this criterion. Here are some examples:

1. *Squirrels as living fossils.* "The recently discovered skeleton of *Protosciurus* (perhaps the oldest squirrel fossil) shows that the earliest recognized sciurid is strikingly similar in its osteology to living *Sciurus*. In the sense that it has evolved very little from what is apparently the primitive squirrel morphotype, *Sciurus* is a living fossil" (Emry and Thorington Jr 1984, p. 23).
2. *Tapirs as living fossils.* "[T]he postcranial skeleton of *Tapirus* remains essentially unchanged from that of an early Eocene perissodactyl, and its skull and dentition are essentially unchanged since the first appearance of the genus in the middle Oligocene" (Janis 1984, pp. 85–86).
3. *Nautiluses as living fossils.* "*Nautilus* … is easily differentiable from the ancestral *Cenoceras* species of the Late Triassic on the basis of suture and cross section. A nonspecialist, however, holding the two shells, would probably find the differences quite slight" (Ward 1984, p. 247).
4. *Crocodilians as living fossils.* "Many living broad-nosed crocodilians are so similar to extinct species that if they were exhibited side-by-side in zoos, the public and many herpetologists would be hard pressed to distinguish them … Although they lived in the Upper Jurassic, more than 140 million years ago, their skulls had only a small difference in outline from several living crocodiles" (Meyer 1984, p. 107).
5. *Gars as living fossils.* "Cretaceous and Eocene gars look like Recent gars" (Wiley and Schultze 1984, p. 163).
6. *Velvet worms as living fossils.* "In general, the onychophorans provide a spectacle of persistent stasis. They have not changed much in a long period of time" (Ghiselin 1984, p. 216).
7. *Teleostean fishes as living fossils.* Colin Patterson (1984) recounts that while attempting to select a suitable candidate for the category of "living fossil," he was astonished to discover countless specimens that merited such a designation:

   > This article is something of a fraud. I began work on it with the intention of showing that at least one Recent teleost, *Chanos chanos* … deserved to be called a living fossil. In the end, I failed to convince myself. The article is an account of that failure, with comments on other potential living fossils among teleosts … The number of candidates is too great to select one or a few as living fossils. (Patterson 1984, p. 132)

In fact, theoretically, any higher taxonomic group undergoes supraspecific stasis once it has appeared in evolutionary history. For example, some examples of supraspeciation reviewed above include the following observations:

8. *Snakes as living fossils.* "Anatomical features common to all recognized fossil and modern snake lizards are present in these jaw fragments" (Caldwell 2020, p. 40).
9. *Bats as living fossils.* "These findings indicate that the bat wing digit proportions have not changed substantially during the past 50 million years of evolution" (Sears et al. 2006, p. 6582).
10. *Chameleons as living fossils.* "These fossil fragments are unquestionably chamaeleonid on the basis of several characters" (Moody and Roček 1980, p. 89).

To be precise, all extant higher taxa could be categorized as living fossils. This is due to a widespread pattern of morphological stability at this level, which could be referred to as *supraestasis*. Once this evolutionary pattern is identified, the next step is to explore the underlying mechanisms that contribute to this morphological stability. Various authors have proposed different mechanisms to explain suprastasis with the chief aim of understanding the phenomenon of living fossils.

For example, in his attempt to explain the phenomenon of "living fossils," Eldredge (1984) proposes that the low rates of change observed at the supraspecific level result from a low speciation rate and the group's ability to tolerate and adapt to a broad range of environmental factors. Meyer (1984) suggests that possible causes include stabilizing selection, environmental stability, and large population sizes that prevent high speciation rates. Lee et al. (2006) assert that these patterns are influenced by speciation rates and genetic constraints, while Louchart et al. (2011) also emphasize the role of stabilizing selection. Lindholm (2014) proposes that significant physiological flexibility in these groups contributes to their stability. Werth and Shear (2014) further assert that stabilizing selection and genetic accumulation maintain morphological equilibrium until disrupted by environmental changes. Brownstein et al. (2024) suggest that slow genetic substitution rates help prevent reproductive isolation, allowing lineages to sustain gene flow and avoid divergence. Conversely, Lidgard and Love (2018) and Bicknell et al. (2022) argue that the persistence of living fossils results from the combined effects of developmental constraints and stabilizing selection. Lastly, while DiFrisco and Wagner (2022) and Galis (2023) do not directly address living fossils, they provide arguments related to developmental constraints to explain the stability of body plans observed at the "phylotypic stage."

Among these options, I concur that developmental constraints are responsible for suprastasis; however, I disagree with those who include stabilizing selection as part of this process. As discussed in the previous chapter, considering these three mechanisms—(1) developmental constraints, (2) developmental bias, and (3) natural selection—I emphasized that developmental constraints operate very differently from natural selection regarding the production and distribution of variation within the morphospace. Developmental constraints (in a negative sense) lock an ancestral structure at a specific position within the morphospace. When these constraints are broken (due to genetic variation or environmental change), a new variation is directed by developmental bias to occupy other regions within the morphospace.

Subsequently, developmental constraints stabilize this variation again. Finally, natural selection determines which variations are retained or eliminated based on their adaptive value (see Chap. 11, Fig. 11.7). Thus, while natural selection may eliminate forms during their origin, once established, developmental constraints can maintain structures almost invariant for millions of years until they eventually disappear through extinction.[15]

12.1.3 Supraextinction

According to the H-Evo-Devo Theory, *supraextinction* represents the evolutionary mode indicating the death of higher taxa. That is, once these higher taxa are established over an extended geological period, they are susceptible to disappearing at any point during their geological existence. This section posits that the causes of supraextinction are mass extinctions and the loss of parts. It is crucial to note that these processes reflect a hierarchical perspective, as discussed below.

Broadly speaking, a higher taxon becomes extinct when the last individual of its last species disappears (Taylor 2004). According to the H-Evo-Devo Theory, the extinction of units of evolution occurs within a hierarchical context. Thus, as observed in species, the risk for a higher taxon is proportional to the number of species (or less inclusive taxa) it contains. For example, a new higher taxon—currently in the process of diversification—or an old one that has lost species (or less inclusive taxa) over time—despite having been diverse in the past—is highly prone to extinction. Considering this, the pivotal question is whether the disappearance of higher taxa is solely a consequence of the natural process of species extinction—which in turn is dependent on subspecies extinction—or if it is driven by specific processes unique to this level. From the perspective of this model, while the survival of higher taxa depends on the number of species and, consequently, the number of subspecies—potentially explainable through microevolutionary extinction—higher-level rules cannot be reduced solely to events occurring at lower levels.

Mass Extinction Although there are different opinions on the causes of mass extinctions (Hallam 2004; Fattorini 2021; Algeo and Shen 2023), Congreve et al. (2018) argue that mass extinctions are characterized by their global impact on taxonomic groups and their large-scale effects compared to background extinctions. According to some authors, mass extinctions exhibit selectivity, i.e., some higher taxa survive due to specific characteristics, while others, whose properties make them more vulnerable, tend to go extinct (e.g., Jablonski 2005, 2017; Friedman 2009; Janevski and Baumiller 2009; Foster et al. 2022; Payne et al. 2023; Liu et al. 2024). For example, González-Suárez and Revilla (2013) demonstrate that large mammals face greater challenges in resisting extinctions due to factors such as lower abundance, slower reproduction, and longer life cycles. Additionally, Payne et al. (2023) emphasize that variations in groups' physiological properties might

[15] See in detail in Chap. 11, Sect. 11.2.2.

influence their tolerance to rapid changes in oxygen levels, temperature, and other variables during mass extinctions. Liu et al. (2024) also demonstrate that trait selectivity varies among different groups, although not all exhibit this pattern. Conversely, Monarrez et al. (2023) state that selectivity in these events could result from a complex interaction between intrinsic and extrinsic properties of higher taxa. In other cases, as Gould (2002, pp. 1316–1320) has noted through his "different rules" model, during global cataclysms, characters that are advantageous under normal conditions may become disadvantageous, while others may become crucial for survival as merely fortuitous exaptations.

However, Allen et al. (2019) show that, at least in the group of archosauromorphs, there is no evidence of selectivity. Similarly, Cole and Hopkins (2021) found no evidence of selectivity during two mass extinctions in the evolution of a group of crinoids. If this pattern holds for all groups, it could be concluded that, during mass extinctions, the death or survival of higher taxa is reduced to mere historical contingencies, suggesting that the extinction or survival of these major groups occurs randomly (Gould 1989; Liang et al. 2023). Nonetheless, whether mass extinction is selective or random, both scenarios result in the sudden disappearance of higher taxa.

Loss of Parts As explored throughout this book, the patterns observed in evolutionary entities reflect a hierarchical perspective; therefore, extinction must be explained in those terms. It is important to note that, according to the H-Evo-Devo Theory, taxonomic structure is arranged in reverse order, i.e., higher taxa originate first, followed sequentially by lower taxa.[16] Moreover, as discussed in the previous chapter, this model suggests that subspecies aid species in adapting to various environmental conditions, thereby reducing their risk of extinction, e.g., the more subspecies, the greater the probability of the species' survival. At a higher level, while the role of species (or genera) is not directly related to adaptation—since body plans can be fitted to different environments—a higher taxon with a large number of species (or less inclusive taxa) has a greater chance of survival (Flessa and Jablonski 1985). Thus, the critical question is whether the causes of subspecies extinction—such as environmental stochasticity, habitat fragmentation, competition, etc.—can also affect the extinction of higher levels over geological time. Although this is generally supported hypothetically, the extinction of higher taxa also depends on stochastic variables that modify their taxonomic structure concerning the rates of origin and extinction of lower levels (Foote 1993; Sclafani et al. 2018; MacDougall et al. 2019; Liang et al. 2023; Liu et al. 2024).

[16] As Michael Foote (1993, p. 202), a pioneer in studying the dynamics of the origin and extinction of higher taxa over geological time, aptly points out: "Nested diversification patterns may differ predictably up and down the taxonomic hierarchy, with the highest-level taxa showing early morphological bursts as major body plans are established, and lower-level taxa showing concordant morphological and taxonomic diversity increases as simple variations on those body plans are generated." For further details on this taxonomic structure, see Sect. 12.1.1.

Interestingly, several authors have noted that disparity and diversity may be decoupled when examining extinction patterns in the fossil record (Foote 1993; Hopkins 2013; MacDougall et al. 2019; Puttick et al. 2020; Beech and Lamsdell 2021; Serra et al. 2023; Bault et al. 2024; Guo et al. 2024; Liu et al. 2024; Xue et al. 2024). For example, even when higher taxa have lost most of their species, it is observed that disparity either increases (Foote 1993; Erwin 2007; Hopkins 2013; Korn et al. 2013; Puttick et al. 2020) or decreases (Bault et al. 2024) in many cases. Indeed, this increase in disparity—and the possible decrease in diversity—is the result of morphological innovations that occur after a mass extinction (Guo et al. 2024; Xue et al. 2024). Given the above, the fossil record clearly shows that dynamics of higher taxa surviving mass extinctions change stochastically (Foote 1993; Erwin 2001, 2008; Korn et al. 2013; Congreve et al. 2018; Sclafani et al. 2018; Guo et al. 2024; Jablonski and Edie 2025). As mentioned earlier, while mass extinctions typically eradicate many higher taxa along with their components, from a hierarchical perspective, a higher taxon may persist with well-defined boundaries within the morphospace, despite the loss of its components (Bault et al. 2024). If less inclusive higher taxa are lost, it is referred to as a loss of disparity, whereas if the species are lost, it is referred to as a loss of diversity.

Therefore, mass extinctions can randomly eliminate taxa across various levels. However, surviving higher taxa can be affected by both disparity and diversity, acting as a bottleneck and altering their taxonomic structure during recovery. For example, MacDougall et al. (2019) note that the disparity of a higher taxon may decrease following a mass extinction (or be low before such an event), but it can increase rapidly afterward. In other cases, disparity may be affected, while diversity remains stable. Conversely, Sclafani et al. (2018) show that the recovery of certain higher taxa can involve diversification into new and previously unoccupied regions of the morphospace, i.e., new groups can be very different from those that were previously dominant, even though the diversification of one group is largely random when compared to another. For example, Congreve et al. (2018) demonstrate that among four brachiopod groups that survived the Late Ordovician mass extinction, only one diversified during the recovery period, while the other three maintained low genus diversity until their eventual extinction. In this case, surviving higher taxa are not guaranteed to persist. Higher taxa with few species are "dead clades walking" (Jablonski 2002), because, despite potentially remaining stable for extended periods (Barnes et al. 2021), they remain vulnerable to extinction in future mass extinction events.

In conclusion, according to the H-Evo-Devo Theory, newly established higher taxa typically exhibit low diversity but tend to accumulate lower-level entities (species or less inclusive taxa) over geological time. However, after a mass extinction, many higher taxa vanish abruptly, and those that survive—having few parts—face serious challenges in recovering and are highly susceptible to extinction. Nevertheless, some higher taxa find the opportunity to regenerate lower-level entities downstream, though they can completely change their initial taxonomic structure.

12.2 Modular-Evolution (Homologue Level)

Modular-evolution is the level that studies the patterns and mechanisms of homologues as developmental modules whose variational dynamics contribute to the construction of the body plans. According to the H-Evo-Devo Theory, homologues correspond to morphological modules, defined as those quasi-autonomous parts of the organism that are highly integrated in developmental terms and maintained phylogenetically over a long time. This perspective on homologues suggests that evolution occurs in mosaics. Some authors have described mosaic evolution as the idea that each of the semi-autonomous units composing an organism exhibits different rates of evolution (Bastir and Rosas 2009; Felice and Goswami 2018; Larouche et al. 2018; Parins-Fukuchi 2020; Larouche et al. 2023). Accordingly, each morphological unit has an evolutionary potential (evolvability) that contributes to either the construction or the modification of the body plan (Brigandt 2007). According to this view, the transition between body plans is explained as a mosaic of characters with variable transformation rates (Lü et al. 2010; Larouche et al. 2018); for example, some characters undergo rapid changes, while others remain stable (Felice and Goswami 2018; Larouche et al. 2023). This perspective contrasts with the theory of "correlated progression," which posits that all parts of the body are interconnected, forming a network of interrelation among different characters. This model suggests that evolutionary transitions occur through gradual and parallel changes in several characters over time (see Kemp 2007).

This section discusses the mechanisms that contribute to the modes of origin and stability of homologues as modules in traditional morphology. I refer to the mode of origin of the modules as *morphological novelty*. Here, I argue that the morphological structures arise from the co-option of developmental mechanisms of preexisting structures. Once a new module is established, it tends to exhibit distinct variational modalities. In this regard, I present an argument that distinguishes between the limits of character identity and character state to clearly differentiate between "novelty" and "innovation," respectively. Conversely, I refer to the mode of stability of homologues as *morphological stability*, which is attributed to developmental constraints acting at the deepest level of the hierarchy. Here, I will emphasize that developmental constraints act hierarchically. At a higher level, they establish the stability of the modules, while at lower levels, they maintain stability within each of their variational modalities.

12.2.1 Morphological Novelty

According to the H-Evo-Devo Theory, the term *morphological novelty* refers to the origin of homologues as developmental modules. In Fig. 12.6, the ancestral Archetype (a) consists of the three modules (1, 2, 3). A new module (4) emerges and integrates into the descendant Archetype (b). Over time, this new module has acquired two different character states (homogenies) corresponding to distinct body plans (c) and (d). In this context, *morphological novelty* refers to the event of the

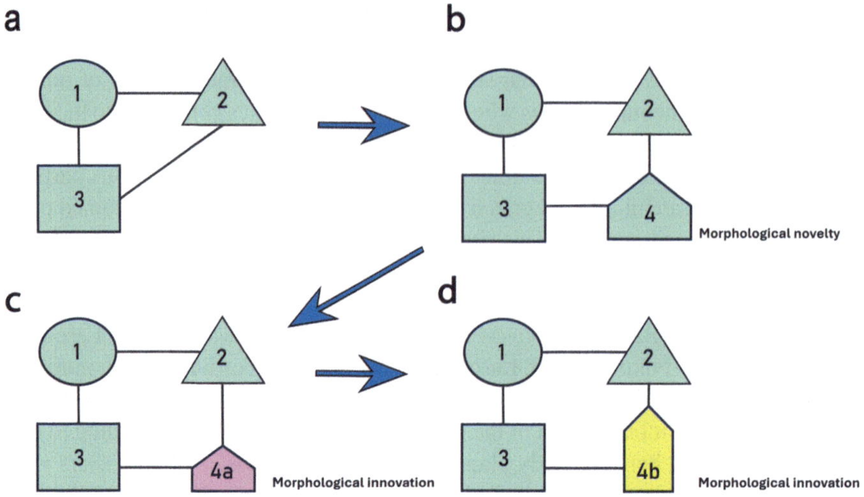

Fig. 12.6 *Modular-evolution*. Each geometric figure represents a module, indicated by a number (1, 2, 3, 4). The arrows show the sequence of generations. (**a**) The ancestral Archetype. (**b**) The descendant Archetype that acquired a new module (4). Although this module may have arisen from the co-option of elements from the module (2), its developmental dynamics changed so radically that it must be considered a distinct module. This is an example of *morphological novelty*. (**c**) and (**d**) Two body plans that altered the character state of the module (4). Since this involves a change in the color and shape of a previous module rather than the addition of a new element, both (4a and 4b) are examples of *morphological innovation*

origin of the module (4). Although module (4) arises from the co-option of elements (e.g., gene networks, cellular tissues, or even submodules) of a preexisting module—e.g., module (2)—its developmental dynamics have changed so radically that it must be considered a distinct module. Later, as module (4) evolves, it can still be identified in body plans (c) and (d), retaining its identity as a "homologue," although it varies in distinct forms and functions. So, the origin of variational modalities (4a) and (4b) is termed *morphological innovation*. Likewise, as these new character states are generally accompanied by a functional change—e.g., from (4a) to (4b)—which allows the lineage to diversify successfully, such innovation can also be referred to as *key innovation*. In order to elucidate this model and examine the mechanisms involved, this section will explore contemporary discussions on novelty in comparative morphology.

As discussed in Chap. 8, authors such as Hall (1999) have noted that the term "homologue" can be applied to different levels of organization (e.g., molecules, organs, behavior). For this reason, Hall and Kerney (2012) employ the concept of "novelty" to any of these levels. However, according to the H-Evo-Devo Theory, the term "homologue" is applied exclusively to morphological modules, which are defined as quasi-autonomous parts of the organism that are closely integrated in developmental terms. In this context, "novelties" refer to the origin of these modules. Indeed, this is precisely the notion of "novelty" that Wagner (2014, 2015) has in mind when discussing the origin of character identity.

Researchers who have discussed this topic recognize at least two distinct types of novelties (Müller and Wagner 1991, 2003; Müller and Newman 2005; Müller 2010; Peterson and Müller 2013; Wagner 2014, p. 126; Erwin 2021; McKenna et al. 2021; Schlosser 2023; Lala et al. 2024, p. 173; Love 2024, p. 43), which are related to the differences between character identity and character state (Brigandt 2007; Wagner 2014; Havstad et al. 2015; Ochoa and Rasskin-Gutman 2015; McKenna et al. 2021; Schlosser 2023) or between the concepts of homology and homogeny, respectively (Ochoa and Rasskin-Gutman 2015). According to the H-Evo-Devo Theory, character identity refers to the module, while character state pertains to a variational modality of that module. For instance, the forelimb of tetrapods represents a module (homologue), whereas its shape as a wing in bats represents one variational modality of that module (homogeny). It is important to note that "morphological novelty" (novelty type I sensu Wagner) refers to the origin of modules, whereas "morphological innovation" (novelty type II sensu Wagner) pertains to the emergence of variational modalities within the module. Wagner (2014) claims that this distinction is crucial and defines different research agendas. While the theoretical separation between these two modes of novelty is clear, making this distinction in practice can be challenging.

From the beginning of this debate, Müller and Wagner (1991, p. 243) proposed that "[a] morphological novelty is a structure that is neither homologous to any structure in the ancestral species nor homonomous to any other structure of the same organism." Yet, subsequent research has revealed that structures appearing novel in descendants actually arise from the reuse of elements (e.g., gene networks, cell types, tissues, submodules) or processes of preexisting structures (Shubin et al. 2009; Brigandt and Love 2012; Hallgrímsson et al. 2012; Havstad et al. 2015; Linz et al. 2020; McKenna et al. 2021). As Havstad et al. (2015, p. 579) note, "nothing genuinely new is generated through evolution; everything that exists is a transformation of something that existed before." More specifically, Brigandt and Love (2012, p. 418) observe that "even for those structures that appear to be qualitatively novel, there are always precursors or homologous features at lower levels, such as tissues, cells, or gene expression patterns."

According to DiFrisco et al. (2023b), "co-option" is a key process in the origin of characters. This "occurs when an ancestral gene, network, or mechanism that is active in the development of one character or region becomes activated in the development of a different character or region at a later point in evolution" (p. 7). In essence, this co-option hypothesis suggests that a descendant module does not appear as a complete novelty at every level of organization but rather arises from the reorganization of different components of an ancestral module. If this is the case, then the definition of "novelty" proposed by Müller and Wagner (1991) would not apply to any character.

Within the framework of the H-Evo-Devo Theory, however, since a "homologue" is equivalent to a module, each distinct element (genes, cell types, tissues, etc.) of an ancestral module should not be strictly termed "homologous." Instead, the term "homologue" refers specifically to the module as an integrated unit across all developmental levels. Although one module originates from another module, they are not considered homologous because their equivalent levels of organization are decoupled. For example, two structures might share a similar development in the early stages but subsequently diverge in the later stages. Indeed, through this phenomenon, the developmental dynamics of the descendant module can diverge so significantly from the original module that it must be regarded as a distinct module. In addition, stressing the origin of modules through embryological or molecular reuse supports the argument for phylogenetic relationships between modules. These genealogical connections among homologous elements are analogous to the connections among lineages. For example, birds are considered a distinct and independent group from theropod dinosaurs, even though they originated from them.

An important aspect of this debate is the concept of serial homology. Generally, serial homology refers to situations where two or more body parts of the same organism result from repetition. Thus, it is emphasized that these repeated characters share a fundamental identity, despite their variations (e.g., vertebrae, fingers, segments in arthropods, etc.). According to Wagner (2014, 2015), the origin of serial homologues is one of the most significant and prevalent modes for understanding character novelty. Nevertheless, the nature of serial homology remains unclear to some researchers.

Another important aspect of serial homologues is that they are paradigmatically viewed as copies of a base structure from the ancestral lineage, which are then individualized and modified, resulting in certain differences. However, it has become clear that serial homologues are not necessarily exact replicas of an ancestral structure. In fact, in most cases, serial homology results from the process of co-option (DiFrisco et al. 2023b). This process implies that instead of merely replicating an ancestral structure, serial homologues emerge from the reuse and modification of preexisting elements, creating new variants within the organism.

Recently, some scholars have distinguished between two concepts of serial homology (Diogo 2020; Siomava et al. 2020; Fusco 2022): (1) *a historical concept*, which suggests that an ancestral structure produces similar structures in descendants, which then diverge and specialize for specific functions; and (2) *a developmental concept*, which posits that descendant serial structures are derived from the same elements or developmental processes of an ancestral structure. According to this concept, serial homology does not imply that ancestral serial structures are similar but rather that they share a common identity (DiFrisco et al. 2023a, b). In fact, based on the paleontological and embryological evidence, serial homology should be conceptualized exclusively under the developmental concept rather than the historical one (Siomava et al. 2020).

According to DiFrisco et al. (2023a), serial homologues are categorized into two types: *homomorphs* and *paramorphs*. Homomorphs are structures that are replicas of others, arising from the same developmental mechanisms or Character Identity

Mechanisms (ChIMs).[17] Examples include thoracic vertebrae, hairs, and leaves. In contrast, paramorphs are individualized serial structures with different ChIMs, yet they derive from a common ancestral ChIM. Examples include the atlas/axis, hair/mammary gland, and petals/sepal. Additionally, they introduce the concept of "paramorph hierarchy," suggesting that paramorphs can occur at different levels of organization. Considering this, they maintain that true novelty is identified when an individualized structure results from the co-option of paramorph elements, rather than from mere repetition of ChIMs as seen in homomorphs. Moreover, recognizing the origin of paramorphs at different levels not only helps identify novel structures but also facilitates the tracing of genealogical relationships among characters.

In this regard, in the context of the H-Evo-Devo Theory, a distinction is made between "novelty" and "innovation." Novelty pertains to the emergence of a new module, which may arise from co-option, i.e., by reusing the elements of a preexisting module such as gene networks, tissues, or submodules. Under this view, the developmental dynamics of the new module change so radically (by generating new processes, removing or adding subparts of the ancestral module, etc.) that this module must be considered distinct from the ancestral one. In contrast, "innovation" refers to the origin of a character state or a variational modality of an existing module, where this variation does not deviate significantly enough from the developmental constraints of the descendant module to warrant identifying it as a distinct module (see Fig. 12.6). Below, I will briefly review some examples to illustrate this distinction.

Insect Wings The origin of insect wings constitutes a morphological module that has sparked intense debate. The emergence of this character can be considered an example of morphological novelty: once the wing module originated in insects, it gave rise to different variational modalities (wing shapes) such as halteres in flies, elytra in beetles, and the distinctive wings of butterflies. The origin of each of these forms (halteres, elytra, etc.) exemplifies morphological innovation. Although this framework is clear, the boundary between "novelty" and "innovation" can appear ambiguous, especially given the paradigm that "all the new comes from the old."

According to various hypotheses about the origin of insect wings, biologists have debated whether they originate from tergal or pleural tissue (Bruce and Patel 2020; Ohde et al. 2022; Ohde and Prokop 2025). However, it is also postulated that they could have arisen from a combination of both tissues (Prokop et al. 2017; Linz and Tomoyasu 2018; Tomoyasu 2021; Clark-Hachtel et al. 2021). Although some suggestions propose that insect wings are homologous to certain crustacean appendage structures (Averof and Cohen 1997; Bruce and Patel 2020; Prokop et al. 2022),

[17] According to DiFrisco et al. (2020), the Character Identity Mechanisms (ChIMs) model represents the most conserved elements of the developmental system that define a morphological structure as a character identity. This model explains the persistence and traceability of a character across generations. According to the H-Evo-Devo Theory, ChIMs are equivalent to developmental constraints operating at the highest level of the hierarchy. See below for more details.

Clark-Hachtel and Tomoyasu (2020) identified that the tergal edge and proximal leg segment of crustaceans correspond to the tergum and pleural plates of insects, respectively. Also, they found that all these tissues share the same gene regulatory network as insect wings. In this regard, they state that this gene network found in both tissues may have evolved serially, i.e., as a duplication of the tissue with the same underlying gene network. Otherwise, this gene network would have had to be either co-opted into new tissue or evolved independently in both tissues.

But the issue of serial homology does not end there. Evidence shows that the tissues and genes that produce wings in insects—in thoracic segments 2 and 3—are present in all thoracic and abdominal segments of insects (Linz et al. 2020), including those without wings (Clark-Hachtel and Tomoyasu 2020). This, of course, supports their paramorphic nature. For example, researchers have recognized that the embryonic materials forming wings are the same ones producing other structures in different segments, such as horns in the first thoracic segment of dung beetles (Hu et al. 2019), helmets in the first thoracic segment of treehoppers (Fisher et al. 2020), and gin traps in the abdominal segments of red flour beetle pupae (Linz and Tomoyasu 2018), etc. (see Fig. 12.7). Indeed, Fisher et al. (2021) assert that these structures share an identity with other structures in other arthropods such as "bilayered cuticular epithelial structures," which exhibit the same developmental gene network.

This example illustrates that novel structures generally arise from the serial co-option of embryonic components from other modules at different levels of organization, emphasizing the phylogenetic relationships between modules (DiFrisco et al. 2023a). In other words, although insect wings, beetle horns, and treehopper helmets are distinct modules, they maintain a genealogical relationship as serial homologues. Moreover, their ancestral node can be traced to the tergal edge and proximal leg segment of crustaceans and a deeper ancestral state as bilayered epithelia of arthropods.

Tetrapod Limbs The origin of tetrapod limbs exemplifies morphological novelty. For instance, once the forelimb module emerged, it underwent different variational modalities, e.g., bat wings, whale fins, lion claws, human arms, etc. The origin of each of these character states serves as a paradigmatic example of morphological innovation. However, the issue arises when considering that tetrapod limbs evolved from paired fish fins. Interestingly, Wagner (2014, p. 126) categorizes tetrapod limbs as a paradigmatic example of morphological innovation (i.e., novelty type II, according to his terminology) rather than morphological novelty (i.e., novelty type I). But are tetrapod limbs truly paradigmatic examples of morphological innovation?

Undoubtedly, there is a "homologous" relationship between the pectoral fin of ancestral fish and the forelimb in tetrapods. For example, the stylopod and zeugopod of sarcopterygian fins are homologous to the metapterygium of chondrichthyan fins, and it is hypothesized that the protopterygium and mesopterygium were lost during the evolution of the forelimb (Wagner 2014; Tanaka 2016, 2017). Nonetheless, many researchers have questioned whether a structure that is

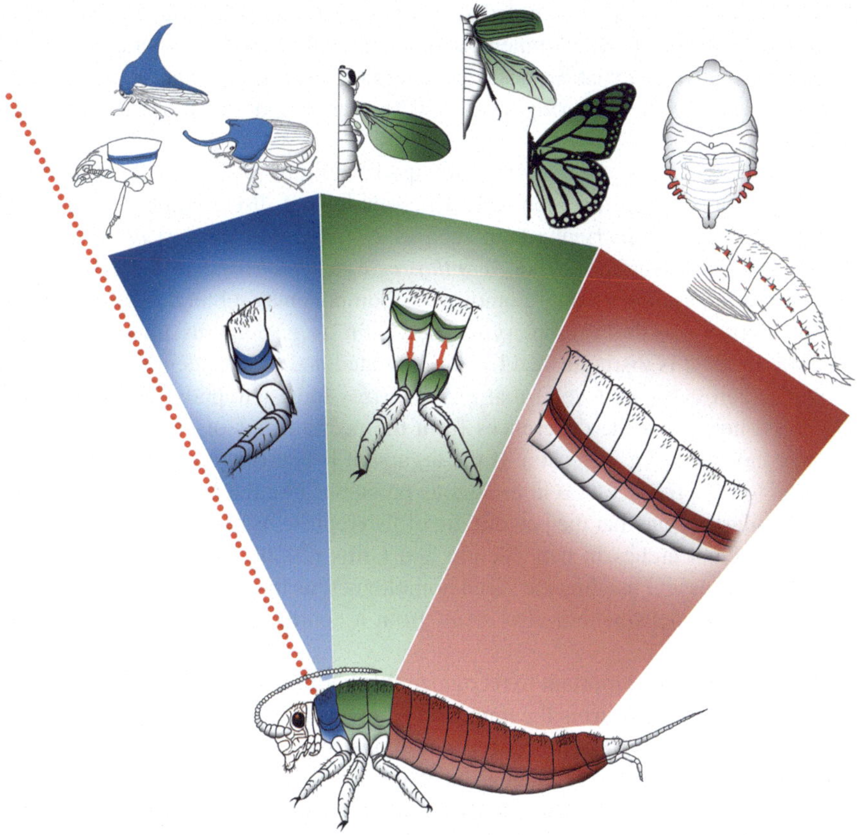

Fig. 12.7 *Serial homology as a source of morphological novelties.* The combination of ancestral tergal and pleural tissues, indicated by red arrows, gave rise to insect wings in T2 and T3 (green). Interestingly, these tissues and their underlying gene regulatory network are also located in other segments (e.g., in the first segment or the abdomen, indicated in blue and red, respectively), which produce other structures such as horns in beetles, helmets in treehoppers, and gin traps in red flour beetles. Although there may be doubts about the use of the term "novelty" considering the paradigm "all the new comes from the old," each structure is considered a morphological novelty, as they arise from the co-option of preexisting elements. (Modified from Linz et al. 2020, with permission from Elsevier)

homologous to the autopod exists in these basal fish. If no homologous structure is found, the autopod could be considered a morphological novelty. Those who advocate for considering the autopod as a "novelty" support this view based on differences in genetic expression observed in studies of teleost fish (e.g., Sordino et al. 1995; Wagner and Chiu 2001). However, some paleontological studies—particularly in fossil fish such as *Panderichthys*, *Tiktaalik*, and *Elpistostege*—suggest that homologous elements are present in the autopod, including digits (Shubin et al. 2006; Boisvert et al. 2008; Cloutier et al. 2020). This hypothesis is further supported by molecular evidence (Davis et al. 2007; Johanson et al. 2007; Nakamura et al.

2016). Even if the autopod is regarded as a novel structure due to its differences in genetic expression, its development is fundamentally based on a straightforward modification of a preexisting basal developmental program (Davis et al. 2007; Shubin et al. 2009; Woltering et al. 2020; Varga and Varga 2022; Fernando et al. 2023).

Regardless, I consider the origin of tetrapod forelimbs as an example of morphological novelty due to the pronounced morphological discontinuity between the pectoral fins of sarcopterygian fish, such as *Tiktaalik* and *Panderichthys*, and the forelimbs of early tetrapods like *Acanthostega* and *Ichthyostega*. While the forelimbs have incorporated some homologous components (submodules) from what can be referred to as the "sarcopterygian pelvic fin" (e.g., humerus, ulna, radius), the forelimb module is distinguished by an integrated and independent embryological dynamic (across all its levels) that characterizes most tetrapods. Furthermore, it should be noted that the differences from the pectoral fins of other fish are even more pronounced (e.g., teleosts). If the argument is that the forelimb is an innovation due to the presence of a homologous precursor like the pectoral fin, then one could similarly argue that the pectoral fin of fish represents a derived character state of the branchial arches in fish. Sleight and Gillis (2020), for example, state that branchial arches and fish fins are serial homologues, as they not only share the same gene regulatory network but also result from a continuous change in embryonic tissues.[18]

Additionally, it is important to recognize that both forelimbs and hindlimbs originated simultaneously during the transition from fins to limbs. This suggests that the similarity between the forelimbs and hindlimbs may stem from the co-option of common tissues and developmental mechanisms. Nevertheless, given the significant differences between pectoral and pelvic fins in early fish—where pelvic fins are hypothetically thought to have originated from a serial duplication of pectoral fins—the "true serial homology" between forelimbs and hindlimbs has been challenged (Diogo et al. 2012; Sears et al. 2015; Miyashita and Diogo 2016; Diogo 2020; Molnar and Diogo 2021).

For example, Sears et al. (2015) discovered that the developmental programs of forelimbs and hindlimbs are more similar to each other than those of the pectoral and pelvic fins. In fact, in ancestral sarcopterygian fish, pectoral fins differ significantly from the pelvic fins, whereas the forelimbs and hindlimbs of early tetrapods exhibit greater similarity (Boisvert 2005).[19] Note that serial homologues are defined by their identity rather than their similarity. Given this, my interpretation is that the forelimbs and hindlimbs emerged simultaneously from a radical transformation of the fin developmental program. Therefore, although there is a relationship between fins and limbs, it is only in terms of their genealogical connection as serial homo-

[18] See Gillis et al. (2009) for further details. Recent paleontological evidence has shed light on the homology between the branchial arch and the pectoral girdle (Brazeau et al. 2023). For a review of the ongoing debates on the evolution of paired fins in vertebrates, see Bayramov et al. 2024.

[19] However, one of the considerations of serial homology is that an ancestral structure is never identical to its descendants; see Siomava et al. 2020.

logues. Nevertheless, since their developmental dynamics changed radically (e.g., by integrating the autopod along with a characteristic phalangeal pattern), they can be considered distinct modules. Hence, limbs are examples of morphological novelty rather than paradigmatic examples of morphological innovation.

Feathers Bird feathers are a morphological module characterized by complex interactions between various submodules at different hierarchical levels (Prum and Dyck 2003; Lin et al. 2020; Terrill and Shultz 2023). Notably, the fossil record indicates that feathers existed before the emergence of birds, as evidenced by some theropod dinosaur fossils. These primitive structures were simple and filamentous, undergoing a series of steps that progressively increased their complexity (Xu 2020). Interestingly, this series of transitions coincides with the steps observed in the development of modern bird feathers from a follicular structure (Prum 1999). All extant feathers share a common and distinctive developmental pathway (Brush 1996; Prum 1999). Since the establishment of this common feather type in the Late Jurassic (Longrich et al. 2020), feathers have undergone a series of variations in length, orientation, and structure, resulting in forms that are silky, curly, colorful, and more (Chen et al. 2015; Bhattacharjee et al. 2016; Boer et al. 2017; Ng and Li 2018; Lin et al. 2020). A single bird can exhibit a great diversity of feather shapes, sizes, and coloration patterns on different parts of its body, each serving different functions (Chen et al. 2015; Terrill and Shultz 2023). In short, the origin of the feather is an example of morphological novelty, and each of its successive modifications is an example of morphological innovation.

Like the previous example, the problem with considering feathers as a novelty primarily lies in recognizing reptilian scales as their "homologues." For instance, Wagner (2014, p. 126) argues that since feathers and scales share a common developmental pathway and feathers appeared after scales, feathers should be considered a paradigmatic example of morphological innovation (type II novelty, under his terminology). In other words, feathers are a character state of the character identity scales. However, despite suggestions that feathers likely formed from the reutilization of a developmental program present in primitive scales (Harris et al. 2002; Prum 2005), some researchers have emphasized that feathers have such a distinctive embryological development that they cannot simply be considered derivatives of scales (Brush 1996, 2000; Prum 1999, 2005; Harris et al. 2002; Dhouailly 2009; Benton et al. 2019). For example, Prum (1999, p. 300) notes: "[F]eathers and scales are apparently homologous as epidermal appendages at the level of the placode and papilla, but not as mature structures. The origin of the follicle created derived developmental mechanisms unique to feathers that grow structures whose details are not homologous with any aspect of a mature scale."

Comparably, while there is a morphological similarity between avian scales (e.g., those observed on the legs) and reptilian scales, the developmental processes of these structures are quite distinct. Furthermore, the same developmental mechanisms regulate both avian scales and feathers. This suggests that avian scales originated from feathers rather than reptilian scales (Dhouailly 2009, 2024; Wu et al.

2018; Cooper et al. 2019). Most importantly, no intermediate form between reptilian scales and feathers, or vice versa, has ever been observed (Dhouailly 2023), not even in fossil specimens that exhibit both types of epidermal structures, where these remain developmentally partitioned (Yang et al. 2024).

In any case, assuming that feathers are "innovations" rather than "novelties" is refuted by considering the paradigm that "all the new comes from the old." For example, recent research has demonstrated that all epithelial appendages in vertebrates (e.g., fish scales, reptile scales, mammalian hair, and bird feathers) stem from a shared anatomical placode, which is developed through a highly conserved gene regulatory network that was likely present in a vertebrate ancestor at least 420 million years ago (Cooper et al. 2017; Benton et al. 2019; Dhouailly et al. 2019; Cooper et al. 2023; Dhouailly 2023). Cooper et al. (2023) mention that early jawed fish possessed epithelial appendages in the form of teeth known as odontodes. Sharks and rays have preserved these structures as dermal denticles on their skin and as teeth in their mouths. Embryological comparisons reveal that denticles and teeth are not only morphologically similar but also share a set of genes. Interestingly, when comparing the development of these denticles with that of feathers, they found that both share an early developmental program as well.

In light of these findings, it can be concluded that despite the wide variety of epidermal appendage forms, all these structures arose from the co-option of a highly conserved preexisting developmental system (see Fig. 12.8). Cooper et al. (2017) and Dhouailly et al. (2019) stress that this homology is not only confined to the genetic level but also extends to the morphological level through the epithelial placode, suggesting that the homology may not be as deep as previously thought. So, while feathers may have evolved from a structure that initially developed reptilian scales, their early developmental program was eventually decoupled from its higher level, leading to the emergence of feathers (similar to how the development of denticles decoupled to give rise to scales). In short, given that feathers are integrated into all levels of organization, regardless of their varieties of form and function, their origin should be considered an example of morphological novelty rather than morphological innovation.

12.2.2 Morphological Stability

According to the H-Evo-Devo Theory, *morphological stability* refers to the constancy of homologues spatiotemporally. That is, once a module originates, it stabilizes at all hierarchical levels; nevertheless, it can exhibit different variational modalities. A key discussion point in this evolutionary mode is the need to distinguish between the mechanisms that establish the identity of the modules and those responsible for their variational modalities. In fact, stasis in the character identity is deeper than that observed in character states. In other words, the mechanisms that establish the morphological stability of the modules can be separated from those that fix their specific forms and functions. Now, I will explore this discussion in detail.

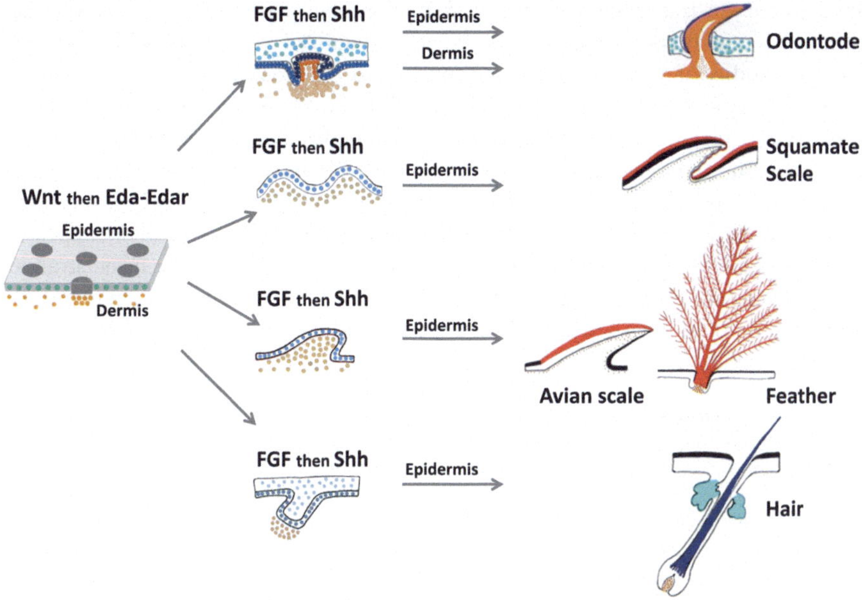

Fig. 12.8 *Embryonic development of denticles, scales, feathers, and hairs*. All these epidermal appendages arise from the co-option of a highly conserved preexisting developmental process, spanning from the genetic to the morphological level through the epithelial placode (gray circles in the epidermis). Similar to the example of insect wings, it is important to emphasize that each of these structures represents a case of morphological novelty arising from the co-option of preexisting elements. (Modified from Benton et al. 2019, with permission from Elsevier)

When I state that homology is the recognition of the same module regardless of its form and function, I am emphasizing that these different forms and functions (variational modalities) represent another level of comparison. The issue lies in identifying the mechanisms that explain the stability of character identity and distinguishing them from those responsible for the stability of character states. Wagner (2014), in his initial attempts to differentiate between character identity and character states, proposed that the most conserved part of the gene regulatory network, known as the Character Identity Network (ChIN), determined the character identity. Subsequently, DiFrisco et al. (2020), expanded this argument to include other levels of organization, which they refer to as Character Identity Mechanisms (ChIMs).

According to DiFrisco et al. (2020), the ChIMs model proposes that homologous characters are recognized spatiotemporally through certain highly conserved mechanisms during ontogeny, as opposed to other mechanisms that exhibit more variability within the system. These conserved mechanisms theoretically account for the stability of the character identities (homologues). However, there are also stability mechanisms for character states (or homogenies), which can be referred to as Character State Mechanisms (ChSMs). In the context of the H-Evo-Devo Theory, when developmental constraints operate at the highest levels of the hierarchy, they can be viewed as ChIMs, which establish the fundamental architecture of the module (as illustrated in Chap. 5, Fig. 5.4). Conversely, when these constraints act at lower hierarchical levels, they can be conceptualized as ChSMs, which account for the stability of a specific form and function of the module.

For example, considering the previously mentioned case of the evolution of epithelial appendages in vertebrates, after reptilian scales emerged during the Triassic, their structure remained largely unchanged for millions of years. The same applies to hair structure, which has stayed relatively stable since the Early Cretaceous (Dhouailly 2023). In bird evolution, while various types of protofeathers appeared in dinosaurs, it is only with *Archaeopteryx lithographica* that both the basic arrangement of modern feathers and the diversification of feather forms—including those specialized for flight—are established.[20] Considering this event, Longrich et al. (2020) comment:

> [M]any aspects of feather morphology and arrangement found in modern birds originated in the Jurassic and are retained, within minimal modification, in modern birds … It would not be fair to say that the feathers and their arrangement have not changed; clearly, major changes did occur between *Anchiornis* and *Archaeopteryx* and crown birds in feather structure … and arrangement … However, these adaptations notwithstanding the feathers of birds represent a remarkable example of stasis. Birds are typically considered highly advanced organisms, and yet in terms of feather morphology, they are every bit as much a "living fossil" as a crocodile or a coelacanth. (Longrich et al. 2020, pp. 137–138)

This illustrates the distinction between the two levels of structural stability. The basic architecture of the feather represents the stability of the character identity, while minor modifications within this basic structure can exhibit stability as a character state. During the origin of the modern feather, developmental constraints would have established its basal architecture, and subsequently, at a different level, they would have acted to stabilize its specific forms, such as those specialized for flight or other functions.

As another example, consider the bird beak as a module defined through comparative methods. The origin of the bird beak is an example of morphological novelty, as the difference between the snout of the ancestral archosaur (probably a dinosaur) and the beak of early birds is so substantial—requiring radical structural changes—that it should be considered a distinct module (Bhullar et al. 2015). This evolutionary emergence likely occurred in the Late Cretaceous, particularly if *Ichthyornis dispar* is considered a key species in the origin of the modern bird beak (Field et al. 2018). Despite the considerable variation in beak shapes and sizes, birds have maintained the characteristic beak pattern due to its high degree of cranial morphological integration (Marugán-Lobón et al. 2022). Nonetheless, these beak variations have remained stable for millions of years across different avian groups (Rombaut et al. 2021). For example, the beak of pelicans has persisted in stasis for approximately 30 million years (Louchart et al. 2011). With this in mind, the H-Evo-Devo Theory posits that developmental constraints operating at the highest levels of

[20] Note that modern feathers emerged very early in the Late Jurassic. Many theropod dinosaurs—along with the primitive arrangement of feathers—reached their peak later during the Cretaceous.

Table 12.1 *Comparison between the two morphological levels within the modular hierarchy: the homologue and one of its variational modalities*. This table highlights the mode of origin, the causes of morphological stability, and examples

Descriptor	Unity	Mode of origin	Causes of morphological stability (stasis)	Examples
Homology	Homologue (module)	Morphological novelty	Developmental constraints (operating at a higher level), or Character Identity Mechanisms (*sensu* DiFrisco, Love, and Wagner)	Insect wings, vertebrate forelimbs, feathers, bird beak
Homogeny	A variational modality of a module	Morphological innovation	Developmental constraints (operating at a lower level), or Character State Mechanisms	Butterfly wings, bird wings, flight feathers, pelican beak

the hierarchy account for the morphological stability of the beak across all birds. At a lower level, developmental constraints work to stabilize specific beak forms within different bird groups. According to Ochoa and Rasskin-Gutman (2015), the recognition of stability in the module is homology, while the recognition of stability in one of the variational modalities is homogeny (see Table 12.1).

Finally, in addition to developmental constraints, it is important to recognize that many researchers have explored other factors to explain morphological stability. For instance, in the example of feathers, Longrich et al. (2020) note that while developmental patterns may constrain feather morphology, stabilizing selection also plays a crucial role by imposing functional limitations on feathers. In this regard, DiFrisco (2023) suggests that developmental constraints alone need not be the only explanation for module stability, suggesting that other mechanisms, including stabilizing selection, may also be involved. It is important to note that, for DiFrisco, stabilizing selection encompasses not only the elimination or preservation of variations advantageous to an environment but also the interdependence of characters in developmental terms. In this context, he references Riedl's (1978) concept of "burden" and Wimsatt's (1986) idea of "generative entrenchment": "both describe a situation where a character is stable in evolution not because of a lack of natural variations, but because all departures from existing variants are non-viable or infertile" (DiFrisco 2023, p. 792).

However, as discussed throughout this book regarding stasis, developmental constraints and natural selection play different roles. For example, natural selection does not aim to limit possible variants but rather to preserve or eliminate them based on their functional utility within the environment. More importantly, at the level of modules, developmental constraints are characterized by maintaining a more stable morphological integration across all levels of organization, even though many levels may become decoupled at different times during their evolutionary history. From this perspective, developmental constraints are not merely a matter of the "lack of natural variations," but rather of the interdependence of generative processes that contribute to forming stable structures. Therefore, developmental constraints can create a kind of "internal selection" by maintaining the boundaries of variation, thus preserving viable organisms while eliminating non-viable ones during development. As Caponi (2008) notes:

> ... these more stable structures should not be considered merely as the simple remnants of a common ancestral form that natural selection has left untouched for now, but rather as the outcome of developmental constraints that actively preserve certain structures by virtue of their organizational significance. (Caponi 2008, p. 210; my translation from Spanish)

If this is the case, the concepts of "burden" and "generative entrenchment" should be viewed as complementary to, rather than opposed to, the arguments of developmental constraints. Furthermore, it should be noted that Riedl's concept of "burden" may not fully apply at the module level, as it primarily addresses the stability of body plans (Rasskin-Gutman and Esteve-Altava 2021). Thus, as argued in this book (see Chap. 5), developmental constraints alone account for the stability of homologues, regardless of their form and function.

12.3 Conclusions

Surprisingly, traditional evolutionary theory has overlooked many significant processes in large-scale evolution. However, the H-Evo-Devo Theory offers theoretical tools that allow for reinterpreting the study of higher levels in the evolutionary hierarchy. For example, higher taxa are no longer viewed merely as historical entities resulting from speciation; instead, they can be conceptualized as dynamic and autonomous entities that shape the taxonomic structure in a particular way. Meanwhile, homologues are no longer perceived solely as evidence of phylogenetic relationships between species but can be interpreted as fundamental units that contribute to the construction of body plans over geological time.

Higher taxa are units of evolution that represent the Natural System through a hierarchy of body plans. Homologues are modules that arise from the co-option of elements from other modules, which, through their modifications as character states, produce different body plans hierarchically. The taxonomic structure proposed by the H-Evo-Devo Theory is built from the top-down, emphasizing the role of major body plans as the initiators of the Tree of Life. Furthermore, this theory provides a scheme that not only elucidates the origin of major multicellular life groups but also explains their stability over geological time, the dynamics of their extinction, and the interconnection among these processes. In summary, the H-Evo-Devo Theory offers a comprehensive framework that effectively unifies taxonomy, morphology, and evolution at all hierarchical levels.

References

Alexander DE (2015) On the wing: insects, pterosaurs, birds, bats and the evolution of animal flight. Oxford University Press, New York

Algeo TJ, Shen J (2023) Theory and classification of mass extinction causation. Natl Sci Rev 11:nwad237. https://doi.org/10.1093/nsr/nwad237

Alhajeri BH, Steppan SJ (2018) Disparity and evolutionary rate do not explain diversity patterns in muroid rodents (Rodentia: Muroidea). Evol Biol 45:324–344

Allaire N, Ginot S, de Baets K, Korn D, Goudemand N, Monnet C, Crônier C (2023) Morphological disparity of early ammonoids: a geometric morphometric approach to investigate conch geometry. Acta Palaeontol Pol 6:193–212

Allen BJ, Stubbs TL, Benton MJ, Puttick MN (2019) Archosauromorph extinction selectivity during the Triassic–Jurassic mass extinction. Palaeontology 62:211–224

Averof M, Cohen SM (1997) Evolutionary origin of insect wings from ancestral gills. Nature 385:627–630

Barnes BD, Sclafani JA, Zaffos A (2021) Dead clades walking are a pervasive macroevolutionary pattern. Proc Natl Acad Sci USA 118:e2019208118. https://doi.org/10.1073/pnas.2019208118

Bastir M, Rosas A (2009) Mosaic evolution of the basicranium in *Homo* and its relation to modular development. Evol Biol 36:57–70

Bault V, Crônier C, Monnet C (2024) Coupling of taxonomic diversity and morphological disparity in Devonian trilobites? Hist Biol 36:473–484

Bayramov AV, Yastrebov SA, Mednikov DN, Araslanova KR, Ermakova GV, Zaraisky AG (2024) Paired fins in vertebrate evolution and ontogeny. Evol Dev 26:e12478. https://doi.org/10.1111/ede.12478

Beech JD, Lamsdell JC (2021) Phylogeny, disparity and mass extinction response in the trilobite order Harpetida. Pap Palaeontol 7:2205–2225

Benton MJ, Dhouailly D, Jiang B, McNamara M (2019) The early origin of feathers. Trends Ecol Evol 34:856–869

Bhattacharjee MJ, Yu CP, Lin JJ, Ng CS, Wang TY, Lin HH, Li WH (2016) Regulatory divergence among Beta-Keratin genes during bird evolution. Mol Biol Evol 33:2769–2780

Bhullar BA, Morris ZS, Sefton EM, Tok A, Tokita M, Namkoong B, Camacho J, Burnham DA, Abzhanov A (2015) A molecular mechanism for the origin of a key evolutionary innovation, the bird beak and palate, revealed by an integrative approach to major transitions in vertebrate history. Evolution 69:1665–1677

Bicknell RD, Kimmig J, Budd GE, Legg DA, Bader KS, Haug C, Kaiser D, Laibl L, Tashman JN, Campione NE (2022) Habitat and developmental constraints drove 330 million years of horseshoe crab evolution. Biol J Linn Soc 136:155–172

Boer EF, Van Hollebeke HF, Shapiro MD (2017) Genomic determinants of epidermal appendage patterning and structure in domestic birds. Dev Biol 429:409–419

Boisvert CA (2005) The pelvic fin and girdle of *Panderichthys* and the origin of tetrapod locomotion. Nature 438:1145–1147

Boisvert CA, Mark-Kurik E, Ahlberg PE (2008) The pectoral fin of *Panderichthys* and the origin of digits. Nature 456:636–638

Brazeau MD, Castiello M, El Fassi El Fehri A, Hamilton L, Ivanov AO, Johanson Z, Friedman M (2023) Fossil evidence for a pharyngeal origin of the vertebrate pectoral girdle. Nature 623:550–554

Brigandt I (2007) Typology now: homology and developmental constraints explain evolvability. Biol Philos 22:709–725

Brigandt I, Love AC (2012) Conceptualizing evolutionary novelty: moving beyond definitional debates. J Exp Zool (Mol Dev Evol) 318B:417–427

Brownstein CD, MacGuigan DJ, Kim D, Orr O, Yang L, David SR, Kreiser B, Near TJ (2024) The genomic signatures of evolutionary stasis. Evolution 78:821–834

Bruce HS, Patel NH (2020) Knockout of crustacean leg patterning genes suggests that insect wings and body walls evolved from ancient leg segments. Nat Ecol Evol 4:1703–1712

Brush AH (1996) On the origin of feathers. J Evol Biol 9:131–142

Brush AH (2000) Evolving a protofeather and feather diversity. Am Zool 40:631–639

Caldwell MW (2020) The origin of snakes morphology and the fossil record. CRC Press, Boca Raton

Caponi G (2008) Selección interna: el control de la filogenia por la ontogenia en una perspectiva variacional. Theoria 62:195–218

Carretero-Paulet L, Librado P, Chang TH, Ibarra-Laclette E, Herrera-Estrella L, Rozas J, Albert VA (2015) High gene family turnover rates and gene space adaptation in the compact genome of the carnivorous plant *Utricularia gibba*. Mol Biol Evol 32:1284–1295

Casane D, Laurenti P (2013) Why coelacanths are not 'living fossils': a review of molecular and morphological data. BioEssays 35:332–338

Cavin L, Alvarez N (2022) Why coelacanths are almost "living fossils"? Front Ecol Evol 10:896111. https://doi.org/10.3389/fevo.2022.896111

Cavin L, Guinot G (2014) Coelacanths as "almost living fossils". Front Ecol Evol 2:49. https://doi.org/10.3389/fevo.2014.00049

Chan BKK, Dreyer N, Gale AS, Glenner H, Ewers-Saucedo C, Pérez-Losada M, Kolbasov GA, Crandall KA, Høeg JT (2021) The evolutionary diversity of barnacles, with an updated classification of fossil and living forms. Zool J Linnean Soc 193:789–846

Chen CF, Foley J, Tang PC, Li A, Jiang TX, Wu P, Widelitz RB, Chuong CM (2015) Development, regeneration, and evolution of feathers. Annu Rev Anim Biosci 3:169–195

Cieslak T, Polepalli JS, White A, Müller K, Borsch T, Barthlott W, Steiger J, Marchant A, Legendre L (2005) Phylogenetic analysis of *Pinguicula* (Lentibulariaceae): chloroplast DNA sequences and morphology support several geographically distinct radiations. Am J Bot 92:1723–1736

Clark JW, Hetherington AJ, Morris JL, Pressel S, Duckett JG, Puttick MN, Schneider H, Kenrick P, Wellman CH, Donoghue PCJ (2023) Evolution of phenotypic disparity in the plant kingdom. Nat Plants 9:1618–1626

Clark-Hachtel CM, Tomoyasu Y (2020) Two sets of candidate crustacean wing homologues and their implication for the origin of insect wings. Nat Ecol Evol 4:1694–1702

Clark-Hachtel C, Fernandez-Nicolas A, Belles X, Tomoyasu Y (2021) Tergal and pleural wing-related tissues in the German cockroach and their implication to the evolutionary origin of insect wings. Evol Dev 23:100–116

Cloutier R, Clement AM, Lee MSY, Noël R, Béchard I, Roy V, Long JA (2020) *Elpistostege* and the origin of the vertebrate hand. Nature 579:549–554

Cohn MJ, Tickle C (1999) Developmental basis of limblessness and axial patterning in snakes. Nature 399:474–479

Cole SR, Hopkins MJ (2021) Selectivity and the effect of mass extinctions on disparity and functional ecology. Sci Adv 7:eabf4072. https://doi.org/10.1126/sciadv.abf4072

Congreve CR, Falk AR, Lamsdell JC (2018) Biological hierarchies and the nature of extinction. Biol Rev 93:811–826

Cooper LN, Sears KE (2013) How to grow a bat wing. In: Adams R, Pedersen S (eds) Bat evolution, ecology, and conservation. Springer, New York, pp 3–20

Cooper RL, Martin KJ, Rasch LJ, Fraser GJ (2017) Developing an ancient epithelial appendage: FGF signalling regulates early tail denticle formation in sharks. EvoDevo 8:8. https://doi.org/10.1186/s13227-017-0071-0

Cooper RL, Lloyd VJ, Di-Poï N, Fletcher AG, Barrett PM, Fraser GJ (2019) Conserved gene signalling and a derived patterning mechanism underlie the development of avian footpad scales. EvoDevo 10:19. https://doi.org/10.1186/s13227-019-0130-9

Cooper RL, Nicklin EF, Rasch LJ, Fraser GJ (2023) Teeth outside the mouth: the evolution and development of shark denticles. Evol Dev 25:54–72

Cui D-F, Hou Y, Yin P, Wang X (2022) A Jurassic flower bud from China. Geol Soc Lond, Spec Publ 521:81–93. https://doi.org/10.1144/sp521-2021-122

Cundall D, Irish F (2008) The snake skull. In: Gans C, Gaunt AS, Adler K (eds) Biology of the Reptilia, Volume 20, morphology H, the skull of Lepidosauria. Society for the Study of Amphibian and Reptiles, New York, pp 349–692

Davis MC, Dahn RD, Shubin NH (2007) An autopodial-like pattern of Hox expression in the fins of a basal actinopterygian fish. Nature 447:473–476

Daza JD, Stanley EL, Wagner P, Bauer AM, Grimaldi DA (2016) Mid-Cretaceous amber fossils illuminate the past diversity of tropical lizards. Sci Adv 2:e1501080. https://doi.org/10.1126/sciadv.1501080

Dhouailly D (2009) A new scenario for the evolutionary origin of hair, feather, and avian scales. J Anat 214:587–606

Dhouailly D (2023) Evo devo of the vertebrates integument. J Dev Biol 11:25. https://doi.org/10.3390/jdb11020025

Dhouailly D (2024) The avian ectodermal default competence to make feathers. Dev Biol 508:64–76. https://doi.org/10.1016/j.ydbio.2024.01.002

Dhouailly D, Godefroit P, Martin T, Nonchev S, Caraguel F, Oftedal O (2019) Getting to the root of scales, feather and hair: as deep as odontodes? Exp Dermatol 28:503–508

Diaz RE Jr, Trainor PA (2015) Hand/foot splitting and the 're-evolution' of mesopodial skeletal elements during the evolution and radiation of chameleons. BMC Evol Biol 15:184. https://doi.org/10.1186/s12862-015-0464-4

DiFrisco J (2023) Toward a theory of homology: development and the de-coupling of morphological and molecular evolution. Br J Philos Sci 74:771–810

DiFrisco J, Wagner GP (2022) Body plan identity: a mechanistic model. Evol Biol 49:123–141

DiFrisco J, Love AC, Wagner GP (2020) Character identity mechanisms: a conceptual model for comparative-mechanistic biology. Biol Philos. https://doi.org/10.1007/s10539-020-09762-2

DiFrisco J, Love AC, Wagner GP (2023a) The hierarchical basis of serial homology and evolutionary novelty. J Morphol 284:e21531. https://doi.org/10.1002/jmor.21531

DiFrisco J, Wagner GP, Love AC (2023b) Reframing research on evolutionary novelty and co-option: character identity mechanisms versus deep homology. Semin Cell Dev Biol 145:3–12

Diogo R (2020) Cranial or postcranial-Dual origin of the pectoral appendage of vertebrates combining the fin-fold and gill-arch theories? Dev Dyn 249:1182–1200

Diogo R, Linde-Medina M, Abdala V, Ashley-Ross MA (2012) New, puzzling insights from comparative myological studies on the old and unsolved forelimb/hindlimb enigma. Biol Rev Camb Philos Soc 88:196–214

Eldredge N (1984) Simpson's inverse: bradytely and the phenomenon of living fossils. In: Eldredge N, Stanley SM (eds) Living fossils. Springer Verlag, New York, pp 272–276

Eldredge N, Stanley SM (1984) Living fossils: introduction to the casebook. In: Eldredge N, Stanley SM (eds) Living fossils. Springer Verlag, New York

Emry R, Thorington RW Jr (1984) The tree squirrel *Sciurus* (Sciuridae, Rodentia) as a living fossil. In: Eldredge N, Stanley SM (eds) Living fossils. Springer Verlag, New York, pp 23–31

Erwin DH (2001) Lessons from the past: biotic recoveries from mass extinctions. Proc Natl Acad Sci USA 98:5399–5403

Erwin DH (2007) Disparity: morphological pattern and developmental context. Palaeontology 50:57–73

Erwin DH (2008) Extinction as the loss of evolutionary history. Proc Natl Acad Sci USA 105:11520–11527

Erwin DH (2021) A conceptual framework of evolutionary novelty and innovation. Biol Rev 96:1–15

Erwin DH, Valentine JW, Sepkoski JJ Jr (1987) A comparative study of diversification events: the early Paleozoic versus the Mesozoic. Evolution 41:1177–1186

Fattorini S (2021) Climate change and extinction events. In: Alderton D, Elias SA (eds) Encyclopedia of geology. 2nd. Academic, pp. 585–595

Felice RN, Goswami A (2018) Developmental origins of mosaic evolution in the avian cranium. Proc Natl Acad Sci USA 115:555–560

Fernando PC, Mabee PM, Zeng E (2023) Protein-protein interaction network module changes associated with the vertebrate fin-to-limb transition. Sci Rep 13:22594. https://doi.org/10.1038/s41598-023-50050-2

Field DJ, Hanson M, Burnham D, Wilson LE, Super K, Ehret D, Ebersole JA, Bhullar BS (2018) Complete *Ichthyornis* skull illuminates mosaic assembly of the avian head. Nature 557:96–100

Fisher CR, Wegrzyn JL, Jockusch EL (2020) Co-option of wing-patterning genes underlies the evolution of the treehopper helmet. Nat Ecol Evol 4:250–260

Fisher CR, Kratovil JD, Angelini DR, Jockusch EL (2021) Out from under the wing: reconceptualizing the insect wing gene regulatory network as a versatile, general module for body-wall lobes in arthropods. Proc R Soc B 288:20211808. https://doi.org/10.1098/rspb.2021.1808

Flessa KW, Jablonski D (1985) Declining Phanerozoic background extinction rates: effect of taxonomic structure? Nature 313:216–218

Foote M (1993) Discordance and concordance between morphological and taxonomic diversity. Paleobiology 19:185–204

Foster WJ, Ayzel G, Münchmeyer J, Rettelbach T, Kitzmann NH, Isson TT, Mutti M, Aberhan M (2022) Machine learning identifies ecological selectivity patterns across the end-Permian mass extinction. Paleobiology 48:357–371

Friedman M (2009) Ecomorphological selectivity among marine teleost fishes during the end-cretaceous extinction. Proc Natl Acad Sci USA 106:5218–5223

Frolov A, Enushchenko I (2022) Monocotyledon-like leaves from the Middle Jurassic of East Siberia (Russia). Acta Geol Sin 96:1884–1896

Fu Q, Sun J, Zheng S, Wang X (2024) Unique Jurassic ovaries shed a new light on the nature of carpels. Plan Theory 13:2239. https://doi.org/10.3390/plants13162239

Funch P, Kristensen RM (1995) Cycliophora is a new phylum with affinities to Entoprocta and Ectoprocta. Nature 378:711–714

Fusco G (2022) Serial homology. Biol Theory 17:114–119

Galis F (2023) Evolvability of body plans: on phylotypic stages, developmental modularity, and an ancient metazoan constraint. In: Hansen HF, Houle D, Pavličev M, Pélabon C (eds) Evolvability: a unifying concept in evolutionary biology? The MIT Press, Cambridge, MA, pp 329–350

Géant E, Mouchel-Vielh E, Coutanceau JP, Ozouf-Costaz C, Deutsch JS (2006) Are Cirripedia hopeful monsters? Cytogenetic approach and evidence for a Hox gene cluster in the cirripede crustacean *Sacculina carcini*. Dev Genes Evol 216:443–449

Ghiselin MT (1984) Peripatus as a living fossil. In: Eldredge N, Stanley SM (eds) Living fossils. Springer Verlag, New York, pp 214–217

Gillis JA, Dahn RD, Shubin NH (2009) Shared developmental mechanisms pattern the vertebrate gill arch and paired fin skeletons. Proc Natl Acad Sci USA 106:5720–5724

Glenner H, Thomsen PF, Hebsgaard MB, Sørensen MV, Willerslev E (2006) Evolution. The origin of insects. Science 314:1883–1884

González-Suárez M, Revilla E (2013) Variability in life-history and ecological traits is a buffer against extinction in mammals. Ecol Lett 16:242–251

Goto R, Takano T, Seike K, Yamashita M, Paulay G, Rodgers KS, Hunter CL, Tongkerd P, Sato S, Hong JS, Endo K (2022) Stasis and diversity in living fossils: species delimitation and evolution of lingulid brachiopods. Mol Phylogenet Evol 175:107460. https://doi.org/10.1016/j.ympev.2022.107460

Gould SJ (1989) Wonderful life: the Burgess Shale and the nature of history. WW Norton, New York

Gould SJ (2002) The structure of evolutionary theory. The Belknap Press of Harvard University Press, Cambridge, MA

Graham LE, Cook ME, Busse JS (2000) The origin of plants: body plan changes contributing to a major evolutionary radiation. Proc Natl Acad Sci USA 97:4535–4540

Gunnell GF, Simmons NB (2005) Fossil evidence and the origin of bats. J Mamm Evol 12:209–246

Guo Z, Benton MJ, Stubbs TL, Chen ZQ (2024) Morphological innovation did not drive diversification in Mesozoic-Cenozoic brachiopods. Nat Ecol Evol 8:1948–1958

Hall BK (1999) Evolutionary developmental biology. Kluwer Academic Publishers, Dordrecht
Hall BK, Kerney R (2012) Levels of biological organization and the origin of novelty. J Exp Zool B Mol Dev Evol 318:428–437
Hallam T (2004) Catastrophes and lesser calamities: the causes of mass extinctions. Oxford University Press, Oxford
Hallgrímsson B, Jamniczky HA, Young NM, Rolian C, Schmidt-Ott U, Marcucio RS (2012) The generation of variation and the developmental basis for evolutionary novelty. J Exp Zool B Mol Dev Evol 318:501–517
Han L, Zhao Y, Zhao M, Sun J, Sun B, Wang X (2023) New fossil evidence suggests that angiosperms flourished in the Middle Jurassic. Life (Basel) 13:819. https://doi.org/10.3390/life13030819
Harris MP, Fallon JF, Prum RO (2002) Shh-Bmp2 signaling module and the evolutionary origin and diversification of feathers. J Exp Zool 294:160–176
Harrison CJ, Morris JL (2017) The origin and early evolution of vascular plant shoots and leaves. Philos Trans R Soc B 373:20160496. https://doi.org/10.1098/rstb.2016.0496
Havstad JC, Assis LCS, Rieppel O (2015) The semaphorontic view of homology. J Exp Zool B Mol Dev Evol 324:578–587
Head JJ, Howard AFC, Müller J (2022) The first 80 million years of snake evolution: the Mesozoic fossil record of snakes and its implications for origin hypotheses, biogeography, and mass extinction. In: Gower DJ, Zaher H (eds) The origin and early evolutionary history of snakes, vol Systematics association special volume 90. Cambridge University Press, Cambridge, pp 26–54
Herman AB (2024) Cretaceous angiosperms: evolutionary, geographical, and paleoclimatic aspects (on S.V. Meyen's scientific legacy). Stratigr Geol Correl 32:317–330
Holman EW (1996) The independent variable in the early origin of higher taxa. J Theor Biol 181:85–94
Hopkins MJ (2013) Decoupling of taxonomic diversity and morphological disparity during decline of the Cambrian trilobite family Pterocephaliidae. J Evol Biol 26:1665–1676
Hu Y, Linz DM, Moczek AP (2019) Beetle horns evolved from wing serial homologs. Science 366:1004–1007
Hughes CL, Kaufman TC (2002) Hox genes and the evolution of the arthropod body plan. Evol Dev 4:459–499
Hughes M, Gerber S, Wills MA (2013) Clades reach highest morphological disparity early in their evolution. Proc Natl Acad Sci USA 110:13875–13879
Jablonski D (2002) Survival without recovery after mass extinctions. Proc Natl Acad Sci USA 99:8139–8144
Jablonski D (2005) Mass extinctions and macroevolution. Paleobiology 31:192–210
Jablonski D (2017) Approaches to macroevolution: 2. Sorting of variation, some overarching issues, and general conclusions. Evol Biol 44:451–475
Jablonski D, Edie SM (2025) Mass extinctions and their rebounds: a macroevolutionary framework. Paleobiology 51:83–96. https://doi.org/10.1017/pab.2024.13
Janevski GA, Baumiller TK (2009) Evidence for extinction selectivity throughout the marine invertebrate fossil record. Paleobiology 35:553–564
Janis C (1984) Tapirs as living fossils. In: Eldredge N, Stanley SM (eds) Living fossils. Springer Verlag, New York, pp 80–86
Johanson Z, Joss J, Boisvert CA, Ericsson R, Sutija M, Ahlberg PE (2007) Fish fingers: digit homologues in sarcopterygian fish fins. J Exp Zool B Mol Dev Evol 308:757–768
Katayama N, Kato M, Nishiuchi T, Yamada T (2011) Comparative anatomy of embryogenesis in three species of Podostemaceae and evolution of the loss of embryonic shoot and root meristems. Evol Dev 13:333–342
Katayama N, Kato M, Yamada T (2013) Origin and development of the cryptic shoot meristem in *Zeylanidium lichenoides* (Podostemaceae). Am J Bot 100:635–646
Kemp TS (2007) The origin of higher taxa: macroevolutionary processes, and the case of the mammals. Acta Zoologica (Stockholm) 88:3–22
Kenrick P, Crane PR (1997) The origin and early evolution of plants on land. Nature 389:33–39

Kin A, Błażejowski B (2014) The horseshoe crab of the genus Limulus: living fossil or stabilomorph? PLoS One 9:e108036. https://doi.org/10.1371/journal.pone.0108036

Korn D, Hopkins MJ, Walton SA (2013) Extinction space–a method for the quantification and classification of changes in morphospace across extinction boundaries. Evolution 67:2795–2810

Kristensen RM, Funch P (2000) Micrognathozoa: a new class with complicated jaws like those of Rotifera and Gnathostomulida. J Morphol 246:1–49

Lala KN, Uller T, Feiner N, Feldman MW, Gilbert SF (2024) Evolution evolving: the developmental origins of adaptation and biodiversity. Princeton University Press, Princeton/Oxford

Larouche O, Zelditch ML, Cloutier R (2018) Modularity promotes morphological divergence in ray-finned fishes. Sci Rep 8:7278. https://doi.org/10.1038/s41598-018-25715-y

Larouche O, Gartner SM, Westneat MW, Evans KM (2023) Mosaic evolution of the skull in labrid fishes involves differences in both tempo and mode of morphological change. Syst Biol 72:419–432

Leal F, Cohn MJ (2018) Developmental, genetic, and genomic insights into the evolutionary loss of limbs in snakes. Genesis 56:e23077. https://doi.org/10.1002/dvg.23077

Lee J, Alrubaian J, Dores RM (2006) Are lungfish living fossils? Observation on the evolution of the opioid/orphanin gene family. Gen Comp Endocrinol 148:306–314

Liang Y, Strotz LC, Topper TP, Holmer LE, Budd GE, Chen Y, Fang R, Hu Y, Zhang Z (2023) Evolutionary contingency in lingulid brachiopods across mass extinctions. Curr Biol 33:1565–1572

Lidgard S, Kitchen E (2023) Revealing the rise of a living fossil menagerie. Front Ecol Evol 11:1112764. https://doi.org/10.3389/fevo.2023.1112764

Lidgard S, Love AC (2018) Rethinking living fossils. Bioscience 68:760–770

Lidgard S, Love AC (2021) The living fossil concept: reply to Turner. Biol Philos 36:13. https://doi.org/10.1007/s10539-021-09789-z

Lidgard S, Love AC (2023) Editorial: new perspectives on living fossils. Front Ecol Evol 11:1250106. https://doi.org/10.3389/fevo.2023.1250106

Lin GW, Li A, Chuong CM (2020) Molecular and cellular mechanisms of feather development provide a basis for the diverse evolution of feather forms. In: Foth C, Rauhut O (eds) The evolution of feathers: from their origin to the present. Springer, Cham, pp 13–26

Lindholm M (2014) Morphologically conservative but physiologically diverse: the mode of stasis in Anostraca (Crustacea: Branchiopoda). Evol Biol 41:503–507

Linz DM, Tomoyasu Y (2018) Dual evolutionary origin of insect wings supported by an investigation of the abdominal wing serial homologs in *Tribolium*. Proc Natl Acad Sci USA 115:E658–E667

Linz DM, Hu Y, Moczek AP (2020) From descent with modification to the origins of novelty. Zoology (Jena) 143:125836. https://doi.org/10.1016/j.zool.2020.125836

Liu X, Song H, Chu D, Dai X, Wang F, Silvestro D (2024) Heterogeneous selectivity and morphological evolution of marine clades during the Permian-Triassic mass extinction. Nat Ecol Evol 8:1248–1258

Liu ZJ, Wang X (2015) A perfect flower from the Jurassic of China. Hist Biol 28:707–719

Longrich NR, Tischlinger H, Foth C (2020) The feathers of the Jurassic Urvogel *Archaeopteryx*. In: Foth C, Rauhut O (eds) The evolution of feathers: from their origin to the present. Springer, Cham, pp 119–146

López-Martínez AM, Magallón S, von Balthazar M, Schönenberger J, Sauquet H, Chartier M (2024) Angiosperm flowers reached their highest morphological diversity early in their evolutionary history. New Phytol 241:1348–1360

Louchart A, Tourment N, Carrier J (2011) The earliest known pelican reveals 30 million years of evolutionary stasis in beak morphology. J Ornithol 152:15–20

Love AC (2024) Evolution and development: conceptual issues. Cambridge University Press, Cambridge

Lü J, Unwin DM, Jin X, Liu Y, Ji Q (2010) Evidence for modular evolution in a long-tailed pterosaur with a pterodactyloid skull. Proc R Soc B 277:383–389

MacDougall MJ, Brocklehurst N, Fröbisch J (2019) Species richness and disparity of parareptiles across the end-Permian mass extinction. Proc R Soc B 286:20182572. https://doi.org/10.1098/rspb.2018.2572

Marugán-Lobón J, Nebreda SM, Navalón G, Benson RBJ (2022) Beyond the beak: brain size and allometry in avian craniofacial evolution. J Anat 240:197–209

McKenna KZ, Wagner GP, Cooper KL (2021) A developmental perspective of homology and evolutionary novelty. Curr Top Dev Biol 141:1–38

Melzer R, Theißen G (2016) The significance of developmental robustness for species diversity. Ann Bot 117:725–732

Meyer ER (1984) Crocodilians as living fossils. In: Eldredge N, Stanley SM (eds) Living fossils. Springer Verlag, New York, pp 105–131

Minelli A (2016) Species diversity vs. morphological disparity in the light of evolutionary developmental biology. Ann Bot 117:781–794

Miyashita T, Diogo R (2016) Evolution of serial patterns in the vertebrate pharyngeal apparatus and paired appendages via assimilation of dissimilar units. Front Ecol Evol 4:71. https://doi.org/10.3389/fevo.2016.00071

Mohan Ram HY, Sehgal A (1992) Podostemaceae-the strange family of aquatic angiosperms. J Palaeosciences 41:192–197

Molnar JL, Diogo R (2021) Evolution, homology, and development of tetrapod limb muscles. Diversity 13:393. https://doi.org/10.3390/d13080393

Molnar JL, Diaz RE Jr, Skorka T, Dagliyan G, Diogo R (2017) Comparative musculoskeletal anatomy of chameleon limbs, with implications for the evolution of arboreal locomotion in lizards and for teratology. J Morphol 278:1241–1261

Monarrez PM, Heim NA, Payne JL (2023) Reduced strength and increased variability of extinction selectivity during mass extinctions. R Soc Open Sci 10:230795. https://doi.org/10.1098/rsos.230795

Moody S, Roček Z (1980) *Chamaeleo caroliquarti* (Chamaeleonidae, Sauria): a new species from the Lower Miocene of central Europe. Věstnik Ústředniho Ustavu Geologického 55:85–92

Mouchel-Vielh E, Rigolot C, Gibert JM, Deutsch JS (1998) Molecules and the body plan: the Hox genes of Cirripedes (Crustacea). Mol Phylogenet Evol 9:382–389

Müller GB (2010) Epigenetic innovation. In: Pigliucci M, Müller GB (eds) Evolution: the extended synthesis. MIT Press, Cambridge, MA, pp 307–332

Müller GB, Newman SA (2005) The innovation triad: an evodevo agenda. J Exp Zool B Mol Dev Evol 304:487–503

Müller GB, Wagner GP (1991) Novelty in evolution: restructuring the concept. Annu Rev Ecol Syst 22:229–256

Müller GB, Wagner GP (2003) Innovation. In: Hall BK, Olson WM (eds) Keywords and concepts in evolutionary developmental biology. Harvard University Press, Cambridge MA, pp 218–227

Nakamura T, Gehrke AR, Lemberg J, Szymaszek J, Shubin NH (2016) Digits and fin rays share common developmental histories. Nature 537:225–228

Ng CS, Li W-H (2018) Genetic and molecular basis of feather diversity in birds. Genome Biol Evol 10:2572–2586

Ochoa C, Rasskin-Gutman D (2015) Evo–devo mechanisms underlying the continuum between homology and homoplasy. J Exp Zool B Mol Dev Evol 324:91–103

Ohde T, Prokop J (2025) The transition to flying insects: lessons from evo-devo and fossils. Curr Opin Insect Sci 68:101332. https://doi.org/10.1016/j.cois.2025.101332

Ohde T, Mito T, Niimi T (2022) A hemimetabolous wing development suggests the wing origin from lateral tergum of a wingless ancestor. Nat Commun 13:979. https://doi.org/10.1038/s41467-022-28624-x

Oyston JW, Hughes M, Wagner PJ, Gerber S, Wills MA (2015) What limits the morphological disparity of clades? Interface Focus 5:20150042. https://doi.org/10.1098/rsfs.2015.0042

Parins-Fukuchi C (2020) Mosaic evolution, preadaptation, and the evolution of evolvability in apes. Evolution 74:297–310

Patterson C (1984) Family Chanidae and other teleostean fishes as living fossils. In: Eldredge N, Stanley SM (eds) Living fossils. Springer Verlag, New York, pp 132–139

Payne JL, Al Aswad JA, Deutsch C, Monarrez PM, Penn JL, Singh P (2023) Selectivity of mass extinctions: patterns, processes, and future directions. Cambridge Prisms: Extinction 1:e12. https://doi.org/10.1017/ext.2023.10

Peterson T, Müller GB (2013) What is evolutionary novelty? Process versus character based definitions. J Exp Zool B Mol Dev Evol 320:345–350

Phillips AJ, Arauco-Brown R, Oceguera-Figueroa A, Gomez GP, Beltrán M, Lai YT, Siddall ME (2010) *Tyrannobdella rex* n. gen. n. sp. and the evolutionary origins of mucosal leech infestations. PLoS One 5:e10057. https://doi.org/10.1371/journal.pone.0010057

Prokop J, Pecharová M, Nel A, Hörnschemeyer T, Krzemińska E, Krzemiński W, Engel MS (2017) Paleozoic nymphal wing pads support dual model of insect wing origins. Curr Biol 27:263–269

Prokop J, Rosová K, Krzemińska E, Krzemiński W, Nel A, Engel MS (2022) Abdominal serial homologues of wings in Paleozoic insects. Curr Biol 32:3414–3422

Prum RO (1999) Development and evolutionary origin of feathers. J Exp Zool 285:291–306

Prum RO (2005) Evolution of the morphological innovations of feathers. J Exp Zool B Mol Dev Evol 304:570–579

Prum RO, Dyck J (2003) A hierarchical model of plumage: morphology, development, and evolution. J Exp Zool B Mol Dev Evol 298:73–90

Puttick MN, Guillerme T, Wills MA (2020) The complex effects of mass extinctions on morphological disparity. Evolution 74:2207–2220

Rabosky DL, Slater GJ, Alfaro ME (2012) Clade age and species richness are decoupled across the eukaryotic tree of life. PLoS Biol 10:e1001381. https://doi.org/10.1371/journal.pbio.1001381

Rasskin-Gutman D, Esteve-Altava B (2021) Concept of burden in evo-devo. In: Nuño de la Rosa L, Müller GB (eds) . Evolutionary developmental biology. Springer, Cham, pp 39–49

Riedl R (1978) Order in living organisms: a systems analysis of evolution (trans: RPS Jefferies). Wiley, Chichester

Romano M (2021) Disparity versus diversity in ankylosaurid dinosaurs: explored morphospace indicates two separate evolutive radiations. Rend Online Soc Geol Ital 53:2. https://doi.org/10.3301/ROL.2020.22

Rombaut LMK, Capp EJR, Cooney CR, Hughes EC, Varley ZK, Thomas GH (2021) Allometric conservatism in the evolution of bird beaks. Evol Lett 6:83–91

Ronshaugen M, McGinnis N, McGinnis W (2002) Hox protein mutation and macroevolution of the insect body plan. Nature 415:914–917

Rubinoff D, Le Roux JJ (2008) Evidence of repeated and independent saltational evolution in a peculiar genus of sphinx moths (*Proserpinus*: Sphingidae). PLoS One 3:e4035. https://doi.org/10.1371/journal.pone.0004035

Rutishauser R (2016) Evolution of unusual morphologies in Lentibulariaceae (bladderworts and allies) and Podostemaceae (river-weeds): a pictorial report at the interface of developmental biology and morphological diversification. Ann Bot 117:811–832

Schlosser G (2023) Rebuilding ships while at sea-character individuality, homology, and evolutionary innovation. J Morphol 284:e21522. https://doi.org/10.1002/jmor.21522

Schwentner M, Combosch DJ, Pakes Nelson J, Giribet G (2017) A phylogenomic solution to the origin of insects by resolving crustacean-hexapod relationships. Curr Biol 27:1818–1824

Sclafani JA, Congreve CR, Krug AZ, Patzkowsky ME (2018) Effects of mass extinction and recovery dynamics on long-term evolutionary trends: a morphological study of Strophomenida (Brachiopoda) across the Late Ordovician mass extinction. Paleobiology 44:603–619

Sears KE (2007) Molecular determinants of bat wing development. Cells Tissues Organs 187:6–12

Sears KE, Behringer RR, Rasweiler JJ 4th, Niswander LA (2006) Development of bat flight: morphologic and molecular evolution of bat wing digits. Proc Natl Acad Sci USA 103:6581–6586

Sears KE, Capellini TD, Diogo R (2015) On the serial homology of the pectoral and pelvic girdles of tetrapods. Evolution 69:2543–2555

Serra F, Balseiro D, Waisfeld BG (2023) Morphospace trends underlying a global turnover: ecological dynamics of trilobite assemblages at the onset of the Ordovician radiation. Palaeogeogr Palaeoclimatol Palaeoecol 615:111448. https://doi.org/10.1016/j.palaeo.2023.111448

Shubin NH, Daeschler EB, Jenkins FA Jr (2006) The pectoral fin of *Tiktaalik roseae* and the origin of the tetrapod limb. Nature 440:764–771

Shubin N, Tabin C, Carroll S (2009) Deep homology and the origins of evolutionary novelty. Nature 457:818–823

Siomava N, Fuentes JSM, Diogo R (2020) Deconstructing the long-standing a priori assumption that serial homology generally involves ancestral similarity followed by anatomical divergence. J Morphol 281:1110–1132

Sleight VA, Gillis JA (2020) Embryonic origin and serial homology of gill arches and paired fins in the skate, *Leucoraja erinacea*. eLife 9:e60635. https://doi.org/10.7554/eLife.60635

Soltis DE, Bell CD, Kim S, Soltis PS (2008) Origin and early evolution of angiosperms. Ann N Y Acad Sci 1133:3–25

Sordino P, van der Hoeven F, Duboule D (1995) Hox gene expression in teleost fins and the origin of vertebrate digits. Nature 375:678–681

Sterner B (2023) Norms of evidence in the classification of living fossils. Front Ecol Evol 11:1198224. https://doi.org/10.3389/fevo.2023.1198224

Suno-Uchi N, Sasaki F, Chiba S, Kawata M (1997) Morphological stasis and phylogenetic relationships in Tadpole shrimps, *Triops* (Crustacea: Notostraca). Biol J Linn Soc 61:439–457

Tanaka M (2016) Fins into limbs: autopod acquisition and anterior elements reduction by modifying gene networks involving 5'Hox, Gli3, and Shh. Dev Biol 413:1–7

Tanaka M (2017) Evolution of vertebrate limb development. In: eLS. Wiley, Chichester. https://doi.org/10.1002/9780470015902.a0002099.pub2

Taylor PD (2004) Extinction and the fossil record. In: Taylor PD (ed) Extinctions in the history of life. Cambridge University Press, Cambridge, pp 1–34

Terrill RS, Shultz AJ (2023) Feather function and the evolution of birds. Biol Rev Camb Philos Soc 98:540–566

Tolley KA, Herrel A (2013) Biology of the chameleons: an introduction. In: Tolley KA, Herrel A (eds) The biology of chameleons. University of California Press, Berkeley, pp 1–5

Tolley KA, Townsend TM, Vences M (2013) Large-scale phylogeny of chameleons suggests African origins and Eocene diversification. Proc R Soc B 280:20130184. https://doi.org/10.1098/rspb.2013.0184

Tomoyasu Y (2021) What crustaceans can tell us about the evolution of insect wings and other morphologically novel structures. Curr Opin Genet Dev 69:48–55

Turner DD (2019) In defense of living fossils. Biol Philos 34:23. https://doi.org/10.1007/s10539-019-9678-y

Valentine JW (1990) The macroevolution of clade shape. In: Ross RM, Allmon WD (eds) Causes of evolution: a paleontological perspective. The University of Chicago Press, Chicago, pp 128–150

Valentine JW (2004) On the origin of phyla. The University of Chicago Press, Chicago/London

Varga Z, Varga M (2022) Gene expression changes during the evolution of the tetrapod limb. Biol Futur 73:411–426

Wagner GP (2014) Homology, genes, and evolutionary innovation. Princeton University Press, Princeton

Wagner GP (2015) Reinventing the organism: evolvability and homology in post-Dahlem evolutionary biology. In: Love AC (ed) Conceptual change in biology: scientific and philosophical perspectives on evolution and development. Springer, Dordrecht, pp 327–342

Wagner GP, Chiu CH (2001) The tetrapod limb: a hypothesis on its origin. J Exp Zool 291:226–240

Wang X (2018) The dawn angiosperms uncovering the origin of flowering plants, 2nd edn. Springer, Cham

Wang X (2019) Origin of angiosperms and their diversification in the Cretaceous. Open J Geol 9:577–580

Ward P (1984) Is nautilus a living fossil? In: Eldredge N, Stanley SM (eds) Living fossils. Springer Verlag, New York, pp 247–256
Watkins A (2021) The epistemic value of the living fossils concept. Philos Sci 88:1221–1233
Werth AJ, Shear WA (2014) The evolutionary truth about living fossils. Am Sci 102:434–443
Wiley EO, Schultze H-P (1984) Family Lepisosteidae (gars) as living fossils. In: Eldredge N, Stanley SM (eds) Living fossils. Springer Verlag, New York, pp 160–165
Wimsatt WC (1986) Developmental constraints, generative entrenchment, and the innate–acquired distinction. In: Bechtel W (ed) Integrating scientific disciplines, Science and philosophy, vol 2. Springer, Dordrecht, pp 185–208
Woltering JM, Irisarri I, Ericsson R, Joss JMP, Sordino P, Meyer A (2020) Sarcopterygian fin ontogeny elucidates the origin of hands with digits. Sci Adv 6:eabc3510. https://doi.org/10.1126/sciadv.abc3510
Wu P, Lai YC, Widelitz R, Chuong CM (2018) Comprehensive molecular and cellular studies suggest avian scutate scales are secondarily derived from feathers, and more distant from reptilian scales. Sci Rep 8:16766. https://doi.org/10.1038/s41598-018-35176-y
Xu X (2020) Filamentous integuments in nonavialan theropods and their kin: advances and future perspectives for understanding the evolution of feathers. In: Foth C, Rauhut O (eds) The evolution of feathers: from their origin to the present. Springer, Cham, pp 67–78
Xue C, Yuan D, Chen Y, Stubbs TL, Zhao Y, Zhang Z (2024) Morphological innovation after mass extinction events in Permian and Early Triassic conodonts based on Polygnathacea. Palaeogeogr Palaeoclimatol Palaeoecol 642:112149. https://doi.org/10.1016/j.palaeo.2024.112149
Yang Z, Jiang B, Xu J, McNamara ME (2024) Cellular structure of dinosaur scales reveals retention of reptile-type skin during the evolutionary transition to feathers. Nat Commun 15:4063. https://doi.org/10.1038/s41467-024-48400-3

The manufacturer's authorised representative in the EU is Springer Nature Customer Service Centre GmbH, Europaplatz 3, 69115 Heidelberg, Germany. If you have any concerns regarding our products, please contact ProductSafety@springernature.com

Printed and bound by CPI Group (UK) Ltd, Croydon, CR0 4YY

30/09/2026

02225596-0012